Geologic History of Utah

Lehi F. Hintze

Brigham Young University Geology Studies
Special Publication 7
Bart J. Kowallis, Editor

DEDICATION

To my father, F.F. Hintze,
who preceded me in the excitement
of geologic discovery.

ABOUT THIS BOOK

This is a completely revised edition of the most popular book on the geology of Utah in print. Originally published in 1973, the first edition has been used by thousands of students, teachers, laymen, and professional geologists as the handiest and most comprehensive introduction to Utah's geology available.

Much new information regarding Utah's geologic past has been discovered in the last fifteen years. Updates have been particularly striking in our knowledge of ancient Precambrian rocks, our concepts of volcanic events in the last 30 million years, and in our understanding of the history of the activity of the Wasatch Fault and Lake Bonneville within the last 30,000 years.

The book is directed primarily toward the beginning student in geology. Professional geologists coming into Utah for the first time will find the stratigraphic charts and the citations of source literature helpful. Other interested persons will find the broad sweep of Utah's geologic history fascinating.

ABOUT THE AUTHOR

Lehi F. Hintze, professor emeritus of geology at Brigham Young University, graduated from the University of Utah, and holds a doctorate from Columbia University. He has done extensive stratigraphic work and published many maps of geologic quadrangles in western Utah in a cooperative program between Brigham Young University, the Utah Geological and Mineral Survey, and the United States Geological Survey. He is author of the 1980 Geologic Map of Utah published by the state, and the earlier Geologic Highway Map of Utah, published by Brigham Young University. He is author of the COSUNA correlation chart of stratigraphic units in the Great Basin published by the American Association of Petroleum Geologists and has written many geological papers and coauthored a textbook on historical geology. His hobbies are music, antique furniture restoration, and grandchildren.

Copies of this book may be obtained from
Publication Sales, Department of Geology
Brigham Young University
Provo, Utah 84602
(801) 378-3918

Cover photo: Canyonlands by David Muench

ISBN — 0-929451-00-7
Library of Congress Catalog Card Number 88-71270
July, 1988

CONTENTS

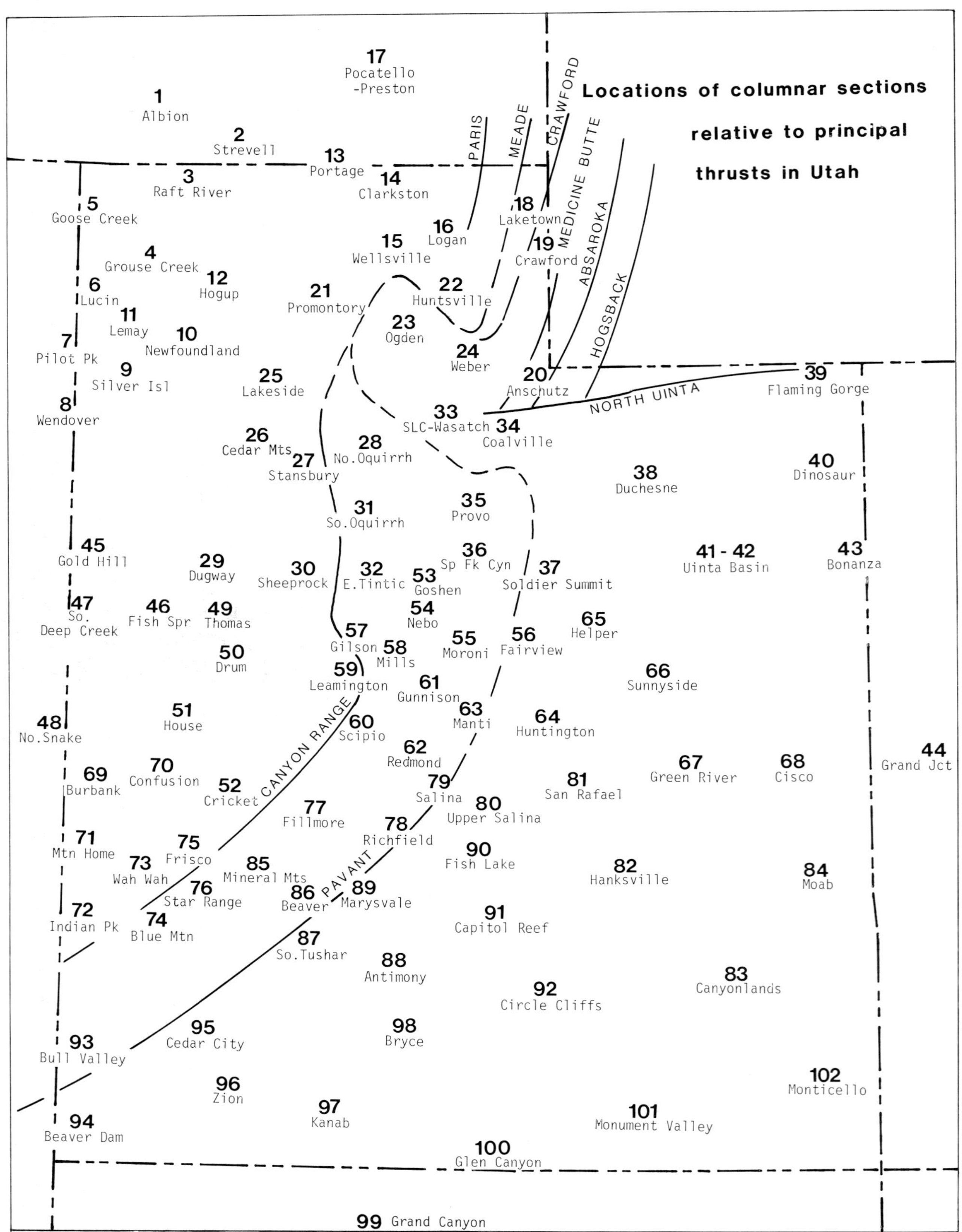

FIGURE 1 — Locations of stratigraphic charts in the back of this book. Trends of leading edges of Cretaceous thrusts are shown in gray. Pre-Cretaceous rocks west of the thrust belt were moved eastward during the Sevier orogeny.

FIGURE 2 — Canyonlands of Utah from Dead Horse Point. Jurassic Wingate Sandstone forms cliffs at upper right of photo. Permian White Rim Sandstone makes thin white ledge in middle distance at right. Lower Permian Elephant Canyon Formation is exposed along the banks of the Colorado River. See Chart 84.

Utah's landscapes are dominated by layered rocks. They come in many colors and configurations and range from rocks formed more than two billion years ago to strata being laid down today. They are the book of stone that has recorded Utah's geologic history.

Many names have been applied to Utah's rock layers and these are shown on the 102 stratigraphic charts in this book; their locations are shown in Figure 1. These charts were prepared as a summary of our knowledge, to date, of Utah's stratigraphy, and represent measurements and descriptions by the great many workers cited on the charts and in the list of references. Interpretations of how and why these rocks were formed, and when and where they were deposited, intruded, or erupted, form the basis for Utah's geologic history. Information from the charts is summarized in two other kinds of diagrams that are scattered through the text: 1) correlation tables, and 2) thickness maps. Correlation tables are designed to show the time-equivalency of rock units that may have different names and lithologies in different areas of the state. Ages for the rocks are based either on fossil evaluation or isotopic dating procedures using radioactive elements found in rocks. Each correlation table shows an isotopic time scale on its vertical margin; these scales are modified from those proposed by Harland et al. (1982), Palmer (1983), and Snelling (1985). The horizontal lines on each table are time lines. Rock units of the same age can be compared horizontally across the correlation table. Unconformities and times of non-deposition and erosion are also shown.

Thickness maps were derived from the stratigraphic charts by adding up the thicknesses of all chart units that belong to a particular time span and then showing this total thickness, to the nearest 100 feet, as a number in its appropriate location on the map. The thickness maps show a considerable variety of patterns. These patterns have been used to identify eight "phases" of Utah's geologic history discussed below. Before discussing these phases, it is important to note that the thickness maps are plotted on two different base maps of Utah. Jurassic and older units are shown on a palinspastic (literally: stretched back) base map that schematically removes the effects of

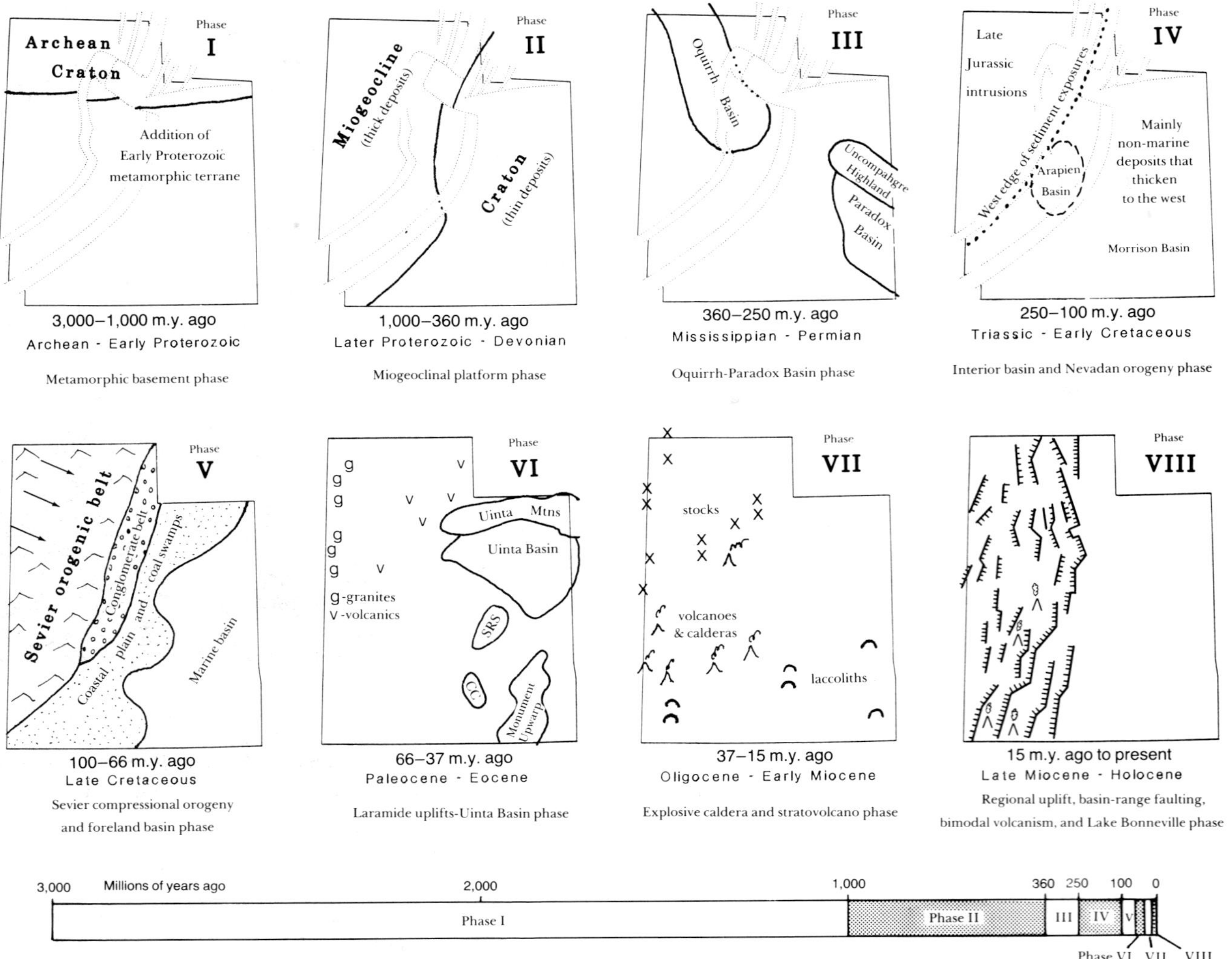

FIGURE 3—In this book Utah's geologic history has been divided into eight phases of differing structural behavior. The length of time that each phase lasted differs greatly, as shown on the bar scale at the bottom where time is linearly represented in millions of years (m.y.).

Cretaceous thrusting in Utah. Rocks deposited west of the front of the thrusts were moved eastward by Cretaceous thrusting 100 to 70 m.y. ago. In general, Triassic and older rocks have overridden Jurassic (and, in a few areas Cretaceous) strata along the frontal thrust zone. Total shortening across the thrust zone is better known in the Utah-Wyoming thrust belt, where it amounts to about 70 miles, than in the thrust belts of central Utah where estimates have not been so well-constrained by seismic work and drilling. Thickness of Cretaceous and later units are plotted on a non-palinspastic base map of Utah. In view of the considerable extension that has occurred in western Utah during basin-range faulting in the last 15 million years, it might have been helpful to represent the pre-extension shape of western Utah as being narrower than at present. But estimates of the amount of extension that has occurred during this latest phase of Utah's history are presently so variable that it did not seem feasible to design an appropriate palinspastic map to show it.

The eight "phases" into which Utah's geologic history has been divided in this book identify spans of time in much the same way that historians do in speaking about the "Civil War era," or the "World War II era." Time is a continuum, but we mortals note its passage by historical signposts. Geologic time is universally identified by the unique nomenclature of geologic eras and periods (Cambrian, Ordovician, etc., see front cover); but geologic events are irregular in their duration, some lasting several time periods, while others last but a fraction. Long or short, all are important in conceptualizing our view of how Utah came to its present being. Note that the times assigned to the end of one phase and the beginning of

FIGURE 4 — City of Rocks about 6 miles north of the Utah-Idaho border near Almo, Idaho. Area is part of the Raft River-Albion Range metamorphic core complex that was intruded by Oligocene Almo granite. Most of the rocky outcrops are Oligocene granite, but the left knob of the Twin Sisters, beyond the main road near the photo center, is Archean Green Creek Complex 2.5 billion years old! Archean rocks underlie the wooded area in the left foreground and the rounded hill in the left middle distance beyond the main road. The California Trail shows faintly at the edge of the trees in the foreground. See Chart 1.

another form a continuum; name-giving events (i.e., Sevier orogeny) do not form a continuum any more than World War I and World War II do; they are episodic.

The text of this book is divided into three main sections: First, a brief overview; second, a time-period by time-period description of geologic events in Utah; and third, a perspective section that discusses plate-tectonic models that have been proposed to account for the various phases of Utah's geologic history. Persons becoming newly acquainted with the geology of Utah will find the overview particularly helpful. Geologists seeking original references on particular aspects of the state's geologic history will find them cited in the latter part of the text.

OVERVIEW

We are so acquainted with Utah's present topography that it is hard for us to imagine how very different it has been in times past. A Sahara desert, a Texas coastline, a Bahamas' limestone bank, an Okefenokee Swamp—has Utah ever looked this way? Yes, and many other ways as well. The rock record testifies that Utah hovered near sea level—slightly above, or slightly below—throughout most of the last billion years. Only in the last tenth of that time has Utah been raised much above sea level; and only in the last 10 million years has it come to resemble anything like the Utah of today.

Just as human history is built on past events, so are present geologic relationships the cumulative product of the past. Certain patterns of tectonic and structural behavior have persisted over long enough segments of geologic time in this part of North America to give a distinctive stamp to these segments in terms of paleogeographic, paleoenvironmental, and paleodepositional patterns. We can better understand any segment of earth history if we know what events have preceded it. Thus it is helpful to divide Utah's geologic history into eight phases as shown in Figure 3.

Phase I—Metamorphic basement phase. It has only recently become apparent that the oldest rocks in Utah, Archean rocks more than 2,500 million years old (Figure 4), are restricted to northern Utah. Granitic and metamorphic rocks south of Salt Lake City are part of a terrane that was added to the southern margin of the Archean craton (a stable part of a continent that has been little deformed for a prolonged period) between 1,600 and 1,800 million years ago. The rocks that compose the southern terrane probably originated as mixed sedimentary-volcanic rock assemblages in island arcs which col-

FIGURE 5 — Precambrian quartzites and argillites of the Uinta Mountain Group in the western part of the high Uinta Mountains. Pleistocene glaciation enlarged the valleys and left moraines on the valley floors. See Chart 38.

lided with the Archean craton in Middle Proterozoic time and became attached as part of the growing North American continent. All rocks of this very long phase of Utah's history are metamorphosed; most are gneisses and schists; some are granitic or pegmatitic; a few are quartzitic, suggesting that they were originally sedimentary sandstone.

Phase II—Miogeoclinal platform phase. This phase marks the first time that a diagonal flexure divided Utah into two parts: a subsiding western half where thousands of feet of shallow-water marine sediments accumulated in a miogeoclinal basin, and an eastern half where similar sedimentary strata a tenth as thick accumulated. Strictly speaking, this pattern held true for only the latter two-thirds of the phase's duration. During the first third an anomalous east-trending basin developed along the trend of the present Uinta Mountains. Some 25,000 feet of sandstone and shale accumulated in the Uinta area (Figure 5) in a basin that has been called an aulacogen. The aulacogenic behavior ceased before Cambrian time, but the main northeasterly diagonal thick-thin pattern persisted throughout lower Paleozoic time. This thick-thin transition zone has been called in recent geological writings "Utah's hingeline." In pre-1970 literature it was known as the "Wasatch line," the term given to it by Kay (1951). Most of the Precambrian and Lower Cambrian deposits of this phase were quartz sandstones, siltstones and mudstones, with rare diabase dikes or flows. Deposits from Middle Cambrian through Devonian time were mostly limestone and dolomite, with some tongues of shale and sandstone, recording fluctuations in sea level.

Several thousand feet of sediments, deposited under shallow-water conditions, accumulated in the miogeocline during this phase. The only way that this could occur is for the miogeoclinal floor to subside at about the same rate as the sediments were laid down. The weight of the sediments themselves contributed to the isostatic subsidence, but the old saying that "you can't sink a rowboat by filling it with sawdust" has some relevance here. The factors that are currently considered most important in causing the miogeocline to subside slowly are loss of heat and thinning of the crust. Warming rocks expand in the earth's interior; cooling ones shrink. What, then, might cause the crust beneath western North America to cool? The best answer suggested so far seems to be splitting or rifting of the proto-North American continent along a zone trending northerly through central Nevada. Miogeoclinal subsidence was initiated when a western part of proto-North America split off and left western Utah and eastern Nevada as the hot trailing edge of the sundered plate. Loss of heat from the raw trailing edge caused it to slowly subside and permitted the accumulation of thousands of feet of shallow water sediments

FIGURE 6 — Pennsylvanian strata of the Hermosa Group exposed in the Goosenecks of the San Juan River. The rocks are interbedded limestone, shale, siltstone and sandstone, deposited on the west side of the Paradox Basin. Nearly 2000 feet of strata are exposed along the canyon walls. Road to the overlook point is at the left center of the photo. See Chart 101.

during the very long period of time (see time bar at the bottom of Figure 3) that this phase lasted.

Phase III—Oquirrh-Paradox Basin phase. Although sedimentation in shallow marine waters continued during this phase of Utah's history, the pattern of where sediment accumulation was thick and where it was thin or absent was strikingly different from preceding patterns. Two deeply subsiding basins with northwesterly trends developed. Sinking of the Paradox Basin in the Four Corners area (Figure 6) was closely related to the uplift of the adjacent Uncompahgre Highland, from which the basin received much arkosic sediment. Thousands of feet of evaporites—rock salt, gypsum, and potash—also accumulated in the Paradox Basin; the potash has been mined commercially near Moab. The Oquirrh Basin was more deeply subsiding than the Paradox Basin, but the sediments that filled it, largely fine-grained, well-traveled siliceous sand, were derived from many sources, some quite distant from the basin. Utah's phase III pattern is part of a belt of yoked basin-uplifts that extended southeastward to Oklahoma. The uplifts are often described as the "Ancestral Rockies," although in fact, they have little to do with the present Rocky Mountains that originated much later. Meager published data suggest that the Uncompahgre Highland formed as a result of combined thrust and left-slip offsets. Plate-tectonic movements that caused the Ancestral Rockies uplifts and basins have not yet been clearly identified. Although some geologists have suggested continental collision as the mechanism, some reconstructions of continental positions in Late Paleozoic time do not support this idea. Slight rotation of one portion of the continental interior with respect to other portions has also been suggested. Northward thrusting of this age is a conspicuous structural element in the Ouachita Mountains of Oklahoma-Arkansas. Whatever the origin, linear intracratonic uplift and subsidence occurred in a belt 1200 miles long between Utah and Oklahoma during this phase.

Phase IV—Interior basin and Nevadan orogeny. Latest Triassic strata spread across the top of the Uncompahgre Highland marking the end of Ancestral Rockies uplift. During the 150 million years (Triassic through Early Cretaceous) of phase IV the thin-thick depositional pattern returned to that of phase II, thickening to the west. However, preservation of deposits of this phase in western Utah is limited to scattered occurrences of Lower Triassic marine beds; all other strata of this phase disappear beneath the leading edge of the Cretaceous thrust

FIGURE 7 — Split Mountain anticline, a Laramide fold subsidiary to the main Uinta Mountain uplift. Arrow near center of right edge of picture points to visitor center at the Dinosaur National Monument. Oldest beds exposed where Green River cuts the anticlinal axis are Mississippian (M). Successively higher formations are Morgan Formation, Weber Sandstone, Park City Formation, Moenkopi Formation, Gartra Member, Chinle Formation, Glen Canyon (Navajo) Sandstone, Carmel Formation, Entrada Sandstone, Curtis Member, Redwater Member, Morrison Formation, Cedar Mountain Formation, Dakota Sandstone, Mowry Shale, Frontier Sandstone, and Mancos Shale. See Chart 40 for thicknesses.

belt, and their subsurface occurrence in western Utah has not yet been reported. Western Utah also includes some granitic intrusions of late Jurassic age, evidence of the Nevadan orogeny. Deposits of this age in eastern Utah include some of the most famous formations of this country. Fossil dunes of the Navajo Sandstone are an important part of the scenic attraction of Zion National Park, forming its spectacular monoliths. The Morrison Formation is world famous for its dinosaurs. Petrified wood from the Chinle Formation is common in Utah, and the basis for a national park in Arizona; and Arches National Park owes its uniqueness to weathering habits of the Entrada Sandstone. Many of these deposits accumulated in gently subsiding basins on the craton. Sources for their sediments lay on all sides of a basinal area which centered on the Four Corners. Marine incursions into Utah occurred during Lower Triassic and Middle Jurassic time but the seas were so shallow and so nearly land-locked most of the time that evaporites (sometimes salt, but more often gypsum) are common in Triassic and Jurassic strata of this interval. Evaporites are particularly abundant in the Arapien Basin of central Utah where they form the basis for a gypsum wallboard industry. Deposits of phase IV were laid down just before the Sevier orogeny of phase V. So it is not surprising that when Sevier thrusts broke toward the surface, the displaced thrust sheets slid across strata just deposited during this phase.

Phase V—Sevier compressional orogeny and foreland basin phase. No event left more widespread effects in Utah than the Sevier orogeny. Western Utah was buckled into mountains by it, and eastern Utah received a thick wedge of sediments eroded from those mountains and deposited in a new marine seaway. Later uplift and erosion has removed much of the original Cretaceous marine wedge from southeastern Utah, and the Cretaceous orogenic belt in western Utah has been broken by later block faults, so that much of its record lies concealed beneath basin fill. Understanding the effects of the Sevier orogeny is critical to an understanding of Utah's geology. It marked the turning point from a long, mostly shallow-water marine history to the later, entirely non-marine, scenar-

FIGURE 8 — Quaternary basalt cone and lava flows a few thousand years old along Utah Highway 18 north of St. George. Basalts are part of late Cenozoic bimodal volcanism. Basalts rest unconformably on Jurassic Navajo Sandstone that shows in right center edge of photo near Snow Canyon State Park. See Chart 96.

ios. A conglomerate belt marks the eastern edge of the Sevier orogenic highlands; these pass eastward into a Texas-like coastal plain where coal swamps were common, thence into the Cretaceous epicontinental interior seaway that extended from the Gulf of Mexico to the Arctic. Compressional thrusting worked its way eastward across Utah; older Sevier thrusts were carried piggy-back on later thrusts that surfaced successively farther to the east along the belt. Compression that produced the orogeny has been ascribed to subduction of an oceanic plate beneath the west coast of North America, which in Cretaceous time was in western Nevada. In addition, mini-continents or remnants of island arcs were rafted across the Pacific basin on the oceanic plates that brought these small land masses into collision with the west edge of North America. Successive collisions may have produced the series of compressive pulses that made up the orogeny.

Phase VI—Laramide uplifts-Uinta Basin phase. The Rocky Mountains are the most conspicuous product of this phase. Rocky Mountain ranges are formed by elongate, asymmetrical anticlinal structures which are bordered on their steep structural side by a low-angle thrust fault. Utah's Uinta Mountains (Figure 7) are a typical example. Uplift of Rocky Mountain ranges is commonly linked to downwarping of an adjacent asymmetrical synclinal basin, such as the Uinta Basin. Erosion of the rising range provided sediment to fill the adjacent basin, the sediment being the lasting record of the event. In addition to the Uinta Mountains-Uinta Basin couplet, other smaller, structurally asymmetrical uplifts were also formed during this phase. These are the San Rafael Swell, the Monument Upwarp, and the Circle Cliffs anticline. Other monoclinal flexures in southern Utah and Arizona may also be part of this activity. Laramide-type folding overlaps slightly in time with Sevier compression. Laramide uplifts are, in fact, compressive features themselves with a moderate amount (few miles) of local horizontal offset. They have been explained as a result of a few degrees of rotation of the Colorado Plateau relative to the continental interior during latest Cretaceous and early Cenozoic time when the eastern Pacific floor was being subducted beneath the North American plate.

Phase VII—Explosive caldera and stratovolcano phase. Igneous activity started in northwestern Utah about 40 million years ago, in late Eocene time, after a long interval (100 million years, since late Jurassic) of volcanic quiescence during Sevier compressional events. Igneous

FIGURE 9 — Wasatch Mountains east of Mapleton. U.S. Highway 89-6 emerges from Spanish Fork Canyon at right edge of photo; Hobble Creek Canyon extends into range at left side. Triangular form of interstream ridges along mountain front (triangular faceted spurs) resulted from episodic uplift of range front along the Wasatch fault. Recent fault splays that cut Quaternary lake deposits follow irregular courses along base of mountains. The regular bench that follows the mountain front just beneath the lowest triangular facet is the highest shoreline of Lake Bonneville. The bedrock of the snow-capped mountain front is entirely Pennsylvanian-Permian Oquirrh Formation; the farmland of the Mapleton bench is mostly underlain by Provo-level Lake Bonneville deltaic materials from Hobble Creek and Spanish Fork Canyons.

action spread southward through the next 20 million years affecting all parts of the state except the Uinta Basin. Three areas are particularly noteworthy for the volume of igneous rocks produced. Stratovolcanoes in the East Tintic area erupted a pile of volcanic rocks more than 2 miles thick; associated intrusive stocks produced important mineral deposits. Near Marysvale, some stratovolcanoes collapsed into calderas after producing a great blanket of lava flows, ash-flow tuffs, and other volcanogenic deposits that still cover the area between Richfield and Panguitch. Perhaps the most spectacular volcanic activity of Oligocene time took place near Indian Peak on the Utah-Nevada border. From 33 to 26 million years ago a caldera complex astride the Utah-Nevada line broadcast 2,500 cubic miles of dacite and rhyolite ash-flows over an area of 10,000 square miles. Compared to this, the small volume of ash erupted by modern volcanoes, such as Mt. St. Helens (about one cubic mile), is insignificant. The Utah Oligocene explosive eruptions were so big and so numerous that the worldwide dust clouds they produced were likely the cause of Oligocene atmospheric cooling, an event recorded in 30 million-year-old sediments on the Pacific ocean floor. In southeastern Utah, dioritic magma bodies of stiff, dough-like consistency were intruded into Mesozoic strata, doming them up to form the world-famous laccoliths of the Henry, La Sal, and Abajo Mountains. In southwestern Utah similar intrusions formed the Iron Mountain-Granite Mountain-Three Peaks laccoliths west of Cedar City. Stocks of granitic rocks congealed at depth in various places in northern Utah; they include the mineral-producing stocks now exposed at Park City, Alta, Bingham, and Gold Hill.

A number of plate-tectonic models have been proposed to explain the widespread igneous activity in Utah 20–40 million years ago. Most models include subduction of an oceanic plate beneath western North America; various angles of descent of the plate, even splitting of the plate into segments has been proposed. Another idea relates to the thickening of continental crustal rocks caused by compressional stacking of the upper crustal layers during the Sevier orogeny in the preceding phase. Delayed reaction of heat flow patterns to the changed crustal thickness may have been a significant factor. To date, no one has proposed a model that everyone accepts. Because of its preceding history, the western U.S. was already geologically complex before the volcanic phase. Interplay of several

South half of Rock Canyon Delta, showing Fault Scarps.

FIGURE 10 — Recent fault scarps (about 15 feet high) that cut Lake Bonneville deltaic deposits (in mid-distance) and modern alluvial fan deposits (nearest scarps) at the mouth of Rock Canyon east of Provo. Sketch from Gilbert's Lake Bonneville Monograph (1890).

FIGURE 11 — Lake Bonneville shorelines (upper benches) and Recent fault scarps that cut the toe of the Lake Bonneville deposits at the base of the Wasatch Mountains near Farmington. Sketch from Gilbert (1884).

factors probably produced the heating event evidenced by the surface igneous activity.

Phase VIII—Regional uplift, basin-range faulting, bimodal volcanism, and Lake Bonneville phase. We are still in this latest and, so far, shortest phase of Utah's geologic history. During this phase most of the geographic characteristics we associate with Utah were formed: The raising of the state (along with much of the western U.S.) to where its valley bottoms are about a mile above sea level, and its mountain tops two miles; the breaking up of western Utah into north-south basins and ranges, debris from the ranges filling adjacent valleys; the rejuvenation by the uplift, of river systems in eastern Utah so that they have carved and are carving the canyonlands, removing incredible volumes of sediment via the Colorado River conveyor belt to fill the Gulf of California basin; the eruption of small basaltic cinder cones (Figure 8) and shield volcanoes between Delta and St. George in southwestern Utah; and finally, the covering of western Utah by Lake Bonneville between 30,000 and 10,000 years ago along with the concurrent glaciation of the Wasatch and Uinta Mountains and the High Plateaus of central and southern Utah. Bonneville shorelines are a conspicuous feature of northern and western Utah (Figure 11).

We are reminded of the ongoing geologic activity of this phase by occasional earthquakes (see Figures 9 and 10 for a view of the Wasatch Fault) and by expensive fluctuation in the level of Lake Bonneville's successor, the Great Salt Lake. Earth's crust in the Great Basin between Salt Lake City and Reno is spoken of by geophysicists as being "hot and soft;" the abnormally high heat flow in the Great Basin is evidenced by hundreds of warm springs scattered around the region, more than anywhere else in North America. The heat, in turn, has caused the rigid part of the Earth's crust to be thinner here than elsewhere, and heating of the crust has made it "soft" or slightly plastic as shown by its reduced ability to transmit waves of earthquake energy. Western North America, between California and Colorado, is an anomalously broad elevated area. Its elevating force is heat. Why heat is produced in greater than average amounts beneath the western U.S. is not completely understood. Whether it is a by-product of west coast continental plate-oceanic plate relationships, or whether it is caused by anomalies deeper in the mantle is still being investigated.

ARCHEAN

The term "Archean" was redefined in 1976 by the Subcommission on Precambrian Stratigraphy of the International Union of Geological Sciences to include all rocks older than 2500 million years [see Sims (1979), Harland et al. (1982), and Snelling (1985), for review of the history of usage of Precambrian time scale terms].

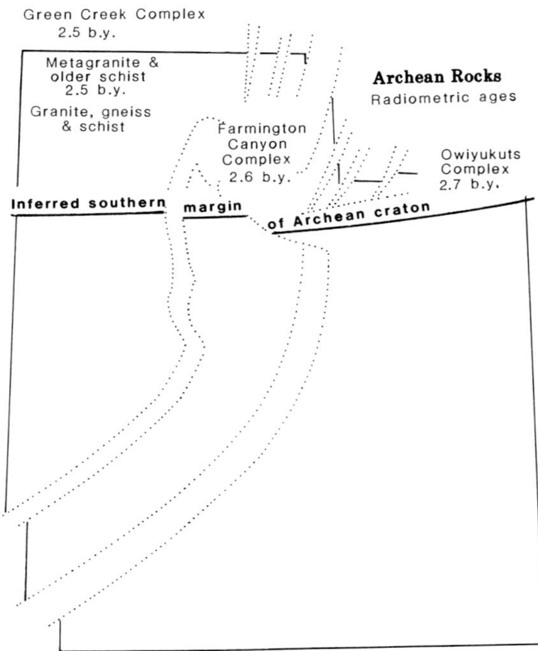

FIGURE 12 — Archean rocks (2.5 billion years and older) are preserved only north of an east-west line through Salt Lake City (Bryant and Nichols, 1988). Oldest rocks are the Owiyukuts Complex and the Farmington Canyon Complex. The latter metamorphic assemblage may have been formed from a protolith of sedimentary and basaltic rocks, possibly as old as 3 billion years, which was first metamorphosed 2.6 b.y. ago. Dotted lines show the trends of Mesozoic thrusts, with western Utah pulled apart schematically to its pre-thrust configuration.

Rocks of this age are limited, in Utah, to the area north of Salt Lake City as shown on Figure 12.

Sears et al. (1982) named the Owiyukuts Complex for exposures in the Red Creek area on the northeastern flank of the Uinta Mountains where the complex consists of granitic gneiss and other types of gneissic rocks from which a Rb-Sr whole-rock isochron age of 2.7 b.y. was obtained.

The Farmington Canyon Complex was mapped by Bryant (1984), who divided the complex into four major units: granitic gneiss; migmatite, gneiss, and schist; schist and gneiss; and schist, gneiss and quartzite. Geochronology of the complex was discussed by Hedge et al. (1983) and Bryant (1988) who concluded that the granitic gneiss is early Proterozoic (1800 m.y.) in age, while the other three groups of rocks are layered Archean rocks. Bryant (1987) suggested the following history for the complex: Detritus containing materials as old as 3600 m.y. was shed from the Archean craton and deposited on oceanic crust, and later intruded by basaltic dikes and sills. About 2600 m.y. ago these sediments and intrusions were metamorphosed. Compositions of the layered Archean rocks suggest that they were derived mostly from a sedimentary

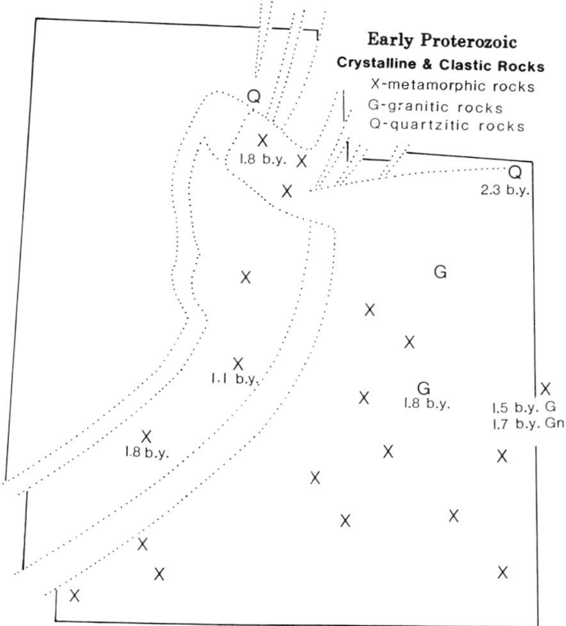

FIGURE 13 — Early Proterozoic rocks consist mostly of gneiss (Gn on map), schist and pegmatite, and lesser amounts of granitic and quartzitic rocks. These rocks represent an addition of a wide belt of rocks to the southern margin of the Archean craton between 1.8 and 1.5 billion years ago. The belt extends southward into Arizona and New Mexico and is part of a northeast-trending belt that extends from Arizona to the Great Lakes region.

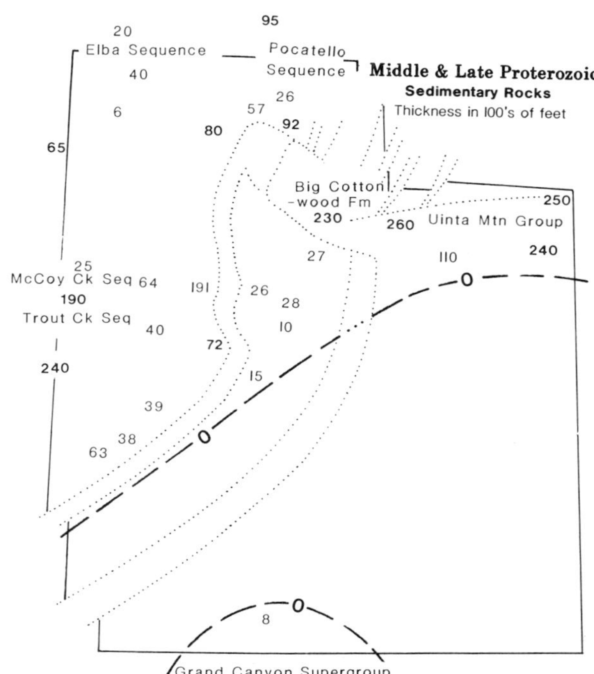

FIGURE 14 — Middle and Late Proterozoic sandstone, shale, and diamictite (now quartzite, argillite, and tillite) accumulated to great thickness in the Uinta trough and the newly-born miogeocline in western Utah. Bold numbers on the map show places where complete sequences, as much as 26,000 feet thick, have been reported. Light numbers represent locations where the Middle and Late Proterozoic sequences are only partially preserved. Sedimentary deposits of this age are absent over a large part of southeastern Utah where Cambrian strata rest directly on Early Proterozoic crystalline rocks.

sequence that consisted of shale, quartz sandstone, and feldspathic sandstone. Thus the sedimentary protoliths of the Farmington Canyon Complex were derived largely from pre-existing continental crust.

The Raft River, Grouse Creek, and Albion ranges of northwestern Utah and Idaho are metamorphic core complex uplifts that have brought Archean rocks to the surface. Compton et al. (1977) summarized the dating problems of these repeatedly heated rocks and concluded that the granitic rocks are about 2500 m.y. old and are intrusive into older gneiss and schist. Armstrong (1968b, 1970b) described similar relationships in the Albion Range, Idaho.

Age relationships of Archean rocks are shown on Figure 15.

PROTEROZOIC

Proterozoic subdivisions shown on Figure 15 were established by the Subcommission on Precambrian Stratigraphy of the International Union of Geological Sciences in 1979 in order to encourage world-wide uniformity of usage. Isotopic dating is the key to the assignment of Precambrian rocks to their proper age bracket, and enough ages have been obtained for Utah's Precambrian rocks to suggest the correlations shown on Figure 15.

Two main groupings of Proterozoic rocks are apparent. Distribution of older Proterozoic rocks is shown on Figure 13. These are "basement" rocks, mostly gneiss and schist, and until radiometric dating and isotopic studies made it apparent that they were younger, they were included with the Archean basement rocks discussed above. Proterozoic crustal history of the western U.S. has been reviewed by Bennett and DePaolo (1987) who suggested that Proterozoic crust was formed in a large part from mantle-derived material, but included about 20 percent of material derived from the Archean craton. Condie (1987) suggested that early Proterozoic rocks represent additions to the craton by collisional accretion of terranes, some of which have lithologic and geochemical characteristics of oceanic island arcs, while other terranes may have originated in continental-margin arcs and back-arc basins. Medaris et al. (1983) presented recent advances in our concepts regarding Proterozoic geology.

Younger Proterozoic rocks are widely exposed in northern and western Utah, as shown on Figure 14. They include several sedimentary suites of considerable thickness that consist mostly of quartzite and argillite, but also

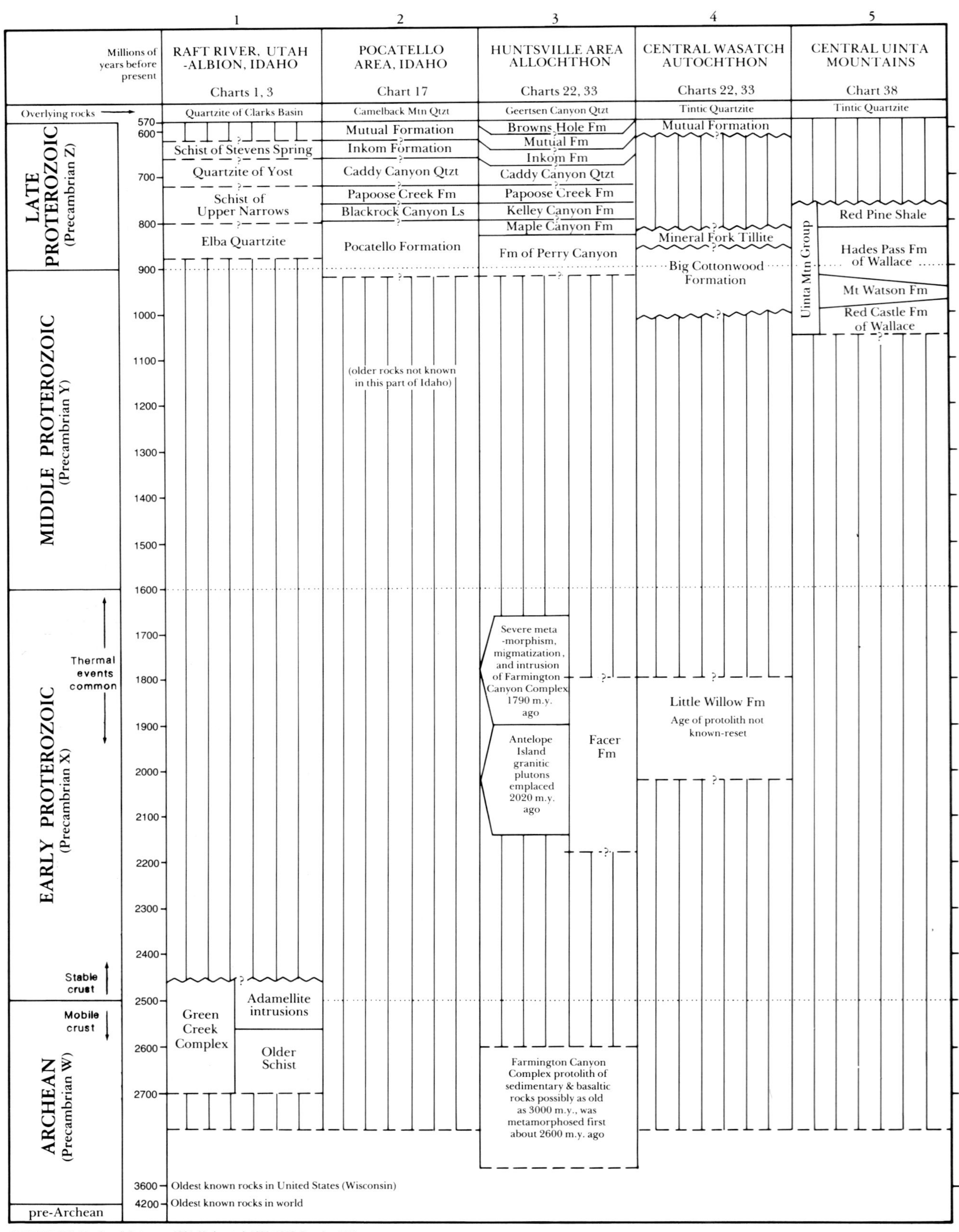

FIGURE 15 — Precambrian correlation table.

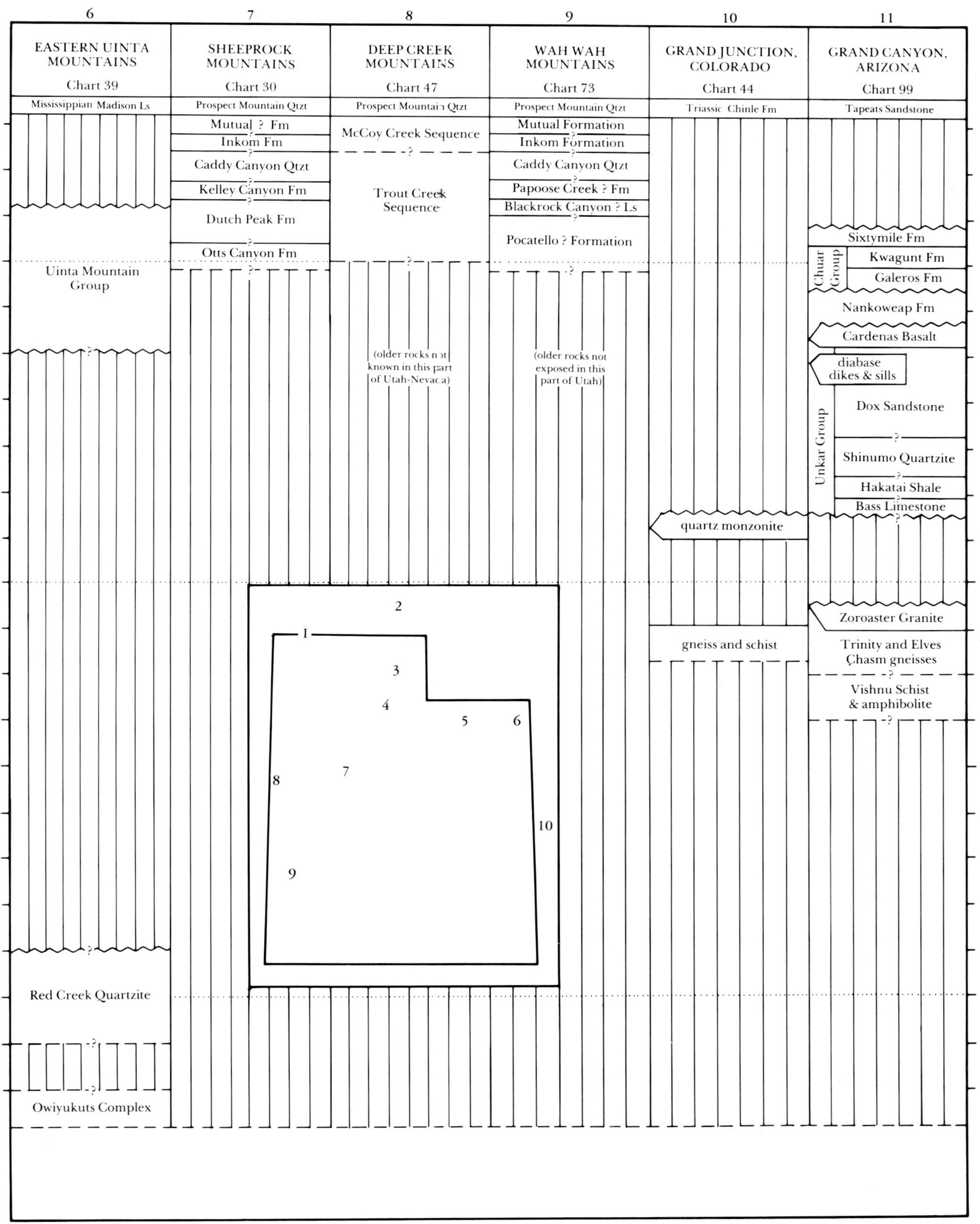
6
EASTERN UINTA MOUNTAINS
Chart 39
Mississippian Madison Ls
Uinta Mountain Group
Red Creek Quartzite
Owiyukuts Complex
7
SHEEPROCK MOUNTAINS
Chart 30
Prospect Mountain Qtzt
Mutual ? Fm
Inkom Fm
Caddy Canyon Qtzt
Kelley Canyon Fm
Dutch Peak Fm
Otts Canyon Fm
8
DEEP CREEK MOUNTAINS
Chart 47
Prospect Mountain Qtzt
McCoy Creek Sequence
Trout Creek Sequence
(older rocks not known in this part of Utah-Nevada)
9
WAH WAH MOUNTAINS
Chart 73
Prospect Mountain Qtzt
Mutual Formation
Inkom Formation
Caddy Canyon Qtzt
Papoose Creek ? Fm
Blackrock Canyon ? Ls
Pocatello ? Formation
(older rocks not exposed in this part of Utah)
10
GRAND JUNCTION, COLORADO
Chart 44
Triassic Chinle Fm
quartz monzonite
gneiss and schist
11
GRAND CANYON, ARIZONA
Chart 99
Tapeats Sandstone
Sixtymile Fm
Chuar Group
Kwagunt Fm
Galeros Fm
Nankoweap Fm
Cardenas Basalt
diabase dikes & sills
Unkar Group
Dox Sandstone
Shinumo Quartzite
Hakatai Shale
Bass Limestone
Zoroaster Granite
Trinity and Elves Chasm gneisses
Vishnu Schist & amphibolite
1
2
3
4
5
6
7
8
9
10

include some quartzite conglomerate, diamictite (tillite), and limestone beds as well as a few basaltic lavas in the basal Pocatello Formation, described by Harper and Link (1986). Deposition of Late Proterozoic strata marks the beginning of major subsidence of the west edge of North America, but the Proterozoic subsidence pattern differs from later miogeoclinal patterns by including an east-west Uinta trough and a less well defined Grand Canyon basin, the north edge of which is identified in wells in southern Utah, as shown on Figure 14.

Very few radiometric or paleomagnetic ages have so far been obtained for Late Proterozoic rocks in Utah. Dating of the volcanic rocks of the Pocatello Formation has not yet been satisfactorily accomplished. Correlation is based on lithology using the less common rock types as keys. Names from the Pocatello area, Idaho have been used as far south as the Wah Wah Mountains in Beaver County, the key beds being the Inkom Formation (a thick argillite), and the Blackrock Canyon Limestone. Christie-Blick et al. (1988) suggested that major lithologic changes, such as that represented by the Inkom Formation, can be correlated on a world-wide basis.

Diamictites are perhaps the most unique rocks in Late Precambrian sequences. Many papers have argued whether they are of glacial origin or not with the current consensus being that they are (Crittenden et al., 1983). As such, they provide the best means of correlation within Late Proterozoic sequences in the western U.S., based on the assumption that a major glacial event would affect a large area at the same time. Crowell (1982) summarized continental glaciation through Precambrian time, suggesting three glaciation peaks at 940, 770, and 615 million years ago. In Utah, in addition to the Mineral Fork Tillite, glacial deposits are included in the Pocatello, Perry Canyon, Dutch Peak, and Trout Creek Formations, as shown on Figure 15 and on charts of individual areas.

The Precambrian-Cambrian boundary has been placed at the base of the Camelback Mountain-Geertsen Canyon-Tintic-Prospect Mountain Quartzites, but the placement is more one of convenience than established on fossil, radiometric, or magnetic data. However, the thickness pattern of Early Cambrian (units listed above) quartzite deposition (Figure 16) is more similar to other early Paleozoic patterns (Figures 18, 24, 29, and 30) than to the rather ill-defined subsidence pattern of Late Proterozoic rocks (Figure 14). Several geologists have suggested that miogeoclinal subsidence of the western margin of North America was a result of rifting, where an as-yet-unidentified land mass broke away from North America along an irregular northerly-trending rift zone in central Nevada. Stewart (1976) and Stewart and Suczek (1977) noted that the subsidence-depositional pattern changed about 850 m.y. ago; Armin and Mayer (1983) using subsidence curves, with the isostatic effects of sediment and water loading removed, concluded that rifting began about 600 m.y. ago. Harper and Link (1986) considered the volcanic rocks of the Pocatello and Perry Canyon formations to be rift-related and argued for starting the rifting earlier than 600 m.y. ago.

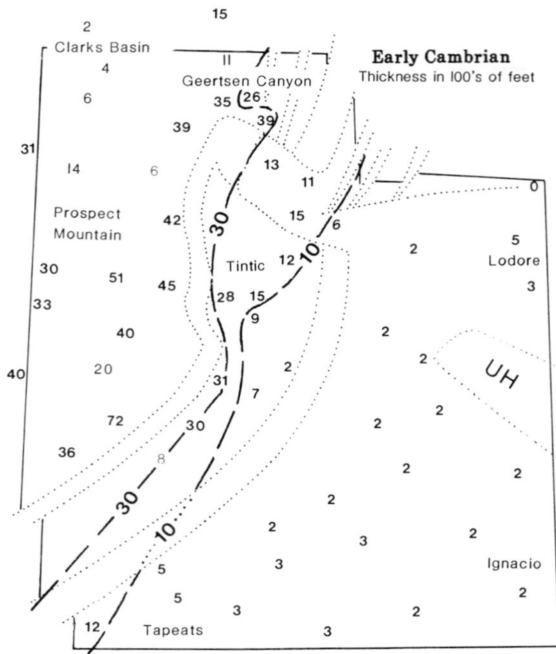

FIGURE 16 — Early Cambrian strata are mostly quartzites that have been given the several names shown. These were the beach sands of a sea that spread eastward across Utah during late Early Cambrian and early Middle Cambrian time. Thus, the quartzites of eastern Utah are slightly younger than those of western Utah. All of the quartzites at the bases of Cambrian sections, including those eastern ones that are early Middle Cambrian, are included on this map. Lightest numbers show places where the sections are only partially preserved. UH indicates the Uncompahgre Highland area, from which Cambrian deposits were removed by later Paleozoic erosion. Eastern Utah, where the basal sandstones are less than 1,000 feet thick, is part of the stable craton. Western Utah, where thicknesses exceeding 4,000 feet are common, is part of the Utah-Nevada miogeoclinal belt.

CAMBRIAN

Middle Cambrian strata in Utah play a key role in North American biostratigraphy. The House Range (Figure 20 and chart 51) contains one of the most complete fossil-bearing sequences of Middle and Upper Cambrian rocks found anywhere (Hintze and Robison, 1987). Trilobites are the guide fossils to Cambrian time and Cambrian correlations are based almost entirely on them, as shown on the left side of the Cambrian correlation table, Figure 19. Agnostoid trilobites enable world-wide correlations; other trilobite zones are more provincial (Robison et al., 1977; Briggs and Robison, 1984). Isotopic age assignments on Figure 19 are taken from Palmer (1983), Snelling (1985), and Harland (1982). Trilobites are

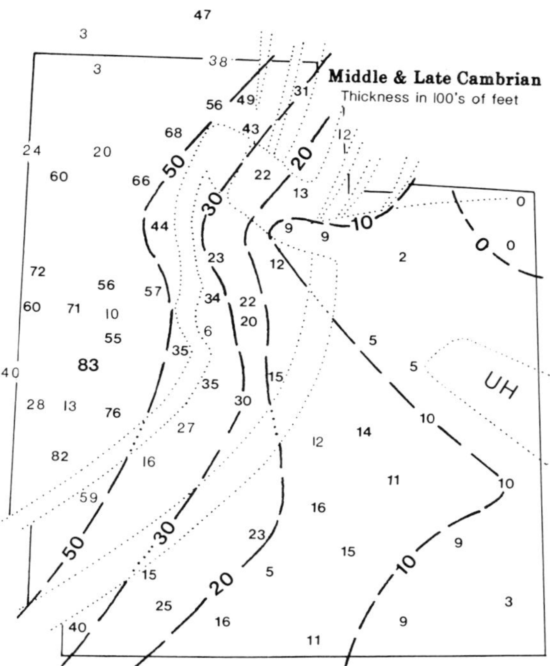

FIGURE 17 — Middle and Late Cambrian deposits are mostly shallow marine limestones and dolomites that are separated into packages by several thin, but widespread, deeper-water shale horizons. Deposits of this interval thicken from zero in northeastern Utah to more than 5,000 feet in the western miogeoclinal belt. The thickest deposits, 8,300 feet, accumulated in the House Range area. Lightest numbers show partial sections or those of uncertain identity. UH indicates the Uncompahgre Highland, an area of late Paleozoic uplift from which Cambrian rocks were eroded. The numerous names that have been applied to rocks of this age are shown on the Cambrian correlation table (Figure 19) and on individual charts.

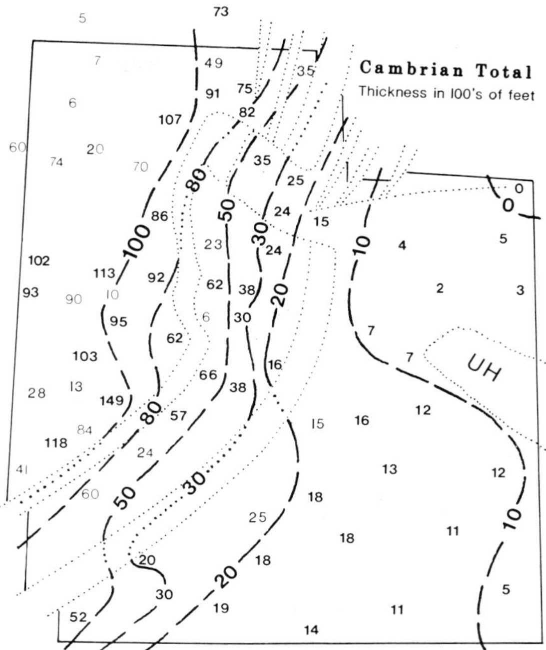

FIGURE 18 — Total thickness of Cambrian deposits exceeds 10,000 feet in western Utah. Cambrian rocks are more widely preserved and distributed in Utah than those of any other period. UH indicates the Uncompahgre Highland which arose in late Paleozoic time and from which all Cambrian rocks have been eroded. The lighter numbers show locations where the Cambrian section is only partially preserved.

uniquely preserved in the Wheeler Shale in the House Range so that whole specimens are abundant. There is likely not a paleontologic museum in the world that does not include *Elrathia kingii* from the Wheeler Shale of Utah in its collections.

Early Cambrian faunas are uncommon in Utah rocks and only uppermost Early Cambrian fossils have been documented (Robison and Hintze, 1972). The Precambrian-Cambrian transition problem (Mount et al., 1983) will not likely be resolved by Utah's rocks.

On the other end of Cambrian time, Utah rocks have much to contribute to the definition of the Cambrian-Ordovician boundary. The International Working Group on the Cambrian-Ordovician Boundary has examined the conodont-trilobite sequence (conodonts are microscopic tooth-like fossils that first appear abundantly in Late Cambrian rocks) in the upper Lava Dam Member of the Notch Peak Formation in western Utah and considers it an excellent reference section that contains a virtually complete record of all trilobite and conodont zones known in uppermost Cambrian and lowermost Ordovician strata (Miller, 1984; Taylor and Miller, 1981).

It has long been apparent that Cambrian strata are more than ten times as thick in western Utah as in the eastern half of the state, as shown on Figures 16, 17, and 18. Lithologies in both areas are similar and were laid down in the shallow waters of an epicontinental sea. The reason why the west half of the state accumulated a very thick stack of shallow-water sediments has recently been explained by the idea that the western edge of the continent subsided as it became the raw trailing edge of North America, when rifting separated it from an unidentified land mass to the west. Slow loss of heat along the newly formed continental margin allowed it to sink at about the same rate as sediments could accumulate atop its foundering surface. Armin and Mayer (1983) suggested that the subsidence began about 600 m.y. ago. This is about the same time as the late Proterozoic Mutual Formation and the Early Cambrian Prospect Mountain and Tintic Quartzites were deposited. These Early Cambrian quartzites include a thin but widespread basalt flow which is taken as partial evidence of a rifting environment. The basalt flow is generally altered and has not yet yielded a satisfactory radiometric or paleomagnetic age.

CAMBRIAN

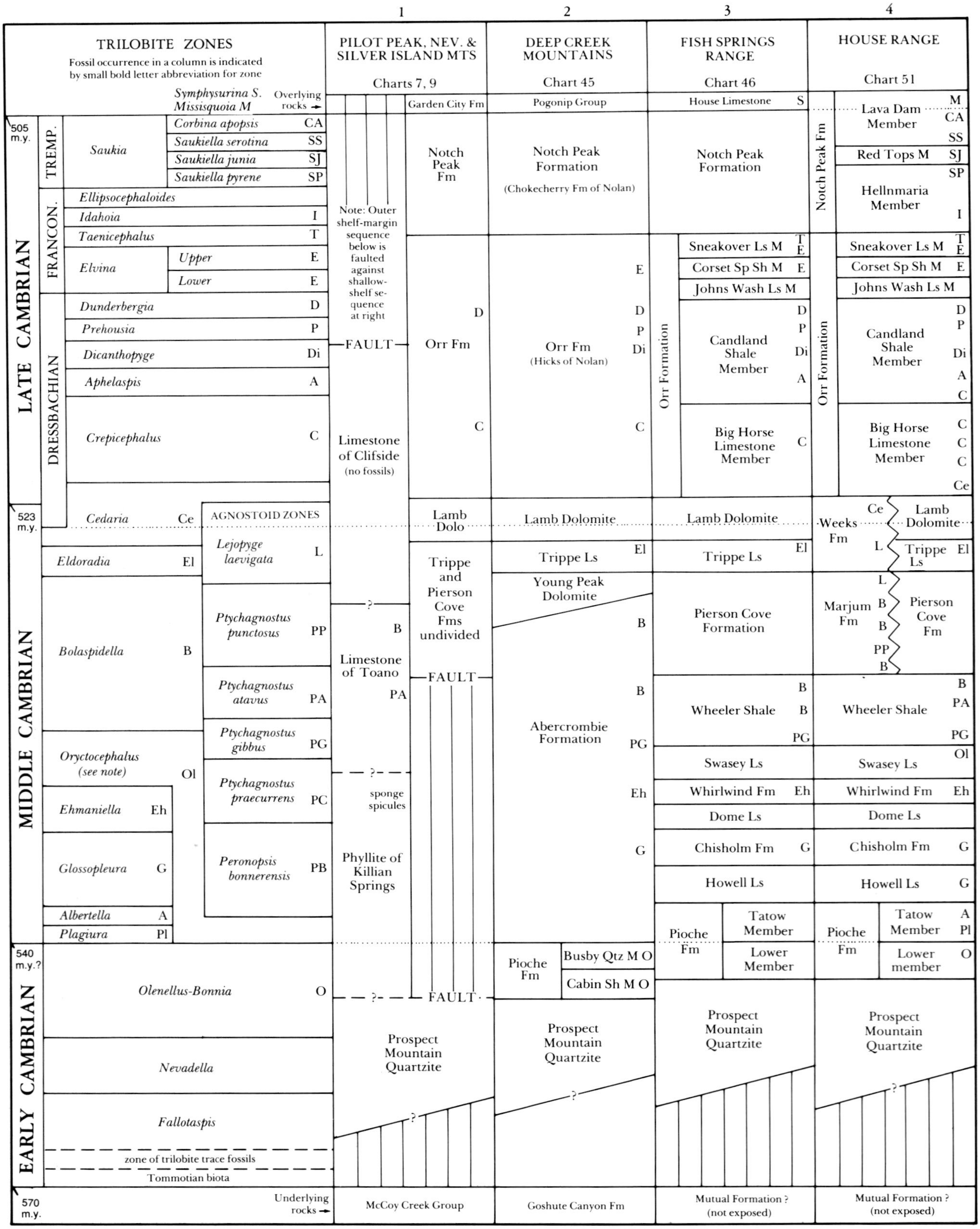

Series / Stage	TRILOBITE ZONES (Fossil occurrence in a column is indicated by small bold letter abbreviation for zone)	Agnostoid Zones	1 PILOT PEAK, NEV. & SILVER ISLAND MTS (Charts 7, 9)	2 DEEP CREEK MOUNTAINS (Chart 45)	3 FISH SPRINGS RANGE (Chart 46)	4 HOUSE RANGE (Chart 51)
	Symphysurina S. / *Missisquoia* M — Overlying rocks →		Garden City Fm	Pogonip Group	House Limestone S	Notch Peak Fm: Lava Dam Member M
LATE CAMBRIAN, TREMP. (505 m.y.)	*Saukia*: *Corbina apopsis* CA		Notch Peak Fm	Notch Peak Formation (Chokecherry Fm of Nolan)	Notch Peak Formation	Lava Dam Member CA
	Saukia: *Saukiella serotina* SS					SS
	Saukia: *Saukiella junia* SJ					Red Tops M SJ
	Saukia: *Saukiella pyrene* SP					Hellnmaria Member SP
FRANCON.	*Ellipsocephaloides*					
	Idahoia I		Note: Outer shelf-margin sequence below is faulted against shallow-shelf sequence at right			I
	Taenicephalus T				Orr Formation: Sneakover Ls M T E	Orr Formation: Sneakover Ls M T E
	Elvina: *Upper* E			E	Corset Sp Sh M E	Corset Sp Sh M E
	Elvina: *Lower* E				Johns Wash Ls M	Johns Wash Ls M
DRESSBACHIAN	*Dunderbergia* D			D	Candland Shale Member D	Candland Shale Member D
	Prehousia P		—FAULT—	Orr Fm	Orr Fm (Hicks of Nolan) P	P	P
	Dicanthopyge Di				Di	Di	Di
	Aphelaspis A					A	A C
	Crepicephalus C		Limestone of Clifside (no fossils)	C	C	Big Horse Limestone Member C	Big Horse Limestone Member C C C Ce
MIDDLE CAMBRIAN (523 m.y.)	*Cedaria* Ce	AGNOSTOID ZONES	Lamb Dolo	Lamb Dolomite	Lamb Dolomite	Weeks Fm Ce / Lamb Dolomite
	Eldoradia El	*Lejopyge laevigata* L	Tripple and Pierson Cove Fms undivided	Tripple Ls El	Tripple Ls El	Weeks Fm L / Tripple Ls El
	Bolaspidella B	*Ptychagnostus punctosus* PP	? B	Young Peak Dolomite; B	Pierson Cove Formation	Marjum Fm L B B PP B / Pierson Cove Fm
		Ptychagnostus atavus PA	Limestone of Toano PA; —FAULT—	Abercrombie Formation B	Wheeler Shale B B	Wheeler Shale B PA
	Oryctocephalus (see note) Ol	*Ptychagnostus gibbus* PG		PG	PG	PG Ol
		Ptychagnostus praecurrens PC	? sponge spicules	Eh	Swasey Ls; Whirlwind Fm Eh	Swasey Ls; Whirlwind Fm Eh
	Ehmaniella Eh				Dome Ls	Dome Ls
	Glossopleura G	*Peronopsis bonnerensis* PB	Phyllite of Killian Springs	G	Chisholm Fm G; Howell Ls	Chisholm Fm G; Howell Ls G
	Albertella A				Pioche Fm: Tatow Member	Pioche Fm: Tatow Member A
	Plagiura Pl					Pl
EARLY CAMBRIAN (540 m.y.?)	*Olenellus-Bonnia* O		? —FAULT—	Pioche Fm: Busby Qtz M O; Cabin Sh M O	Lower Member	Lower member O
	Nevadella		Prospect Mountain Quartzite	Prospect Mountain Quartzite	Prospect Mountain Quartzite	Prospect Mountain Quartzite
	Fallotaspis		?	?		?
	zone of trilobite trace fossils					
	Tommotian biota					
(570 m.y.)	Underlying rocks →		McCoy Creek Group	Goshute Canyon Fm	Mutual Formation ? (not exposed)	Mutual Formation ? (not exposed)

Note: In Middle Cambrian, agnostoid trilobite zones are shown at right side of zone list; Oryctocephalus is a long-ranging open-shelf group that extends from the base of the Middle Cambrian to the base of the Bolaspidella Zone.

REFERENCES - Robison, 1976, 1984; Palmer, 1979, 1981; Mount et al, 1983.

FIGURE 19 — Cambrian correlation table.

CAMBRIAN

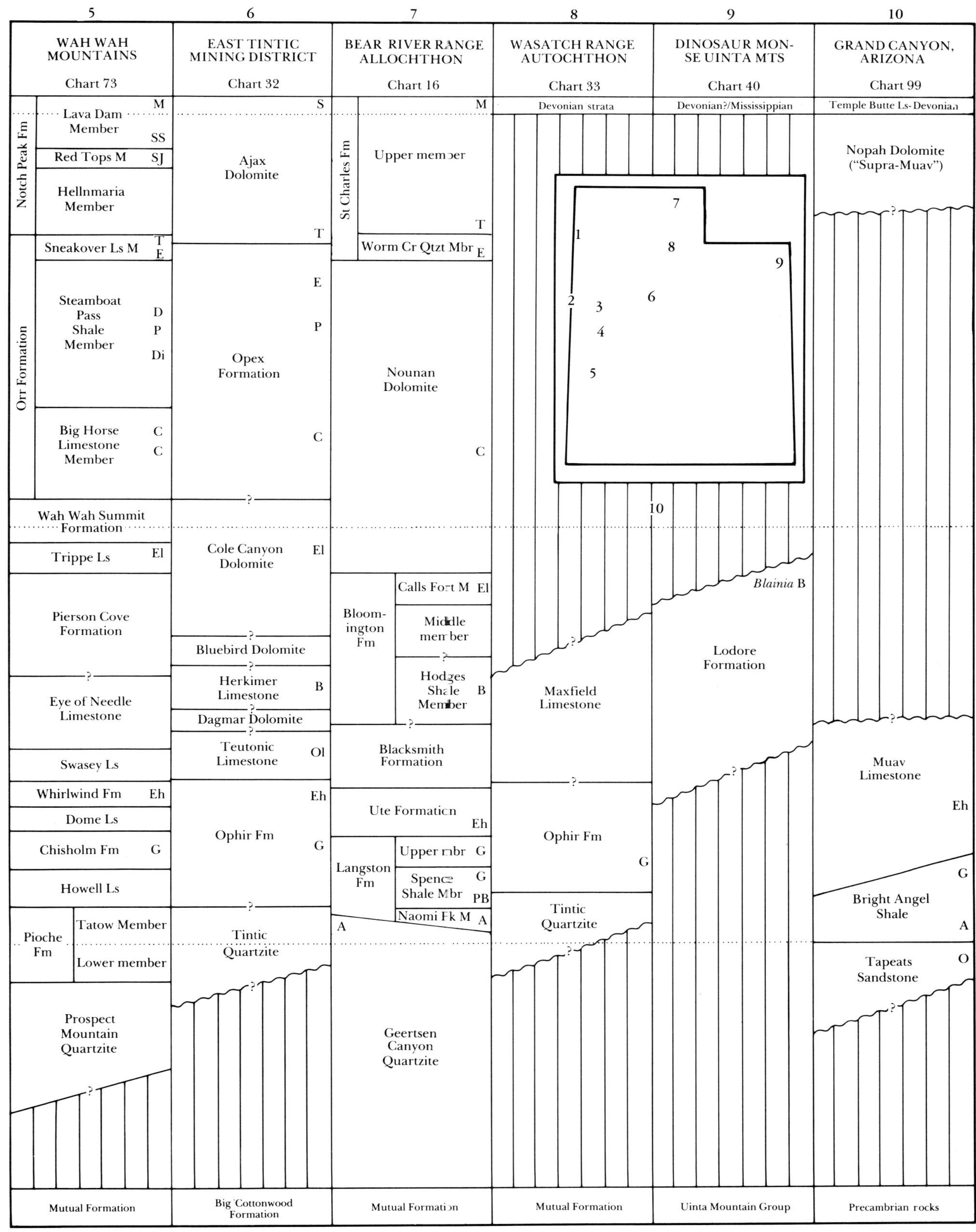

5
6
7
8
9
10
WAH WAH MOUNTAINS
Chart 73
EAST TINTIC MINING DISTRICT
Chart 32
BEAR RIVER RANGE ALLOCHTHON
Chart 16
WASATCH RANGE AUTOCHTHON
Chart 33
DINOSAUR MON-SE UINTA MTS
Chart 40
GRAND CANYON, ARIZONA
Chart 99
Devonian strata
Devonian?/Mississippian
Temple Butte Ls-Devonian
Notch Peak Fm
Lava Dam Member
Red Tops M
Hellnmaria Member
Orr Formation
Sneakover Ls M
Steamboat Pass Shale Member
Big Horse Limestone Member
Wah Wah Summit Formation
Trippe Ls
Pierson Cove Formation
Eye of Needle Limestone
Swasey Ls
Whirlwind Fm
Dome Ls
Chisholm Fm
Howell Ls
Pioche Fm
Tatow Member
Lower member
Prospect Mountain Quartzite
Ajax Dolomite
Opex Formation
Cole Canyon Dolomite
Bluebird Dolomite
Herkimer Limestone
Dagmar Dolomite
Teutonic Limestone
Ophir Fm
Tintic Quartzite
St Charles Fm
Upper member
Worm Cr Qtzt Mbr
Nounan Dolomite
Bloomington Fm
Calls Fort M
Middle member
Hodges Shale Member
Blacksmith Formation
Ute Formation
Langston Fm
Upper mbr
Spence Shale Mbr
Naomi Pk M
Geertsen Canyon Quartzite
Maxfield Limestone
Ophir Fm
Tintic Quartzite
Lodore Formation
Blainia B
Nopah Dolomite ("Supra-Muav")
Muav Limestone
Bright Angel Shale
Tapeats Sandstone
M
SS
SJ
T
E
D
P
Di
C
C
El
Eh
G
S
T
E
P
C
El
B
Ol
Eh
G
M
T
E
C
El
B
Eh
G
G
PB
A
A
G
Eh
G
A
O
?
1
2
3
4
5
6
7
8
9
10
Mutual Formation
Big Cottonwood Formation
Mutual Formation
Mutual Formation
Uinta Mountain Group
Precambrian rocks

Figure 20 — Cambrian strata on the west face of the House Range west of Delta, Utah. Upper cliff with cave is Dome Limestone; tree-covered bench beneath it is shaly Chisholm Formation; light-banded cliff underneath Chisholm is upper member of Howell Limestone with dark Millard Member at its base. Shale slope beneath Howell cliffs is Tatow Member of the Pioche Formation; dark ledges at center of photo are quartzites of the lower member of the Pioche Formation. Light and medium gray bedrock at the base of the slope is down-faulted Dome Limestone, Whirlwind Formation and Swasey Limestone. Light terrace near base of slope is Lake Bonneville beach deposit. White playa deposits of Tule Valley show in foreground. See Chart 51.

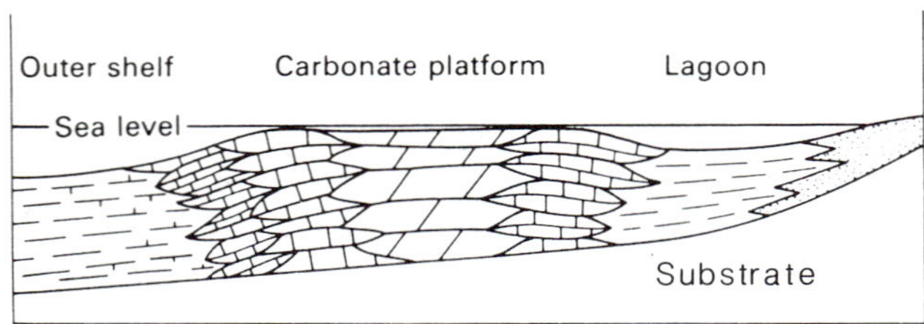

FIGURE 21 — Generalized west-east reconstruction of Cambrian depositional environments in western Utah showing a landward inner detrital belt, a shallow-water carbonate bank, and an outer slope detrital belt facing the open ocean to the west. Free-swimming agnostoid trilobites are found mostly in deposits of the outer belt. Bottom-dwelling polymeroid trilobites occur in deposits of both the inner detrital belt and along the shallow edge of the outer belt. The middle carbonate bank is generally barren except for algal-deposited limestone.

Most Cambrian sequences in Utah consist of a basal quartzite (Figure 16), overlain by a thin shale (Pioche, Ophir, Bright Angel), which is succeeded by a thick carbonate blanket with some shaly interbeds. Many paleoenvironmental studies (Buterbaugh, 1982; Deputy, 1984; Gardiner, 1974; Hay, 1982; Podrebarac, 1976; Kepper, 1976; Lohmann, 1976; Koepnick, 1976; Brady and Rowell, 1976; Brady and Koepnick, 1979; White, 1973; Rees, 1986; Rogers, 1984; McCollum, 1987; to name some) have concluded that environments represented by Utah Cambrian lithologies range from a broad subtidal platform, shoaling to the east, to deeper water slope deposits towards the west. Figure 21 depicts the general relationships. In a regional study, Rees (1986) suggested that the platform-slope transition was controlled by a northeasterly trending fault. Its direction runs perpendicular to a Cambrian rift system envisioned by Larson et al. (1985). Kepper (1981) proposed Cambrian syndepositional faulting that parallels the trends of Larson et al.

Regional syntheses of Cambrian paleoenvironmental patterns have been given by Palmer and Hintze (1988), and by Ross (1975). Location of the western margin of the North American continent in central Nevada in Cambrian time was discussed by Rowell et al. (1979), and Kepper

FIGURE 22 — Ibex, Utah, site of Jack Watson's one-shack "ranch" during the first third of this century. Potholes in Ordovician Eureka Quartzite ledges in the foreground caught rainwater for bare subsistence needs. Shack burned down about 1960. Fossil Mountain, in left distance, is peaked by Eureka Quartzite underlain by band of darker Crystal Peak Dolomite; light ledges below that are Watson Ranch Quartzite, about 240 feet thick here. Gullies in slopes at base of Fossil Mountain expose the very fossiliferous Lehman Formation and Kanosh Shale. See Chart 70.

(1981). Ziegler et al. (1979) concluded that during Late Cambrian time world land areas were limited to low latitudes, the higher latitudes were entirely occupied by polar oceans. Heat budget calculations indicate that this particular distribution of land and sea would result in less absorbed solar radiation, suggesting that Cambrian climates were cooler than most other non-glacial times in earth history.

Cambrian deposits originally blanketed almost all of Utah and they have been fortuitously preserved for us by other early Paleozoic strata which protected them from mid-Paleozoic and later erosion. Because they are the most uniform and widespread pre-Cretaceous-thrusting strata in Utah they make the best rocks to use to estimate amounts of relative horizontal displacement of upper and lower plates across thrusts. In two locations in Utah, Cambrian sequences of contrasting thickness and formational nomenclature are juxtaposed by thrusting. In the Canyon Range, Cambrian rocks of House Range aspect (Chart 59) are on the upper (Canyon Range) plate structurally emplaced above Cambrian rocks of East Tintic aspect (Chart 60). Similarly, near Ogden the allochthonous Cambrian (Chart 22) carries Logan-Bear River Range nomenclature and the autochthonous Cambrian (Chart 23) uses central Wasatch names for the most part. Systematic evaluation of the distance of thrust transport necessary to obtain the present relations has not been published.

ORDOVICIAN

Spanning 70 million years, the Ordovician is the longest Paleozoic period. Rapid evolutionary radiation during the period created many new life forms, several of which are useful as guides to time. In addition to trilobites, the main Cambrian guide-fossil group, the Ordovician also contains graptolites, conodonts, brachiopods, and cephalopods that are ubiquitous and diverse enough to be divided into faunal zones. The International Subcommission on Ordovician Stratigraphy has published a correlation chart with explanatory notes (Ross et al., 1982) that summarizes fossil zonation schemes and radiometric dating findings for the Ordovician System. Utah Ordovician rocks play a prominent role in defining fossil ranges within the system; no place in the world has a more diverse assemblage of early Ordovician fossils than western Utah. The name Ibexian, for the lowest part of the Ordovician, as shown on Figure 26, comes from a fossili-

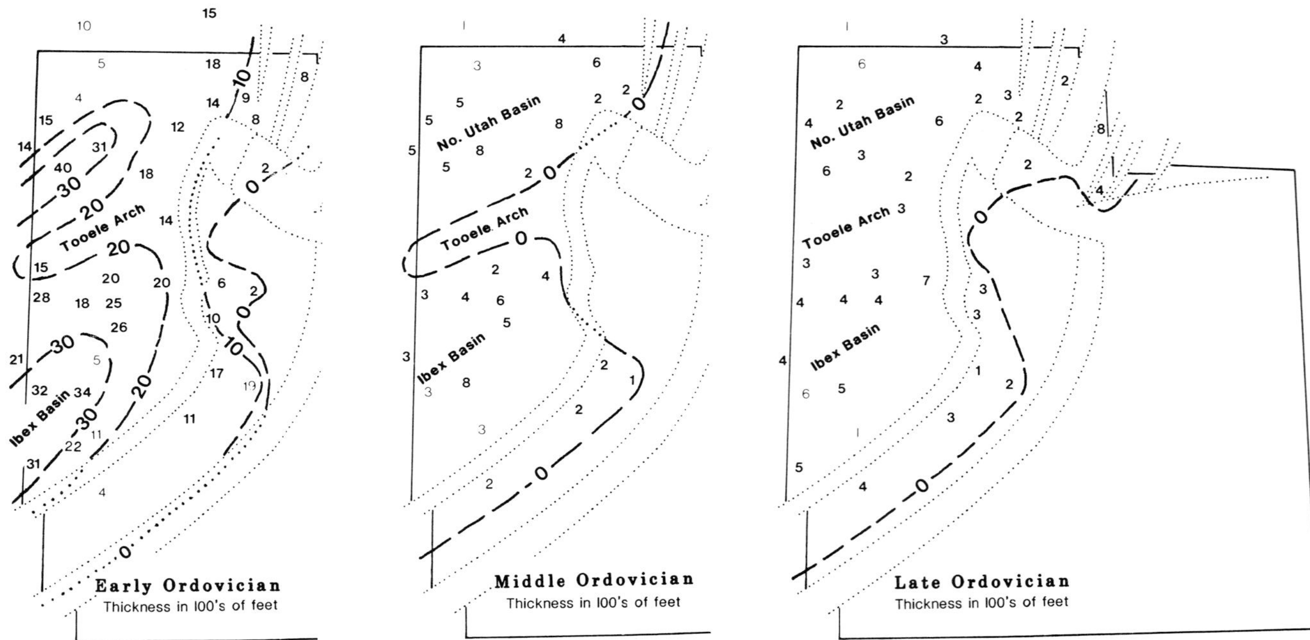

FIGURE 23 — Ordovician strata fall naturally into three parts. Early Ordovician strata are called the Pogonip Group in west-central Utah and the Garden City Formation in northern Utah. These beds have yielded exceptionally diverse fossil faunas that serve as a North American reference standard. Middle Ordovician beds include the Eureka and Watson Ranch quartzites in the Ibex Basin and the Swan Peak Quartzite in northern Utah. Late Ordovician carbonates are the Fish Haven Dolomite of northern Utah and the Ely Springs Dolomite elsewhere. The Tooele Arch was a non-depositional positive area in mid-Ordovician time; but Late Ordovician dolomite spread across the arch. Lighter numbers show locations where only parts of the original thicknesses are preserved.

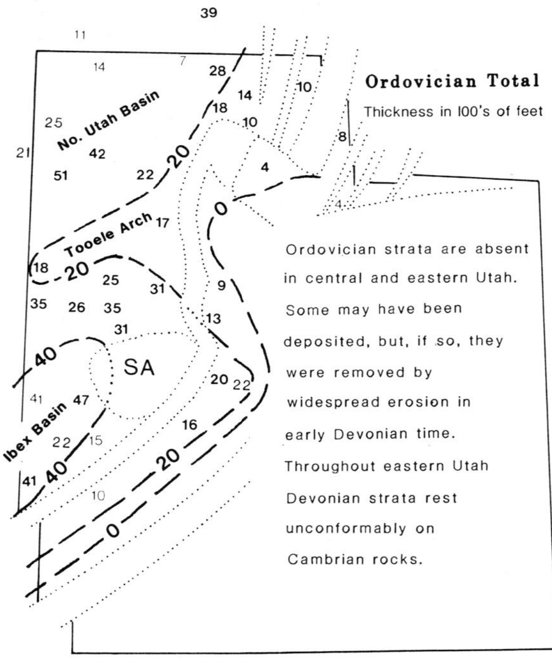

FIGURE 24 — Total Ordovician thickness exceeds 4,000 feet in the Ibex and Northern Utah basins. SA indicates the Sevier Arch, a Mesozoic uplift from which all Ordovician strata have been removed. Lighter numbers show locations where Ordovician strata are incompletely preserved.

ferous locality (Figure 22) in the south part of the Confusion Range in western Utah (Hintze 1982, 1987). Precise boundaries for the upper and lower limits of the Ordovician System are still in the process of being set by International Working Groups on the Cambrian-Ordovician and Ordovician-Silurian boundaries. Again, the western Utah strata are useful because they include a nearly complete succession of trilobite and conodont zones across the Cambrian-Ordovician boundary interval (Miller, 1984). The Ordovician-Silurian boundary will not be established using Utah rocks because they are not fossiliferous enough. Only recently has the Late Ordovician-Silurian chronostratigraphic zonation been worked out for Utah's strata using conodonts (Leatham, 1985).

Ordovician strata are absent from the craton east of the Mesozoic thrust belt but are thick in the western part of the state. As shown on Figures 23 and 24 the thrust belt nearly coincides with the Ordovician zone of abrupt thickening (hinge-line). The miogeocline is divided into two subbasins during Early and Middle Ordovician time by a positive feature called the Tooele Arch, as shown on Figures 23 and 25. This arching documents differential subsidence within the miogeocline along a northeasterly trend, similar to that suggested for Middle Cambrian faulting postulated by Rees (1986). Some have suggested that the Tooele Arch may be a westward continuation of an east-west Uinta Mountain positive trend, but its direction is different.

Ordovician strata consist of a three-fold sequence: Lower Ordovician rocks are mostly sandy bioclastic limestones; Middle Ordovician rocks include thick quartz sandstones; Upper Ordovician rocks are dolomites, as shown on Figure 26. Lower Ordovician strata in Utah commonly include intraformational flat-pebble conglomerate beds as their most distinctive rock type. Such rocks are believed to have formed in an environment approaching that of a tidal flat, where newly deposited silty limestone layers could be ripped apart by currents and redeposited as "pebbles" almost *in situ*. A few intraformational conglomerate surfaces display ripple marks truncating the pebbles, attesting to their softness at the time of redeposition. Lower Ordovician limestones of Utah generally include a great percentage of fine quartz sand and silt, and the limestone itself is commonly clastic. Trilobites, brachiopods and echinoderms lived in great numbers in these shallow marine waters, but their remains are usually fragmented and water-worn. A broad shallow coastal shelf, across which marine waters deepened westward, is visualized for Early Ordovician time in Utah and Nevada (Ross et al., 1987; Palmer and Hintze, 1988). The equator stretched across Utah during Ordovician time, as shown on Figure 28. Thus, we might picture Ordovician Utah as being something like present-day Florida with warm shallow waters extending many tens of miles offshore. Organisms capable of building local hard-grounds or patch-reefs appeared during Early Ordovician time. Sponge-algal reef mounds a few feet high and a few tens of feet in circumference interrupt the thinner bedded bioclastic deposits in the Fillmore, Wah Wah, and Juab formations of the Ibex area (Chart 70 and Figure 25). Thicknesses of Early Ordovician shallow-water limestones range up to 4000 feet; they make up the lower two-thirds of most Ordovician stratigraphic sections in Utah.

Middle Ordovician quartzites form one of the most conspicuous stratigraphic horizons in the Paleozoic sections of western and northern Utah, appearing as light orange cliffs above gray limestone slopes and below black dolomite cliffs. Middle Ordovician deposits are less widespread than older and younger Ordovician beds, being entirely absent over the Tooele Arch. In northern Utah, quartzite makes up only the upper part of the Swan Peak Formation; the lower Swan Peak is a shale which is equated with the Kanosh Shale of western Utah (see Figure 26). Webb (1958) concluded that the upper Swan Peak and lower Eureka Quartzites were deposited while the Ordovician shoreline was regressing westward into central Nevada. The uppermost part of the Eureka was thought by Webb to have been deposited during the ensuing eastward marine transgression. Oaks et al. (1977) presented the most detailed summary of Swan Peak-

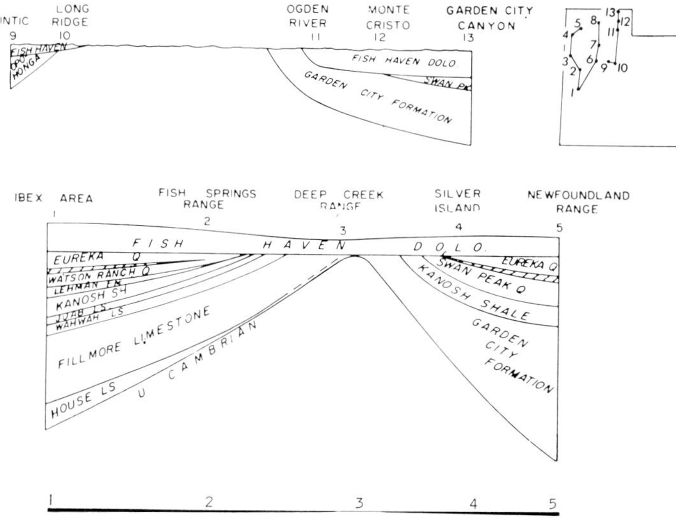

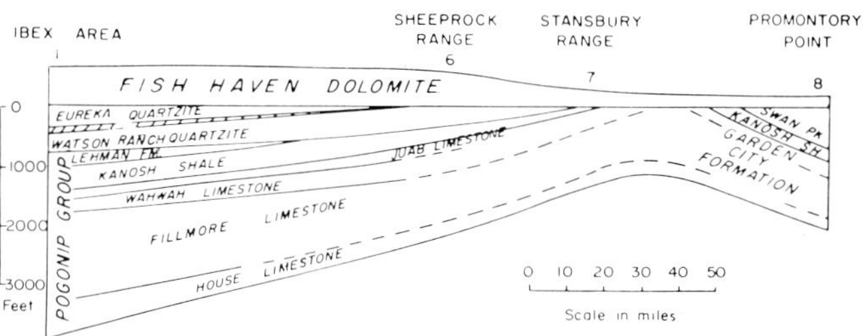

FIGURE 25 — Restored cross-sections through the Ordovician Tooele Arch (after Hintze, 1959). Upper Ordovician Fish Haven Dolomite rests with regional unconformity across Lower and Middle Ordovician strata.

Eureka paleogeographic relationships available. They suggested that the Eureka Quartzite and its Idaho equivalent, the Kinnikinic, were deposited on a broad shallow-marine shelf swept by predominantly southwest-flowing high-energy currents. Ross et al. (1988) reviewed Middle Ordovician paleogeographic and depositional patterns for the entire western U.S.

Upper Ordovician (Fish Haven-Ely Springs) dolomite (Figure 27) is the most widely distributed Ordovician rock in Utah; it is also the most uniform in thickness and in lithology. Typically, it is a dark gray, cliff-forming, cherty dolomite. It is usually overlain by dolomite of Silurian age which is so similar that they are sometimes lumped as undivided Ordovician-Silurian dolomite on maps. The most obvious fossils in Upper Ordovician dolomites are corals and brachiopods (Sheehan, 1971, 1982), which are better for paleoenvironmental interpretations than for age determinations. Leatham (1985) has zoned the Fish Haven Dolomite on the basis of conodonts.

Poole and Sandberg (1988) noted that during Late Ordovician time the North American continent became flooded by shallow marine waters to a greater extent than at any previous Paleozoic time; these waters drowned the source areas of quartz sand that characterized the under-

FOSSIL ZONATION	1 IBEX AREA CONFUSION RANGE Chart 70	2 SPOR MTN, THOMAS RANGE Chart 49	3 EAST TINTIC MINING DISTRICT Chart 32	4 BEAR RIVER RANGE Chart 16
Ordovician strata in Utah are well-zoned with conodonts, trilobites and brachiopods. Silurian strata are partially zoned with brachiopods.				
Overlying strata →	Sevy Dolomite	Sevy Dolomite	Upper Bluebell Dolomite	Water Canyon Fm

Fossil zonation (time scale):

408 m.y.

SILURIAN: PRIDOLI, LUDLOW, WENL'K, LLANDOVERY

Note: Brachiopods and corals are the only fossils commonly found in the Silurian Laketown Dolomite. Conodonts, cephalopods, and trilobites are very scarce. Sheehan, 1971, identified the fossil communities listed below in the Laketown Dolomite.

- *Spirinella* (C6)
- *Verticilipora* (C4–C5)
- *Pentamerus* (C1–C3)
- *Virgiana*

Ordovician conodont-based chronozones of Sweet, 1984, below

435 m.y.

ORDOVICIAN:

Series	Stage	Ross-Hintze trilobite-brachiopod zones	Conodont chronozones
ASHGILL	CINCINNATIAN: GAMACHIAN		*A. shatzeri*
ASHGILL	CINCINNATIAN: RICHMONDIAN		*A. divergens*
ASHGILL	CINCINNATIAN: RICHMONDIAN		*A. grandis*
ASHGILL	CINCINNATIAN: MAYSVILLIAN		*O. robustus*
ASHGILL	CINCINNATIAN: MAYSVILLIAN / EDENIAN		*O. velicuspis*
CARADOC	CINCINNATIAN: EDENIAN		*B. confluens*
CARADOC	MOWHAWKIAN: SHERMANIAN		*P. tenuis*
CARADOC	MOWHAWKIAN: KIRKFIELDIAN		*P. undatus*
CARADOC	MOWHAWKIAN: ROCKLANDIAN		*P. undatus*
CARADOC	MOWHAWKIAN: BLACK RIVERAN		*B. compressa*
CARADOC	MOWHAWKIAN: BLACK RIVERAN		*E. quadridactylus*
CARADOC	MOWHAWKIAN: BLACK RIVERAN		*P. aculeata*
LLAND	WHITEROCKIAN	Ross-Hintze trilobite-brachiopod zones listed below	*P. sweeti*
LLANVIRN	WHITEROCKIAN	O *Kirkina - Eofletcheria*	*P. friendsvillensis*
LLANVIRN	WHITEROCKIAN	O *Kirkina - Eofletcheria*	*P. "Pre-flexuosus"*
LLANVIRN	WHITEROCKIAN	N *Illaenus - Desmorthis*	*H. holodentata*
ARENIG	WHITEROCKIAN	N / M	*H. sinuosa*
ARENIG	IBEXIAN	M *O. michaelis - Orthidiella*	*H. altifrons*
ARENIG	IBEXIAN	L *Ectenonotus*	*M. flabellum*
ARENIG	IBEXIAN	K *H. minor*	
ARENIG	IBEXIAN	J *Pseudocybele nasuta*	
ARENIG	IBEXIAN	I *Presbynileus ibexensis*	
TREMADOC	IBEXIAN	H *Trigonocerca typica*	
TREMADOC	IBEXIAN	G2 *Protopliomerella contracta*	
TREMADOC	IBEXIAN	G1 *Hintzeia celsaora*	
TREMADOC	IBEXIAN	F *Rossaspis superciliosa*	
TREMADOC	IBEXIAN	E *Tesselacauda*	
TREMADOC	IBEXIAN	D *Leiostegium - Kainella*	
TREMADOC	IBEXIAN	C *Paraplethopeltis*	
TREMADOC	IBEXIAN	B *Symphysurina*	
TREMADOC	IBEXIAN	Mi *Missisquoia*	

505 m.y.

Underlying Cambrian rocks →

1 — Ibex Area, Confusion Range (Chart 70):

- Sevy Dolomite
- Laketown Dolomite: Decathon M; ? Jack Valley Member; ? Gettel Mbr; ? Tony Grove Lake Mbr (see note)
- Ely Springs Dolomite: Floride Mbr (see note); ? Lost Canyon Member; ? Barn Hills Member; ? Ibex Member
- Eureka Quartzite
- Crystal Peak Dolomite (O)
- Watson Ranch Quartzite
- Pogonip Group: Lehman Fm (N); Kanosh Shale (M); Juab Ls (L); Wah Wah Ls (J, K); Fillmore Formation (I, H, G2, G1, F, E, D); House Limestone (C, B)
- (Mi) Lava Dam Mbr, Notch Peak Fm (see note)
- Ross-Hintze trilobite & brachiopod zones

2 — Spor Mtn, Thomas Range (Chart 49):

- Sevy Dolomite
- Laketown Dolomite: ? Thursday Member; ? Lost Sheep Mbr; ? Harrisite Mbr; ? Bell Hill Mbr (see note)
- Ely Springs Dolomite: Floride Mbr; ? Lower member
- ? Eureka Quartzite
- Pogonip Group: Kanosh Shale (M); Juab Ls; Wah Wah Ls (J); (H); Fillmore Fm (G2); House Ls (B)
- Notch Peak Fm

3 — East Tintic Mining District (Chart 32):

- Upper Bluebell Dolomite
- Middle part of the Bluebell Dolomite (See note)
- Lower part of the Bluebell Dolomite
- Fish Haven Dolomite
- Opohonga Limestone (J, G, B, Mi)
- Ajax Dolomite

4 — Bear River Range (Chart 16):

- Water Canyon Fm
- ? Laketown Dolomite: Decathon Mbr; ? Jack Valley Member; ? Portage Cany M; ? High Lake Mbr; ? Tony Grove Lake Mbr (see note)
- Fish Haven Dolomite: Bloomington Lake Member (see note); Deep Lakes Member; Paris Peak Member
- ? Swan Peak Fm: Qtzt mbr; Shale mbr (M)
- Garden City Fm: Cherty member (L, K, J); Lower Member (I, H, G2, G1, F, E, D, C, B, A)
- (Mi) St. Charles Formation
- ← Ross trilobite & brachiopod zones

Index map of Utah showing locations 1, 2, 3, 4.

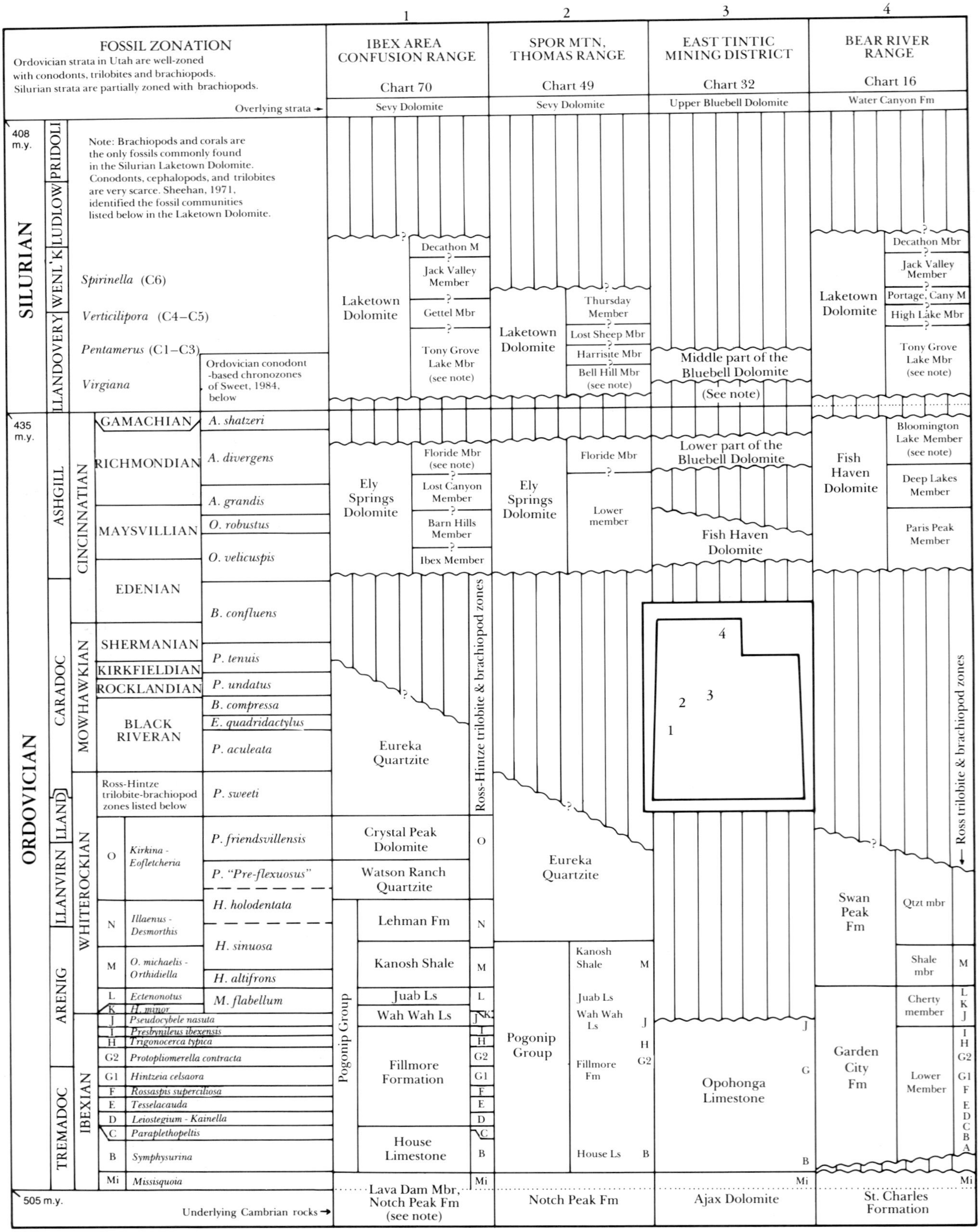

REFERENCES - Ross et al, 1982; Ross & Naeser, 1984; Hintze, 1979; Sweet, 1984; Sheehan, 1971; Berry & Boucot, 1970; Boucot, 1979; Jaanusson, 1979.

Note: Budge & Sheehan, 1980, subdivisions of the Laketown and Ely Springs have yet to be established as valid mappable members. Assignment of parts of the Bluebell Dolomite is based on Budge & Sheehan, 1980, p. 62. Age assignment of Fish Haven subdivisions in Bear Lake Range based on Leatham, 1985. Laketown correlations in Thomas Range based on Budge & Sheehan, 1980, p. 65–66. Miller, 1984, outlines conodont occurrences at the Cambrian-Ordovician boundary.

FIGURE 26 — Ordovician-Silurian correlation table.

FIGURE 27 — Ordovician, Silurian, and Devonian strata in the southern part of the Confusion Range 50 miles west of Delta, Utah on U.S. Highway 6-50. Light cliffs at right side of photo are Eureka Quartzite; black ledges above the Eureka are Upper Ordovician Ely Springs Dolomite. Light gray rocks on skyline in right third of photo are Devonian Sevy Dolomite; banded ledges between light Sevy and black Ely Springs are Silurian Laketown Dolomite. Nonresistant strata in foreground are Lower Ordovician Fillmore Formation. U.S. Highway 6-50 from Delta enters left middle edge of photo and is hidden behind low hill in right third of photo. Playa is about 5 miles long. See Chart 70.

lying Middle Ordovician rocks. Generally, a major unconformity separates the Middle (Mohawkian) and Upper (Cincinnatian) Ordovician strata, as shown on Figure 26. Upper Ordovician rocks thin slightly over the Tooele Arch (Figure 23) and are absent east of Utah's hingeline. Carpenter et al. (1986) suggested that the transition between the Utah shelf and the Nevada basin of Late Ordovician time was likely a gentle ramp rather than a steeper margin.

Ross (1975, 1976) has reviewed Utah's paleogeographic setting throughout Ordovician time, as part of a larger regional and world-wide analysis, and has postulated North America's position relative to circulation of ocean currents. He described North America as lying slightly south of the equator and subject to southern equatorial currents deflected by the continent into gyres.

SILURIAN

Current isotopic age assessments have made the Silurian, embracing less than 30 million years, the shortest Paleozoic period on everybody's time scales (Palmer, 1983; Harland et al., 1982; Snelling, 1985). Silurian stratigraphy is Utah's simplest: it consists of one lithology, dolomite, assigned mostly to one formation, the Laketown Dolomite. The type area for the Laketown Dolomite is in northeastern Utah, where it is generally a light-gray unit; elsewhere in Utah (Figure 32) it is banded light-gray, dark-gray, and pinkish-gray; it is commonly cherty. Budge and Sheehan (1980) subdivided the Laketown into four or five members, as shown on Figure 26, but these have yet to be established as valid mappable units. The only places in Utah where Silurian strata have been subdivided on maps is in the Thomas Range (Chart 49), where Staatz and Osterwald (1959) divided the Silurian into four formations in order to delineate a complex fault pattern, and in the Fish Springs Range (Chart 46). In the East Tintic mining district (Chart 32) Silurian rocks are included as part of the Ordovician-Devonian Bluebell Dolomite.

In older American literature most Silurian fossils were referred to as "Niagaran." In this sense the term "Niagaran" is essentially synonymous with Silurian. Berry and Boucot (1970) have proposed that American Silurian faunas be compared with the British-Czechoslovakian standard series. The most complete zonation of the Silurian is based on graptolites. Cocks et al. (1971) subdivided the lower and middle Silurian into 23 zones, based on *Mono-*

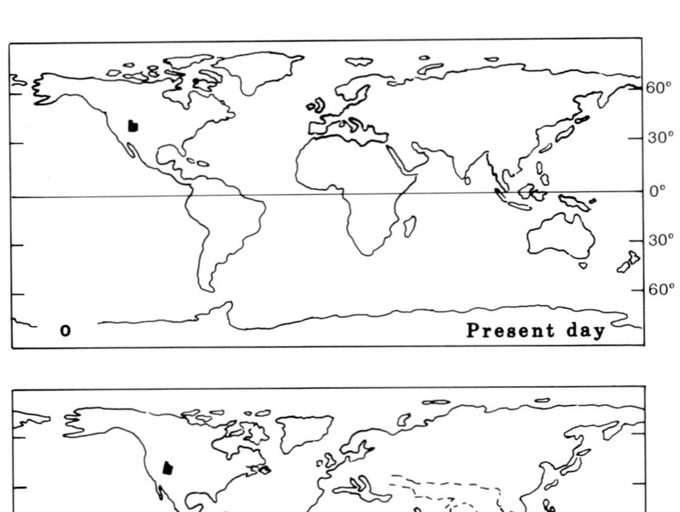

FIGURE 28 — Utah's position at 40 million year intervals back through Cenozoic, Mesozoic, and Paleozoic time is shown on these maps adapted from Smith, Hurley, and Briden (1981). Unlike many parts of the world, Utah has remained no farther from the equator than it is today; during Paleozoic time it crossed the equator into the southern hemisphere.

Because we are used to recognizing continents based on their present coastlines, these have been retained to show past positions despite the fact that coastlines are ephemeral, changing

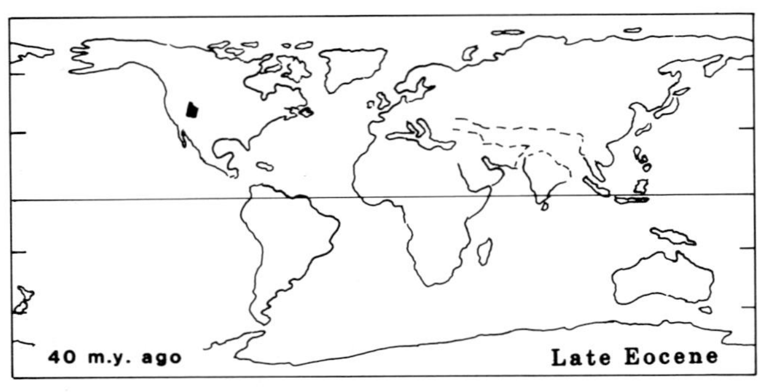

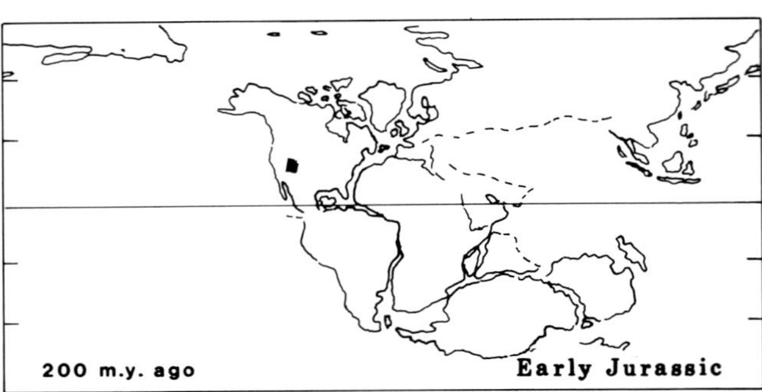

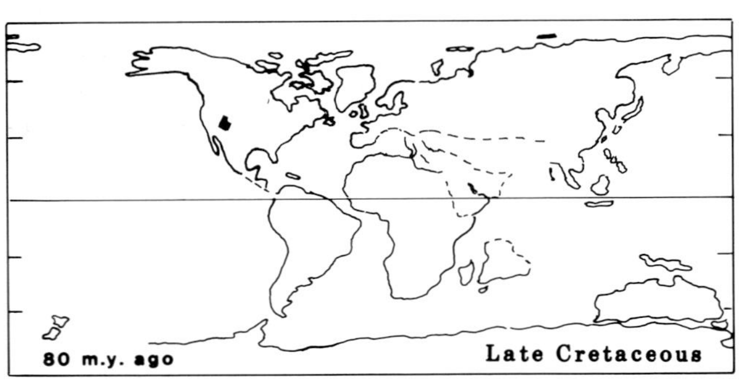

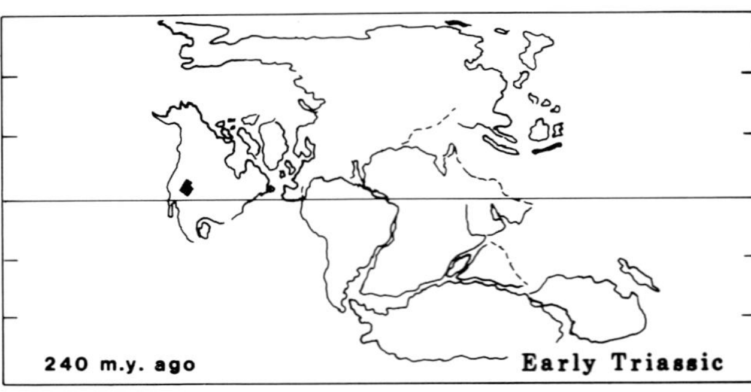

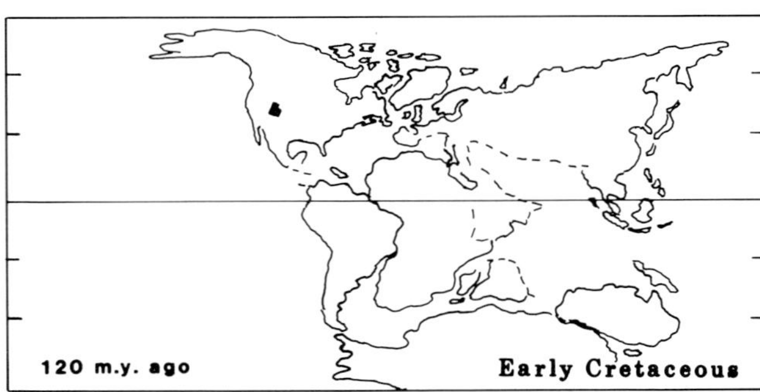

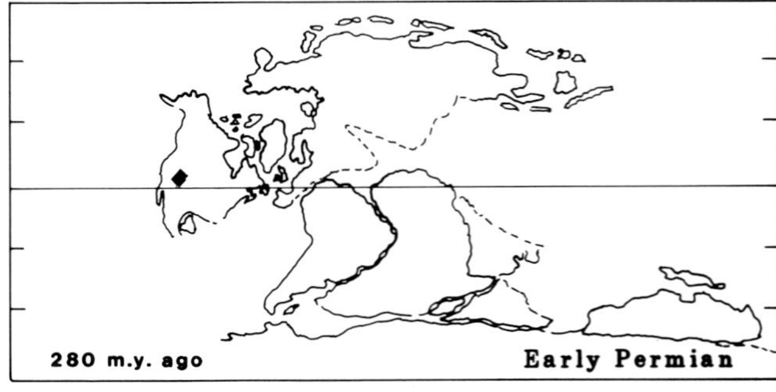

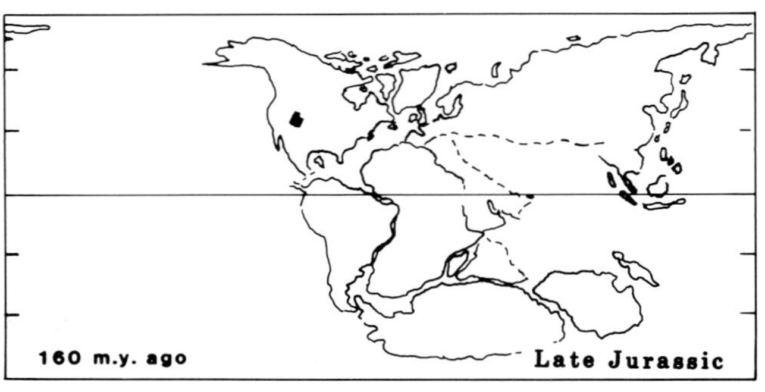

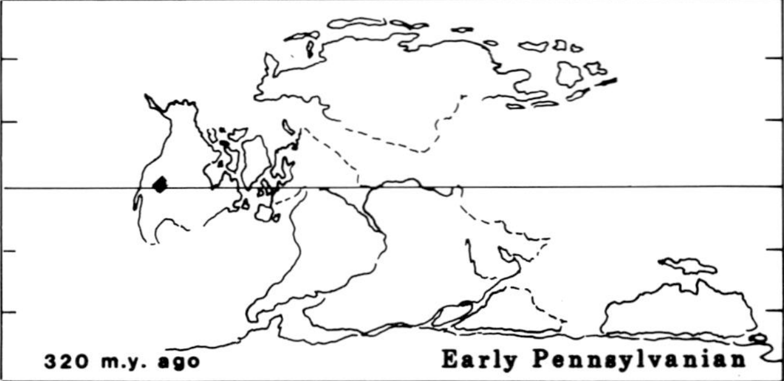

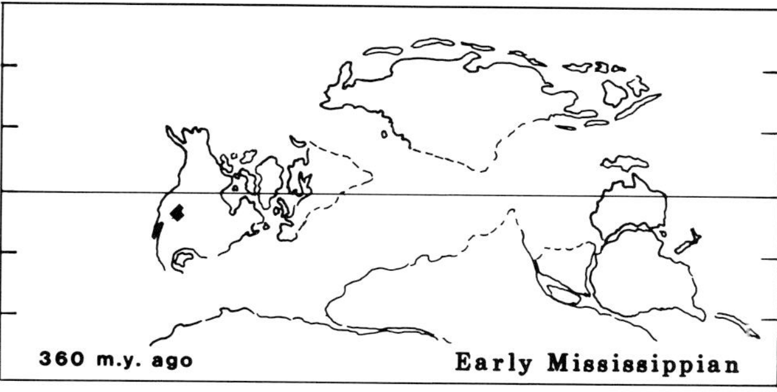

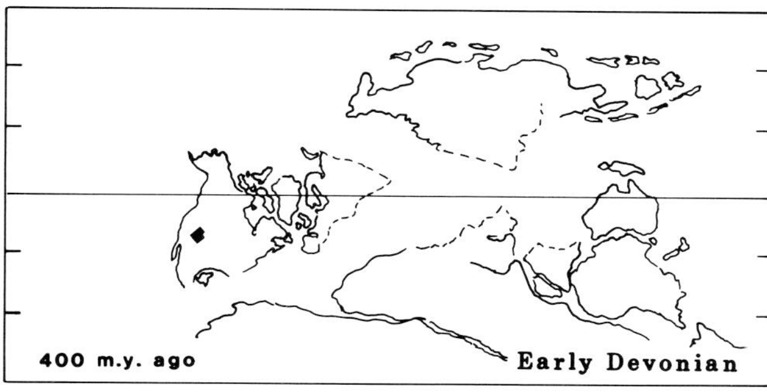

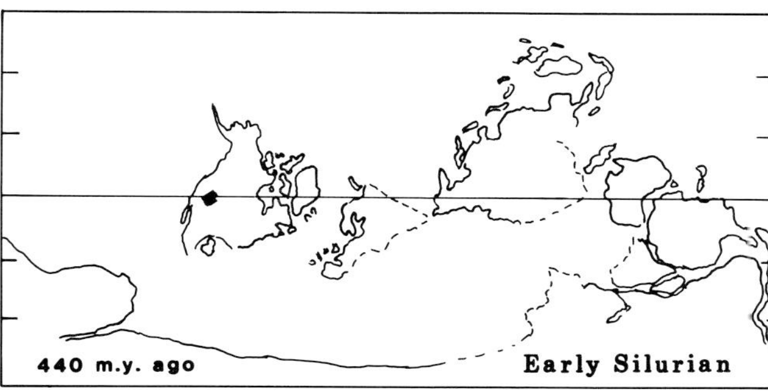

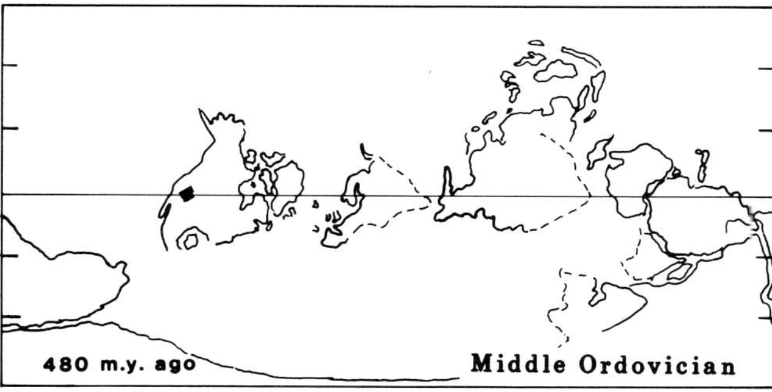

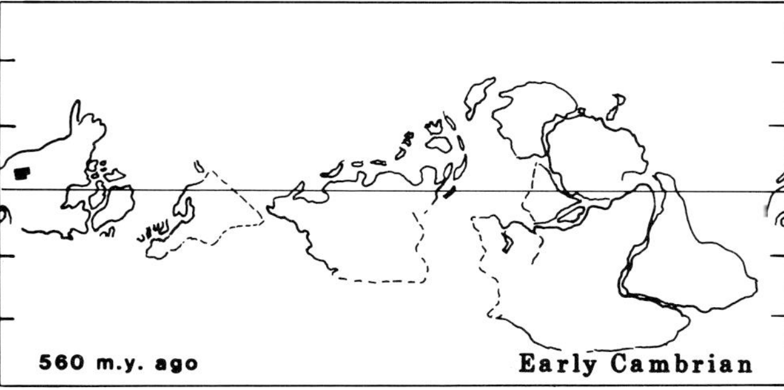

greatly as sea-level rises and falls. Utah, in fact, was submerged beneath shallow seas during most of Paleozoic time, as were much of other continents.

The maps are shown on a cylindrical base which distorts the shapes of continents when they lie in high latitudes. The shape of Greenland, for example, appears quite different when it is near the pole, as at present, than when it is near the equator.

graptus species. Berry and Boucot (1970) listed five additional upper Silurian zones. Unfortunately, Silurian dolomites in Utah have not yielded any graptolites. Conodonts, another useful fossil group, are so sparse in Utah's Laketown Dolomite that no one has yet processed enough rock to know what conodont zones may be present (David L. Clark, 1987, personal communication). Silurian coral and brachiopod faunas from Utah have been described by Budge (1972) and Sheehan (1980b, 1982). These faunas permit correlation with the British Llandovery and Wenlock Series, as shown on Figure 26. Sheehan (1982b) noted that brachiopods suffered a major extinction at the Ordovician-Silurian boundary; the extinction was associated with glacio-eustatic lowering of sea level which destroyed most shallow water, epicontinental habitats. After the glaciation the shallow water habitat re-

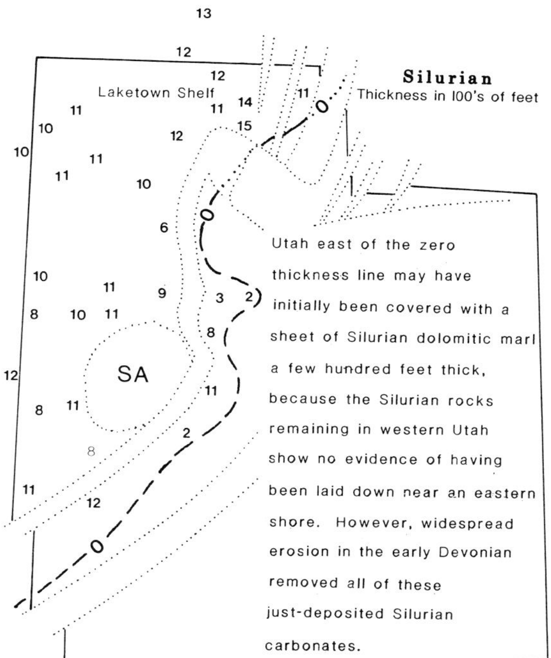

FIGURE 29 — Silurian rocks are almost entirely dolomitic and form a sheet about 1,000 feet thick in western Utah. SA indicates the Sevier Arch, a Mesozoic uplift across which Silurian strata are absent.

turned and new groups of brachiopods appeared in early Silurian time. Sheehan (1980a) presented a series of maps that delineated areas occupied by various brachiopod communities. He concluded that pentamerid brachiopod communities represented shallow, rough water, and dasyclad algal communities lived in shallow calm water.

Berry and Boucot (1970) suggested that a Silurian carbonate blanket extended across the entire North American craton. Absence of Silurian strata in eastern Utah (Figure 29), Colorado, and Wyoming was, in their view, occasioned by widespread erosion in Devonian time during which the soft, newly deposited Silurian carbonates were readily removed. Sheehan (1979) and Poole and Sandberg (1988) presented regional summaries of Silurian thicknesses and lithologic variations for western North America.

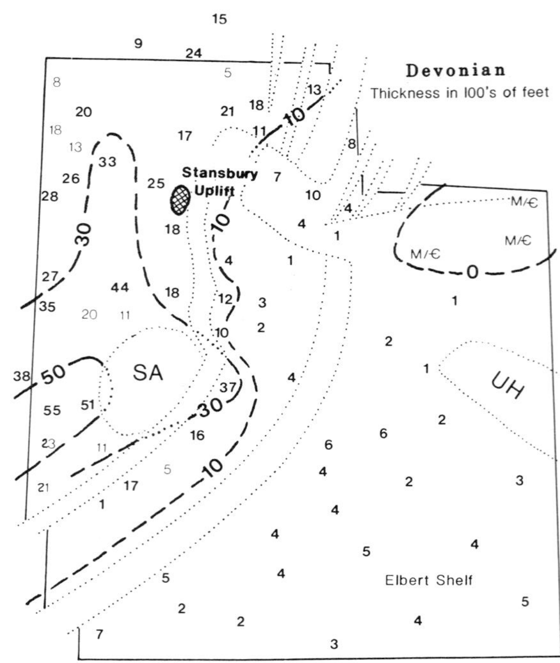

FIGURE 30 — Devonian strata are more than 5,000 feet thick in west-central Utah, and they follow the typical early Paleozoic pattern: thin on the craton in eastern Utah, but thousands of feet thick in the miogeocline. Devonian rocks are chiefly dolomites and limestones, but quartz sandstone and silty shale beds appear in the late Devonian as a result of uplift in the Stansbury area, Utah, and in the Antler area of central Nevada. SA shows the location of the Sevier Arch, a Mesozoic positive area from which Devonian strata have been removed. UH shows the Uncompahgre Highland, a late Paleozoic uplift. Mississippian rocks rest unconformably on Cambrian strata in the eastern Uinta area. Lighter numbers indicate partial thicknesses. Devonian strata of the Elbert Shelf area in southeastern Utah are oil and gas reservoir rocks.

DEVONIAN

Devonian time embraced about 50 million years (Palmer, 1983; Harland et al, 1982; Snelling, 1985) which has been best subdivided on the basis of conodont zones, as shown on Figure 31. Conodont zonation has permitted provincial North American stage names used in older literature to be replaced by European stages. Figure 31 shows that four sets of formation names are used to cover Utah's Devonian strata.

Devonian rocks are thick in western Utah and thin to the east (Figure 30); Devonian strata are more widespread than are Ordovician or Silurian, and are almost as extensive as Cambrian deposits. Breakup of the simple miogeocline-craton pattern which had persisted in Utah since late Precambrian time, began in the Late Devonian with a mini-orogeny called the Stansbury Uplift. One of the effects of the uplift was the development of an unconformity surface in north-central Utah that beveled downward across beds as old as Early Cambrian in the Stansbury Mountains and to the Precambrian in the western Uinta Mountains, as recognized by Rigby (1959), and Morris and Lovering (1961) (Figure 33). Another effect was deposition of a coarse conglomerate, the Stansbury Formation, in the immediate vicinity of the uplift (Stokes and Arnold, 1958) and the more widespread dispersal of Late Devonian sands such as the Victoria and Beirdneau sandstones. Quartz sandstones and shales are found in Late Devonian rocks in all areas of Utah. The source of these sands has never been certainly identified; many are frosted and cross-bedded suggesting wind transport. Commonly they occur interbedded with carbonates, as in the upper part of the Guilmette Formation.

The most common fossils in Utah's Devonian rocks are stromatoporoids, whose small tubular forms are so abundant in some layers that they are called "spaghetti" beds. Other stromatoporoids resemble small cabbage heads; neither "spaghetti" nor "cabbage" forms are useful for fossil zonation. Useful zone fossils are brachiopods, or better yet, conodonts. Sandberg (1976), Sandberg and Dreesen (1984), Sandberg and Gutschick (1979), Sandberg and Poole (1977), and Sandberg, Poole, and Gutschick (1980) have developed conodont zonations which permit detailed correlations of Utah's Middle and Late Devonian strata. Paleoecology and paleogeography have been outlined for Early and Middle Devonian (Johnson and Sandberg, 1977; Poole and Sandberg, 1988) and Late Devonian (Sandberg and Poole, 1977, 1988; Sandberg et al, 1983).

Maximum eustatic transgression of Devonian seas onto the North American craton occurred during Frasnian and early Famennian time, partly coincidental with early phases of the Antler orogenic uplift in Nevada and the Stansbury uplift in Utah. A general regression in late Famennian (post-*Palmatolepis trachytera*) time is reflected in the narrow distribution of rocks of this interval and in unconformities shown in the upper part of the Devonian correlation table, Figure 31. Following the

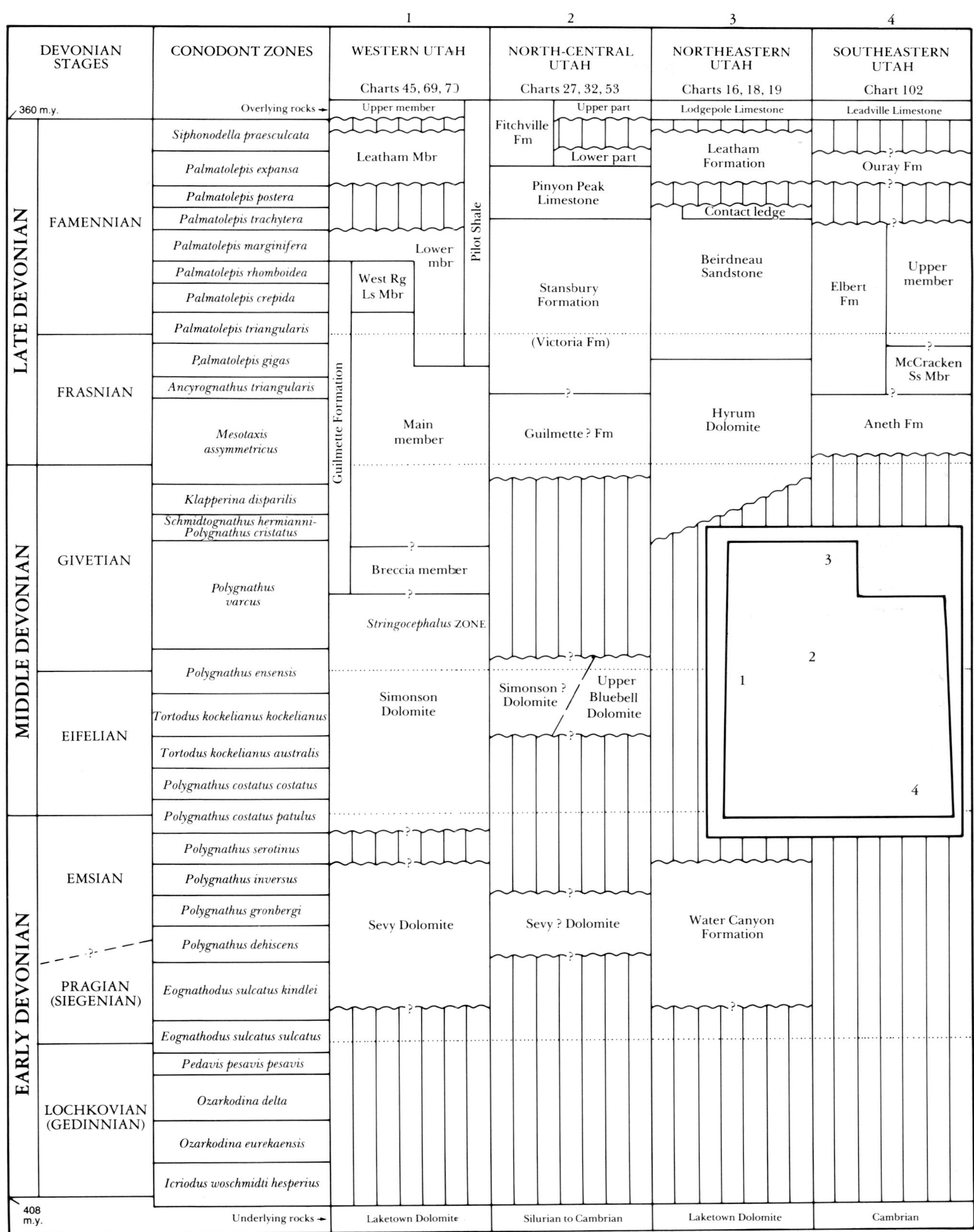

FIGURE 31 — Devonian correlation table.

FIGURE 32 — Silurian Laketown Dolomite in the southern Confusion Range 50 miles west of Delta Utah. See Figure 27 for distant view of the same hills and playa. Light gray rocks capping the hills from photo-center to right edge are Sevy Dolomite. Banded rocks to base of central hill are all Laketown Dolomite. Hill just beyond playa is down-faulted block of upper Laketown. Upper Ordovician Ely Springs Dolomite forms darkest ledges at base of hill in the right quarter of the photo. See Chart 70 for thicknesses.

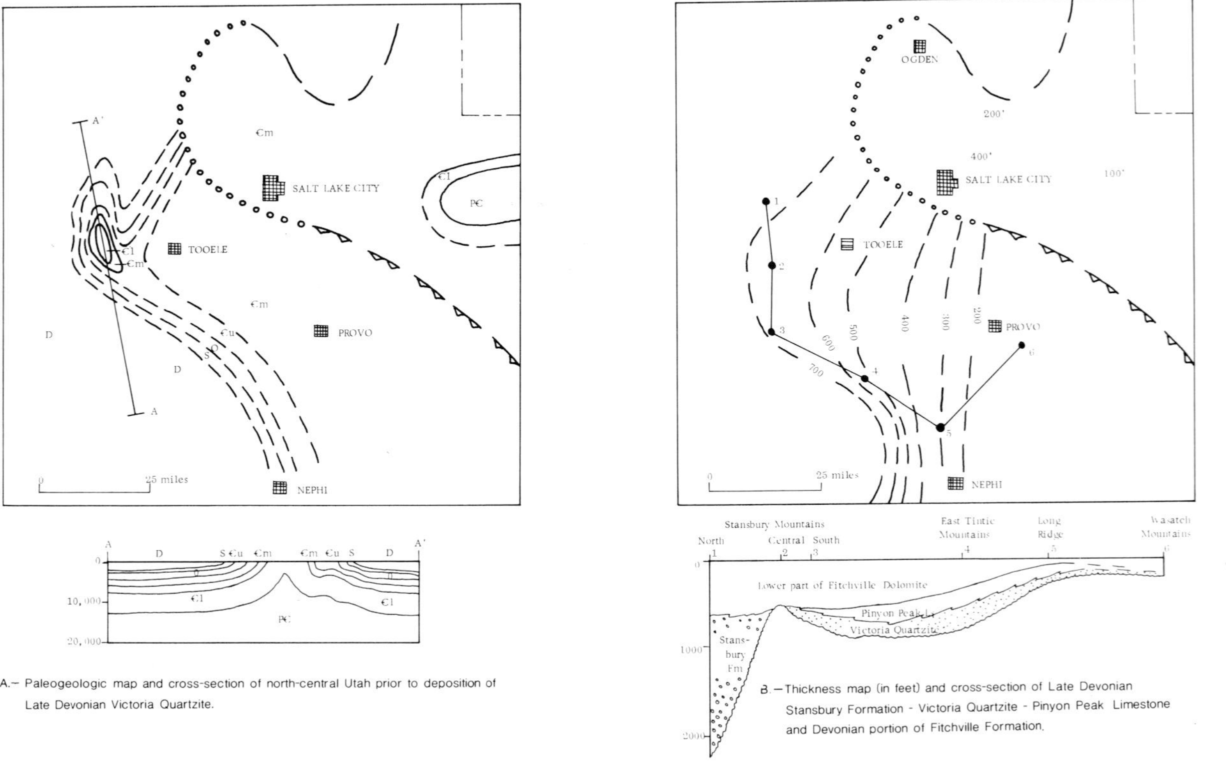

A.– Paleogeologic map and cross-section of north-central Utah prior to deposition of Late Devonian Victoria Quartzite.

B.–Thickness map (in feet) and cross-section of Late Devonian Stansbury Formation - Victoria Quartzite - Pinyon Peak Limestone and Devonian portion of Fitchville Formation.

FIGURE 33 — Late Devonian Stansbury uplift, map and cross-sections (after Rigby, 1959, and Morris and Lovering, 1961). The Stansbury uplift is Utah's expression of the Antler orogeny which affected central Nevada. It probably resulted from collision of the west coast of North America (then in central Nevada) with an island arc advancing eastward on a plate in the Pacific region.

FIGURE 34 — Wasatch Mountains east of Provo. Frontal hills at left edge of photo include Precambrian strata (pC) in the mouth of Rock Canyon, overlain by Cambrian (C) and Mississippian (M) rocks, mostly carbonates. Dirt road behind frontal hills follows a strike valley in the Manning Canyon Shale (Mm). Higher rocks beyond the dirt road are Pennsylvanian-Permian limestones and sandstones of the Oquirrh Group (PPo). See Chart 35 for thicknesses. Wasatch fault separates bedrock of the mountains from alluvial and lake deposits of Utah Valley.

Famennian regression, a 3 million-year-long period of continental stability ensued and most of the western North American craton was emergent (Sandberg and Poole, 1988).

MISSISSIPPIAN

The Mississippian Period was about 40 million years long (Harland et al., 1982; Palmer, 1983), and in Utah its complex depositional and diverse paleoenvironmental patterns mark a pronounced change from the simpler patterns of earlier Paleozoic time. Mississippian formational nomenclature is correspondingly diverse, as shown on Figure 37. The general pattern of rock thickness continued from the earlier Paleozoic (thin on the craton in eastern Utah and thicker in the miogeocline in western Utah), but superimposed on this was the development of a new feature, the Oquirrh Basin, wherein as much as 7000 feet of Mississippian rocks were deposited (Figure 35). "Basin," as used here, means depositional center, or place where sediments accumulated to greater thickness than in adjacent areas. It does *not* mean a topographic basin filled with deep water. In fact, most basin-filling deposits accumulated under shallow-water conditions; only rarely do deep-water conditions develop in epicontinental seas, and even then the water depth is generally less than 1000 feet. Within the Oquirrh Basin, near Provo, a virtually continuous record of fossiliferous Mississippian deposits is preserved (Figure 34); thus, Utah has one of the most complete sequences of Mississippian rocks and their contained fossils to be found anywhere. Mississippian strata are more fossiliferous than most other Utah rocks. Traditionally, the Mississippian has been zoned on endothyrid and fusulinid foraminifera, corals, brachiopods, and ammonoids. But recently conodonts have proved to be very useful, enabling excellent time resolution, especially when combined with forams and corals, as shown on Figure 37. Conodonts, being microscopic, have the disadvantage that they cannot be recognized directly during field collecting, but they have the advantage that they are not damaged by oil-well drilling and can be recovered from well cuttings. Source references for fossil zonations are listed at the bottom of Figure 37; additional zonation schemes for fusulinids, brachiopods, bryozoans, and ammonoids are shown in Dutro et al. (1979), and Welsh and Bissell (1979, p. Y4). Brachiopods are probably the most commonly seen fossils, and many genera are present. Unfortunately, the brachiopod

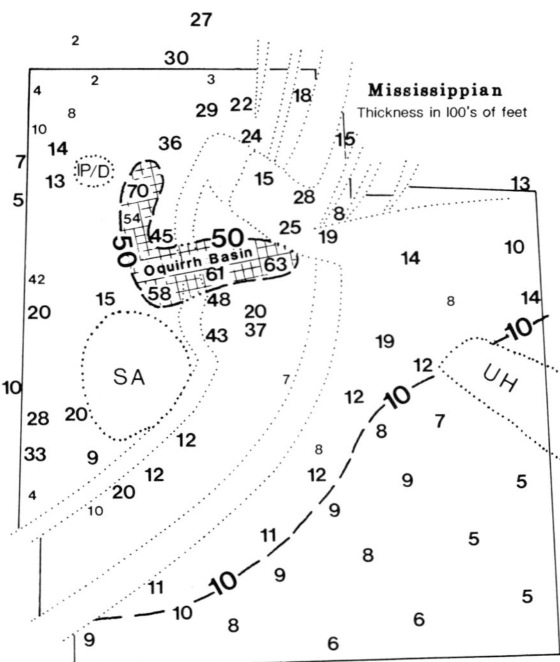

FIGURE 35 — Mississippian rocks, mostly limestones, are widely distributed in Utah and show a change in thickness pattern from that of earlier Paleozoic deposits because the late Paleozoic Oquirrh Basin made its first appearance. Mississippian rocks are as much as 7,000 feet thick in the basin and include deposits representing the entire span of Mississippian time, one of the most complete records of this interval in North America. P/D on the map indicates the area in the Newfoundland Mountains where Mississippian rocks are absent between Devonian and Pennsylvanian strata. The lighter numbers represent locations where Mississippian rocks are only partially preserved. SA indicates the Sevier Arch, a late Mesozoic positive area. UH indicates the Pennsylvanian Uncompahgre Highland.

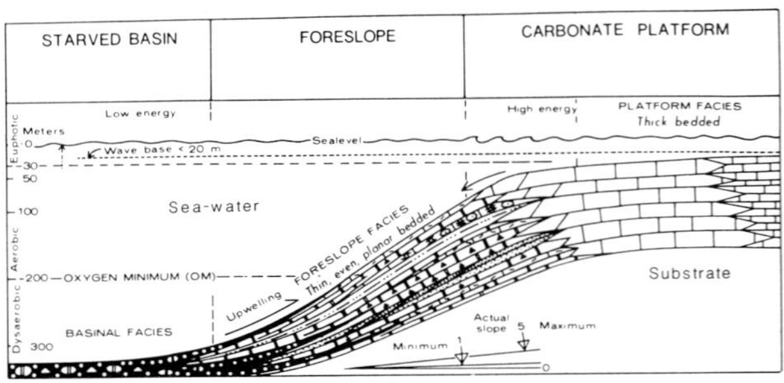

FIGURE 36 — Schematic cross-section, with vertical scale greatly exaggerated, to show environmental relationships between starved basin, foreslope, and platform on the east side of the Mississippian Oquirrh Basin during deposition of the Deseret Limestone. Kinds of corals, brachiopods, conodonts, algae, and other fossils are controlled by their living environment; some fossil forms are especially useful as paleoenvironmental indicators. Adapted from Gutschick and Sandberg (1983).

faunas of Mississippian-Pennsylvanian-Permian time look alike to most people, so it takes an expert to identify them. Furthermore, faunal lists in many published reports classify Mississippian brachiopods under certain stereotyped names which may be too "catchall" in usage. Hence, brachiopod listings are frequently suspect, and despite their abundance, brachiopods are not the most specific time indicators. Karklins (1986) has recently documented zonal ranges of bryozoans from Late Mississippian strata in western Utah.

Mississippian sedimentation began in western Utah with deposition of the Pilot Shale, fine grained detritus derived from the Antler orogenic belt in central Nevada, and the dolomitic Fitchville Formation, a shelf deposit in central Utah. The most widespread shallow marine incursion in Utah is represented by deposition of the Joana-Gardison-Lodgepole-Redwall fossiliferous limestones of late Kinderhookian-early Osagean age; Joana Limestone is thin to absent locally in the northern Confusion Range. Limestone deposition was followed by a brief interval in middle Osagean time during which subsidence in western Utah exceeded the rate of sedimentation and a starved basin developed (Figure 36), as described by Gutschick and Sandberg (1983). Phosphatic siltstone and shale of the Delle Member (Sandberg and Gutschick, 1984) were deposited in waters that may have been more than 1000 feet deep. The last half of the Mississippian was dominated by cyclic sedimentation. Simple sandstone-limestone cyclothems characterize the Humbug Formation. More complex cyclothems have been described from the Manning Canyon Shale by Moyle (1959) and Prince (1963). Veevers and Powell (1987) suggested that cyclic sedimentation in North America is related to glacial episodes elsewhere. Influence of Nevada's Antler orogenic belt resulted in influx of clastics, forming the Chainman Shale in western Utah and Diamond Peak conglomerates in northwestern Utah. Clastic muds of eastern provenance are included in the Doughnut Formation and Great Blue Limestone. Welsh and Bissell (1979, p. Y11) presented a paleogeographic map of Utah that outlined the Doughnut trough in northeastern Utah, the Great Blue carbonate bank of central Utah, and the extensive platform in southeastern Utah that was subjected to subaerial erosion in late Mississippian time, developing a lateritic soil. The Manning Canyon and Chainman Shales and the Doughnut and Great Blue Formations were regarded by Swetland et al. (1978) as favorable petroleum source beds. The Manning Canyon Shale is about 1500 feet thick in the Wasatch Mountains between Provo and Nephi, and wherever it is exposed it is prone to slumping and landsliding. Its instability has destroyed homes in northeast Provo and has caused highway maintenance problems wherever it is encountered.

MISSISSIPPIAN

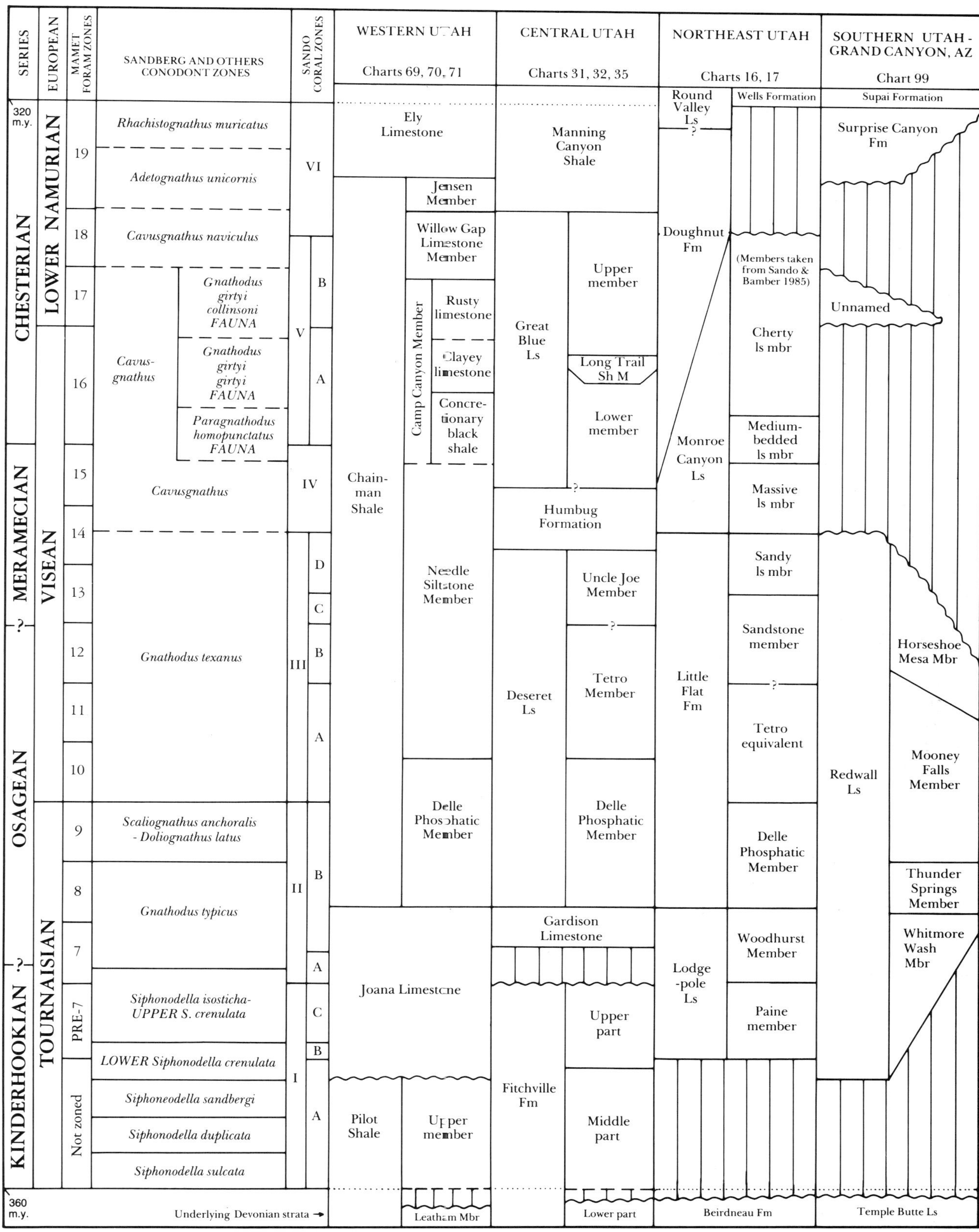

REFERENCES - Sando & Bamber, 1985; Sandberg, Poole & Gutschick, 1980; Tynan, 1980; Sandberg, 1979; Webster et al, 1984; Welsh & Bissell, 1979.

FIGURE 37 — Mississippian correlation table.

FIGURE 38 — Goosenecks of the San Juan River near Mexican Hat, Utah, are incised into marine limestones, sandstones, and shales of the Pennsylvanian Hermosa Group. The lower massive cliffs are limestone equivalents to evaporite facies in the subsurface of the Paradox Basin to the northeast. Equivalents of the Ismay, Desert Creek, Akah, and Barker Creek zones that produce oil in the Paradox Basin are exposed here. The Akah and Barker Creek equivalents contain numerous bioherms at Goosenecks. See Chart 102.

During Mesozoic folding and thrusting, the Chainman-Manning Canyon shales commonly served as a glide-plane for thrust sheets. In such areas the apparent thicknesses of these shales vary greatly. Upper Mississippian shales have been mined for brick clays west of Utah Lake. In these clay pits and wherever else they occur, the Upper Mississippian shales are fossiliferous, and collecting fossils from them is fun.

Regional paleogeographic analyses of the Mississippian System in the U.S. were presented by Craig and Varnes (1979), Skipp (1979), Mallory (1979), Welsh and Bissell (1979), and Sandberg et al. (1982). The reader anxious for more detailed accounts is referred to these more comprehensive summaries.

PENNSYLVANIAN

Although the Pennsylvanian is one of the shorter Paleozoic periods, spanning about 30 million years, it included the principal development of two unusual depositional basins that dominate the regional pattern of Pennsylvanian deposition in Utah: the Paradox Basin of southeastern Utah, and the Oquirrh Basin of northwestern Utah, shown on Figure 39. Although Pennsylvanian rocks are concealed, for the most part, beneath the surface in the Paradox Basin (Figure 38), oil exploration has made them much better understood than the Oquirrh Basin strata that are magnificently exposed in the basin-ranges from Nephi northwestward.

Paradox Basin — Discovery of oil and potash in southeastern Utah has led to many publications on the Paradox Basin. Comprehensive summaries are available in Cater (1970), Hite and Cater (1972), Baars (1983), and in guidebooks published in 1963, 1975, 1978 and 1979 by the Four Corners Geological Society. The Paradox Basin contains a thick sequence of evaporites included within the Paradox Formation (see Charts 84 and 102, Figures 9, 40 and 46, columns 10 and 11), and a thinner complementary facies of shelf carbonates, the Pinkerton Trail and Honaker Trail Formations. The Paradox evaporites, chiefly anhydrite and rock salt, but also including potash layers, are interbedded with black shale; some 29 evaporite-shale cycles have been recognized in the thicker parts of the basin. Each evaporite cycle can be correlated with an equivalent carbonate cycle on the southwest shelf of the basin near the Four Corners. Many of these cyclic carbonates are petroleum reservoirs. These producing intervals, spanning several depositional cycles, have been grouped into petroleum zones known as Ismay, Desert Creek, Akah, Barker Creek, and Alkali Gulch, as shown on Chart 102 and Figure 46.

The Paradox Basin lies on the flank of the Uncompahgre Uplift, a highland which shed arkosic clastic rocks into the basin (Mack and Rasmussen, 1984). Drilling and seismic data have shown that the Uncompahgre Uplift is bounded on the southwest by a high-angle reverse fault, which accommodated several thousand feet of vertical displacement from Pennsylvanian to Early Triassic time. This "yoked" basin-uplift pattern extended from Utah southeastward to Oklahoma forming a paleogeographic feature called the "Ancestral Rockies." Larson et al. (1985) showed that Late Cambrian and Early Ordovician dike swarms in Colorado are aligned in the same direction as the later Paradox Basin faults. Baars (1966, 1972) suggested that the Uncompahgre structural block moved repeatedly—during Late Cambrian, Devonian, Mississippian and late Paleozoic time. Stevenson and Baars (1986) indicated that at least some of the Paradox fault movement was strike-slip.

The Paradox Basin is unique among Ancestral Rockies basins because of the great amount of salt that accumulated in it. The original thickness of salt that was deposited in the Paradox Basin is difficult to estimate because of the widespread effects of later plastic flowage of the lower density, and therefore buoyant, salt, but certainly several thousand feet of bedded salt were present in the thickest part of the basin. Fifteen diapiric salt anticlines (Figure 42) have subsequently formed above the subsurface salt (Doelling, 1983). They are aligned in a southeasterly trend that extends from the Salt Valley anticline of Arches National Park, Utah, to the Gypsum Valley anticline north of Cortez, Colorado. The northwest end of the Uncompahgre paleohighland and the Paradox paleobasin are buried beneath later sedimentary deposits of the Uinta Basin. The Paradox Basin is separated from the Oquirrh Basin by a paleopositive feature called the Emery High. It should be noted that the Paradox Basin subsided only partly in isostatic response to its sedimentary loading; subsidence of the basin was initiated and lasted only a finite time as a result of subcrustal intracratonic behavior which is not yet completely understood, but which may be related to late Paleozoic continental sinistral rifting (Stevens and Stone, 1988). Klein and Hsui (1987) discussed the origin of cratonic basins and noted that even many of the larger ones such as the Illinois, Michigan, and Williston basins were initiated by faulting. A nice balance between subsidence and sedimentation permitted the surface of the Paradox Basin to remain in a sabkha environment as the basin floor dropped. Cyclical repetitions of sedimentation patterns in this environment were caused by world-wide eustatic sea-level changes which created Pennsylvanian cyclothems on all continents.

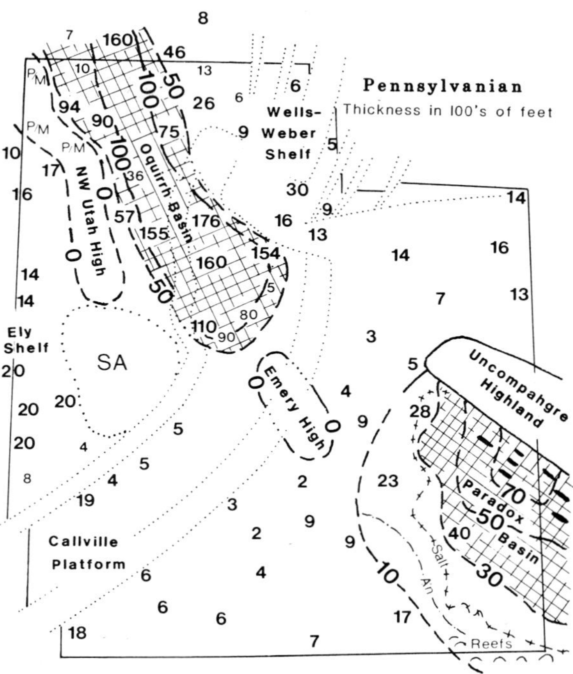

FIGURE 39— Pennsylvanian thickness patterns reflect activity at the northwest end of the Ancestral Rockies belt, which extended southeastward into Oklahoma. The fault-bounded Uncompahgre Highland shed tongues of arkosic debris into the adjacent Paradox Basin where they intertongue with marine sabka deposits in a basin restricted at its southern margin by algal reefs. Limits of basinal salt and anhydrite (An) deposits are indicated; the salt quickly migrated into salt anticlines (black ovals) that continued to be active later. Pennsylvanian strata are absent over the Emery High and the Northwest Utah High where Permian rocks lie unconformably over the Mississippian (P/M). The Oquirrh Basin, where as much as 17,600 feet of fine quartz sand and sandy limestone accumulated, follows the Ancestral Rockies trend; the Emery High sits atop a northeasterly positive trend between the fault-bounded Paradox and Oquirrh basins. SA shows the Sevier Arch, a late Mesozoic positive feature; whether Pennsylvanian rocks were ever present over the Sevier Arch is a moot question; it may have been a part of the NW Utah High. The area southwest of the Emery High where Pennsylvanian strata are less than 1000 feet thick was called the Piute Platform by Mallory (1972). Lighter numbers indicate locations where Pennsylvanian stratigraphic sections are incompletely preserved.

Oquirrh Basin—Pennsylvanian marine strata more than three miles thick accumulated in the Oquirrh Basin (Figures 39 and 41). These rocks cap Mount Timpanogos and Mount Nebo and make up most of the Oquirrh Mountains on the west side of Salt Lake City. Despite these grand exposures, it is still not entirely clear where Pennsylvanian rocks of the Oquirrh Group (Charts 28, 31, 35) came from and why they accumulated to such an outrageous thickness, double that of the Paradox Basin. Unlike that in

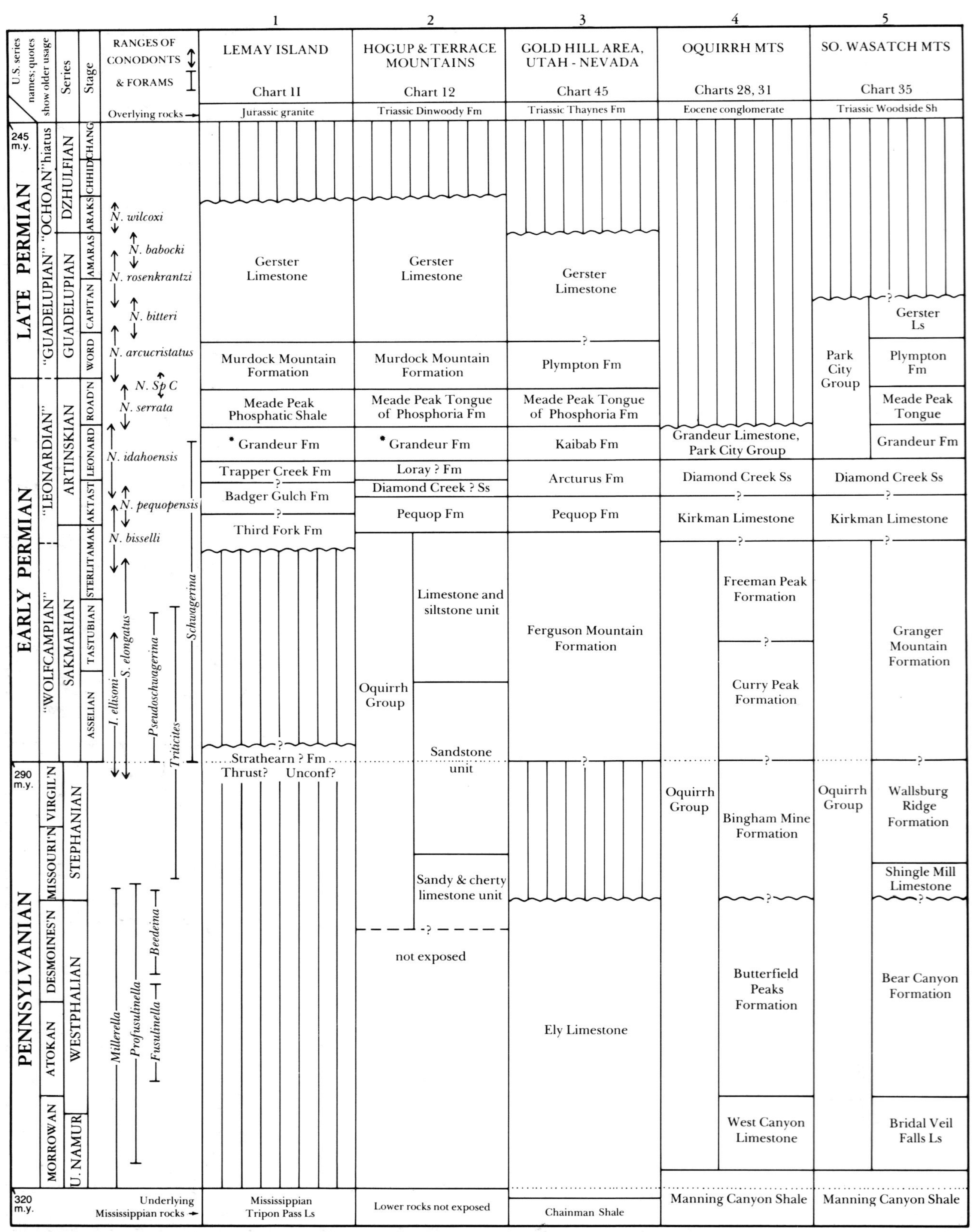

REFERENCES - Ross, 1979; Ross & Ross, 1979; Clark et al, 1979; Furnish, 1973; Welsh & Bissell, 1979; Baars, 1979.

NOTE—*Grandeur is a term from autochthonous section at Park City; elsewhere in western Utah allochthonous strata of this interval have been called Kaibab.

FIGURE 40 — Pennsylvanian-Permian correlation table.

PENNSYLVANIAN-PERMIAN

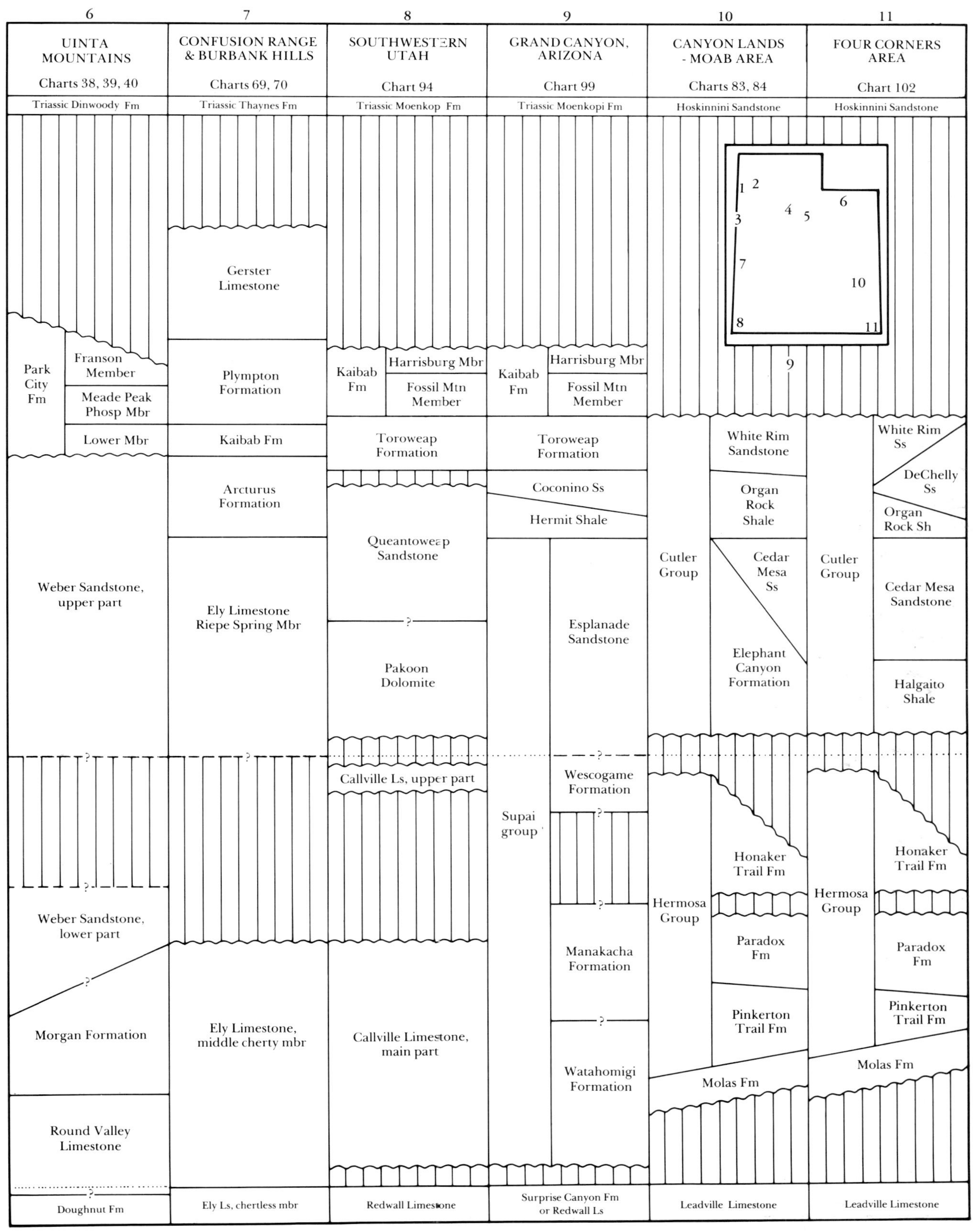

the Paradox Basin, there has been little drilling and no petroleum production from the Oquirrh Basin. If it took 30 million years to deposit 16,000 feet of sediment in the Oquirrh Basin, that gives an average rate of deposition of about one foot per 2000 years.

The Oquirrh Basin sediments later became part of the allochthon, or overthrust sheet, that was transported southeastward during Mesozoic time. The northeast edge of the Oquirrh Basin presently coincides with the Charleston thrust or tear fault of Cretaceous age exposed in the Wasatch Mountains between Salt Lake City and Provo; this probably reflects the mechanical influence that rocks of the older basin had in localizing the breakthrough and trend of the Mesozoic thrust. Some have argued that because Oquirrh thicknesses are ten-fold greater than those of equivalent rocks on the Wells-Weber shelf (Figure 39), they must therefore indicate a considerable distance of transport on Mesozoic thrust sheets in order to juxtapose such contrasting thicknesses. But Oquirrh Basin rocks thin abruptly on *all* sides of the basin so that thickness trends alone cannot serve to tell us the distance of thrust transport.

FIGURE 41 — Pennsylvanian strata of the lower part of the Oquirrh Group on the east side of Mt. Timpanogos in the Wasatch Mountains near Provo. Quaternary glacial moraines in the foreground.

Formational nomenclature for the Oquirrh Group has been slow to become established, although a number of schemes have been proposed. Earlier nomenclature was time-stratigraphic, and depended on the occurrence of fusulinids for its validity. The formation names shown in columns 4 and 5 of Figure 40 are lithologic units based on geologic mapping. The basal limestone unit of the Oquirrh Group (West Canyon-Bridal Veil Falls) is the easiest Oquirrh unit to trace, not only because it is mostly limestone, but also because it bears a prolific fauna of brachiopods, corals, bryozoans, and other organisms that make it readily identifiable. Davis (1987) has proposed that Morrowan limestones at the north end of the Oquirrh Mountains could serve as a stratotype for the Mississippian-Pennsylvanian boundary, based on conodonts. Conodonts are being increasingly used in Pennsylvanian biostratigraphic studies (Driese, Carr and Clark, 1984). Post-Morrowan Oquirrh rocks are mostly unfossiliferous sandstones, except for thin fossil-bearing limestones in the Desmoinesian part of the Butterfield Peaks-Bear Canyon Formations, and the key limestone beds, the Shingle Mill, and its correlative limestone beds at the base of the Bingham Mine Formation (Charts 28, 31). The Late Pennsylvanian part of the Oquirrh Group is mostly unfossiliferous sandstone, but it is overlain by the Permian part of the Oquirrh that contains abundant Wolfcampian fusulinids.

Most Oquirrh sandstones are very fine- to medium-grained, subarkosic (75–90 percent quartz, 5–25 percent feldspar) packstones, grainstones, or wackestones with a calcareous matrix. The best analysis of depositional environments in the Oquirrh Basin is that of Jordan and Douglass (1980), and the following briefly summarizes some of their findings. Middle Pennsylvanian shallow-water limestone and sandstone gave way in Late Pennsylvanian time to great quantities of subarkosic sand whose trace fossils indicate increased water depth and changing basin morphology (Chamberlain and Clark, 1973). Water depths may have reached more than 1000 feet, resulting in stagnation in the northwest part of the basin. Coarse conglomerates, composed largely of older clasts from the Oquirrh Group itself, occur in uppermost Pennsylvanian and earliest Permian deep-water deposits. They record rapid lithification, uplift, and erosion of the margins of the basin, and imply nearby active faults. Faulting ceased in late Wolfcampian time and the uppermost Oquirrh rocks record passive filling of the remnant trough. Figure 39 plots a Northwest Utah High (NWUH) on the west side of the Oquirrh Basin, based on absence of Pennsylvanian rocks along its trend. But this high was not nearly so substantial an uplift as the Uncompahgre since it did not lift Precambrian, or even early Paleozoic, rocks to the surface; nor has it been surely identified as a source for Oquirrh sediments. Jordan and Douglas (1980, p.234) showed a Pequop Uplift in the Nevada extension of the NWUH. The northwestern part of the Oquirrh Basin was the area studied by Jordan and Douglass (1980); an earlier study that emphasized trace fossils and conodonts as evidence for changes in water depths (Chamberlain and Clark, 1973) reached somewhat similar conclusions for the southeastern part of the Oquirrh Basin.

The quartzose sands that make up the bulk of Oquirrh sediments are part of a great Pennsylvanian sand sheet that includes the Weber-Wells-Tensleep sandstones of the Utah-Wyoming shelf. The quantity of silica included in the Oquirrh and related sandstones is so considerable that Bissell (1959) suggested that some of it may have come from the Antler belt of central Nevada. Detailed diagenetic history of the Weber Sandstone in western Colorado, discussing the interplay of tectonic and water circulation conditions was presented by Koelmel (1986).

Pre-Wolfcampian unconformity—Superficially the Pennsylvanian and Permian strata of Utah seem to form a continuum of sedimentation controlled by the same basic influences. However, detailed tracing of fusulinid (Welsh and Bissell, 1979) and conodont (Ritter, 1987) zones has shown that Early Permian rocks rest unconformably on Desmoinesian strata over the southwestern quarter of Utah (see Figure 40), and as shown also on paleogeographic maps by Welsh and Bissell (1979, p. Y14–Y15).

Regional analysis of the Pennsylvanian System in the western U.S. was presented by Mallory (1975), accompanied by paleogeologic maps, thickness maps, and correlation tables.

PERMIAN

Figure 43 shows the same general thickness pattern for the Permian as Figure 39 does for the Pennsylvanian, with slight changes: locus of thickest deposition shifted southward in the Paradox Basin; the Emery High and Northwest Utah High were gone; subsidence and deposition was more rapid in the northern Oquirrh Basin where 19,000 feet of Permian sediments occur.

Worldwide, the Permian was a time of changing environments, more so than most periods; in Utah this is reflected in marked lateral variation in rock types, called facies changes, which can be noted on the correlation table (Figure 40) in the equivalencies of various carbonate and sandstone formations.

Early Permian strata are widely distributed in Utah; Late Permian strata much less so, and there is an unconformity at the top of the Permian that represents an erosion interval more than five million years long between Permian and Early Triassic deposits. The Permo-Triassic unconformity is found on all continents and is the subject of a symposium volume (Logan and Hills, 1973) in which various authorities discussed radiometric, paleontologic and nomenclatorial aspects of the Permian and Triassic systems and their mutual boundary.

The common and obvious fossils in most Permian carbonate rocks in Utah are brachiopods, bryozoans, and sponges; fusulinids are also common and some of the Permian schwagerinid fusulinids are quite large, half the size of a dime. Ammonoids, the standard international guide fossils, are rare in Utah's Permian rocks. Conodonts, microscopic tooth-like fossils, have increased in utility in dating Permian rocks as a result of much new biostratigraphic work (Clark, 1979; Clark et al., 1979; Ritter, 1986, 1987). However, even taken all together—

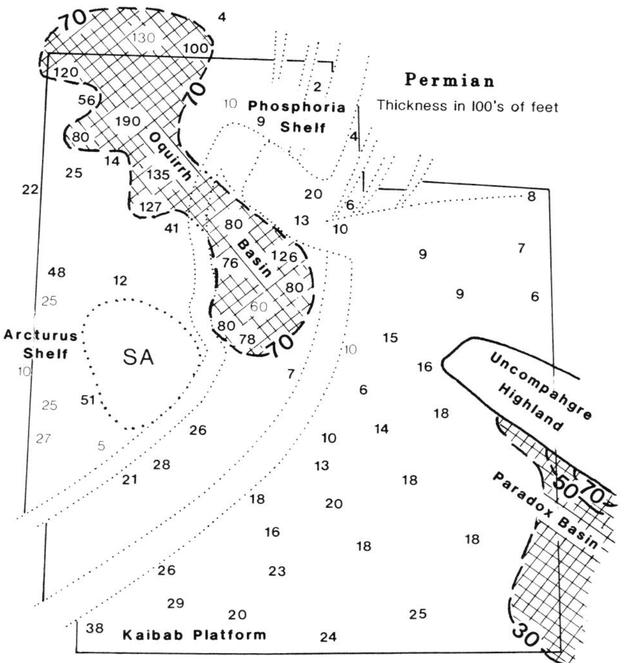

FIGURE 43 — Permian thicknesses reflect a subdued continuation of the Ancestral Rockies trends established in Pennsylvanian time. In the Paradox Basin, clastic rocks of the Cutler Group were derived from the active Uncompahgre Highland. In the Oquirrh Basin an incredible 19,000 feet of Permian deposits have been reported for the Hogup Mountains area, and elsewhere thicknesses in excess of 7,000 feet are typical. Early Permian rocks of the Oquirrh Basin are largely dolomitic fine-grained quartz sandstones; later Permian rocks include fossiliferous and cherty limestones and phosphatic shale. SA indicates the Mesozoic Sevier Arch where Permian rocks have been removed by later erosion; lighter numbers show locations where Permian sections are incompletely preserved.

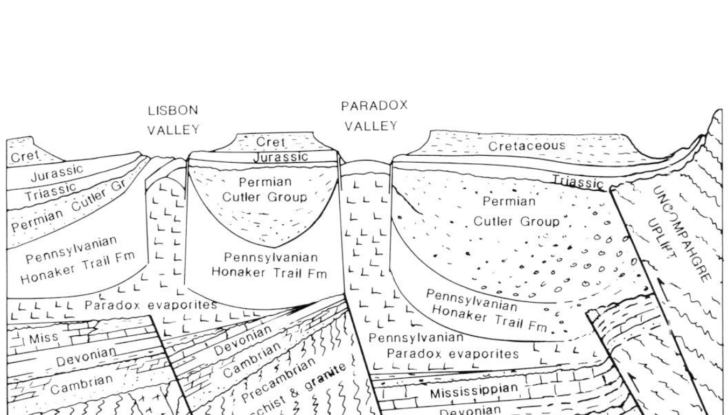

FIGURE 42 — Schematic cross-section through the Paradox Basin showing deformation and thinning of Permian and Mesozoic strata caused by movement of Pennsylvanian Paradox salt into anticlines. Salt anticlines are believed to have been localized above earlier faults that cut pre-salt bedrock. Adapted from Baars (1983) and White and Jacobson (1983).

FIGURE 44 — Comb Ridge Monocline, looking northward from the San Juan River; Abajo Mountains in distance. Oldest rocks exposed are Pennsylvanian Honaker Trail Formation, overlain successively by the Halgaito, Cedar Mesa, Organ Rock and DeChelly Formations of the Permian Cutler Group. The Triassic Moenkopi Formation is concealed by alluvium along its strike valley. The Chinle Formation crops out at the base of the Wingate Sandstone strike ridge. On the backslope of Comb Ridge the Navajo, Carmel, Entrada, Wanakah, and Bluff (cliff-forming) Formations are exposed. See Charts 101 and 102.

fusulinids, ammonoids, brachiopods, and conodonts—fossils do not occur in Utah rocks widely enough to permit us to establish unequivocal correlation between rocks of differing facies; they are especially lacking in sandstones and red shales of southeastern Utah.

Kaibab Limestone, which forms the rimrock of the Grand Canyon, is the southern Utah representative of the greatest epicontinental marine transgression of Permian time. In the type Permian in Russia, this transgression is represented by the Artinskian Series, the most abundantly fossiliferous of the Russian Permian sequence. In northern Utah it is represented by Park City and Phosphoria strata.

Phosphoria and Park City are old names which have been applied to post-Wolfcampian Permian strata in northern Utah and adjacent Wyoming, Idaho, and Nevada. In their current usage they are applied to intertonguing facies, the Park City (Franson and Grandeur Formations) being dominantly shallow-water marine limestone, and the Phosphoria (Tosi and Rex Cherts, and Retort and Meade Peak Phosphatic Shales) representing unusual deeper water sediments deposited in places on the Utah-Wyoming-Idaho shelf by upwelling currents (Sheldon et al., 1967).

Chamberlain and Clark (1973) used trace fossils and conodonts as evidence for Wolfcampian deep-water (1000 feet plus) deposits in the southern Oquirrh Basin. They suggested that much of the basin fill represents fluxoturbidites deposited at bathyal depths. Jordan and Douglass (1980), using lithofacies, trace fossils and body fossils as paleoenvironmental indicators, concluded that the northern part of the Oquirrh Basin also had deep-water deposition in the Wolfcampian. They suggested that faulting, which accompanied Oquirrh Basin subsidence, became inactive in latest Wolfcampian so that shallower water siltstones capped the deeper water sequence in latest Wolfcampian to Leonardian time. Post-Oquirrh Permian rocks in the ranges rising out of the Salt Flats (Charts 11 and 12) reach great thicknesses. Wardlaw et al. (1979) established new nomenclature to accommodate the later Permian deposits of northwestern Utah and adjacent Nevada.

Faulting that created the Uncompahgre Highland ceased major activity during Permian time, but erosion of the highland continued into the Early Triassic as evidenced by granite cobbles in Moenkopi sediments near the uplift (H. H. Doelling, personal communication, 1988); sandstones and shales on the shelf west of the

FIGURE 45 — Monument Valley Navajo Tribal Park at the Utah-Arizona border. Big Indian in foreground. Massive cliff is Permian DeChelly Sandstone. Above the DeChelly on the monument behind Big Indian is the less resistant Moenkopi Formation capped by the Shinarump Conglomerate Member of the Chinle Formation. Thin bedded redrocks at the base of the monuments is the Permian Organ Rock Formation; light-colored beds below are Cedar Mesa Sandstone. See Chart 101.

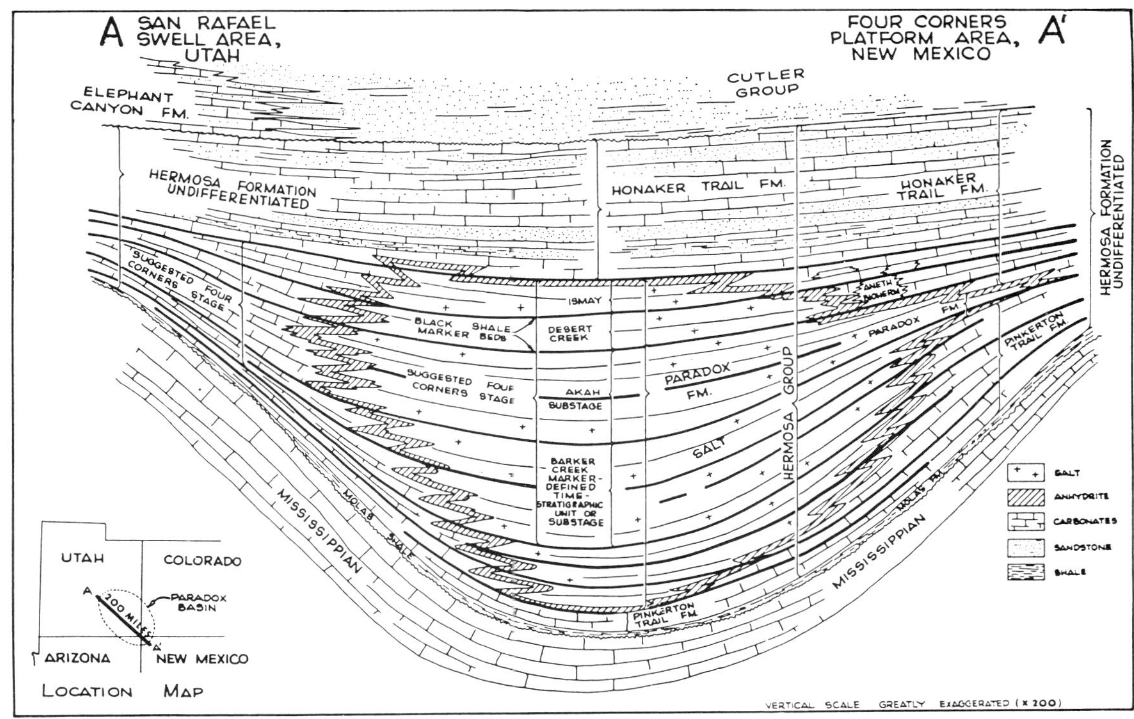

FIGURE 46 — Schematic cross-section through subsurface of Paradox Basin showing facies relationships. After Baars (1983).

Paradox Basin (Figures 44 and 45) interfinger with marine carbonates to the west; review of stratigraphic relationships here can be found in McKee (1934, 1982) Blakey (1980), Rascoe and Baars (1972), Rawson and Turner-Peterson (1979), and Stevens (1977, 1979).

Regional analysis of the Permian System was presented by McKee and Oriel (1967) and has been updated by Peterson (1980), and Peterson and Smith (1986).

TRIASSIC

Triassic, meaning "threefold," is divided into Early, Middle, and Late; Middle Triassic fossil-bearing rocks are absent in Utah. We can, thus, rather handily make the generalization that Utah's Early Triassic strata have marine affinities, that Middle Triassic is missing, and that Late Triassic rocks are continental (Figure 49). The thickness pattern of Triassic deposits (Figure 48) is reminiscent of early Paleozoic patterns: thin in eastern Utah, and thickening markedly westward. Original extent of Triassic strata in western Utah will never be known because these deposits were removed from all but a few locations during later Mesozoic uplift and erosion. In eastern Utah the Late Paleozoic Uncompahgre Highland became completely buried by Late Triassic sediments: Chinle redbeds rest directly on Precambrian metamorphic rocks in the center of the uplift (Chart 44). Even so, the buried Uncompahgre is reflected on Figure 48 in the curved form of the 1000-foot thickness line.

Moenkopi Formation—The Early Triassic Moenkopi Formation lends its reddish-brown colors to the Painted Desert and to the sculptured Canyonlands of southern Utah (Figures 47 and 50). The formation is predominantly a mudstone that was deposited on a broad flat coastal plain that sloped gently westward towards the miogeocline in southern Nevada. Thin tongues of marine limestone (Timpoweap, Sinbad, Virgin, Shnabkaib of Figure 49) indicate times when the seas spread out of Nevada across the Moenkopi flats. Westward, these limestone tongues

FIGURE 47 — East side of San Rafael Swell near Green River, Utah. Interstate I-70 descends through Spotted Wolf Canyon cut in resistant Jurassic sandstones. Oldest rocks in this view from the rest area at Mile 141 are Permian Black Box Dolomite (B) in photo center. In earlier reports this dolomite was called "Kaibab" (Welsh et al., 1979). S marks Sinbad Limestone Member of the Moenkopi Formation; M, Moss Back Member of the Chinle Formation; W, Wingate Sandstone. See Chart 67 for thicknesses.

merge into nearly continuous limestone deposits near Las Vegas (Bissell, 1973). Stewart et al. (1972b) presented a comprehensive study of the Moenkopi Formation, including numerous measured sections; Blakey (1974) also presented measured sections of the Moenkopi in southeastern Utah and established new member names shown on the charts herein. Irwin (1971, 1976) focused on the potential of Permian and Early Triassic rocks in east-central Utah to serve as petroleum reservoirs. Mitchell (1985) constructed detailed thickness maps for Permian and Triassic strata in southeastern Utah, based on both surface and well data.

Moenkopi environments, as interpreted from fossil sedimentary features, include stream channels, flood plains, fresh or brackish ponds, playas, and shallow seas. Beds of gypsum and casts of salt cubes preserved in the red beds indicate evaporation. Large reptiles, represented by tracks, and big amphibian skeletons in the stream deposits suggest warm climates.

Thaynes and Dinwoody Formations—Early Triassic rocks in northern and western Utah generally represent slightly deeper-water deposition than the Moenkopi and are known as the Dinwoody and Thaynes Formations. Near Salt Lake City the red Woodside Shale is present, but it passes westward into limestones of the Thaynes Formation. Koch (1976) and Collinson and Hasenmueller (1978) published thickness and lithofacies maps for Early Triassic rocks in western Utah and Nevada. Carr and Paull (1983), Carr, Paull, and Clark (1984), and Clark and Carr (1984) summarized Early Triassic paleogeography of northern Utah with lithofacies maps and cross sections showing inter-tonguing relationships in the inner shelf-outer shelf-basin transition, noting that conodont distribution is controlled by environment.

Ammonoid and conodont zones in the Thaynes Formation and in limestone members of the Moenkopi, shown on Figure 49, enable sure dating and correlation of Utah's Early Triassic strata with Triassic standard sequences elsewhere. Marine clams, snails, and brachiopods, are, as usual, more obvious and abundant in these limestones than the conodonts and ammonoids but are not as distinctive in indicating time. Utah's Late Triassic rocks lack marine fossils and cannot be correlated inter-continentally with the same degree of confidence.

Chinle continental deposits—Petrified Forest National Park in Arizona features thousands of fossilized conifer logs that came to rest as driftwood on Chinle river floodplains in Late Triassic time. Petrified wood fragments can be found elsewhere in most Chinle exposures, but not in such abundance. In some instances the cellular tissues of the wood have been replaced by silica in exquisite detail. Silicification of wood in the Chinle Formation was probably related to bentonitic volcanic ash in the formation, which provided a ready source of silica.

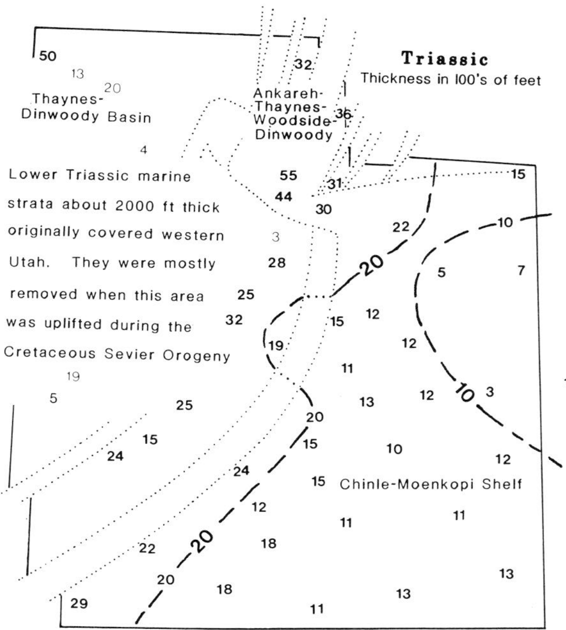

FIGURE 48—The Triassic depositional pattern resembles that of early Paleozoic periods: thinner on the craton of eastern Utah and thicker in the western miogeocline. Early Triassic (Moenkopi) strata are marine and tidal flat mudstone, siltstone, sandstone and platy limestone. Late Triassic (Chinle) rocks are continental clastics that include a substantial amount of reworked volcanic ash. Latest Triassic rocks covered the area of the Uncompahgre Highland, active from Pennsylvanian time into the earliest Triassic. Lighter numbers show locations where Triassic stratigraphic sections are incompletely preserved.

Stewart et al. (1986), in discussing the dilemma of the source for volcanic detritus in the Chinle, noted that although Chinle sediment transport was from the south, southern Arizona does not have volcanic sources of the appropriate age, whereas Nevada and California do.

Although conifers are the most common plants preserved in the Chinle Formation, it also yields fossil cycads, ferns, and horsetails (Ash, 1975). Scott (1982) and Fisher and Dunay (1984) described fossil spores and pollen from the Chinle Formation. Other fossils from the Chinle include fresh-water clams, snails, ostracods, and fish, as well as amphibians and reptiles. The most common reptile was the river-dwelling phytosaur, which resembles modern crocodiles but has a slender snout.

Arrows on Figure 51 indicate direction of stream transport during Chinle time. Gravels and sands which make up the Shinarump and Moss Back Members were derived chiefly from the east and south. Chinle stream-channel deposits are common sites of later localization of uranium mineralization. Regional relationships of Triassic strata are summarized on lithofacies and thickness maps by

TRIASSIC

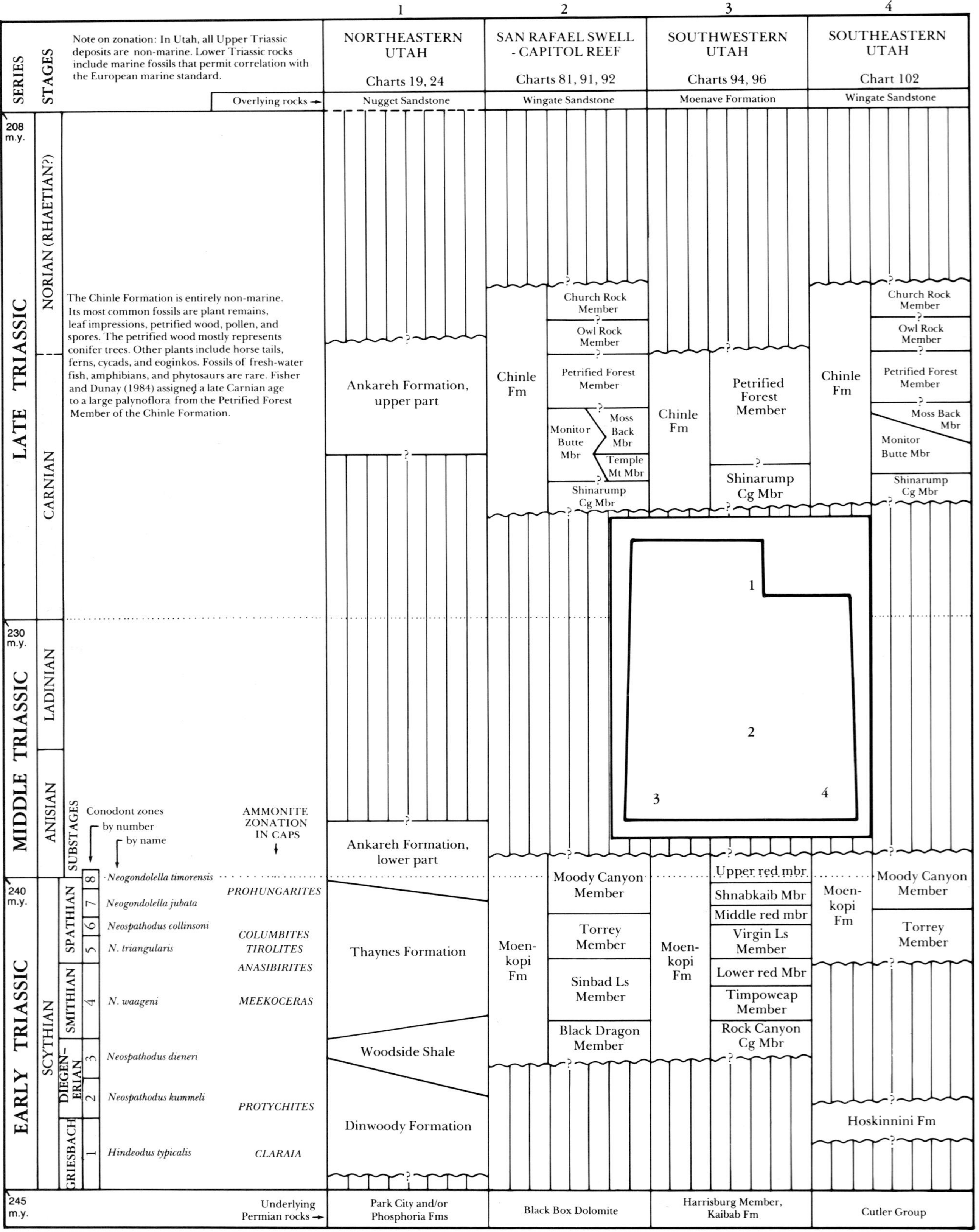

REFERENCES - Carr & Paull, 1983; Clark & Carr, 1984 et al, 1979; Ash, 1975; Blakey, 1974; Blakey & Gubitosa, 1983; Kummel, 1979, Stewart et al, 1986; Scott, 1982; Fisher and Dunay, 1984.

FIGURE 49 — Triassic correlation table. Along the south flank of the Uinta Mountains (not represented on this table), the Gartra Formation, conglomerate and sandstone, unconformably overlies the Ankareh (Chart 38) and Moenkopi (Chart 40) Formations. It conformably underlies the Chinle Formation which is there called the Popo Agie Formation by Uyger and Picard (1985).

FIGURE 50 — Dead Horse Point, view southwestward across the north end of the Monument Upwarp. Permian Elephant Canyon Formation (Pec) at river level is overlain by arkosic redbeds of the Cutler Formation (Pu) capped by the White Rim Sandstone (Pwr). Triassic Moenkopi Formation and Chinle formations form a retreating slope above the White Rim and below the cliff-forming Wingate Sandstone (Jw), which is capped by the Kayenta Formation. See Charts 83 and 84. Entrenchment of the Colorado River within the last 10 million years has stripped more than a mile of post-Kayenta strata from the Monument Upwarp.

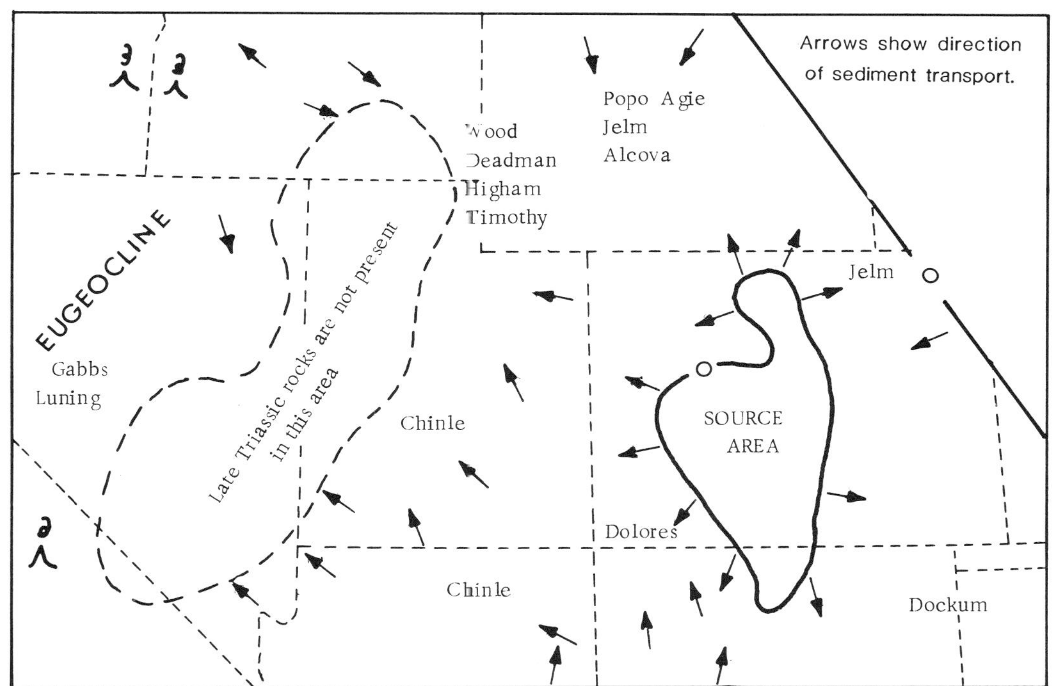

FIGURE 51 — Chinle Formation (Upper Triassic) directions of sediment transport during Phase IV of Utah's geologic history. Relict Uncompahgre Uplift remained as low area in Colorado. No western area had developed as a significant source yet.

FIGURE 52 — Checkerboard Mesa in Zion National Park shows crossbedding of the Navajo Sandstone cut by surficial cracks. Larger vertical fractures, called joints, have controlled weathering and erosion in the development of Zion; all major valleys follow joints, some of which can be traced for miles. The Navajo Sandstone is made of fossil sand dunes, stacked one upon another to a cumulative thickness of about 2000 feet.

McKee et al. (1959) and MacLachlan (1972). More detailed information on the Chinle Formation in Utah may be found in Stewart et al. (1972), Lupe (1984), and Blakey and Gubitosa (1983), who concluded that Chinle sedimentation occurred in an enclosed continental basin and represents a complex of alternating fluvial and lacustrine depositional systems.

JURASSIC

Jurassic strata are among Utah's most scenic; they include the Navajo Sandstone that makes the great monoliths at Zion Park (Figure 52) and, along with the Wingate and Kayenta Sandstones, forms the reef at Capitol Reef and the Lake Powell dam abutments at Glen Canyon; they also include the Entrada Sandstone that makes the arches at Arches National Park and the goblins at Goblin Valley; the world's most famous dinosaur quarry is in the Morrison Formation at Dinosaur National Monument. Jurassic rocks are exposed over wide areas in eastern Utah, and their multicolored, variously eroded layers are an open textbook to the knowledgeable traveler, and a scenic delight to all.

Worldwide, the Jurassic System is zoned on the basis of ammonites; in Utah this works only for the marine part of the Middle Jurassic, as shown in the left column of Figure 55. Kent and Gradstein (1985) reviewed radiometric, magnetic polarity, and paleontologic zonation schemes for the Jurassic System; Imlay (1980) summarized a life's work on paleontologic aspects of Jurassic rocks in the U.S., and his zonations that apply to Utah rocks are shown on Figure 55.

Jurassic strata in Utah fall naturally into three packages: Early Jurassic rocks are chiefly non-marine sandstones; Middle Jurassic rocks record changing conditions as an epicontinental seaway advanced from Canada, then retreated; Late Jurassic deposits were laid down in the interior non-marine Morrison basin, replete with lakes and streams that attracted dinosaurs.

Early Jurassic Sandstones—Over much of eastern Utah, Early Jurassic sandstones occur in an easily recognizable three-part sequence: at the base, the eolian Wingate Sandstone erodes to vertical-walled cliffs; in the middle, the fluviatile Kayenta Formation (sandstone-siltstone-mudstone) forms juniper-covered ledge-slope benches; at the top, the eolian Navajo Sandstone forms rounded cliffs that show magnificent cross-bedding features etched into their surfaces. At Glen Canyon and also near Dinosaur Monument, Kayenta-type features are missing or so sub-

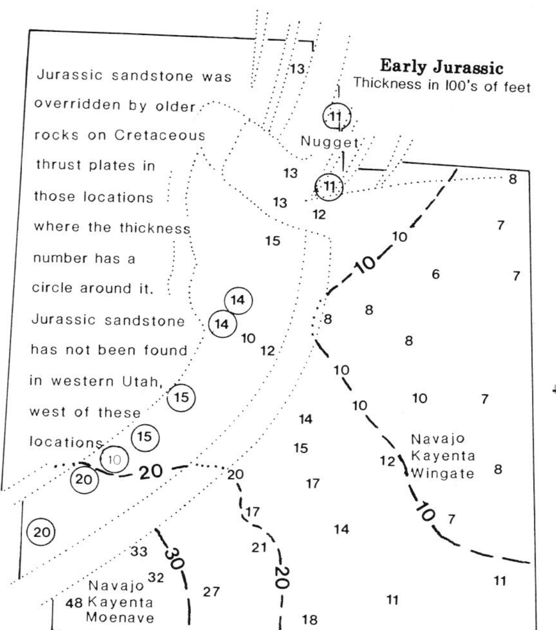

FIGURE 53 — Early Jurassic deposits include the Nugget Sandstone north of Salt Lake City, and the Navajo, Kayenta, Wingate, and Moenave Formations (Glen Canyon Group) in the rest of the state. Thicknesses increase towards southwestern Utah where 4,800 feet are present in the Beaver Dam Mountains area west of St. George. Jurassic sandstones seemed to form a convenient base for overriding Cretaceous thrust plates; the thinner-bedded later Jurassic strata were commonly peeled away during advance of the thrust plate. Lighter numbers indicate locations where Early Jurassic sandstones are incompletely preserved.

dued that the entire package merges into one formation, the Glen Canyon Sandstone or Group; the same is true in northern Utah where the term Nugget Sandstone is used.

The Early Jurassic sandstone sequence thickens westward as shown on Figure 53, and passes beneath overthrust plates along Utah's hingeline. It has not been identified in the subsurface in western Utah and its western subsurface extent there remains unknown. In southern Nevada it continues westward as the Aztec Sandstone. Navajo-Nugget sandstones are believed by most geologists to represent a coastal to inland dune field, although Freeman (1976) argued for a shallow-marine origin. Kocurek and Dott (1983, p. 107) suggested that Navajo eolian dunes may have been the largest recorded in earth's history. Figure 54 shows that the amazing Navajo-Nugget-Aztec sand sheet now covers a great part of Utah, and was originally more extensive to the west; dip directions of Navajo cross-beds indicate wind directions from the northwest. Analysis of cross-bedding types found in the Navajo was given by Rubin and Hunter (1987).

Age of the Glen Canyon Group has been difficult to determine; the Navajo and Wingate Sandstones are unfossiliferous, and vertebrate fossils from the Kayenta and Moenave Formations have been variously assigned by some to the Late Triassic and by others to the Early Jurassic. The problem has been reviewed by Clark and Fastovsky (1986) and Olsen and Galton (1977), who favor a Jurassic assignment. Peterson and Pipiringos (1979, p.31) cited palynostratigraphic data which supports an Early Jurassic age for the Moenave Formation. Additional unpublished information from Ronald J. Litwin (letter, February, 1986), who studied palynomorphs from the Dinosaur Canyon Member, the lowest member of the Moenave Formation, in the Kanab, Utah, area supports the assignment of the Moenave to the Jurassic. Fish scales and bone fragments found in the Whitmore Point Member of the Moenave Formation near St. George suggest that this member was deposited in fresh-water lakes (Wilson, 1967).

Middle Jurassic Twin Creek-Carmel marine rocks— Marine fossil-bearing limestones and shales of the Twin Creek (Figure 57) and Carmel Formations attest to a shallow seaway that extended from Canada southward to

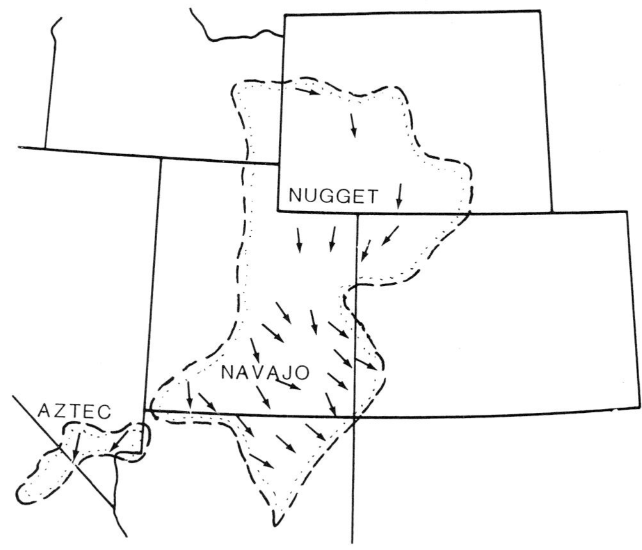

FIGURE 54 — Dashed lines outline present extent of Early Jurassic Navajo-Nugget-Aztec sandstones. Utah lay about 15 degrees north of the equator at this time (see Figure 28). Paleowinds blew mostly from the north and northwest as shown by the mean dip of foreset beds (arrows). After Kocurek and Dott (1983).

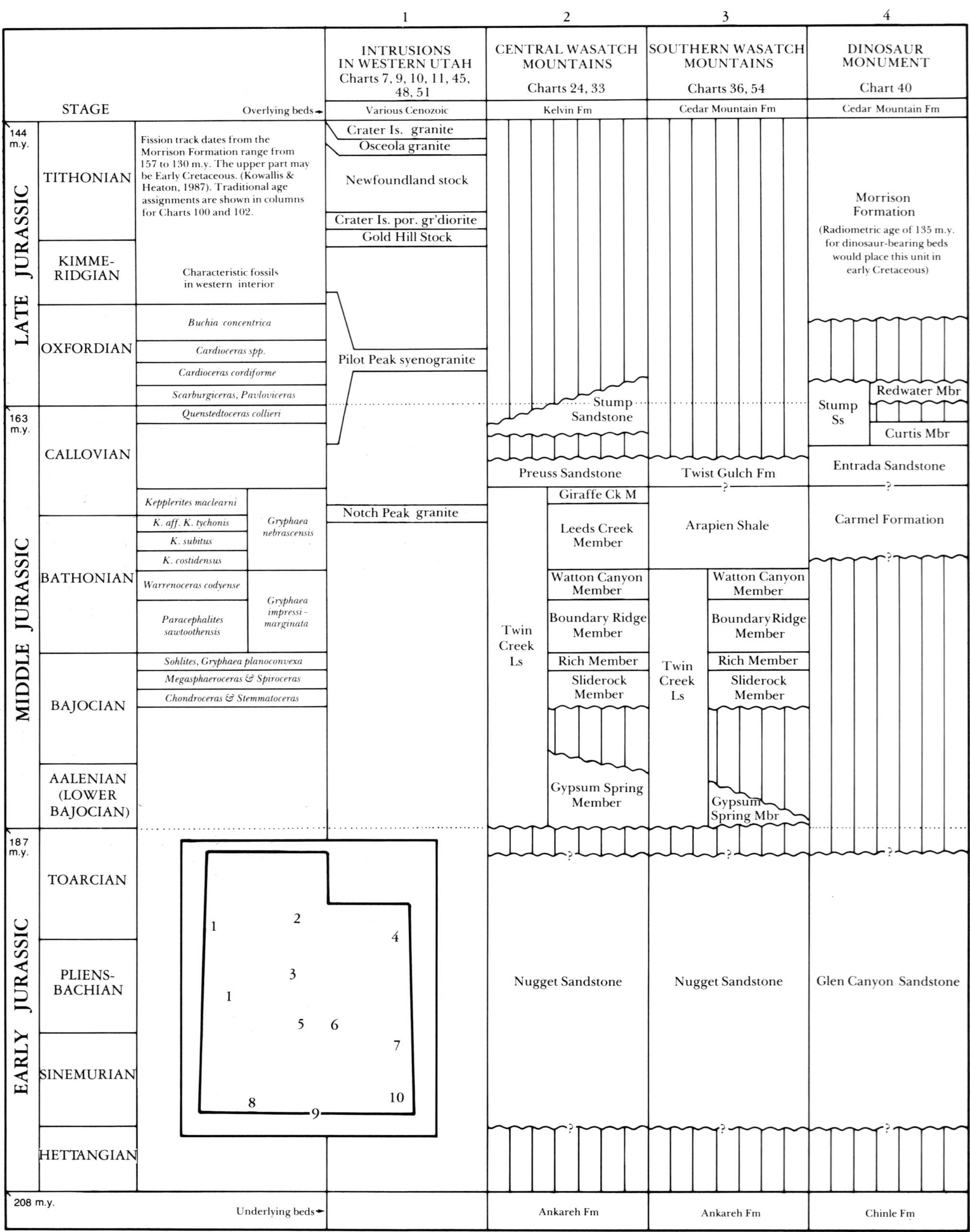

REFERENCES - Imlay, 1980; Kent & Gradstein, 1985; Stewart et al, 1986; Peterson, F., 1987, in press.
Notes-"Henrieville Sandstone" is the bleached upper part of the Entrada (Fred Peterson, 1987, per. comm)
"Kolob Limestone" Member of Carmel Fm (Thompson & Stokes, 1971) is preoccupied. Equivalent to Judd Hollow Member of this chart

FIGURE 55 — Jurassic correlation table.

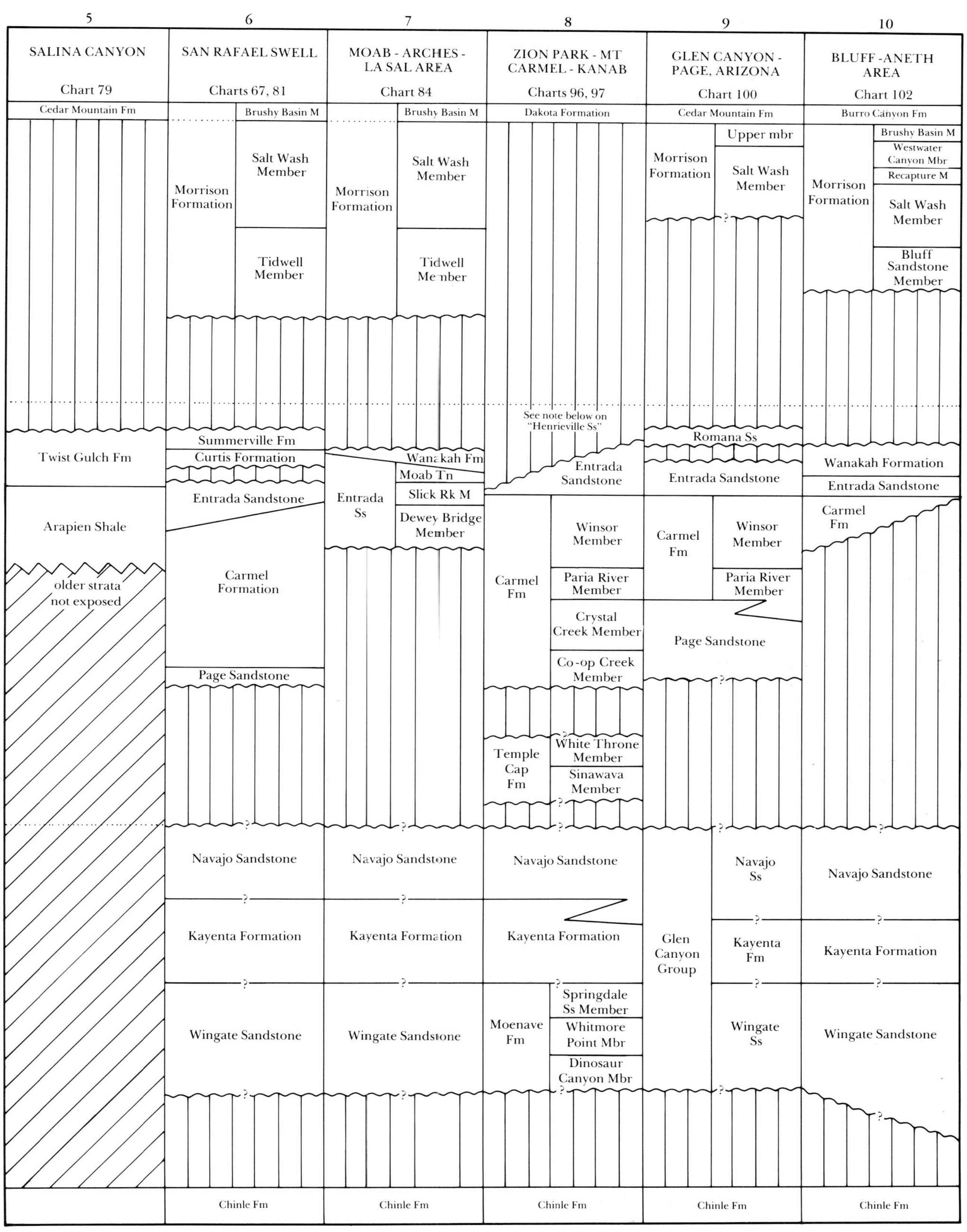

5
6
7
8
9
10
SALINA CANYON
Chart 79
SAN RAFAEL SWELL
Charts 67, 81
MOAB - ARCHES - LA SAL AREA
Chart 84
ZION PARK - MT CARMEL - KANAB
Charts 96, 97
GLEN CANYON - PAGE, ARIZONA
Chart 100
BLUFF -ANETH AREA
Chart 102
Cedar Mountain Fm
Brushy Basin M
Brushy Basin M
Dakota Formation
Cedar Mountain Fm
Burro Cañyon Fm
Upper mbr
Brushy Basin M
Westwater Canyon Mbr
Recapture M
Salt Wash Member
Morrison Formation
Tidwell Member
See note below on "Henrieville Ss"
Summerville Fm
Curtis Formation
Wanakah Fm
Moab Tn
Slick Rk M
Dewey Bridge Member
Romana Ss
Wanakah Formation
Entrada Sandstone
Entrada Ss
Twist Gulch Fm
Arapien Shale
older strata not exposed
Carmel Formation
Carmel Fm
Winsor Member
Paria River Member
Crystal Creek Member
Co-op Creek Member
Page Sandstone
Temple Cap Fm
White Throne Member
Sinawava Member
Bluff Sandstone Member
Navajo Sandstone
Navajo Ss
Kayenta Formation
Kayenta Fm
Glen Canyon Group
Wingate Sandstone
Wingate Ss
Moenave Fm
Springdale Ss Member
Whitmore Point Mbr
Dinosaur Canyon Mbr
Chinle Fm

FIGURE 56 — Jurassic strata exposed along the east side of the San Rafael Swell near Green River, Utah. Interstate I-70 enters the Swell here. Light-colored flatirons are Wingate (W), Kayenta (K) and Navajo (N) Sandstones. Carmel Formation (Ca) forms low, dark flatirons as well as light gypsum beds one-half inch to left of "Ca." Entrada Sandstone makes largely covered terrain near rest-stop pads and "E." Curtis Formation (Cu) forms a light gray (green, actually) strike ridge. Summerville Formation (S) and Morrison Formation at left edge of photo. Chinle (Ch) and Moenkopi (Mo) Formations at right edge. See Chart 67 for thicknesses.

Carmel Junction, near Zion Park. Bajocian-age limestones are the most widespread marine Jurassic rocks in Utah; later Bathonian and Callovian strata include evaporites and red beds that indicate marginal marine conditions. By the time the Entrada-Preuss sandstones were deposited, marine waters had retreated northward out of Utah. Another Jurassic marine invasion is recorded by marine fossils in the Curtis Formation (Brenner, 1983). These marine waters penetrated only to the San Rafael Swell and Capital Reef areas of central Utah where the Curtis Formation (Figure 56) interfingers with marginal marine gypsiferous chocolate shales of the Summerville and Entrada Formations. Peterson (1972) and Kocurek and Dott (1983) have traced the marine Jurassic invasions northward to the more continuously limy deposits in Wyoming and Montana as shown on Figure 59. Intricate facies relationships of Middle Jurassic rocks in southern Utah have been reviewed by Blakey et al. (1983). Thickness patterns (Figure 58) show the continued positive influence of the buried Uncompahgre Uplift in eastern Utah, and development of the Arapien Basin in central Utah where more than 6000 feet of Middle Jurassic mudstone, siltstone, evaporite, and limestone were deposited. Nomenclature of Middle Jurassic strata in the

FIGURE 57 — Thistle, Utah was destroyed in April, 1983, by a mudslide that covered the main highway and rail route in Spanish Fork Canyon, and dammed the river to form a lake whose shoreline shows on the bleached lower hillslopes this side of the dam. This picture was taken in August, 1984, after the railroad was rerouted in a tunnel through Billies Mountain, the hill in the right foreground. The new highway cuts through the top of the mountain. The temporary lake was drained by drilling a tunnel through Billies Mountain. The old road junction at Thistle in the foreground is about at the contact between the Navajo Sandstone (Jn) and the Twin Creek Limestone (Jtc). Older Triassic and Permian beds are exposed downstream beyond the slide. The slide was activated by heavy spring rains that saturated the slide-prone North Horn Formation which unconformably overlies older bedrock out of view to the left. See Chart 36.

Arapien Basin has been modified by geologists working in that area over the years (Sprinkel, 1982). The charts and correlation tables, herein, follow nomenclature recommended by Witkind and Hardy (1983).

Although thin beds and veinlets of gypsum are common in the Carmel and Summerville Formations over much of southern and eastern Utah, commercial gypsum and salt of Jurassic age have been claimed only from the Arapien Basin between Nephi and Richfield. Depositional thickness of Middle Jurassic gypsum and salt in this area is not known because the plasticity of these evaporites and their enclosing mudstones has brought about much distortion (Witkind, 1983, 1987), but certainly several hundred feet of evaporites are indicated by available data. These evaporites may have had a key role in the development of many of the structural features of the Sanpete Valley area (Witkind and Page, 1984; Witkind, 1987; Willis, 1986, 1988).

Volcanic ashes occur in several horizons in the Carmel Formation and their correlation potential is just begin-

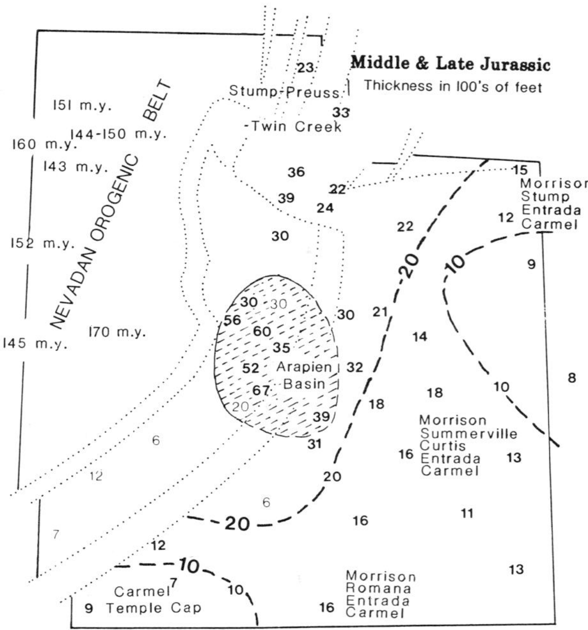

FIGURE 58 — Thickness numbers on this Middle and Late Jurassic map include an increment for the Morrison Formation, part of which may actually be of Lower Cretaceous age. Late Jurassic intrusive rocks, shown by radiometric ages, predated compressional tectonism of the Sevier Orogeny in western Utah. Names of formations elsewhere in Utah are shown. Thickest deposits accumulated in the Arapien Basin where more than 6,000 feet of mudstone, siltstone, evaporite, and limestone were laid down. Shading on the map shows the area of maximum gypsum and salt deposition. Mobility of Jurassic evaporites and mudstone has affected structural development in central Utah. Several thin volcanic ash beds are included in the Carmel and Morrison Formations; these serve as key correlation horizons and radiometric age markers. Lighter numbers indicate where deposits of this age are incompletely preserved.

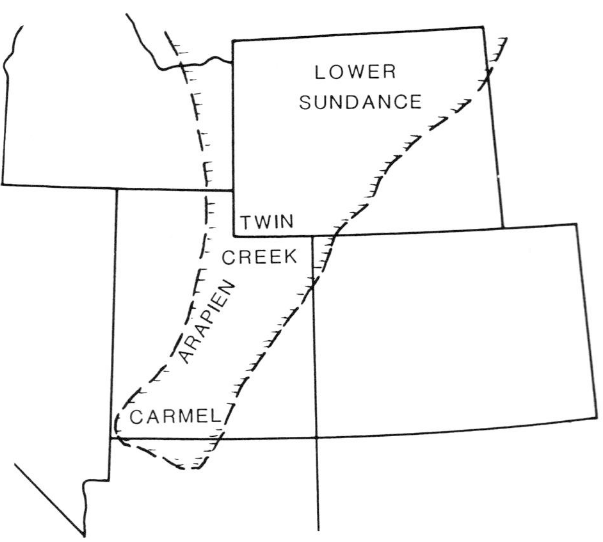

FIGURE 59 — Middle Jurassic marine invasion of Sundance-Twin Creek-Carmel seas from the north. After Kocurek and Dott (1983). Dashed lines show present extent of marine strata.

ning to be realized (Hintze, 1986b; Nielson, 1988). B. J. Kowallis (personal communication, 1986) is currently undertaking fission track dating of these bentonities under a study sponsored by the Utah Geological and Mineral Survey.

Late Jurassic-Early Cretaceous Morrison Formation— Varicolored mudstones, stream channel sandstones and conglomerates, and lacustrine limestones make up the distinctive lithologies of this unique formation. The Morrison has yielded more dinosaur bones than any other formation in the world, chiefly from quarries at Como Bluff in central Wyoming, Dry Mesa near Grand Junction, Colorado, Dinosaur National Monument in eastern Utah (Figure 62), and the Cleveland area south of Price. Bones occur both in claystones and in coarse sandstones, and, although whole skeletons are rare, bone fragments are common in many places. Considering that the Morrison Formation is entirely continental in origin, it was deposited in a basin that covered a surprisingly large area (Figure 60) rather uniformly. Some Morrison conglomerate beds contain clasts that were derived from central Arizona (Peterson, 1986); others certainly came from western Utah and eastern Nevada (Peterson, 1987).

Morrison sediments were likely deposited by intermittent shifting streams under semi-arid conditions (Hallam, 1982). A large playa-lake complex may have played an important role in dinosaur-related paleoenvironments during deposition of the Brushy Basin Member of the Morrison Formation in eastern Utah, Colorado, and New Mexico (Peterson and Turner-Peterson, 1987). Morrison mudstones include several horizons of reworked volcanic ash which have been recently dated by Kowallis and Heaton (1987). They show that part of the Brushy Basin Member in the upper Morrison may be Lower Cretaceous in age. In addition, rocks that were previously assigned to the Morrison (?) Formation in central Utah have been shown by Witkind et al. (1986) and Willis and Kowallis (1988) to be better regarded as belonging to the Lower Cretaceous Cedar Mountain Formation, thus lessening the western extent of the known Morrison basin of deposition.

Morrison stream channel sand deposits became the locus of uranium mineralization and have been widely prospected and mined in the Colorado Plateau. Sanford (1982) has modeled the groundwater conditions in the Morrison basin that led to the transportation and deposition of uranium.

FIGURE 60 — East-central Utah between Capitol Reef and the south end of the San Rafael Swell. See Chart 81. Jurassic Entrada Formation, Je, (earthy facies) is oldest unit exposed, and is overlain by Curtis Formation, Jc; Summerville Formation, Js; Morrison Formation, Jm; Cedar Mountain Formation, Kcm; Dakota Sandstone, Kd; Tununk Shale, Kt; Ferron Sandstone, Kf; Blue Gate Shale, Kb.

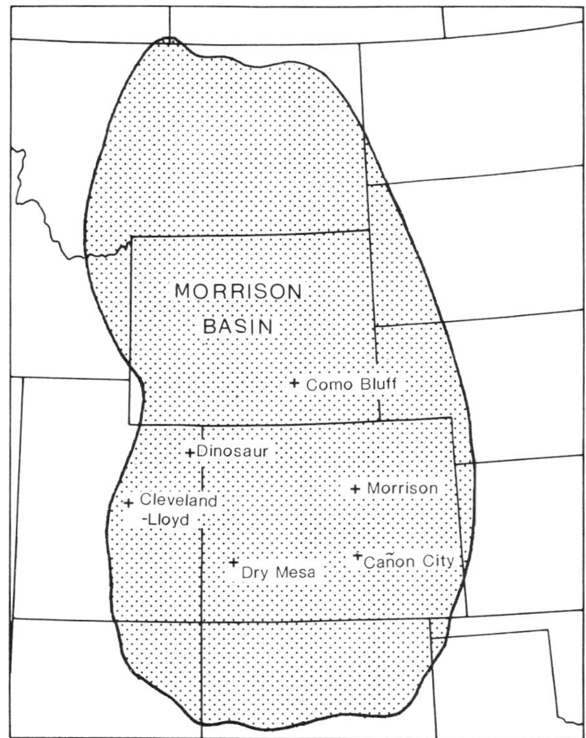

FIGURE 61 — Extent of the interior basin in which the Upper Jurassic-Lower Cretaceous Morrison Formation was deposited. Dinosaur bones are found throughout the basin but significant quarries have been developed only at the locations named on the map.

FIGURE 62 — Dinosaur National Monument quarry face, a continually changing display. Workmen remove sandstone to expose bones in the Morrison Formation. See Chart 40 and Figure 7.

FIGURE 63 — Cretaceous strata at the southern end of the Kaiparowits Plateau, west of Page, Arizona, looking northwest near Warm Creek. See Chart 100. Kt, Tropic Shale; Kst, Tibbet Canyon Member; Kss, Smoky Hollow Member; Ksj, John Henry Member; Ksd, Drip Tank Member; Kwl, lower Wahweap Formation; Kwu, upper Wahweap Formation.

Nevadan orogeny, and metamorphic core complexes— Figure 58 shows radiometric ages for Late Jurassic granitic intrusions in western Utah. Allmendinger and Jordan (1981, 1984) showed that rocks in the Newfoundland Mountains (Chart 10) were compressed about 150 m.y. ago and that this sector of western Utah, thus, records the earliest thrusting in the Sevier orogenic belt. Later, during mid-Cretaceous to Early Tertiary thrusting, the rocks of the Newfoundland Mountains were rafted passively eastward on the upper plate of a deep-seated decollement. Clasts from western Utah that were deposited in the Morrison Formation (Peterson, 1987) confirm the beginning of uplift in the hinterland in western Utah. Heller et al. (1986) preferred not to include this Jurassic tectonism under the term Sevier orogeny. Armstrong's (1968a) original presentation of the Sevier orogeny showed it slightly overlapping in time with the classic Nevadan orogeny. Little (1987) described Late Jurassic folding and thrusting in north-central Nevada, and Schweichert et al. (1984) refined the dating of the Nevadan orogeny in the Sierra Nevada as occurring 155 m.y. ago. They ascribed the orogeny to collision of a western island arc with an andean-type margin of North America. Kistler (1983) confirmed major displacement of crust in western Nevada and California at about 150 m.y. ago using isotope geochemistry.

Miller et al. (1987) summarized geochronological data relating to metamorphic core complexes along the Utah-Nevada border. They showed that local amphibolite-facies and widespread greenschist-facies metamorphism peaked about 160 m.y. ago, accompanied by emplacement of plutons. The metamorphic and intrusive rocks emplaced at depth in the crust during Late Jurassic time were domed to the near-surface in Cretaceous or early Cenozoic time to form the cores of several metamorphic core complexes in this area. Perhaps the emplacement of granitic rocks and the Late Jurassic compressive features in northwestern Utah may be best included with the Nevadan orogeny.

Regional summary papers —Summaries of Jurassic stratigraphy may be found in works by J. A. Peterson (1972), Kocurek and Dott (1983), F. Peterson (1986), and Peterson and Turner-Peterson (1987). Imlay (1980) summarized Jurassic marine fossil zonations and presented a series of maps outlining the extent of marine deposits at various times in the Jurassic.

CRETACEOUS

The Cretaceous Period was the time of the last epicontinental sea in Utah. Marine waters extended from arctic Canada to Texas, separating North America into an Appalachian mega-island of low relief and a western Cordilleran mountain chain, the Sevier orogenic belt. Throughout Late Cretaceous time, erosion in the active Sevier belt in western Utah provided clastic materials to central and eastern Utah in such volume that a geosynclinal wedge of sediment, three miles thick in central Utah, and thinner eastward as shown on Figure 64, spread across eastern Utah, Colorado, and into Kansas. Cretaceous deposition and accompanying subsidence in eastern Utah was such that the thickness of Cretaceous strata, alone, equaled or exceeded the combined thicknesses of all preceding Paleozoic and Mesozoic deposits. Much Cretaceous rock has been stripped from eastern and southern Utah through erosion by the Colorado River and its tributaries in late Cenozoic time; Figure 64 outlines the areas where Cretaceous strata remain. Utah's most imposing erosional escarpments—the Book Cliffs, the east front of the Wasatch Plateau, and the Straight Cliffs—are all held up by resistant sandstones of the Late Cretaceous Mesaverde Group. The underlying Mancos Shale forms broad low areas that have attracted traveller's pathways, ancient and modern.

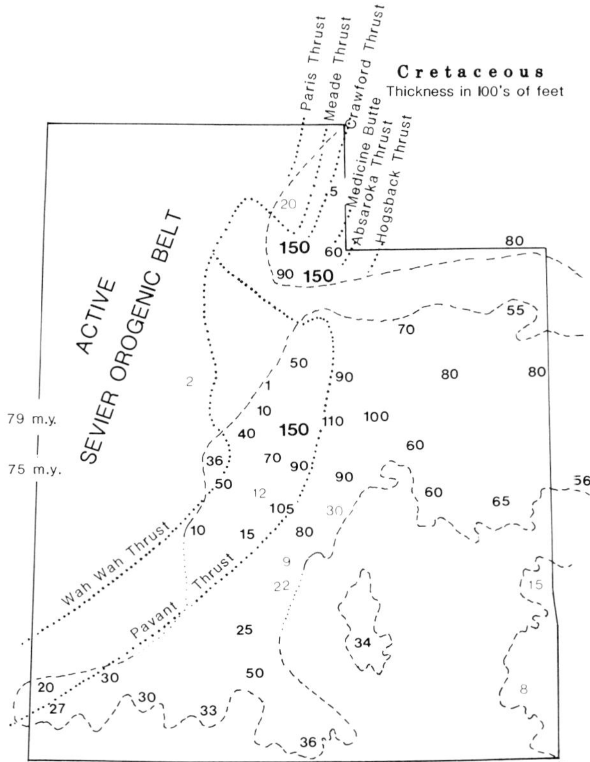

FIGURE 64 — Cretaceous deposits are largely clastic rocks derived from the mountainous Sevier orogenic belt and laid down along a coastal plain where coal-bearing non-marine sandstones and conglomerates interfinger with marine shales of the mid-continental seaway. As much as 15,000 feet of Cretaceous deposits have been reported for the Cedar Hills area in central Utah and the thrust belt of northeastern Utah. Thrusting was partly concurrent with deposition; later Cretaceous thrusts overrode Cretaceous deposits derived from earlier thrusts. The thrusts named above are younger on the east; trends of the thrusts are shown by heavier dotted lines. The dashed lines and light dotted lines show the present-day erosional edge of Cretaceous strata. They have been largely removed over southeastern Utah because of later uplift and erosion by the Colorado River and its tributaries. Lighter numbers show locations where Cretaceous strata are partially preserved. Two-mica granites, shown near the Nevada border by their radiometric dates, appear to represent melting in the thickened hinterland of the Sevier thrusts.

The Cretaceous is one of the longest post-Proterozoic periods, encompassing 78 million years (Kent and Gradstein, 1985). Its paleogeography and depositional patterns in Utah changed several times in response to thrust activity in the Sevier orogenic belt. Hence, numerous

FIGURE 65 — Willard thrust exposed above the town of Willard, Utah. Light band is Cambrian Tintic Quartzite. It rests unconformably on Archean Farmington Canyon Complex metamorphic rocks that form the ragged outcrops on the right side of the photo. Just above the Tintic Quartzite several hundred feet of Cambrian Maxfield Limestone (dark outcrops on photo) rest on the Tintic on a bedding-parallel fault (the arrow points down the fault surface) that cuts out most of the Ophir Shale. The main Willard thrust fault places early Proterozoic Facer Formation and the late Proterozoic Huntsville sequence (Chart 22) on the Cambrian limestone. The Willard thrust is part of the Paris thrust complex shown in Figure 64 as the westernmost thrust in the Sevier overthrust belt.

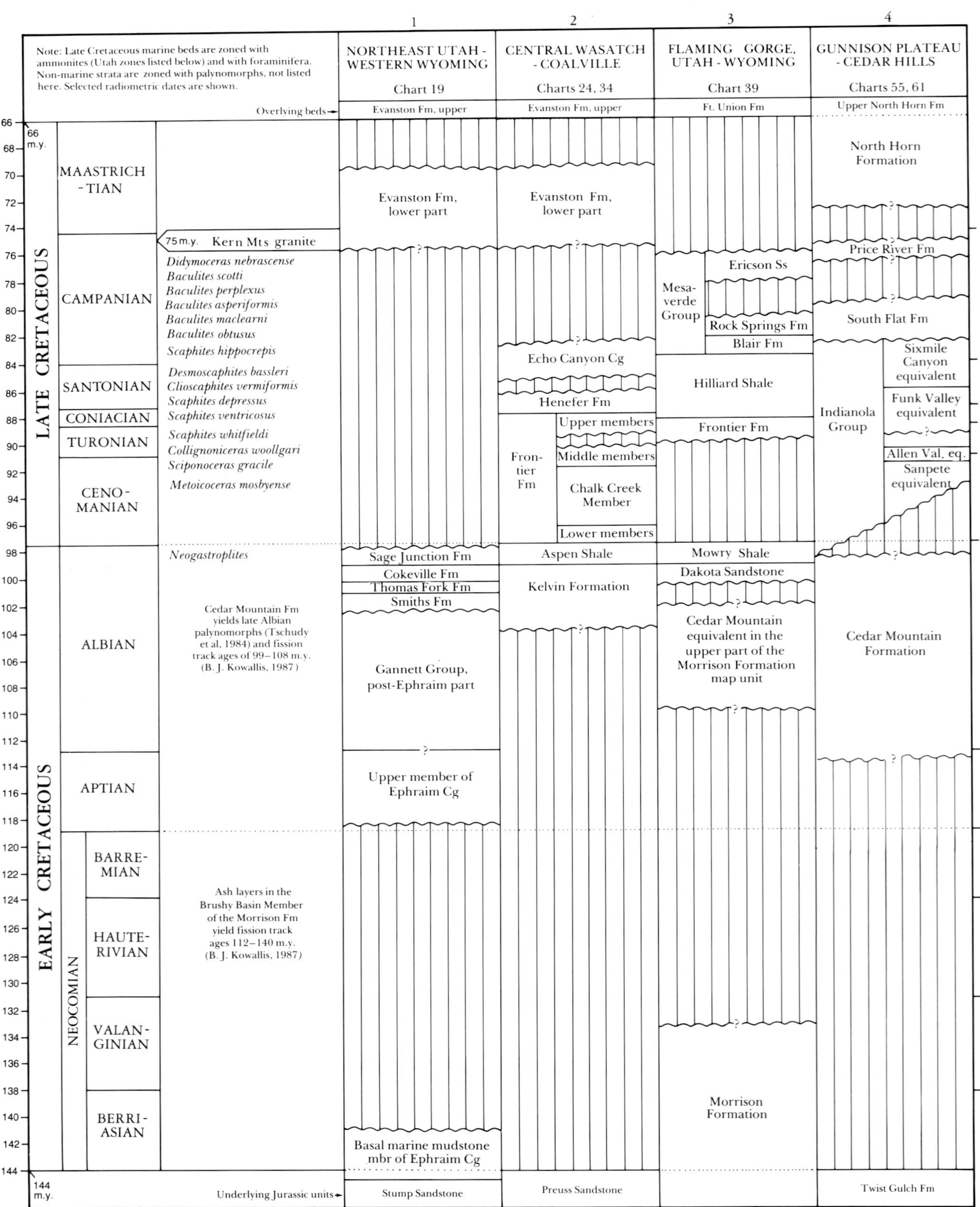

FIGURE 66 — Cretaceous correlation table.

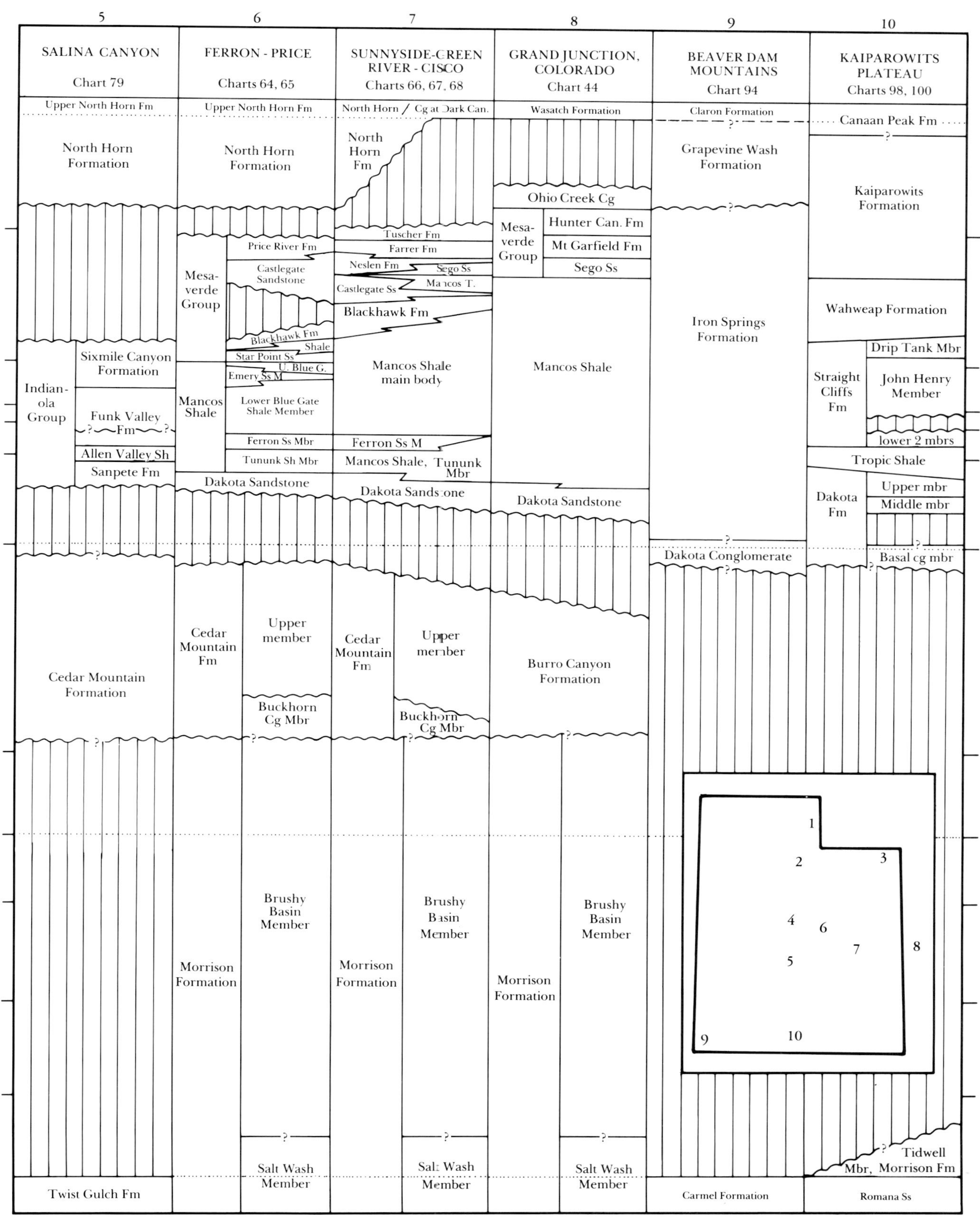
5
6
7
8
9
10
SALINA CANYON
Chart 79
FERRON - PRICE
Charts 64, 65
SUNNYSIDE-GREEN RIVER - CISCO
Charts 66, 67, 68
GRAND JUNCTION, COLORADO
Chart 44
BEAVER DAM MOUNTAINS
Chart 94
KAIPAROWITS PLATEAU
Charts 98, 100
Upper North Horn Fm
Upper North Horn Fm
North Horn / Cg at Dark Can.
Wasatch Formation
Claron Formation
Canaan Peak Fm
North Horn Formation
North Horn Formation
North Horn Fm
Grapevine Wash Formation
Kaiparowits Formation
Ohio Creek Cg
Mesa-verde Group
Hunter Can. Fm
Mt Garfield Fm
Sego Ss
Tuscher Fm
Farrer Fm
Price River Fm
Neslen Fm
Sego Ss
Castlegate Sandstone
Castlegate Ss
Mancos T.
Blackhawk Fm
Mesa-verde Group
Iron Springs Formation
Wahweap Formation
Blackhawk Fm
Shale
Star Point Ss
U. Blue G.
Emery Ss M
Sixmile Canyon Formation
Mancos Shale main body
Mancos Shale
Drip Tank Mbr
Straight Cliffs Fm
John Henry Member
Indian-ola Group
Funk Valley Fm
Mancos Shale
Lower Blue Gate Shale Member
lower 2 mbrs
Allen Valley Sh
Ferron Ss Mbr
Ferron Ss M
Tropic Shale
Sanpete Fm
Tununk Sh Mbr
Mancos Shale, Tununk Mbr
Dakota Sandstone
Dakota Sandstone
Dakota Sandstone
Dakota Fm
Upper mbr
Middle mbr
Dakota Conglomerate
Basal cg mbr
Cedar Mountain Formation
Cedar Mountain Fm
Upper member
Cedar Mountain Fm
Upper member
Burro Canyon Formation
Buckhorn Cg Mbr
Buckhorn Cg Mbr
Morrison Formation
Brushy Basin Member
Morrison Formation
Brushy Basin Member
Morrison Formation
Brushy Basin Member
1
2
3
4
6
7
8
5
9
10
Salt Wash Member
Salt Wash Member
Salt Wash Member
Tidwell Mbr, Morrison Fm
Twist Gulch Fm
Carmel Formation
Romana Ss

FIGURE 67 — Cretaceous strata near Helper, Utah. M, Mancos Shale; P, Panther Sandstone Tongue of the Star Point Sandstone; S, Storrs Sandstone Tongue of the Star Point Sandstone; SC, Spring Canyon Sandstone, basal unit of the Blackhawk Formation. See Figure 69 and Chart 65. Slopes between the labeled sandstones are tongues of marine Mancos Shale.

formational names have been coined to identify the interfingering rock bodies that formed in this active environment. These are shown on Figure 66, which also lists those fossil ammonite zones that pertain to marine Cretaceous strata in Utah. Cretaceous exposures in Utah fall naturally into northern, central, and southern sectors, as may be seen on a geologic map of Utah (back cover). The northern sector, north of the 40th parallel, includes columns 1–3 on Figure 66; the central sector, between the 38th and 40th parallels, includes columns 4–8 on Figure 66; the southern sector includes columns 9 and 10.

Northern Utah—Cretaceous rocks in the overthrust belt of northern Utah (Figure 65), Idaho, and Wyoming include several important non-marine conglomeratic units that have been difficult to date precisely. Biostratigraphic work using spores and pollen (Nichols and Jacobson, 1982a, 1982b; Nichols et al., 1982; and Jacobson and Nichols, 1982) has resulted in the correlations shown on Figure 66, columns 1 and 2. Royce (1975) suggested that the Paris thrust fault system might be related to deposition of the Ephraim Conglomerate, the Crawford-Meade system to the Echo Canyon Conglomerate, the Absaroka fault to the Lower Evanston Formation, and the Hogsback fault to the upper Evanston of early Tertiary age. Models of thrust fault relationships in this area have been reviewed by Blackstone (1977) and Link (1982). Applicability of critical-wedge thrust models have been discussed by Woodward (1987). Bruhn et al. (1986) discussed the influence of the Uinta Arch on thrust boundaries.

Northeastern Utah is host to the earliest marine Cretaceous deposits in Utah, the fish-scale bearing Aspen-Mowry Shale and the underlying Dakota beds. Sandstones called "Dakota" range somewhat in age in Utah, as shown on the correlation table, because they are the basal deposit of a transgressing sea. Alternating marine-non-marine beds of the Frontier Formation are best exposed in Utah near Coalville, where they are succeeded by coarse clastics of the Henefer Formation and Echo Canyon Conglomerate. These rocks pass eastward into the Hilliard Shale and Mesaverde Group sandstones, whose environments of deposition have been interpreted by Kiteley (1983) as representing repetitive regressions and transgressions of Campanian seas.

Coalville was the first successful coal field in Utah (Chart 34); coal was discovered there in 1859 and has been mined sporadically since then, principally from the Coalville Member (Doelling and Graham, 1972b). Coal in the Frontier and Mesaverde Formations near Vernal served that area's fuel needs until the middle part of this century when oil and gas took over the market. Ryer and Lovekin (1986) speculated on the existence of an Upper Cretaceous Vernal delta.

Central Utah—Uinta Basin Tertiary deposits conceal Cretaceous beds in the transitional zone between northern and central Utah Cretaceous outcrops. The central Utah sector includes Early Cretaceous deposits of the Brushy Basin Member of the Morrison Formation, and the somewhat similar Cedar Mountain-Burro Canyon Formations. Kowallis and Heaton (1987) gave fission-track dates for the Brushy Basin Member that indicate that at least the upper part is Early Cretaceous in age, rather than Late Jurassic as had previously been supposed. Witkind et al. (1986) showed that beds of conglomerate, sandstone, mudstone and fresh-water limestone exposed in central Utah that had been assigned, in the past, to the Morrison (?) Formation were better assigned

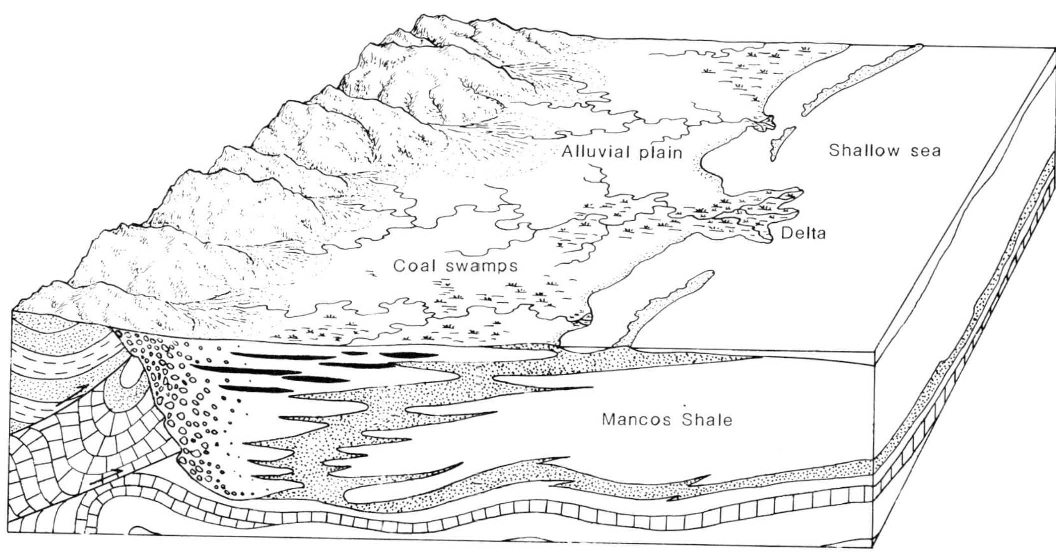

FIGURE 68 — Block diagram showing depositional environments of Late Cretaceous rocks in the foreland basin of eastern Utah.

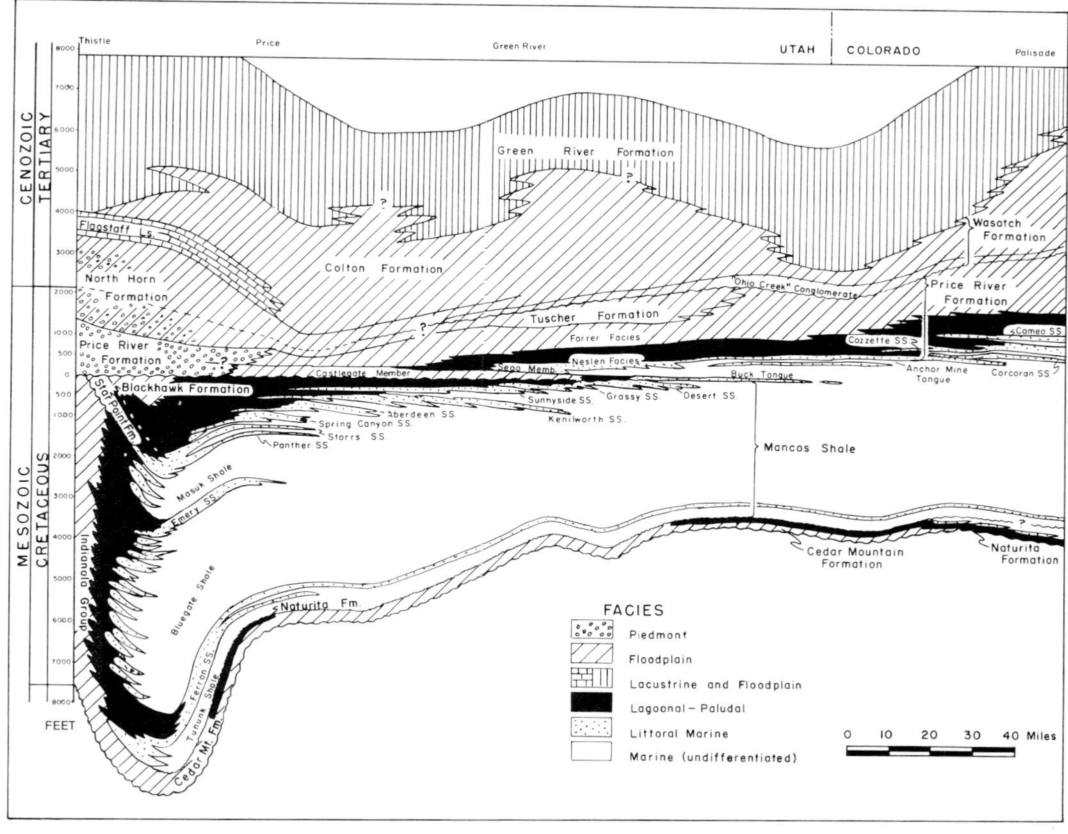

FIGURE 69 — Cretaceous and early Cenozoic strata between Thistle, Utah (Chart 36) and Grand Junction, Colorado (Chart 44), emphasizing intertonguing relations between marine Mancos Shale and Mesaverde Group sandstones. Vertical scale greatly exaggerated. From Young (1966).

FIGURE 70 — View northwestward along Straight Cliffs (Fiftymile Mountain) towards Escalante, Utah (at far distant right edge). Lowest continuous sandstone ledges on Fiftymile Mountain are Tibbet Canyon Member of Straight Cliffs Formation. Tree-covered bench in lower third of escarpment is Smoky Hollow Member. Upper two-thirds of Fiftymile Mountain face exposes John Henry Member of the Straight Cliffs Formation. Colluvium conceals the Tropic Shale on slopes beneath the Straight Cliffs. Low escarpments along right side of photo are held up by resistant Dakota Sandstone, Morrison Formation, and Romana Sandstone. Non-resistant beds of Entrada Sandstone form the valley bottom in foreground. See Chart 100.

to the late Early Cretaceous Cedar Mountain Formation, as shown on Figure 66. These findings give to central Utah a depositional record of the first 40 million years of Cretaceous time, a record that had heretofore been nearly blank on correlation tables. Yingling and Heller (1987) averred that the first evidence of Cretaceous thrusting in the Sevier orogenic belt is recorded by fine-grained clastics in the upper part of the Cedar Mountain Formation. This may have signaled the initial movement on the Canyon Range thrust.

Two sets of nomenclature are used for Late Cretaceous strata across central Utah: west of the Wasatch Plateau the Indianola Group names are used; east of the plateau the Mancos-Mesaverde terminology reigns. Key to the correlation between the two is the equivalency of the Tununk and Allen Valley Shales. Relationships west of the Wasatch Plateau are well summarized by Lawton (1985). Schwans (1988) proposed a new name for the Cedar Mountain equivalents in west-central Utah; Villien and Kligfield (1986) differed markedly from other authors in their interpretation of thrust dating. Witkind (1987) outlined interpretive uncertainties related to incomplete exposures of the Charleston-Nebo thrust. Easternmost proven extent of this thrust is documented by the Paleozoic-over-Jurassic sequence encountered in Placid Oil Daniels Land No. 1 drilled near Wallsburg in 1983. Lawton (1985, 1986) suggested that thrusting began in western Utah in late Albian time and moved episodically successively eastward, transporting older thrusts piggyback on younger thrusts and culminating in deformation east of the Wasatch Plateau by Maastrichtian-Paleocene time.

A great many papers cover various aspects of Cretaceous stratigraphy on the east side of the Wasatch Plateau where the principal coal mines are located (Figure 67). These papers are cited at the bottom of the charts herein; only summary papers follow. Ryer and McPhillips (1983) reconstructed paleogeography associated with the Frontier-Ferron Vernal delta, and the Ferron Sandstone Last Chance delta. Fouch et al. (1983) outlined facies changes of clastic wedges in the foreland basin, as related to thrust-fold events in the Cretaceous hinterland. Lawton (1986) concluded that synorogenic nonmarine strata in the upper Mesaverde and North Horn record late Ca-

FIGURE 71 — Angular unconformity exposed along Interstate I-70 near the mouth of Salina Canyon, about 4 miles upstream from Salina city. Unconformity records the end of deformation of the Sevier orogeny with Laramide basinal deposits overlying the near-vertical beds of the Jurassic Twist Gulch Formation (Jt) exposed at the base of the hill. Flat ledge just above the unconformity has been identified by Willis (1986) as a thin lens of Flagstaff Limestone (Tf). Banded clayey beds above Flagstaff ledge are early Eocene Colton Formation (Tc). Ledge in middle of hillside marks the base of the Green River Formation (Tg) which extends up to the horizon. See Chart 79.

panian transition, in the foreland around the San Rafael Swell area, from Sevier thrust-type deformation to Laramide basement-cored uplift, indicating a late Paleocene minimum age for the San Rafael Swell.

Cretaceous coal is a vital economic resource for east-central Utah. Doelling's monograph (1972) on central Utah coal fields summarized occurrence, quality, production, and reserves, field by field. The Book Cliffs coal field is the most important producer in Utah; all of its coal is contained in the Blackhawk Formation where high-volatile bituminous coals are mined from a number of coal beds. Regardless of age, Utah coals are usually associated with sandstones and were deposited in coal swamp areas behind barrier beach or deltaic sands along Cretaceous coastlines (Figures 68 and 69). Visualize, if you can, eastern Utah at sea level, with a broad, flat, marshy belt along the shoreline of the coastal plain. Dinosaurs roamed in the coal marshes, as indicated by thousands of their footprints left in coal beds. The proportion of different plants in the coal swamps is not well known, but conifers were abundant, and deciduous trees such as gum, maple, walnut, fig, and cottonwood were represented.

Oil has not yet been produced in central Utah. Britt and Howard (1982) reviewed the petroleum potential for the central Utah hingeline-thrust belt region and provided oil maturation curves suggesting that potential source beds, ranging in age from Devonian to Cretaceous, exist in the area.

Southern Utah—In the Kaiparowits Plateau area the term Dakota Formation applies to a thicker sequence of rocks than in other parts of Utah. Coal occurs in the Dakota in the Kolob, Harmony, and Alton fields (Doelling and Graham, 1972a; Ryer, 1982, 1983). Equivalent Iron Springs Formation sandstones in the Beaver Dam Mountains contain a little fossil wood, but lack coal. Gustason (1987) traced elements of the Greenhorn cyclothem across Dakota exposures in southern Utah using high-resolution stratigraphic event techniques to demonstrate transgressive-regressive strand-plain cycles. Dakota stream channels show current directions uniformly toward the southeast.

A regional summary of formational correlations and depositional cycles of Late Cretaceous strata of southeastern Utah and adjacent states has been compiled by Molenaar (1983); Cobban and Hook (1984) related molluscan biostratigraphy to paleogeography for the same interval; a field guide to these rocks was prepared by Eaton et al. (1987).

Largest coal reserves in southern Utah occur in the Straight Cliffs Formation in the Kaiparowits Plateau (Figure 70). As much as 15 billion short tons of coal, in seams more than 4 feet thick, are estimated to be present in the Kaiparowits Plateau coal field (Doelling and Graham, (1972a).

Eaton (1987) concluded that the Cenomanian through late Campanian rocks in the Kaiparowits Plateau contain the most complete Late Cretaceous record of land life known in the world, including spores and pollen, nonmarine mollusks, lower vertebrates, and mammals. All formations listed in column 10 of Figure 66 have yielded terrestrial fossils except the marine Tropic Shale (Figure 63). Diverse terrestrial faunas have been recovered from the Wahweap and Kaiparowits Formations.

Western Utah Hinterland—Compressive structural features in western Utah and eastern Nevada are generally ascribed to the Sevier orogeny but more precise dating of particular structures is limited by the wide difference in ages of the bracketing datable rocks—Early Triassic marine beds and late Eocene to Oligocene volcanic rocks. In the northern Utah-Idaho-Wyoming overthrust belt extensive stratigraphic work on rocks capable of giving more restricted age brackets has shown that thrust fronts progressed from west to east through Cretaceous time (Royce, 1975). Thus the Paris thrust on Figure 64 is the oldest of the six thrusts shown. It is with much less certainty that thrusts in western Utah are correlated from range to range and then identified as the likely orogenic impetus to generate the various clastic wedges present in Late Cretaceous rocks in eastern Utah, as discussed above.

Two-mica granites, dated at 75 and 79 m.y., in the Snake Range and Kern Mountains (Figures 64 and 66) may have formed by melting of the thrust-thickened crust in the Sevier hinterland. A more extensive review of structural features in the Sevier orogenic hinterland can be found in the discussion of Phase V in later pages of this book.

Regional summary papers—Stratigraphy, biogeography, tectonics and other regional aspects of the Cretaceous System in the Western Interior of the U.S. have been summarized by McGookey (1972), Kauffman (1984), Weimer (1986) and Merewether and Cobban (1986).

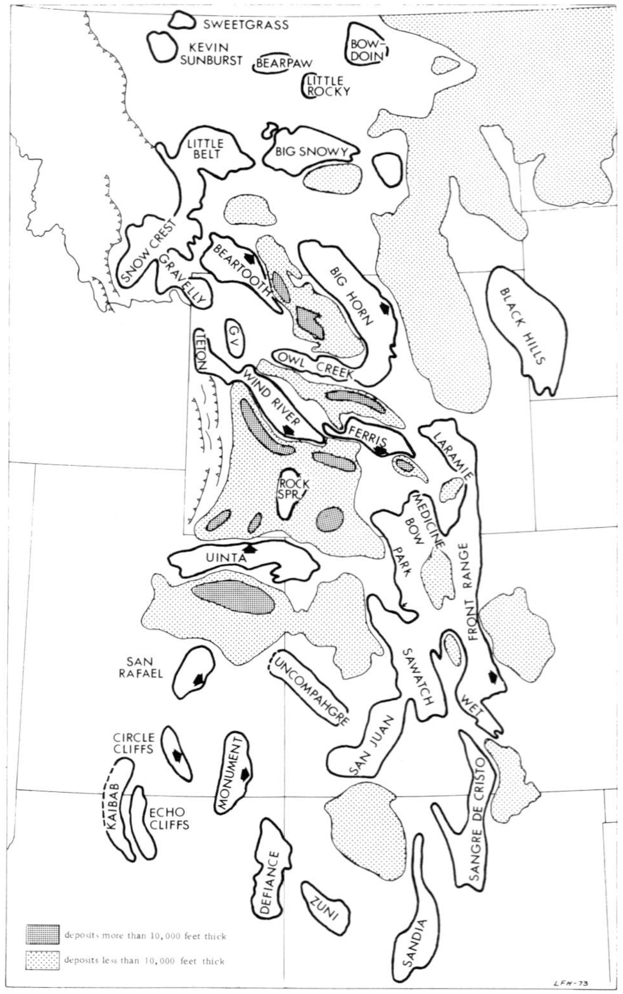

FIGURE 72 — Laramide uplifts and basins of Phase VI. Basins are shaded, darker shading shows more than 10,000 feet of early Cenozoic deposits. Blunt arrows show steep or overthrust limb on uplifts.

FIGURE 73 — Starvation Reservoir near Duchesne in the Uinta Basin. Dam abuts beds of the Eocene Uinta Formation. Flat-topped ridge in middle distance beyond reservoir exposes Duchesne River Formation. Faint ridges on skyline are Uinta Mountains. Duchesne lies near the axis of the Uinta Basin where Laramide deposits are more than 3 miles thick. See Charts 38, 41, and 42.

PALEOCENE-EOCENE

Cenozoic geochronology—Authoritative consensus has recently been reached on the ages of the boundaries between Cenozoic subdivisions. Berggren et al. (1985) presented a geochronology based on magnetostratigraphic, biostratigraphic and radiometric data that has been accepted as the standard and is used herein on Figure 74. The Berggren proposal bases Cenozoic epoch boundaries on European marine sequences, which have long been hard to relate directly to the North American terrestrial mammal ages that are used in Utah chronology. However, even this difficulty has been partly resolved: Woodburne (1987) presented a recommendation for mammal age chronology embracing all of the Cenozoic; Sloan (1987) focused on latest Cretaceous and Paleocene mammal ages. In Utah the mammal ages apply only to Paleocene-Eocene rocks because this is the only portion of the Utah Cenozoic record that has numerous enough mammal fossil occurrences. Oligocene and later rocks in Utah are dated primarily by isotopic methods, although magnetic reversal chronology is locally helpful.

North Horn Formation, Sevier-Laramide transition—Just as Late Cretaceous sedimentation in Utah was impelled by uplift in the Sevier orogenic belt, so Eocene sedimentation is dominated by Laramide uplifts that developed in eastern Utah. The Uinta Mountain-Uinta Basin couplet is the largest Laramide structure in Utah, and is similar to other such Laramide couplets in New Mexico, Colorado, Wyoming and Montana, as shown on Figure 72. The Uinta Basin, however, was not exclusively the site of the earliest Paleocene deposition, rather, a long depositional trough extended southwestward from the west side of the Uinta Basin as shown on Figure 75. This trough follows the eastern margin of the Cretaceous Sevier orogenic belt, as well as coinciding with the present High Plateaus of Utah. In the trough, clastic deposits of the North Horn Formation and its temporal equivalents rest almost everywhere with great angular unconformity on beds that were tilted during the last phases of the Sevier orogeny (Figure 71). Lawton (1986) traced latest Cretaceous and Paleocene deposits from the Wasatch Plateau eastward to the west edge of the Uinta Basin and concluded that Sevier thrust-related deposition in upper Mesaverde sandstone (Tuscher Formation) began to be replaced by sediments derived from nearer sources such as the San Rafael Swell, Circle Cliffs, and Monument uplifts, signaling the beginning of Laramide activity. The North Horn Formation is, thus, a post-Sevier caprock consisting of fluvial and some lacustrine strata. In addition to its distribution in central Utah, the North Horn has been identified as the basal basin-filling deposit in the Uinta Basin (Fouch, 1976; Ryder et al., 1976; Johnson, 1985), as shown on Chart 42.

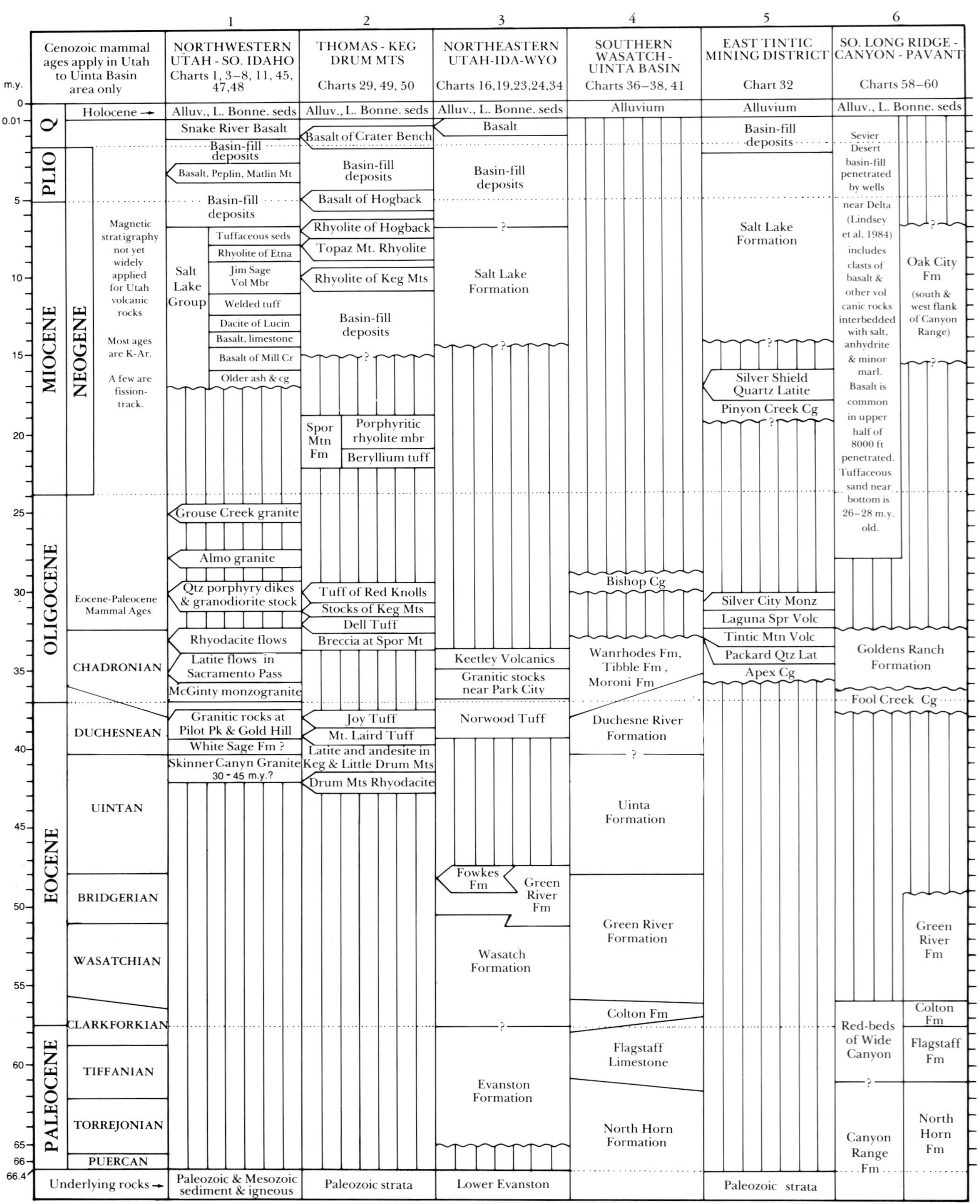

FIGURE 74 — Cenozoic correlation table.

CENOZOIC

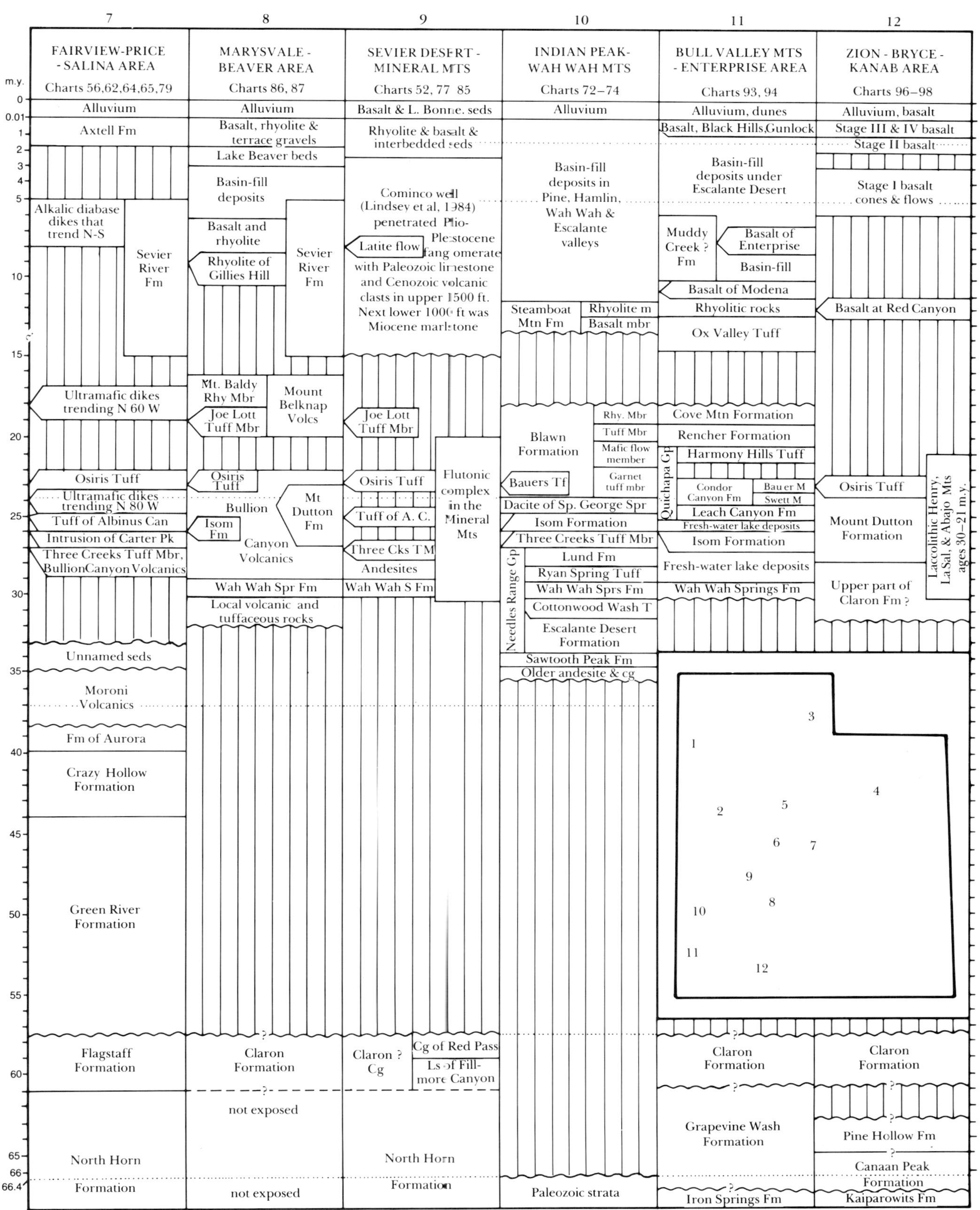

7
8
9
10
11
12
FAIRVIEW-PRICE - SALINA AREA
Charts 56,62,64,65,79
MARYSVALE - BEAVER AREA
Charts 86, 87
SEVIER DESERT - MINERAL MTS
Charts 52, 77 85
INDIAN PEAK- WAH WAH MTS
Charts 72–74
BULL VALLEY MTS - ENTERPRISE AREA
Charts 93, 94
ZION - BRYCE - KANAB AREA
Charts 96–98
m.y.
0
0.01
1
2
3
4
5
10
15
20
25
30
35
40
45
50
55
60
65
66
66.4
Alluvium
Axtell Fm
Alkalic diabase dikes that trend N-S
Sevier River Fm
Ultramafic dikes trending N 60 W
Osiris Tuff
Ultramafic dikes trending N 80 W
Tuff of Albinus Can
Intrusion of Carter Pk
Three Creeks Tuff Mbr, BullionCanyon Volcanics
Unnamed seds
Moroni Volcanics
Fm of Aurora
Crazy Hollow Formation
Green River Formation
Flagstaff Formation
North Horn Formation
Alluvium
Basalt, rhyolite & terrace gravels
Lake Beaver beds
Basin-fill deposits
Basalt and rhyolite
Rhyolite of Gillies Hill
Sevier River Fm
Mt. Baldy Rhy Mbr
Joe Lott Tuff Mbr
Mount Belknap Volcs
Osiris Tuff
Isom Fm
Bullion Canyon Volcanics
Mt Dutton Fm
Wah Wah Spr Fm
Local volcanic and tuffaceous rocks
?
Claron Formation
?
not exposed
not exposed
Basalt & L. Bonne. seds
Rhyolite & basalt & interbedded seds
Cominco well (Lindsey et al, 1984) penetrated Plio-Ple:stocene fanglomerate with Paleozoic limestone and Cenozoic volcanic clasts in upper 1500 ft. Next lower 1000 ft was Miocene marlstone
Latite flow
Joe Lott Tuff Mbr
Osiris Tuff
Tuff of A. C.
Three Cks TM
Andesites
Wah Wah S Fm
Flutonic complex in the Mineral Mts
Claron ? Cg
Cg of Red Pass
Ls of Fill-more Canyon
North Horn Formation
Alluvium
Basin-fill deposits in Pine, Hamlin, Wah Wah & Escalante valleys
Steamboat Mtn Fm
Rhyolite m
Basalt mbr
Blawn Formation
Rhy. Mbr
Tuff Mbr
Mafic flow member
Garnet tuff mbr
Bauers Tf
Dacite of Sp. George Spr
Isom Formation
Three Creeks Tuff Mbr
Needles Range Gp
Lund Fm
Ryan Spring Tuff
Wah Wah Sprs Fm
Cottonwood Wash T
Escalante Desert Formation
Sawtooth Peak Fm
Older andesite & cg
Paleozoic strata
Alluvium, dunes
Basalt, Black Hills,Gunlock
Basin-fill deposits under Escalante Desert
Muddy Creek ? Fm
Basalt of Enterprise
Basin-fill
Basalt of Modena
Rhyolitic rocks
Ox Valley Tuff
Cove Mtn Formation
Rencher Formation
Quichapa Gp
Harmony Hills Tuff
Condor Canyon Fm
Bau er M
Swett M
Leach Canyon Fm
Fresh-water lake deposits
Isom Formation
Fresh-water lake deposits
Wah Wah Springs Fm
?
Claron Formation
?
Grapevine Wash Formation
?
Iron Springs Fm
Alluvium, basalt
Stage III & IV basalt
Stage II basalt
Stage I basalt cones & flows
Basalt at Red Canyon
Osiris Tuff
Mount Dutton Formation
Upper part of Claron Fm ?
Laccolithic Henry, La Sal, & Abajo Mts ages 30–21 m.y.
Claron Formation
?
?
Pine Hollow Fm
?
Canaan Peak Formation
Kaiparowits Fm
1
2
3
4
5
6
7
8
9
10
11
12

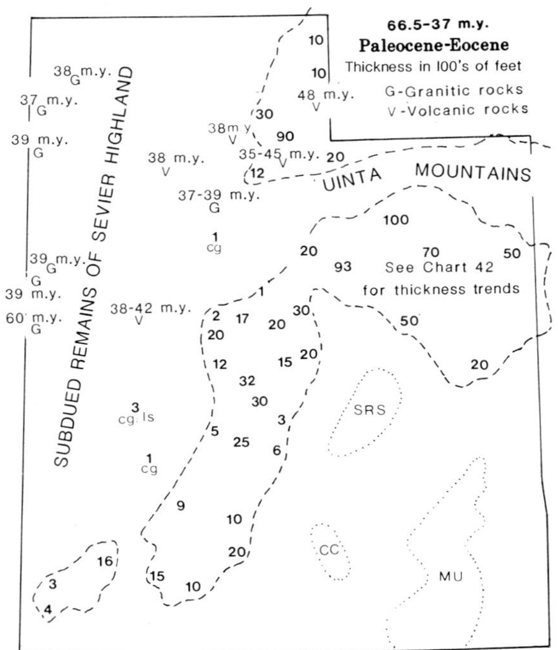

FIGURE 75 — Paleocene and Eocene deposition in the Uinta Basin and elsewhere occurred concurrent with and following uplift of the Uinta Mountains. More than 10,000 feet of alluvial and lake deposits accumulated near Duchesne in the asymmetric Uinta Basin. Dashed lines show the present-day erosional edge of Paleocene-Eocene deposits. Dotted lines outline uplifts active at this time: SRS-San Rafael Swell, CC-Circle Cliffs Uplift, MU-Monument Upwarp. Igneous activity of this interval is shown by radiometric ages of granitic (G) and volcanic (V) rocks in the northwest quarter of the state. Isolated occurrences of conglomerate (cg) and limestone (ls) are shown for the southern Oquirrh, Cricket, and Mineral Mountains of west-central Utah.

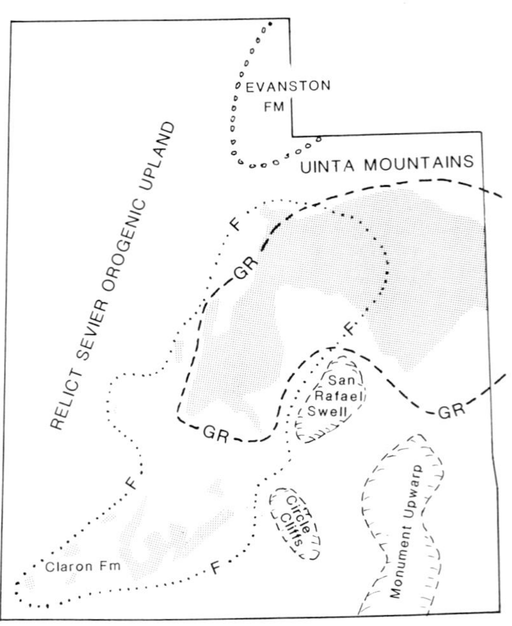

FIGURE 76 — Paleocene-Eocene paleogeography and preserved extent map. Inferred original extents of Paleocene Flagstaff and Eocene Green River lakes are outlined.

Early Cenozoic lakes—Two successive large lakes left extensive Early Cenozoic freshwater deposits in Utah, as shown on Figure 76. The older, Paleocene, Lake Flagstaff left "bird's eye" algal limestones now found in the mountains from Provo to Richfield and found in the Uinta Basin subsurface. Exposed Flagstaff Limestone is fossiliferous and well dated; it has been correlated, somewhat uncertainly, with the almost unfossiliferous Claron Formation that makes the multihued escarpments at Bryce Canyon (Figure 77) and Cedar Breaks in southern Utah. The younger lake left its record as the Eocene Green River Formation, which does not extend south of Richfield. In the Uinta Basin the Green River Formation forms one of the world's few, major, non-marine oil sources.

Uinta Basin—The Uinta Mountains are flanked on the north and south by major depositional basins which subsided as the mountains rose during early Cenozoic time. Formational names of the deposits that filled the Uinta Basin (Figure 73) are shown on Chart 42; this nomenclature was originally set up to apply to layer-cake relations in surface exposures in the Wasatch Plateau, but interfingering of alluvial and lacustrine deposits in the Uinta Basin subsurface turned out to be rather complex and required modification in the usage of the established terms (Fouch, 1975, 1976). The Uinta Basin produces oil, so its subsurface is well known from drilling and seismic data. The best subsurface marker bed is the Mahogany oil-shale zone in the Green River Formation. Volcanic tuffs associated with the Mahogany bed give radiometric ages of 42–46 m.y. (Mauger, 1977; Sheliga, 1980, unpublished thesis, Ohio State). Details of subsurface interfingering relationships in Uinta Basin deposits are summarized by Johnson (1985). Eocene sedimentation in the Bridger and Washakie basins in Wyoming north of the Uinta Mountains was outlined by Surdam and Stanley (1979, 1980).

Laramide uplift of the Uinta Mountains ended after deposition of the Uinta and Duchesne River Formations (Anderson and Picard, 1972). Hansen (1984) noted that the range then lapsed into quiescence during a period when the Gilbert Peak erosion surface formed.

Because of its economic significance, the Uinta Basin has been the subject of many papers and eight guidebooks by the Utah Geological Association and its predecessors (1950, 1957, 1959, 1964, 1969, 1974, 1983, 1985). Surface geology of its northeastern part was best shown by Rowley et al. (1985).

FIGURE 77 — Claron Formation at Bryce Canyon National Park. Interpretation of age and original depositional environment of the Claron Formation has been hindered by destruction of primary depositional features and body fossils during pedogenesis and diagenesis. The Claron is presumed to be equivalent to the Paleocene Flagstaff Formation of central Utah. Fossils and sedimentary structures are well-preserved in the Flagstaff, in marked contrast to the Claron's lack of preserved structures. Vertical joints cut the Claron beds at Bryce, making the scenic pillars or hoodoos.

Western Utah Paleocene-Eocene deposits—Throughout most of western Utah, Oligocene volcanic rocks rest unconformably on older rocks deformed by the Sevier orogeny. Green River equivalents are entirely absent, and Flagstaff equivalents nearly so, as shown on Figures 74 and 75. Late Eocene signaled the onset of igneous activity in northwestern Utah that included emplacement of granitic intrusions and volcanic tuffs and flows. Sources for the older tuff horizons in the Green River Formation have not been identified in Utah and probably lay to the northwest.

Granitic intrusions of latest Eocene age are found in northwestern Utah and include the Immigrant Pass (Chart 4), the McGinty (Chart 6), Bettridge Creek (Chart 7), the Gold Hill and the Ibapah (Chart 45)plutons. Near Salt Lake City, intrusion of the Bingham and Last Chance granitic stocks in the Oquirrh Mountains (Chart 28 and Figure 78) was followed by hydrothermal alteration and mineralization that produced the world-famous Bingham porphyry-copper deposit (Bray and Wilson, 1975). The Clayton Peak stock near Brighton has an age of 37–41 m.y., according to Crittenden et al. (1973), and the various stocks between Heber and Park City (Chart 34) are about 37 m.y. old.

Oldest volcanic activity that produced enough tuff to be given a formal name is that of the Fowkes Formation (48 m.y.) in the Crawford Mountain-Green River Basin area (Chart 19). These tuffs may be related to tuffs in the Green River Formation that are mostly slightly younger, 46–40 m.y. (Mauger, 1977). The Norwood Tuff (38 m.y.) in the Wasatch Mountains near Ogden has yielded a small vertebrate fauna (Charts 23,24). Ages of Late Eocene tuffaceous deposits near Salt Lake City (Chart 33) and the Traverse volcanics near Camp Williams are not tightly constrained (37–45 m.y.). Andesitic lavas of the Grayback Hills (Chart 25) are 38 m.y. old. Similar andesites in the Stansbury Mountains (Chart 27) may also be late Eocene. Late Eocene Joy Tuff (38 m.y.), Mt. Laird Tuff (39 m.y.), Keg Spring Andesite and latite (39 m.y.), and Drum Mountains Rhyodacite (42 m.y.) record the southernmost volcanic activity of this age in Utah (Charts 49, 50).

FIGURE 78 — Bingham copper mine in the Oquirrh Mountains southwest of Salt Lake City. Although copper is the most abundant metal at Bingham, the great volume of rock processed has made Bingham a significant producer of molybdenum, gold, silver, and lead. Other by-products include bismuth, platinum, selenium, rhenium, and sulfuric acid. The Bingham and Last Chance quartz monzonite stocks (late Eocene, 38–39 m.y. old) intruded Pennsylvanian-Permian strata of the Oquirrh Group. Ore minerals disseminated in the Bingham stock and adjacent sedimentary rocks are chiefly chalcopyrite and bornite, with lesser chalcocite and molybdenite. Present ore grade is less than one percent. Bingham, opened in 1904, was the first open-pit copper mine in the world. It has produced nearly 12 million tons of copper, an unequaled record.

OLIGOCENE-EARLY MIOCENE

Recent progress in field mapping and isotopic dating of volcanic rocks in the western U.S. has clarified their spatial, temporal and compositional patterns. In Utah, those patterns changed strikingly in mid-Miocene time with the onset of Basin and Range extension. Therefore we break our discussion of later Cenozoic history at 15 million years into a pre-Basin and Range segment, which includes the Oligocene and early Miocene events, and a late Miocene to Holocene segment, which includes Basin and Range events. To avoid undue map clutter we have shown Oligocene-Early Miocene igneous rocks on two maps: Figure 79 shows only intrusive rocks; Figure 80 shows layered rocks which are entirely sedimentary in eastern Utah and mostly volcanic in western Utah.

Intrusive rocks—Granitic intrusions of this age in the vicinity of northwestern Utah include the Almo (28 m.y.), stocks in the central Grouse Creek Mountains (25 m.y.), intrusions and dikes in the Pilot Range (30–37 m.y.), and granodiorite near Wendover (30 m.y.?). Near Salt Lake

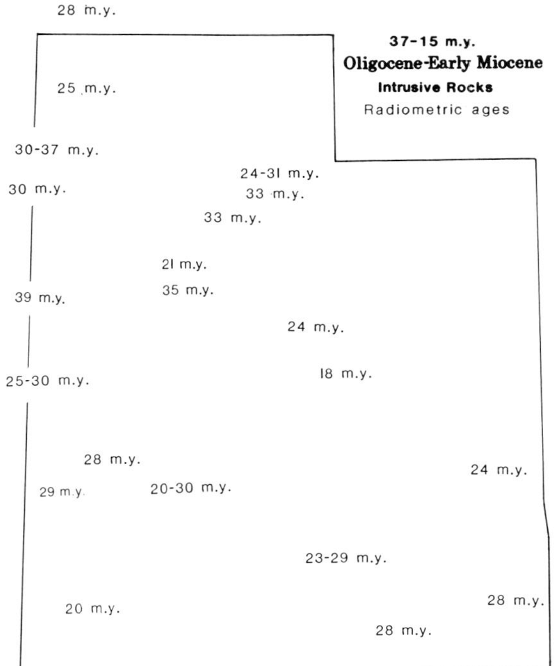

FIGURE 79 — Location and radiometric ages of Oligocene and early Miocene (37–15 m.y.) intrusive rocks. Intrusions include granitic and dioritic stocks, and some dikes and plugs. Note that igneous activity of this age extends farther to the south in Utah than did Paleocene-Eocene activity shown on Figure 75.

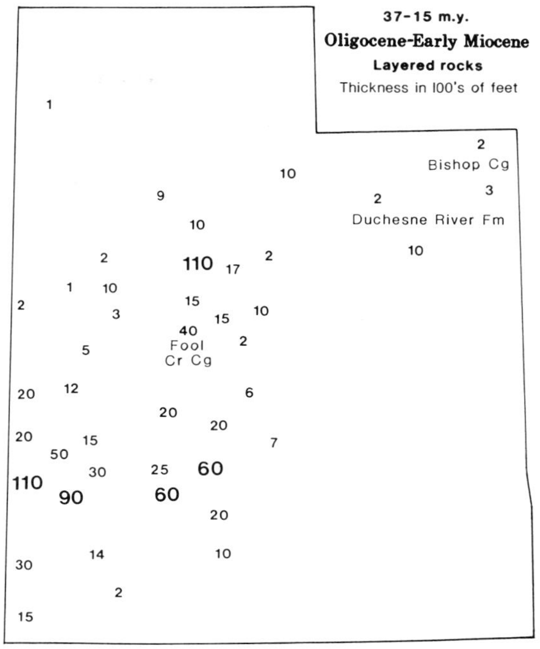

37-15 m.y.
Oligocene-Early Miocene
Layered rocks
Thickness in 100's of feet

Bishop Cg
Duchesne River Fm
Fool Cr Cg

FIGURE 80 — Oligocene and early Miocene layered rocks include sedimentary rocks whose names are shown in eastern and central Utah, and widespread ash-flow tuffs and intercalated conglomeratic rocks in central and western Utah. Thickest accumulations occur around three volcanic centers: 11,000 feet around the Tintic volcanic center, 9,000–11,000 around the Indian Peak center, and 6,000 feet near Marysvale.

City are the Little Cottonwood stock (24–31 m.y.), the Alta stock (32–33 m.y.), and the Shaggy Peak rhyolite plug (33 m.y.). In west-central Utah, the Sheeprock Mountains granitic pluton (21 m.y.), the Desert Mountain pluton (35 m.y.), the Silver Creek two-mica granite (25–30 m.y.?), the Cactus (28 m.y.), and the plutonic complex in the Mineral Mountains (20–30 m.y.) are included in this age group. In east-central Utah mica-peridotite sills and dikes (24 m.y.) occur in the northern Wasatch Plateau, and alkalic diabase dikes and sills (18 m.y., 24 m.y.) occur near Huntington. The famous laccolithic intrusions of southern Utah are dated as follows: Henry Mountains, 21–29 m.y.; La Sal Mountains (Figure 84), 28 m.y.; Abajo Mountains, 23–29 m.y.; and Iron Springs, 20 m.y. Of these, only in the La Sal Mountains was the laccolithic roof breached by the roots of volcanoes (Hunt, 1983). A lamprophyre dike in Monument Valley is 28 m.y. old.

Extrusive rocks—Oligocene-Early Miocene extrusive rock thickness and distribution are shown on Figures 80 and 81. Northernmost occurrence is rhyodacite (33 m.y.) near Lemay Island (Chart 11). Keetley Volcanics near Heber (Chart 34) are 34–35 m.y. old, and include rhyodacite and andesite flows and breccia. In the West Traverse Mountains at the south end of Salt Lake Valley (Chart 31) latite flows and lahars are older than 32 m.y. rhyolite flows and plugs. In central Utah the Moroni and Goldens Ranch formations may be equivalents; they yield K-Ar ages scattered between 30–40 m.y. but are most likely 36–37 m.y. old (Witkind and Marvin, 1988).

The greatest volume of Oligocene-Early Miocene extrusive rocks in Utah accumulated around four principal centers: East Tintic Mountains volcanic center, Thomas-Keg-Drum Mountains volcanic center, Marysvale volcanic field, and the volcanic field of southwest Utah.

In the East Tintic mining district (Chart 32) igneous rocks were derived from an Oligocene composite volcano complex, active mostly between 32–33 m.y. ago. Although the volcanic edifice itself has long since been eroded away, its root system is exposed in the many dikes, plugs, and sills that cut the area. The last volcanic activity in the East Tintic district was emplacement of a large dike and flow unit, the Silver Shield quartz latite, about 18 m.y. ago. The East Tintic mining district has produced a quarter-billion dollars worth of lead, zinc, silver, gold, copper, cadmium, and manganese from mineralization that accompanied the volcanism.

In the Thomas-Keg-Drum mountains area the volcanic rocks that erupted 42–39 m.y. ago (Late Eocene) are dacitic to andesitic in composition; those formed between 38–32 m.y. ago (Oligocene) are rhyolitic; those 21 m.y. old (Early Miocene) are high-silica, topaz-bearing alkalic rhyolites that contain beryllium. The world's largest beryllium mine presently produces ore from open pits in a tuff beneath rhyolite flows on the west side of the Thomas Range. The youngest volcanic rocks in the Thomas-Keg-Drum area range from 8–5 m.y. in age (Late Miocene) and include basalts as well as the topaz-bearing rhyolite of Topaz Mountain and younger rocks at Smelter Knolls and the Honeycomb Hills.

The Marysvale mining district came to life in 1865 when placer gold was discovered there. Over the years the Marysvale district has produced a little bit of several metals: gold, silver, lead, copper, mercury, manganese, aluminum, and uranium. Yet with all this activity Marysvale has never ranked with Utah's major mining districts. However, it has attracted considerable geological study. The latest group to investigate the Marysvale area was the U.S. Geological Survey's Richfield quadrangle CUSMAP team (Hintze, 1986c) led by Thomas A. Steven. Team efforts extended the field mapping to the south and west and resulted in much better understanding of regional stratigraphy of the volcanic rocks (Anderson and Rowley, 1975; Cunningham and Steven, 1979; Steven et al., 1979; Rowley et al., 1979, 1981). Although the rocks in the Marysvale volcanic field are mostly local accumulations of lava and volcanic debris flows that formed several stratovolcanoes (Bullion Canyon, Mt. Dutton, Mt. Belknap), interlayered with these are a few distinctive ash-flow sheets (Isom, Osiris, Three Creeks, and others) that are keys to correlations within these volcanic piles. Recognition of the source calderas for some of these ash layers (Steven, Rowley and Cunningham, 1984) was no mean feat because the Oligocene-Miocene volcanic mountains, craters, and calderas no longer exist in their original form, having been greatly modified by later faulting and erosion. The main calderas that have been identified are shown on Figure 81.

Oligocene-Miocene igneous activity in the volcanic field of southwestern Utah is dominated by widespread ash-flow tuffs that emanated from a caldera complex astride the Utah-Nevada border. From 33 to 26 m.y. ago this magma system broadcast 2,500 cubic miles of dacite and rhyolite ash flows over an area exceeding 10,000 square miles; these eruptions were among the most voluminous of this type known anywhere (Figure 82). By comparison, the 4 cubic miles erupted by Krakatau in 1883 seems almost insignificant. Best and Grant (1987) summarized the stratigraphy of the Oligocene Needles

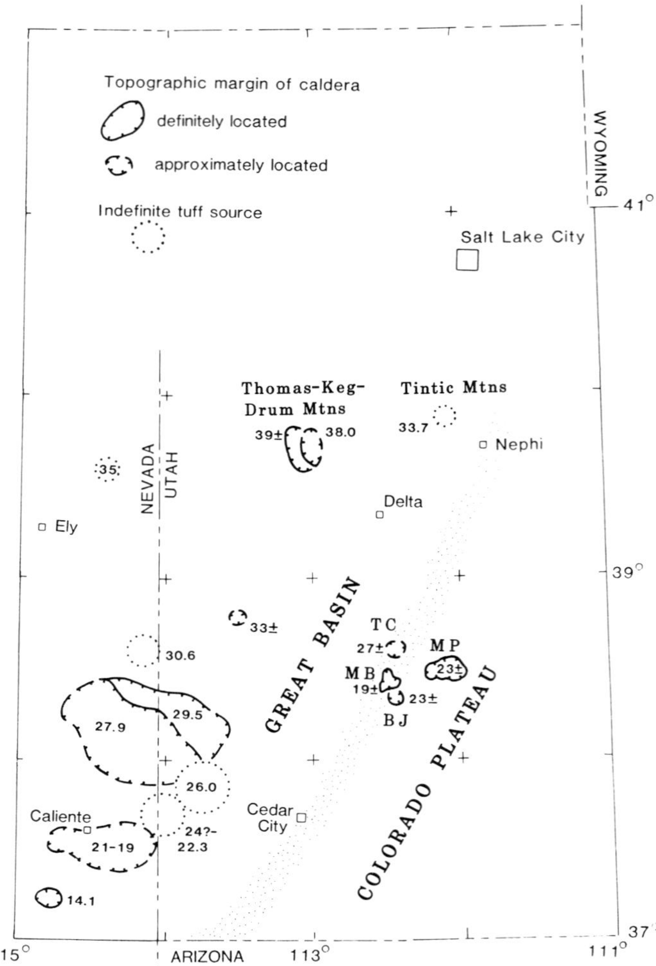

FIGURE 81 — Volcanic fields of Oligocene and early Miocene age in western Utah and eastern Nevada. Small dot pattern shows Colorado Plateau-Great Basin boundary. Calderas within the Marysvale volcanic field are TC, Three Creeks caldera; MB, Mount Belknap caldera; BJ, Big John caldera; and MP, Monroe Peak caldera. The large caldera complex astride the Utah-Nevada border is the Indian Peak caldera, source of the incredibly voluminous Needles Range Volcanic Group. After Best et al. (1988).

FIGURE 82 — Crystal Peak in the southern Confusion Range western Millard County is composed of 33 m.y. old rhyolite tuff, rich in small doubly-terminated quartz crystals. This unusually thick accumulation of Tunnel Spring Tuff (as shown on Chart 70) was probably deposited in an east-west Oligocene paleovalley. Crystal Peak rises about 800 feet above its base. Dwarf ponderosa pines grow in joint crevices here.

Range Group (Figure 83) and noted that eruptions alternated between crystal-poor rhyolite and a larger volume of crystal-rich dacite tuff. Stratovolcano-type eruptions played no role in this volcanism, nor are any large mineral deposits associated with the ash-flow tuffs. This type of volcanism extends westward across Nevada; its space-time-composition patterns have been summarized by Best et al. (1988). They noted that, regionally, tuffs in the 39–33 m.y. age bracket are largely crystal-rich rhyolite; 32–26 m.y., crystal-poor dacite; 25–26 m.y., crystal-poor trachyte (Isom Formation); and during the 25–16 m.y. interval calcalkaline is again predominant. Lava flows were widely scattered but not significant in volume.

Other regional analyses of patterns in Cenozoic igneous rocks in the western U.S. have been written by Snyder et al. (1976), Stewart et al. (1977), Burke and McKee (1979), Leeman (1982), and Chadwick (1985). Plate-tectonic interpretations that bear on Cenozoic volcanism in Utah are considered in a later section of this book.

Sedimentary rocks—Oligocene-Early Miocene sedimentary deposits include the upper part of the Duchesne River Formation in the Uinta Basin (Anderson and Picard, 1972), and the Bishop Conglomerate in the Uinta Mountains. Hansen (1984) noted that the Uinta Mountains consist of two domes aligned on an east-west axis. Following Laramide uplift the Gilbert Peak erosion surface developed across the Uinta Mountain area as an Oligocene pediment, capped by the Bishop Conglomerate in many places. In late Oligocene time the east Uinta dome subsided, deforming the Gilbert Peak surface and the Bishop Conglomerate, which became deeply dissected, as seen today.

In the Canyon Range of central Utah, Campbell (1979) mapped a thick sequence of unfossiliferous sandstones and coarse conglomerate as Fool Creek Conglomerate and concluded that its deposition preceded Basin and Range faulting and post-dated the North Horn Formation; he assigned it, therefore, to the Oligocene. The Fool Creek was locally derived, and laid down on a surface of considerable relief.

Elsewhere in western Utah, sedimentary rocks are not an important component of Oligocene-Early Miocene layered accumulations. In some places, conglomerates a few hundred feet thick are sandwiched within the volcanic sequence; but in most places ash-flow tuffs and related volcanic rocks lie atop one another and constitute the whole layered sequence.

FIGURE 83 — Unconformity between flat-lying Oligocene volcanic and sedimentary rocks, and gently inclined Ordovician strata, 5 miles south of Ibex in the southern Confusion Range of western Utah. Rimrock is a basaltic andesite of the Needles Range Group. Light-colored beds just above the unconformity are Oligocene Cottonwood Wash Formation composed of conglomerates and ash-flow tuff. Tilted Ordovician beds at left are Fillmore Formation. Tilted ledgy beds in upper right are Wah Wah Limestone. See Chart 70.

Metamorphic core complexes—A cluster of metamorphic core complexes cuts Utah's northwest corner and fringes its western border, as shown on Figure 85. Simply described, a metamorphic core complex is an exposure of rocks that were once ductile lower crust, over which shallow brittle extended rocks of the upper crust have been emplaced. Armstrong (1982) summarized the chronology and deformation of the Albion-Raft River-Grouse Creek complex (location shown on Figure 85) as follows: The metamorphic protolith was Archean basement with a Proterozoic, Paleozoic, and early Triassic cover. Regional metamorphism reached amphibolite grade in Jurassic time about 160 m.y. ago. This was followed by a Cretaceous period of imbrication and ductile deformation under conditions of decreasing pressure and temperature. Cretaceous biotite-muscovite granites were emplaced in the northwestern parts of the complex. Profound further modification took place in late Eocene and late Oligocene time, accompanied by additional emplacement of granite. Near the granitic injection complex, temperatures rose to sillimanite grade, and rock strength was greatly reduced when the area underwent regional extension. The several basement-cored domes may have risen at this time, but this is not well- dated and their growth may have been largely Cretaceous. Strong mylonitic foliation and lineation, with lineation trends nearly parallel to Cretaceous fold axes and lineations, developed in areas deformed at this time. Effects of Cenozoic deformation dominate the structural story in the Raft River-Grouse Creek Mountains in Utah, but the Albion Mountains in Idaho retain evidence of Mesozoic deformation. Late features are brittle denudation and detachment faults, breccia sheets, and far-traveled gravity slide masses. A similar structural chronology has been worked out in the Ruby Mountains area, Nevada (Snoke and Howard, 1984; Snoke and Lush, 1984).

Miller et al. (1987) presented detailed geochronological evaluation of metamorphic and intrusive events in the Pilot Range core complex near Wendover and showed that local amphibolite-facies and widespread greenschist-facies metamorphism peaked about 160 m.y. ago in the Late Jurassic, accompanied by early emplacement of

FIGURE 84 — La Sal Mountains, east of Moab, are representative of Oligocene laccolithic clusters in the Colorado Plateau. Diorite porphyry laccoliths invaded Jurassic and Cretaceous country rocks 29 million years ago creating local domes. The uplifts are now more than 12,000 feet above sea level and were glaciated during the Pleistocene ice ages.

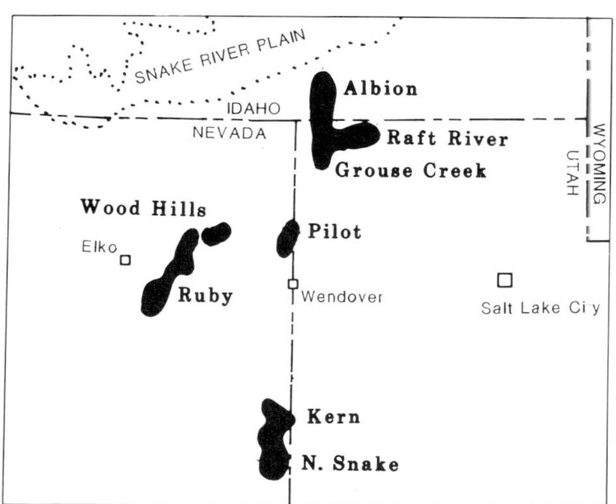

FIGURE 85 — Metamorphic core complexes in Idaho-Utah-Nevada. Core complexes were initiated during the Nevadan orogeny in the late Jurassic and brought to the surface as domal uplifts during the mid-Cenozoic when they underwent denudation of their cover rocks. After Armstrong (1982) and Wust (1986).

granitic plutons. In late Eocene time plutons and dikes cut folds and faults, including a major detachment fault that placed unmetamorphosed on metamorphosed rocks.

Gans et al. (1985), as a result of mapping and seismic data, suggested that the northern Snake Range decollement, near the Utah-Nevada border, originated in Oligocene time as a 12-mile-wide, 4-mile-deep zone of decoupling between ductile stretching below and denudation faulting above. One of the most striking aspects of a metamorphic core complex is the juxtaposition, on a low angle decollement, of the unmetamorphosed, brittlely deformed cover against a plastically stretched and metamorphosed basement. Wust (1986) noted that the directions of Tertiary extensional deformation, in core complexes from Mexico to Canada, fall into regional groups in which the directions are similar throughout each group. This implies a regional control of extension in metamorphic core complexes. Wust showed that the Nevada-Utah-Idaho complexes have been extended northwestward and southeastward.

LATE MIOCENE-HOLOCENE

Events during the past 15 million years have shaped the Utah we know today. Superimposed on a complex foundation of past happenings, Late Miocene to Holocene actions include the raising of the state to its mile-high elevation, the even further uplift of new ranges in the west half of the state, the sculpting of the new Wasatch Mountains and the newly rejuvenated Uinta Mountains by glaciers in the last million years, and the astonishing expansion of fresh-water Lake Bonneville to cover valleys from Idaho to southwestern Utah 16,000 years ago. Although Utah lies 500 miles inland from the west coast where the Pacific plate interacts with the North American plate, Utah's geological and geographical status is fundamentally controlled by plate interactions as well as the nature of the crust in the western United States. The discussions that follow in this section are arranged in order of decreasing age, realizing that most phenomena have been active throughout the entire 15 million years, but the glacially-related events happened only during the last million years.

Bimodal volcanism—The reason for selecting 15 m.y. as the dividing time boundary for late Cenozoic events is primarily the change in the mode and composition of volcanism in Utah at that time. A lull in volcanic activity occurred during the switchover. Late Eocene to mid-Miocene volcanic rocks are mostly voluminous ash-flow tuffs and lava flows of rhyolite-andesite-dacite composition. Late Miocene to Holocene volcanic rocks are commonly less voluminous rhyolites and basalts which are, in a word, bimodal. Stewart and Carlson (1978, plate 11-2), Best et al. (1980), Leeman (1982), Luedke and Smith (1984), and Christiansen et al. (1986) have presented regional summaries of this basaltic and rhyolitic (bimodal) volcanism.

Working our way from north to south: Armstrong et al. (1975) presented radiometric ages for late Cenozoic silicic and basaltic volcanism in the Snake River Plain, an area that impinges on northernmost Utah. Several summary papers concerning bimodal volcanism in southern Idaho are in Idaho Bureau of Mines Bulletin 26 (dated 1982 but actually distributed in 1984). In northwestern Utah bimodal volcanic rocks of the Salt Lake Group are widely

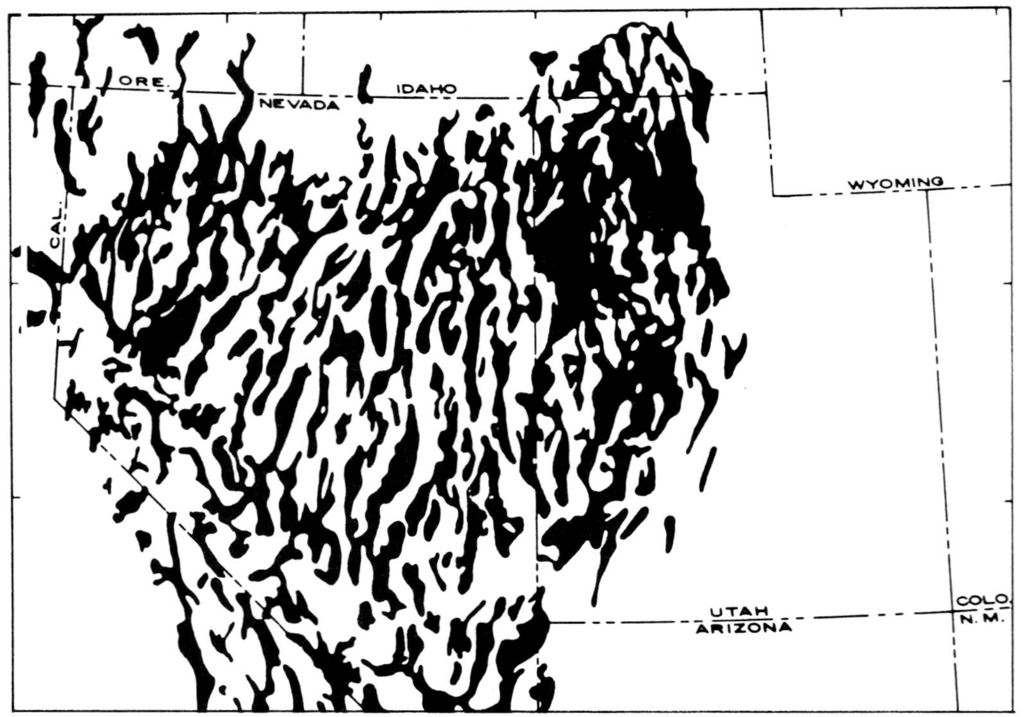

FIGURE 86— Late Miocene to Holocene basins (black) in the western U.S., from Best and Hamblin (1978). Note offset of general north-south trend of basins along northwest transverse faults that parallel the Nevada-California border in western Nevada. Some of these basins contain more than 10,000 feet of late Miocene to Holocene valley fill.

FIGURE 87 — Quaternary basalt cones and lava flows north of St. George along Utah Highway 18. Pine Valley Mountains, an Oligocene laccolith, on skyline. Light outcrops in foreground are Navajo Sandstone, here more than 2500 feet thick. Light-colored strata in the middle distance just above the two basalt cones are Jurassic Carmel Formation. Ragged basalt flow in left foreground is younger than table-capping basalt that Highway 18 runs on in the right foreground.

distributed and have recently been mapped by the Utah State University students identified on Charts 4–6. In west-central Utah, Lindsey (1982) and Christiansen et al. (1986) have recorded the renewal of volcanism in late Cenozoic time in the Thomas-Keg-Topaz Mountain area; a new Cenozoic volcanic center, the Honeycomb Hills, burst into action at the west end of this belt. Rhyolitic domes less than a million years old erupted atop the Mineral Mountains near Milford recently enough to have left deep residual heat for Utah's largest geothermal resource (Lipman et al., 1978). Nearby basalt flows of the same age (Hoover, 1974; Machette, 1985) are not heat sources. In southwestern Utah, Best and Hamblin (1978) have used the basaltic cones and flows (Figure 87) which range in age from a few thousand years to about 6 m.y., to date faulting and to compute rates of regional uplift and erosion (Hamblin et al., 1981; Hamblin, 1984).

Bimodal volcanism is not randomly scattered across the Great Basin. It is concentrated along the southwestern and southeastern margins in a bilaterally symmetrical fashion (Best and Hamblin, 1978, fig. 14-7).

Basin-range Faulting—From the Wasatch Mountains westward, Utah lies in the Great Basin. This basin, which also includes most of Nevada, is that part of the U.S. that has no streams that drain to the sea. The Great Basin is part of a larger segment of North America, the Basin and Range Province, that is characterized by a series of elongate ranges separated by alluvial-fan dominated valleys. Within the Great Basin the valleys (basins) and ranges trend irregularly northward. The largest individual basin within the Great Basin is that occupied by the Great Salt Lake Desert, commonly known as the salt flats, as can be seen on Figure 86.

Basin-ranges are usually asymmetric in cross section, having a steep slope on one side and a gentle slope on the other (Figure 88). The steep slope is an erosionally modified fault scarp and the range a tilted fault block. Stewart (1980) described regional tilt patterns of basin-range fault blocks (Figure 89) and noted that the average tilt is about 20 degrees and that 85 percent of measurable tilts of block in Utah and Nevada were less than 32 degrees. He equated this to indicate extension of 20 to 30 percent for the Great Basin region.

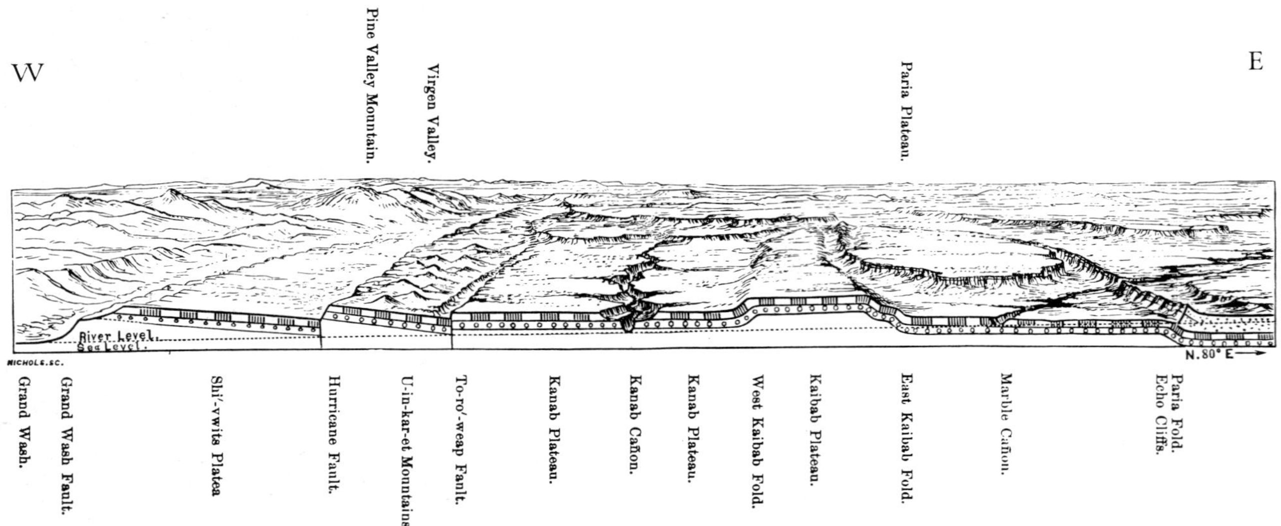

FIGURE 88 — John Wesley Powell's (1875) cross-section near the Utah-Arizona border between the Grand Wash Fault (now seen at Cedar Pockets rest area along Interstate I-15 in the Virgin Gorge west of St. George) and Echo Cliffs south of the Glen Canyon Dam at Page, Arizona. The Toroweap Fault is called the Sevier Fault on modern maps.

In considering basin-range structure it is important to realize that extensional structures within the Great Basin are not limited in time to the 15-million-year interval being discussed here; earlier extensional features of different form and orientation are important in some areas of Nevada (see Stewart, 1983, for a review) but much less so in Utah as indicated earlier in the Oligocene-early Miocene discussion. As used herein, basin-range structure refers only to the development of our present-day basins and ranges in western Utah and to the concurrent normal faulting of similar orientation that cuts the western part of the Colorado Plateaus in central and southern Utah.

Age of inception of basin-range faulting is deduced from the age of the youngest pre-basin-range rocks that are cut. In general, all of the pre-17 m.y. rocks shown on Figure 74 have been cut by basin-range faults, and rocks younger than this occur mostly as fill within modern basins. The latter have also been cut by basin-range faults that have continued to be intermittently active from their inception up to the present. Certain basin-range faults seem to follow older structures that have controlled their location; basin-range faults which follow earlier Laramide structures include the Hurricane (Figures 88 and 90) and Sevier faults of southwest Utah. Zoback and Thompson (1978) concluded that basin-range rifting in northern Nevada began 17–14 m.y. ago, based on dike swarms whose extension direction differs 45 degrees from present extension. Chronology of block faulting in the Great Basin was reviewed by Best and Hamblin (1978, p. 328), who noted

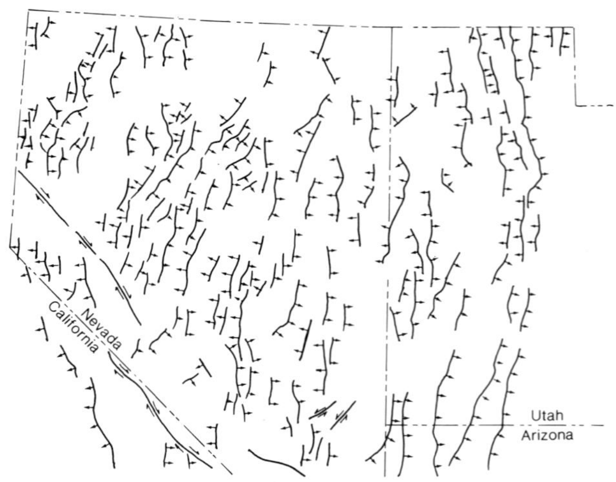

FIGURE 89 — Tilt patterns of late Cenozoic basin-range fault blocks in Utah and Nevada. After Stewart (1980).

FIGURE 90 — Hurricane, Utah, in foreground; La Verkin in center middle distance. Hurricane fault runs along base of escarpment at right edge of town. Stratigraphic displacement along Hurricane fault here is more than 6000 feet, placing Cretaceous strata on the downthrown (town) side of this fault against Permian Toroweap Formation at the base of the escarpment on the upthrown block. Well-bedded layers in the right middle distance are Moenkopi Formation, the white band being the Shnabkaib Member. Above it, the dark slope is the upper red member of the Moenkopi; the mesa is capped by the Shinarump Member of the Chinle Formation. Cliffs of Navajo Sandstone show along the right skyline.

that activity has not been uniform since its inception but has shifted location throughout time. Stewart (1978, 1983) and Zoback et al. (1981) suggested that modern topography probably began to form about 10 m.y. ago in the Great Basin. In northwestern Utah the mid-Miocene Salt Lake Group seems to be an early basin-filling assemblage of basalts, rhyolites and related sedimentary rocks, but the form and extent of its depositional basin is not clear. Lindsey et al. (1981) identified latest Oligocene sedimentary and volcanic rocks at the bottom of more than 7000 ft of basin-fill in the Sevier Desert near Delta. In Utah Valley, the Gulf Oil-Banks well bottomed in beds containing upper Miocene pollen at a depth of 13,000 feet. More age data are needed from deep wells in western Utah valleys before we can be sure when the valley-filling process (and basin-range faulting) began. That basin-range faulting continues to be active up to the present day is documented on the Quaternary Fault Map of Utah (Anderson and Miller, 1979).

Subsurface form of basin-range faults has elicited considerable attention. Most basin-range faults are poorly exposed, buried at the base of mountain fronts in alluvial and colluvial debris. Where fresh faulting, erosion, or man-made cuts have exposed fault surfaces their valley-

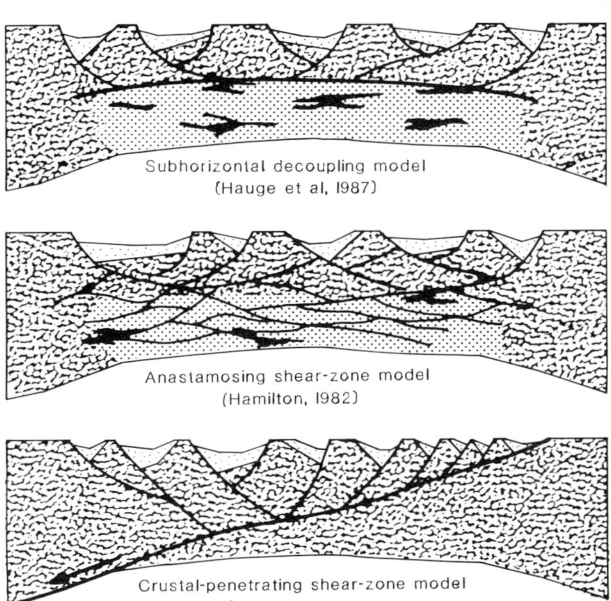

FIGURE 91 — Conceptual models of basin-range faults showing relation of surface blocks to deeper crustal adjustment. Faults are shown as listric with steeper near-surface dips flattening with depth. After Allmendinger et al. (1987).

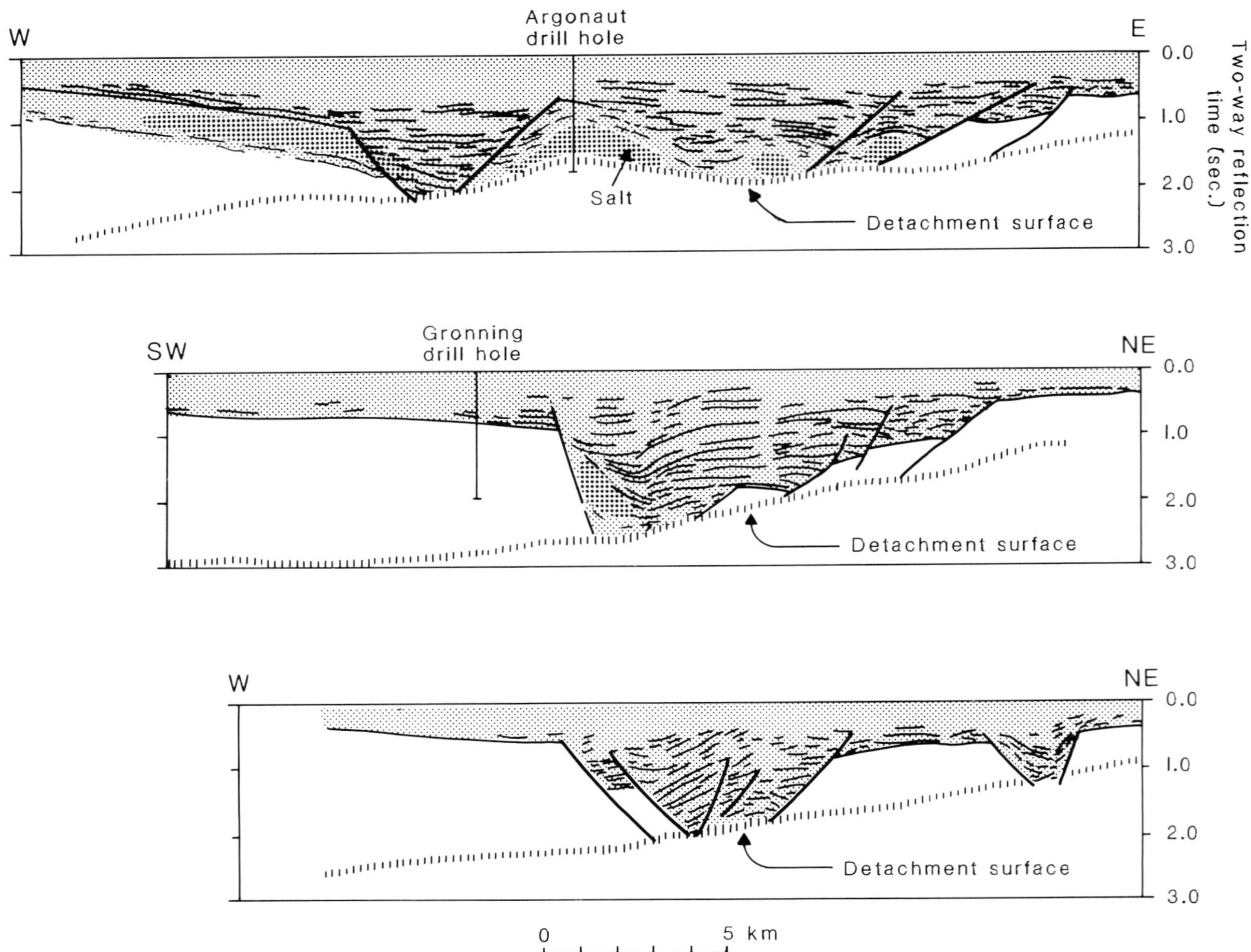

FIGURE 92 — Interpretive cross-sections generalized from seismic reflection data for the eastern Sevier Desert. Valley-filling sedimentary, evaporitic, and volcanic rocks are shaded. Adapted from Anderson et al. (1983).

ward dip angles range between 45 and 90 degrees but are commonly around 60 degrees, and they show dip-slip slickensides. Exposures are almost always limited to a few feet or tens of feet in extent so that projecting them downward is speculative. One of the few places where a basin-range-type fault can be followed through a large vertical distance was studied by Hamblin (1965) in the Grand Canyon area. He showed that the fault surface was curved, its dip angle decreasing with depth. Thus, as the hanging wall block moved down relative to the footwall, the upper part of the hanging wall block moved horizontally away from the footwall block creating a gap into which hanging wall materials rotated or collapsed. A curved downward-flattening fault is called a listric fault and it is believed that many basin-range faults are listric, flattening with depth as shown on Figure 91. Anderson et al. (1983) showed listric faults terminating at a detachment surface in their interpretation of deep seismic reflection data from the Sevier Desert near Delta (Figure 92). Because basin-range fault surfaces are inclined, fault displacement has both horizontal and vertical components. The vertical component is easier to estimate, being the difference in elevation between the downdropped pre-fault topography preserved by burial beneath a valley, and the upraised (and now eroded) pre-fault topography, whose original height may be judged from the thickness of valley fill derived from its destruction. One cannot look at the surface of a Basin and Range valley and tell much about the thickness of valley fill or depth to its pre-fault basement. But gravity surveys provide a reasonable basis for estimating, because valley fill, being less dense than most bedrock, yields lower gravity values. Of the basins in Utah that have been actually drilled to date, Utah Valley has one of the thickest recorded fillings; the Gulf Oil Banks well bottomed in Miocene valley fill at 13,300 feet. Figure 98 shows only half that amount, averaging the thickest value with those from other parts of Utah Valley where the basement is nearer to the surface. Thickness amounts shown on Figure 98 are derived from a variety of sources shown on the individual charts. In northwestern Utah they include thickness estimates for the Salt Lake Group; elsewhere they include much volcanogenic material as shown on the correlation table, Figure 74. Parry and Bruhn (1987) estimated 35,000 feet of vertical offset on the Wasatch Fault near Salt Lake City. Estimates of horizontal displacement (extension) related to basin-range faulting have ranged greatly from extremes of 400 percent to as little as 20 percent. Very large estimates may have been based partially on misinterpretation of gravity slides as listric faults although the two are similar except for scale. Because we have no clear way to fix the original location of points in a horizontal framework at some past time we have no easy basis for knowing how much those points have been horizontally displaced relative to one another through time. Boyer and Allison (1987) suggested that analysis of well, surface, and seismic data suggests that actual extension in the Basin and Range Province is probably more than 20 percent but generally less than 50 percent.

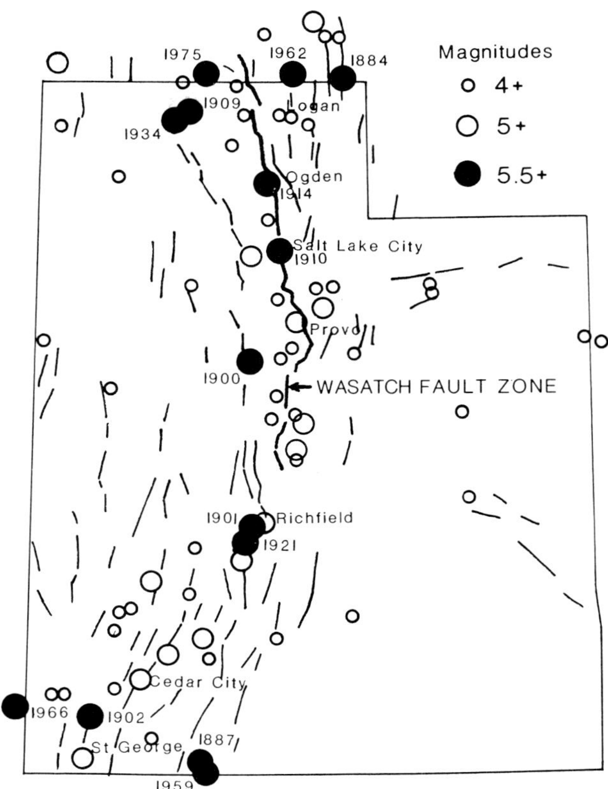

FIGURE 93 — Earthquakes of magnitude four or more in Utah 1850–1986. Earthquakes greater than magnitude 5.5 are indicated with date. From Arabasz et al. (1987).

Wasatch Fault—The Wasatch Mountains, which lie on the east side of the Great Basin, are an upraised fault block that has been tilted eastward by movement on the Wasatch fault, which lies along the west base of the range. Because it cuts through Utah's population centers (Logan-Ogden-Salt Lake City-Provo) the Wasatch fault is the most significant and most studied fault in the state. Although a seismic network to monitor earthquake activity in the region was established earlier, the first serious attempt to map the entire fault zone did not occur until low-sun-angle photography (Figures 94 and 95) made this feasible (Cluff et al., 1970, 1973, 1974). Since then, the Wasatch Fault zone has been the focus of many kinds of studies aimed at understanding the characteristics of the fault so as to implement better planning and zoning procedures that will reduce future earthquake damage and loss of life along the Wasatch Front urban corridor. These studies have been funded primarily by the federal government under the U.S. Geological Survey's earthquake hazard reduction program. The latest published reports of these activities make up two thick volumes edited by Gori and Hays (1987).

Historic earthquakes along the Wasatch fault have been relatively minor (Figure 93). But evidence of prehistoric (pre-1847) fault activity is so extensive that the fault is deemed dormant, not dead. By comparison with other faults of similar size, the Wasatch fault is judged capable of generating an earthquake of magnitude 7.0 to 7.5 at some unknown time in the future. Modern seismic monitoring of earthquakes in Utah did not begin until about 1950 but the seismograph network has been improving since then. In order to extend the earthquake information data base into the pre-pioneer past, a program of digging trenches up to 25 feet deep and 200 feet long across the fault has been conducted since about 1975. More than a score of trenches have yielded data on fault activity of the last few thousand years; some trench data suggest that large earthquakes have struck Utah within the last 1000 years. Deeper trenches that will extend the earthquake record a little further back are being planned. Looking at the faceted spurs (Figure 9) along the Wasatch mountain front, Hamblin (1976) suggested that large Cenozoic movement on the Wasatch fault was episodic; movement occurred along the fault until 500 to 1000 feet of vertical displacement were produced; this movement was followed by a period of quiescence. Hamblin recognized eight periods of movement separated by short periods of stability during the uplift of the southern Wasatch Mountains. Hamblin (1984) showed that a Lake Bonneville terrace near American Fork Canyon had been displaced upwards by recent faulting to above the regional elevation

FIGURE 94 — Low-sun-angle vertical aerial photograph of the Wasatch fault scarp that cuts glacial moraine deposits at Bells Canyon just south of Little Cottonwood Canyon at the south end of Salt Lake Valley. The scarp records multiple movements on the Wasatch Fault, the most recent occurring about 1500 years ago.

of the Lake Bonneville shoreline, indicating that the mountain block moved actively upward while the valley block remained stationary. Taking an even longer view of the uplift of the Wasatch Range, Naeser et al. (1983) reported that apatite fission-track ages indicated an uplift rate of 0.4 mm/yr for the last 10 m.y. for the Wasatch north of Salt Lake City.

The Wasatch fault zone can be divided into segments (Figures 96 and 97) bounded by offsets or salients in its surface trace. It is believed that each segment may behave as a semi-independent entity and that an individual fault rupture is not likely to propagate into adjoining segments, although its earthquake shaking effects may be widely felt. Prediction of future earthquake activity along the Wasatch fault zone is based on assessment of many kinds of information: seismicity, segmentation, recurrence rates based on trenching, maximum expectable magnitude based on comparison with historic earthquakes or similar faults. Because there is no way of telling exactly when the next earthquake will occur, the predictions are couched in probabilistic terms. Youngs et al. (1987) presented probabilities of earthquake ground shaking hazards along the Wasatch Front for valley sites and for mountain (bedrock) sites. They prepared maps showing contours of maximum ground acceleration for probabilities exceeding 10 percent in 10 years, 10 percent in 50 years and 10 percent in 250 years. Stated more simplistically, most experts believe there is a great likelihood that the Wasatch Front area will be shaken by a severe earthquake sometime within the next few hundred years.

FIGURE 95 — Oblique aerial photograph of the same fault scarp shown on Figure 94. Little Cottonwood Canyon is the only glaciated canyon in Utah where glacial ice extended into Lake Bonneville.

Regional uplift—Utah's surface lay near sea-level throughout most of Paleozoic and Mesozoic time, rising locally during the Sevier and Laramide orogenies. Fossils in Eocene Green River lake beds suggest that in early Cenozoic time most of the Wyoming-Utah area was less than 2000 feet above sea level. Regional uplift of the western U.S. began in late Cenozoic time, at about the same time as basin-range block faulting was initiated (Hunt, 1956). The effect of this uplift was to rejuvenate all the major river systems in the Rocky Mountains and Colorado Plateaus (Figure 99), enabling them to cut major gorges, such as the Grand Canyon of the Colorado, the Canyonlands of Utah (Figure 100), and the Royal Gorge. In the Great Basin the uparched area between Salt Lake City and Reno also extended significantly allowing the collapse of portions of the arch, thus producing the basins of the Basin and Range Province. Gable and Hatton (1983, map B) indicated that the central part of the Great Basin has been vertically uplifted more than 10,000 feet in the last 10 million years. Most of Utah that lies east of the maximum uplift has been elevated between 7,000 and 10,000 feet in the same period. The area affected by late Cenozoic uplift is very broad; it extends from the west coast to the Mississippi River but the slope of the uplift from Denver eastward is very gentle, as reflected in modern topography.

Best and Hamblin (1978) and Hamblin et al. (1981) measured rates of vertical uplift of faults along the southwestern Colorado Plateau to range between 75 ft/m.y. and 1200 ft/m.y., based on displacement of radiometrically dated basalt flows. Basalt cobbles in late Cenozoic gravels on northeastward-flowing streams near Flagstaff bracket the major uplift there between 10 and 5 m.y. Numerous localities exist in southern Utah and northern Arizona where age of faulting can be determined by its interrelationships with datable basalts, as indicated by Hamblin (1987).

Valley-fill deposits—Figures 74 and 98 show that western Utah is the site of most of the late Cenozoic deposition in the state. Because the basins of deposition (valleys) are still accumulating sediments and are nearly everywhere covered with recent alluvial, lacustrine, or eolian deposits, subsurface techniques are necessary to ascertain the thickness and form of the valley fill. Water wells give information on the upper few hundred feet of valley fill and reports summarizing these data are referenced on individual charts. Gates (1987) presented an overview of groundwater movement in western Utah that included data from test wells drilled about a decade ago by the Defense Department's preliminary feasibility studies for the MX missile siting program.

In northern Utah the Salt Lake Formation or Group includes basalts and rhyolites, as well as diverse sediments. McClellan (1981) reported a sequence of diverse carbonates about 200 feet thick within the Salt Lake Formation in northern Utah; this is one of the thickest fresh-water limestone sequences of this age so far identified in the state. Patton and Lent (1980) and Bortz (1983) summarized the results of Amoco's oil exploration program (1978–1981) in the Great Salt Lake, offshore from the Rozel Point oil seeps. Fifteen wells were drilled into a faulted anticline; most wells encountered shows of heavy oil and gas in Pliocene basalts, limestones and tuffaceaous shales at depths ranging from 2000 to 5000 feet, but economics for developing this field were unfavorable and it was abandoned. No rocks older than Miocene were penetrated and the late Cenozoic rocks themselves were believed to be the oil source rocks. Bortz (1983) reported that Miocene to Holocene sedimentary and volcanic rocks are up to 15,000 feet thick in the basin beneath the Great Salt Lake, west of the south end of Promontory Point.

In southern Utah Valley, Gulf Oil drilled the Banks well to a depth of 13,000 feet in 1977. Miocene spores and pollen were recovered from sediments at the bottom of the hole. Unconsolidated gravel, clay, silt and sand predominated in the upper few thousand feet; altered tuffs, pyroclastics and volcanogenic sediments were encountered. Beds were better lithified with depth; thin beds of fresh-water limestone were common and thin coal beds were found in the lowest part of the hole. The Banks well was drilled just west of Spanish Fork in the center of Utah Valley where late Cenozoic rocks were believed to be the thickest. Minor gas shows were recorded in the drilling mud but none warranted testing.

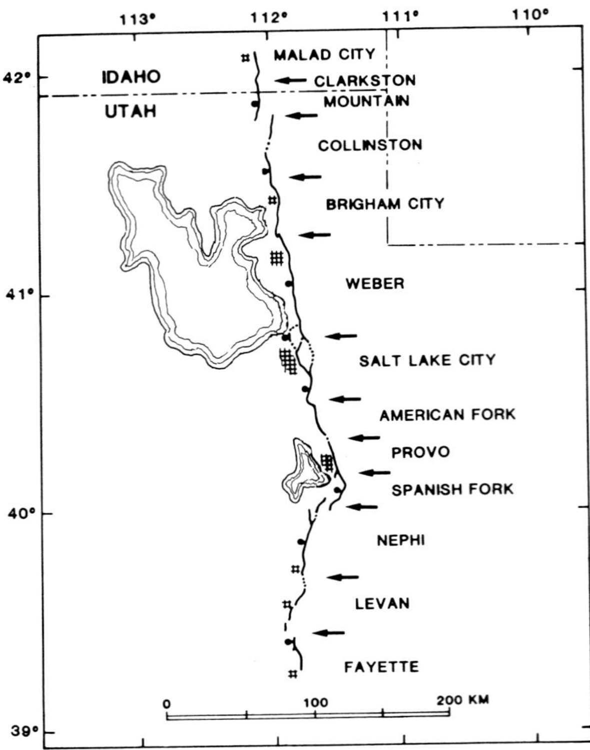

FIGURE 96 — Segments of the Wasatch Fault as identified by Machette et al. (1987). Segment boundaries are placed at offsets or discontinuities in the trend of the fault. It is thought that surface ruptures are not likely to propagate from one segment to another, even in a strong earthquake.

Seven significant wells have penetrated middle to late Cenozoic valley fill in the Sevier Desert basin near Delta, Utah, as summarized by Mitchell and McDonald (1987). Thicknesses of valley fill encountered in these wells range from 1700 to nearly 10,000 feet. The valley fill consists of a heterogeneous assemblage of alluvial, lacustrine and volcanogenic sedimentary rocks, as well as volcanic tuffs and flows. The most unusual sequence in any of the wells is Cenozoic salt and gypsum almost a mile thick encountered in the Argonaut Energy Federal No. 1 described by Mitchell (1979). Alternative interpretations of seismic lines published by McDonald (1976) through the Sevier Desert show several such Cenozoic salt masses (Baer and Hintze, 1987). Lindsey et al. (1981) dated evaporite-bearing volcaniclastic sediments near the bottom of the Gronning well near Delta as late Oligocene, but most of the Cenozoic valley fill in the Sevier Desert basin is middle Miocene and younger.

Near Cedar City a well was drilled in 1975 by Odessa-Southland to a depth of 11,700 feet. It encountered 900

FIGURE 97 — The Salt Lake salient juts out from the Wasatch Front and marks the boundary between the Salt Lake City segment and the Weber segment of the Wasatch Fault, as shown on Figure 96.

feet of alluvial valley fill near the surface, then penetrated almost 900 feet of Miocene and Oligocene ash-flow tuffs and related flows, below which were about 700 feet of fresh-water limestones, shales and conglomerates of probable Oligocene age, then 500 feet of Claron Formation with a basal conglomerate that rests on Cretaceous Iron Springs Formation sandstones. West of Cedar City, in the Escalante Desert basin near Lund, Hunt Oil drilled Table Butte Unit No. 1 in 1984 to a depth of 18,500 feet. The upper 500 feet were alluvial, fresh water and playa lake deposits that overlaid 1200 feet of ash-flow tuffs and varicolored tuff beds; these rested on 600 feet of valley alluvial materials rich in volcanic clasts; these were underlain by Claron (?) Formation sandstone, shale, freshwater limestone and conglomerate which rested unconformably on Cretaceous Iron Springs Formation shale, siltstone, sandstone and conglomerate. In summary, alluvial valley fill is not as thick in the southern Utah sector of the Great Basin

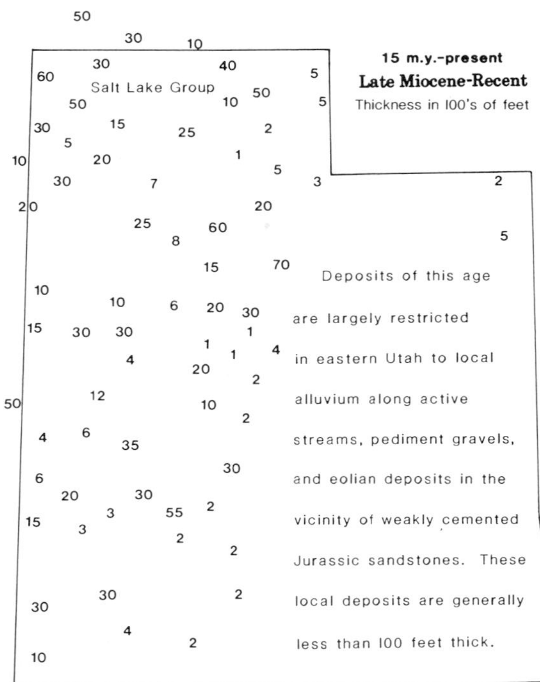

FIGURE 98 — With the onset of major crustal extension in western Utah about 15 m.y. ago, clastic and volcanic rocks began to accumulate in subsiding local basins. Information on thicknesses of these deposits is incomplete, but a few deep wells have penetrated valley-filling deposits of this age more than 5,000 feet thick. Thicknesses of valley-filling deposits are variable from valley to valley, as well as locally within valleys. The deposits are mostly alluvial sands and gravels but also include volcanic lava flows (chiefly basalts), playa and lake beds, and locally evaporites. In northwestern Utah an extensive accumulation of sediments and rhyolitic volcanic rocks is called the Salt Lake Group.

FIGURE 99 — Factory Butte at the north end of the Henry Mountains Basin near Hanksville. Caprock is the Muley Canyon Sandstone (formerly erroneously called Emery Sandstone) which overlies the Blue Gate Member of the Mancos Shale which is about 1200 feet thick here. Factory Butte is an erosional remnant of the once-continuous Cretaceous deposits that blanketed eastern Utah. This view conveys the feeling for the tremendous amount of material that has been removed from eastern Utah by the never-ending work of the Colorado River and its tributaries, many of which only work immediately after a storm. There is hardly any vegetation here to protect the land from rapid erosion.

FIGURE 100 — Rainbow Bridge National Monument. The arch is made of Jurassic Navajo Sandstone which also makes up the massive cliffs above it. The base of the arch rests on Kayenta Formation whose thinner bedded layers are exposed in the bottom of the canyon. This view portrays the active erosion done by small tributaries to the Colorado River conveyor belt.

as it is in central and northern Utah and the southern valleys contain a substantial section of Oligocene-Miocene layered volcanic rocks.

Quaternary alluvial and soil sequences have been scrutinized recently because they are offset by basin-range faults and may provide a means of dating the faulting. Christenson and Purcell (1985) discussed the means by which basin-range alluvial fan sequences may be correlated, relating them to Quaternary climatic changes. Machette (1985) reviewed the development of calcic soils in the southwest U.S. and included a stage classification for soils near Beaver, Utah. Harder et al. (1985) examined Quaternary deposits and soils in Spanish Valley near Moab and established a soil chronosequence for the area. Richmond (1962) mapped Quaternary units in the La Sal Mountains, but for the most part the limited areas of late Cenozoic deposits in eastern Utah have not received special study. Current policy of the Utah Geological and Mineral Survey requires that Quaternary geologic units be shown on quadrangle maps; good examples of this from eastern Utah are the maps of Willis (1985), Morton (1984) and Doelling (1985).

QUATERNARY

Ashfalls and other Quaternary dating methods—Because Quaternary events have so recently been impressed on our surroundings, they are of particular importance to many of our environmental decisions and often require a degree of age discrimination not needed for older happenings. Consequently, a variety of dating techniques have been developed that apply especially to Quaternary time. The Quaternary Period was established to include the glacial deposits that have so significantly modified the landscapes of the northern hemisphere. Many early attempts were made to establish chronologies based on advances and recessions of continental glaciers in Eurasia and North America, but none of them were satisfactory. Oceanographic research in the last half of this century has shown that the complete record of Quaternary climatic fluctuations is preserved in sea-floor sediments (Berggren et al., 1985). Isotope studies from ocean bottom sediments have indicated more than 20 Quaternary cooling events, which have been keyed to the record of magnetic reversals. Accordingly, the base of the Quaternary has been set just above the Olduvai Normal Polarity Subchron at 1.6 m.y. (Berggren et al., 1985).

Cyclicity of Quaternary glacial advances was attributed more than eighty years ago by Milutin Milankovich to interacting cyclic variations of degree of tilt of the earth's axis, precession of the equinoxes, and eccentricity of earth's orbit around the sun. Verification of the Milankovich cycle awaited the detailed record of Quaternary climatic fluctuations available in cores from the sea floor; the earth-sun cyclic variations are now widely used to explain, not only Quaternary glacial cycles, but also cyclic deposition throughout the geologic record.

Extending a Quaternary time scale established on the sea floor to events in Utah requires some bridging correlation techniques. The most straightforward of these involves the use of widespread ashfalls as time lines. Volcanic ashes blown out of eruptive centers may become widely distributed as thin blankets that record an "instant" of geologic time. Each layer bears its own compositional fingerprint and each can be radiometrically dated (Izett, 1981). Ash layers have been identified in deep-ocean sediments and, thus, link the sea-floor record with deposits inland. Izett and Wilcox (1982) compiled maps showing the distribution of several ashes ranging in age from 0.6 to 2.0 m.y. from the Yellowstone caldera complex. Other ash falls that have been studied in considerable detail include those from Mount St. Helens, Crater Lake (Mazama), Glacier Peak, and the Bishop ashes from California (Figure 101).

In addition to ash beds, several other dating methods can be used, depending on availability of suitable materials. Carbon-14 is widely applied to organic materials that are less than 50,000 years old. Uranium-series decay products may be used to date mollusks or bone. Potassium-argon and fission-track techniques are directly applicable only to igneous rocks. Other relative-age methods include lichen growth studies, obsidian hydration, weathering rinds on quartzite clasts, thermoluminescence, amino acid racemization, soil and caliche development, landform modification, and geomorphic position (Pierce, 1986; Coleman and Pierce, 1977).

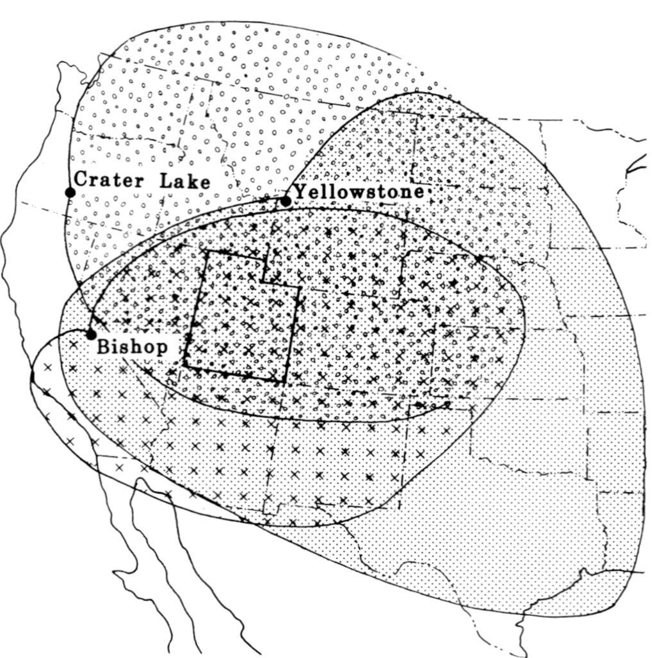

FIGURE 101 — Ash from volcanic eruptions can be distributed over large areas downwind. Map shows major sources and fallout pattern for Pleistocene volcanoes. Such ash layers serve as time markers when they are incorporated into sedimentary deposits. The most voluminous ashes came from three eruptions in Yellowstone 0.6, 1.2, and 2.0 million years ago. The Bishop ash was erupted 0.7 m.y. ago, and the Crater Lake (Mazama) ash only 6000 years ago.

Pleistocene glaciation—"Cookie-sheet" or "biscuit-board" landforms result where glacial cirque basins are plucked out of the smooth crest of a mountain range. Atwood (1909) described such topography in the Uinta Mountains (Figure 102); he also mapped glacial features of the central Wasatch Mountains, as shown on Figure 103. All mountains in Utah above 10,000 feet have been glaciated, as shown on Figures 104, 105, and 106, and glacial histories of several have been described: Uinta Mountains, Bradley (1936); Boulder Mountain, Flint and Denny (1958); La Sal Mountains (Figure 84), Richmond (1962); Little Cottonwood Canyon, Richmond (1964), McCoy (1977), Madsen and Currey (1979). Pre-1970 reports attempted to correlate Rocky Mountain glaciations with midconti-

FIGURE 102 — Glaciated "biscuit-board" topography in the western Uinta Mountains. The Wasatch Mountains show on the horizon. Bedrock in the Uinta Mountains is Proterozoic quartzite and argillite.

nent continental ice ages known as the Wisconsin, Illinoisan, Kansan and Nebraskan glaciations. We consider ourselves to be in an interglacial warm period at present; the last major interglacial warm spell (Sangamonian) peaked 122,000 years ago and is identifiable both in marine cores and in interglacial soils on land. Since that time there have been two glacial maxima during the Wisconsin glaciation. In the Rocky Mountain area these have usually been termed the Pinedale Glacial (30,000–12,000 years ago) and the Bull Lake Glacial (60,000 plus years ago). Because of uncertainties of correlation, however, it is common in current studies to use local names to record minor advances and retreats. The part of Quaternary glacial history that is best-dated on land is that part that can be dated with radiocarbon, the last 50,000 years. Thus the most recent restudy of glaciation in Little Cottonwood Canyon (Madsen and Currey, 1979) deals only with the latest glacial activity: a soil, dated at 26,000 years, developed on earlier glacial deposits and was covered by the Bells Canyon till (Figures 94 and 95) about 19,000 years ago; a mid-canyon deglacial pause about 13,000 years ago, and an upper canyon deglacial pause about 8000 years ago marked the recession of the ice in Little Cottonwood Canyon. No later glacial activity was identified in the canyon during Neoglacial time (about 3000 years ago) although glaciation of this age has been noted elsewhere in the Rocky Mountains. Anderson and Anderson (1981) reported similar findings on Mount Timpanogos (Figure 105) where weathering rinds on quartzite clasts in rock glaciers indicate that no true glaciers existed in the Wasatch Range during the Neoglacial. These findings modify the late glacial history of the Wasatch Range as visualized by Richmond (1964).

Little Cottonwood Canyon is an especially important locality for the Pleistocene history of the Great Basin because it is the best place to observe intertonguing of Lake Bonneville deposits with those of contemporaneous glaciers. Richmond's (1964) correlation of the Bonneville Formation deposited by Lake Bonneville with a Bull Lake glacial age is no longer believed valid; as discussed below, the Bonneville Formation is now regarded as much younger.

Lake Bonneville — Shorelines of Lake Bonneville are one of the most conspicuous geological features in western Utah (Figure 108). A century ago Grove Karl Gilbert

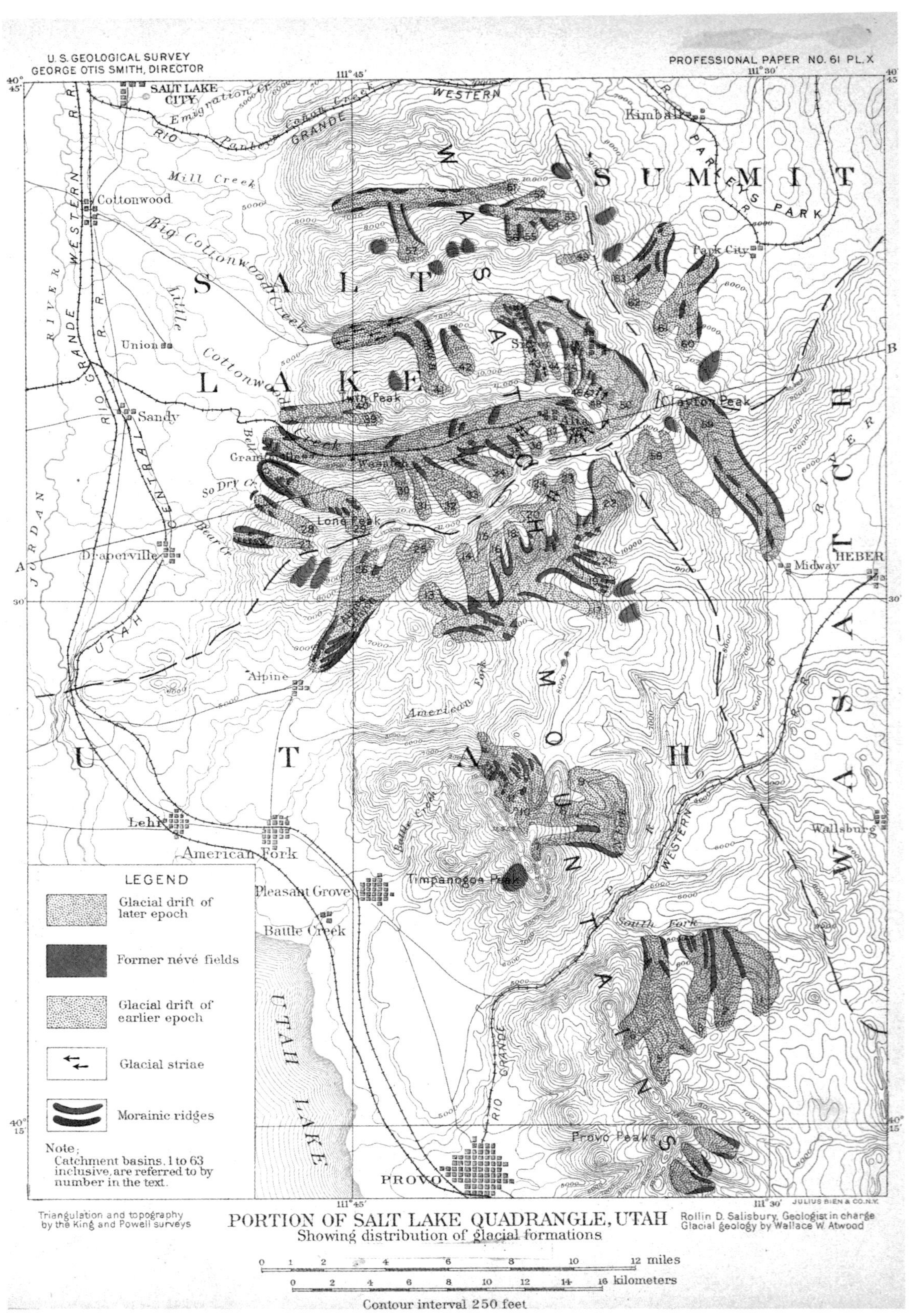

FIGURE 103 — Glaciation of the Wasatch Mountains between Salt Lake City and Provo. From Atwood (1909).

FIGURE 104 — Glaciated topography near Brighton in Big Cottonwood Canyon east of Salt Lake City. Lake Mary near left center of photo is a typical glacially formed shallow lake. Bedrock around the lake is granodiorite of the Alta stock. It intruded Paleozoic strata which form the ridges on the left horizon. The Big Cottonwood glacier extended down canyon only to "The Stairs." It did not reach the canyon mouth.

recorded the features associated with this ice-age lake in his monumental Monograph 1 of the U.S. Geological Survey. Later geologists (Hunt et al., 1953; Williams, 1962; Bissell, 1963) mapped in greater detail the sediments that accumulated in the lake in the form of deltas, beaches, and lake-bottom deposits.

Lake Bonneville was the largest late Pleistocene (Wisconsin) pluvial lake in western North America. At its maximum, shown on Figure 106, it covered nearly 20,000 square miles and was nearly 1000 feet deep in the area of present Great Salt Lake. Most of the shorelines we see were formed during the Bonneville Lake Cycle between 30,000–10,000 years ago, as shown on Figure 107. Certain hillsides that were exposed to vigorous wave action show multiple shorelines, but rigorous tracing of shorelines by many workers has led to the naming of just a few that can be identified basin-wide. The Bonneville Shoreline is easy to pick because it is the highest (5090 feet); it formed between 16,000–14,500 years ago when freshwater Lake Bonneville found the Zenda outlet into the Snake River drainage near present Red Rock Pass between Preston and Downey in southeastern Idaho. The

FIGURE 105 — Glaciated basins on the east side of Mt. Timpanogos in the Wasatch Mountains near Provo. Bedrock is mostly fine-grained sandstones of the Pennsylvanian Oquirrh Group.

Provo Shoreline is easy to identify because it forms the broadest benches; it formed between 14,500–12,500 years ago, after the Bonneville Flood, 14,500 years ago, cut a gorge from the Zenda threshold to a bedrock threshold at Red Rock Pass, lowering the lake level 350 feet catastrophically (Malde, 1960, 1968; Jarret and Malde, 1987). The Stansbury and Gilbert shorelines are less prominent but can be traced widely at their elevations of 4500 and 4250 feet respectively.

Another shoreline that is easy to see in Utah Valley and other places in northern Utah has been called the Alpine Shoreline (Little, 1988). It is older than shorelines of the Bonneville Lake Cycle and may be related to the pre-Wisconsin Little Valley Lake Cycle. It lies just below the highest Bonneville level and is characteristically cut by many gullies, conspicuously more so that any of the later Bonneville Lake Cycle shorelines. Little's (1988) study showed that the Alpine Shoreline was formed before the Bear River drainage was captured into the Bonneville basin. The term Alpine was coined by Hunt et al. (1953) to apply to the oldest lake beds (Alpine Formation) that he could recognize in Utah Valley. Subsequent workers extended the term to apply to similar-appearing silty beds near Leamington that were later shown to be much younger. Accordingly, Oviatt et al. (1987) preferred to abandon the term Alpine Formation because it seemed to have become a catch-all term for silty lake deposits of various ages. Nonetheless, the term Alpine Shoreline seems to be a useful designation for the high-level gullied shoreline prominent in parts of the Bonneville Basin.

The Lake Bonneville history summarized above differs from interpretations described in pre-1980 literature. The present interpretations have been made as a result of better dating methods (McCoy, 1987; Thompson, 1987) and more detailed stratigraphic correlation and mapping of critical exposures of lake beds in the Bonneville Basin (Currey, 1982; Scott et al., 1983; Currey and Oviatt, 1985; Oviatt et al., 1987; Oviatt, 1987). A modern map of Lake Bonneville levels and a popular summary is presented by Currey, Atwood, and Mabey (1984), who trace lake history from Bonneville time up to the fluctuations of the present Great Salt Lake.

Pre-Bonneville lakes existed in the Bonneville basin and have been identified in well cores (Eardley et al., 1973). Gravity surveys of the Great Salt Lake basin show numerous buried bedrock fault blocks in a north-south horst and graben arrangement suggesting that pre-Pleistocene lake basins were individually smaller than the present Great Salt Lake Desert. History of pre-Pleistocene lakes is poorly known.

Deformation of Lake Bonneville shorelines, because of post-Bonneville isostatic crustal rebound following removal of the load of lake waters, originally noticed by Gilbert (1890), has been verified by Crittenden (1963) and Currey et al. (1983). Basically, the deeper the lake, the

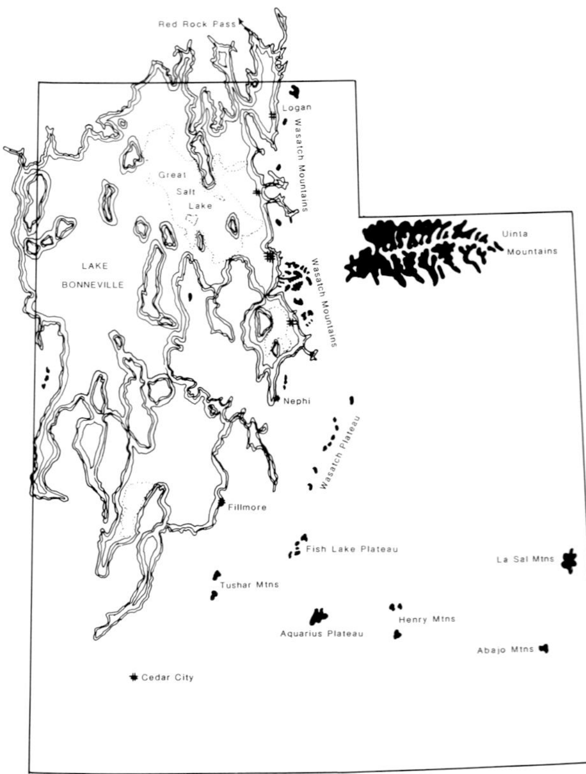

FIGURE 106 — Maximum extent of late Quaternary Lake Bonneville and contemporaneous glaciers (in black).

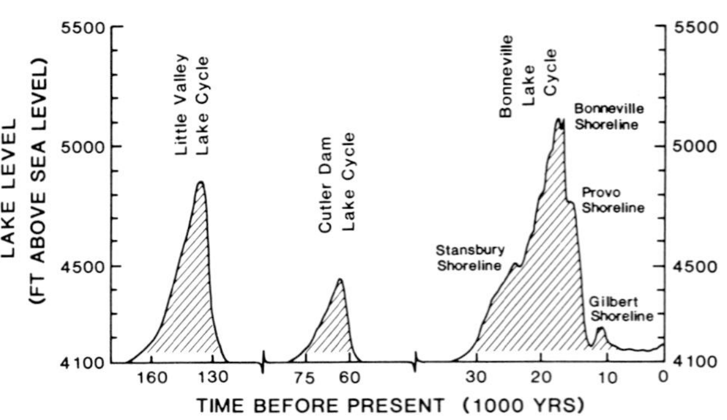

FIGURE 107 — Diagram showing probable levels of Lake Bonneville during the past 150,000 years. Modified from Currey and Oviatt (1985), extended beyond 30,000 years on the basis of stratigraphic studies of older lake deposits by Scott et al. (1983), McCoy (1987), and Oviatt et al. (1987). From Machette et al. (1987).

FIGURE 108 — Lake Bonneville shorelines on the north end of the Oquirrh Range, Utah. From Gilbert (1890).

greater the depression and, hence, the greater the rebound. The Bonneville Shoreline has rebounded more than 200 feet in the area where the lake was 1000 feet deep in the central Great Salt Lake basin. The Provo Shoreline shows a maximum rebound of about 150 feet. Rebound deformation of the shorelines diminishes to zero at the outer limits where lake waters were shallow. It is important to know how elevations of the same shoreline vary when assessing shoreline deplacement caused by faulting. McCalpin et al. (1987) discussed the interplay between Bonneville shoreline deformation and recent faulting in the Hansel Valley area in the northern Bonneville basin and suggested that water loading during Lake Bonneville high stands triggered more frequent large earthquakes than we have experienced during the past 100 years.

PRESENT DAY

Contemporary seismicity—As part of the U.S. Geological Survey's Earthquake Hazards Reduction program, a seismic network that monitors earthquakes in Utah has been operated by the University of Utah for the past few decades. A seismicity map of the state (Stover et al., 1986) lists all felt earthquakes, or those with computed magnitudes greater than 2.5, that originated within Utah; damage reports are also listed. Assessment of earthquake risk has been the subject of numerous reports, the most recent of which was compiled by Gori and Hays (1987).

In addition to aiding in risk assessment, study of earthquake energy transmission has improved our understanding of crustal structure and modes of deformation within the Basin and Range, Colorado Plateau, and Rocky Mountain provinces and their transitional boundaries. Smith et al. (1988) presented a crustal model (Figure 109) that depicts the fundamental differences between the Basin and Range and the Colorado Plateau. Smith and Bruhn (1984) incorporated oil industry seismic reflection data along with earthquake data to develop their model of regional tectonics and the brittle-ductile crustal layering in the eastern basin-range area. Eddington et al. (1987) reviewed kinematics of basin-range extension. Arabasz et al. (1987) applied earthquake data to evaluating future earthquake parameters in the Wasatch Front area and presented a model (Figure 110) that incorporates current understanding of crustal structure in this area.

Fault-plane solutions for earthquakes in the Basin and Range-Colorado Plateau transition area of central and southwestern Utah were presented by Arabasz and Julander (1986). They found no evidence for seismic slip on either downward-flattening (listric) or low-angle normal faults in this region, although such faults are shown on geologic maps in the area. However, low-angle discontinuities play a role in separating locally intense upper-crustal seismicity less than 6–8 km in depth from less frequent earthquakes at depths down to 16 km. Industry reflection seismic data suggest two different forms of basin-range normal faults: listric faults that flatten between 4–6 km depth (the flat portion having never produced any historic earthquake records), and large normal faults that extend in planar form to depths of 15–16 km and have generated major historic quakes (Borah Peak, Idaho; Hebgen Lake, Montana; Dixie Valley, Nevada).

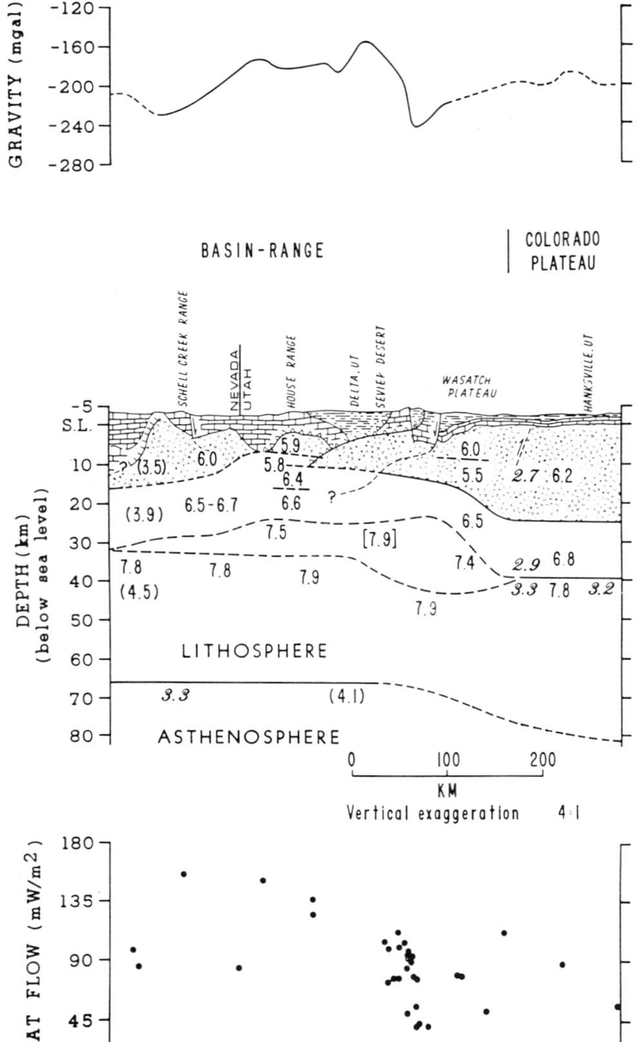

FIGURE 109 — West-east cross-section of lithosphere from east-central Nevada, through Delta, Utah to Hanksville. Upper: Bouguer gravity anomaly profile. Middle: two-dimensional velocity cross-section. Surface geology generalized from Saleeby, 1986. P-wave velocities are in km/sec; S-wave velocites (km/sec) are in parentheses; densities (gm/cubic cm) are in italics. One P-wave velocity number in brackets represents interpretation of Pakiser (1985). Lower: Heat flow data points. After Smith et al. (1988).

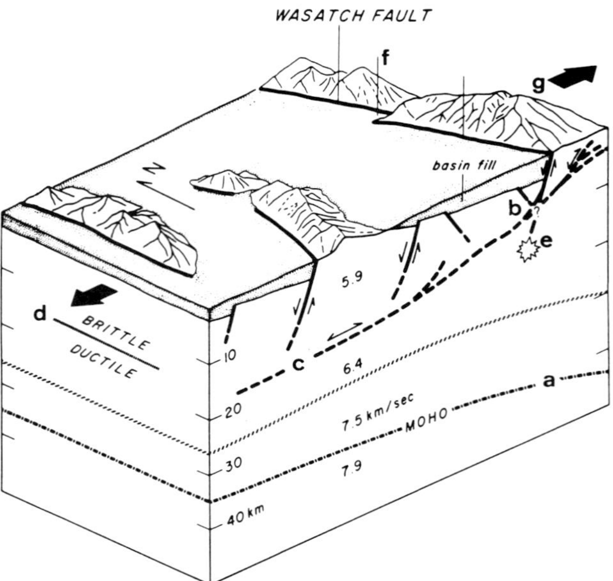

FIGURE 110 — Block diagram cartoon showing some aspects of the seismotectonic framework of the Wasatch Front. a, Moho, marking base of crust, descends eastward; b, basin-range faults have steep dips at surface but flatten with depth to merge into near-horizontal detachment surface (c) near the depth of the brittle-ductile transition (d) which may range between 8–15 km in depth; the Wasatch Fault would likely break (nucleate) (e) at about 15 km depth, based on comparison with other large historic earthquakes in this region; surface rupture from such an event would not likely propagate between Wasatch Fault segments (f). Ongoing extension (g) in this area is about 1–2 mm/year. After Arabasz et al. (1987).

Although most basin-range faults are regarded as normal faults with dip-slip movement, Arabasz and Julander's (1986) fault-plane solutions documented abundant strike-slip movement in the Sevier Valley area. Anderson and Barnhard (1987) described exposed left-lateral and right-lateral strike slip-faults in this area and concluded that the strike-slip movement formed along laterally displaced blocks in a basic extensional framework. They concluded that although strike-slip microseismicity was currently active, the potential for major earthquakes along the strike-slip faults was small when compared to the destructive potential of major normal faults such as the Sevier fault.

One of the enigmas of contemporary seismicity in Utah is that it does not, in general, coincide with surface fault scarps or with faults identified in seismic reflection data. Faults that produce the active seismicity appear to represent internal adjustments within blocks bounded by major basin-range faults.

COCORP—The deep-seismic reflection program funded by the National Science Foundation (COCORP: Consortium for Continental Reflection Profiling) ran its 40th

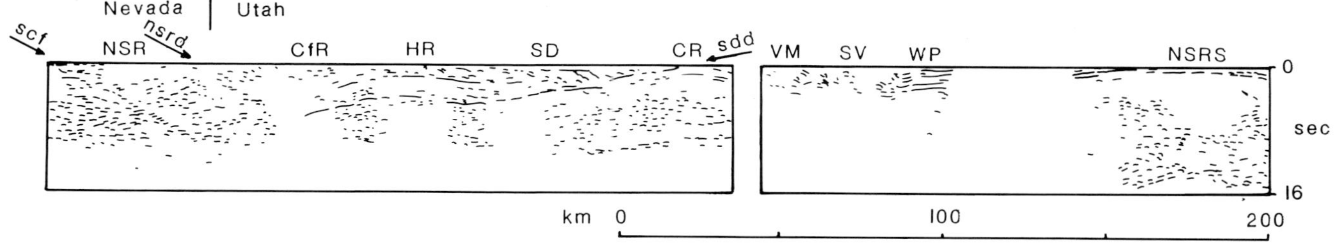

FIGURE 111 — COCORP deep-seismic-reflection profile from eastern Nevada to central Utah. Compare with Figure 112. Structural features: Schell Creek fault, scf; Northern Snake Range detachment, nsrd; Sevier Desert detachment, sdd. Topographic features: Northern Snake Range, NSR; Confusion Range, CfR; House Range, HR: Sevier Desert, SD; Canyon Range, CR; Valley Mountains, VM; Sevier Valley, SV; Wasatch Plateau, WP; North end San Rafael Swell, NSRS. After Allmendinger et al. (1987).

Parallel transect extending from the Sierra Nevada, west of Reno, to the San Rafael Swell near Green River, Utah. Allmendinger et al. (1987b) summarized the results; more detailed accounts of the findings were presented by Allmendinger et al. (1983, 1986, 1987a), and Hauser et al. (1987a). A simplified line drawing of the eastern Nevada and Utah portion of the transect is shown in Figure 111; a schematic comparison of the seismic fabrics with certain known geologic features across the entire transect is shown in Figure 112.

The base of the reflective crust is consistently marked by high-amplitude reflections. It is identified on Figure 112 as the reflection Moho (Moho: discontinuity that separates the Earth's crust from the subjacent mantle) and seems to agree in general form with the refraction Moho (Klemerer et al., 1986). Across Nevada and western Utah the reflection Moho is continuous beneath terranes that experienced very different tectonic histories in the Paleozoic and Mesozoic, and thus it appears to be a young feature, probably taking its present form in late Cenozoic time. In Arizona, Hauser et al. (1987b) identified a 3 km vertical offset in the reflection Moho, which suggested recent offset along the Colorado Plateau transition zone. Average depth of the reflection Moho in the Great Basin is about 30 km (18 miles); in the Colorado Plateau it is more than 45 km (27 miles) deep, but a 30 km horizon of unknown origin also appears as a reflector in the Colorado Plateau data.

Above the Moho in the Colorado Plateau other intermediate-depth reflections or diffractions dip both easterly and westerly but do not persist laterally. According to Allmendinger et al. (1987b) they could be produced by dipping intrusive or lithologic contacts; Precambrian, late Paleozoic, or Mesozoic structures; steep metamorphic gradients; zones of hydration; or reflection artifacts. Seismic fabric of this nature is typical of old, cold cratons; the fabric in eastern Utah probably reflects Proterozoic features.

West of Utah's hingeline (which Interstate-15 approximately follows), COCORP profiles show a prominent

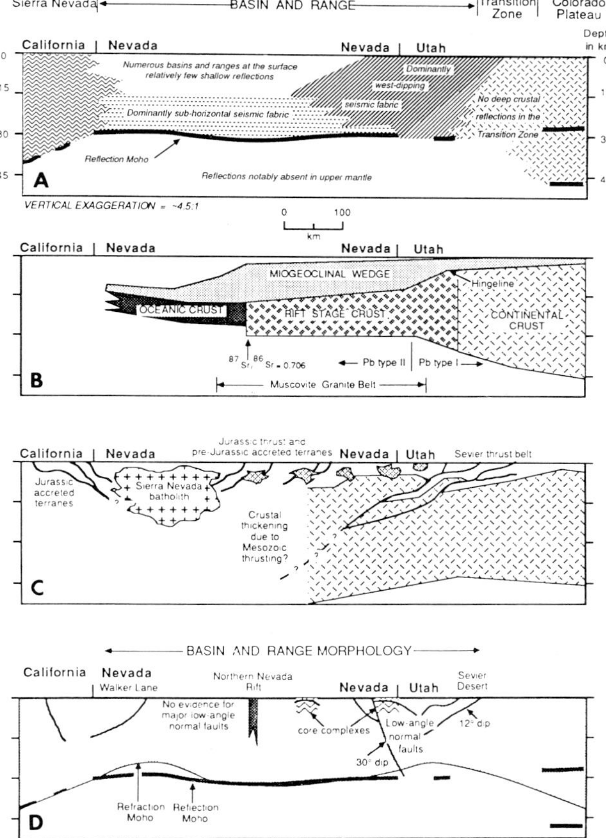

FIGURE 112 — Generalized cross-sections comparing COCORP fabrics with early Cenozoic and early Paleozoic relationships across the Great Basin. A, present-day seismic fabrics; B, structures identified in COCORP profiles; C, early Cenozoic (Phase VI) crustal features (post Mesozoic thrusting, pre-basin-range faulting); D, inferred early Paleozoic (Phase II) continental margin. Not plotted on palinspastic base; vertical exaggeration x 4.5. From Allmendinger et al. (1987). Reflection Moho is a layered zone, about 1 km thick, of horizontal but discontinuous reflections generated by interaction of unknown nature between the crust and the mantle.

reflector that dips westward at about 12 degrees that is called the Sevier Desert detachment. Normal faults that cut upper Cenozoic reflectors in the Sevier Desert basin do not cut this detachment surface which has been interpreted as either a reactivated Mesozoic thrust or a new Cenozoic normal fault by Allmendinger et al. (1986). The Sevier Desert detachment can be traced from shallow depths on the east side of the Sevier Desert to depths of about 15 km (9 miles) on the west side of the desert over a distance of more than 40 miles. It is paralleled by shallower reflectors, the origin of which is speculative, above its western half.

In eastern Nevada a prominent east-dipping reflector called the North Snake Range decollement clearly truncates west-dipping reflectors beneath the Utah-Nevada border. Forty km (25 miles) farther west an even more prominent east-dipping boundary, called the Schell Creek fault, extends almost to the reflection Moho, as shown near the left edge of Figure 111.

COCORP data have provided a new view of the earth's crust. In Utah its fuzzy deep look into the crust has helped to confirm some previous structural interpretations; but it has also showed us problems that need further clarification.

Aeromagnetic data—Regional magnetic patterns in Utah have been portrayed in color on maps at the 1:2,500,000 scale (Zietz, 1982), and the 1:1,000,000 scale (Zietz et al., 1976). Mabey et al. (1978, 1983) have presented interpretations of the anomalies shown on magnetic maps of the state and the following is summarized from their discussions. Magnetic anomalies in Utah mostly reflect the interaction between magnetic properties of Precambrian basement rocks, and the superimposed effects of later (chiefly Cenozoic) igneous bodies. East of the heavy line that cuts northward through Utah on Figure 113 the magnetic anomalies are governed largely by variations in rock types in the Proterozoic metamorphic and igneous basement complex; local magnetic highs are produced by mid-Cenozoic laccoliths in the Henry, La Sal, and Abajo mountains.

West of the heavy line, in western Utah and eastern Nevada, magnetic anomalies from basement rock are inconspicuous in what has been called the "basement quiet zone" except in northwestern Utah where the alignments marked D and E may be basement related. Western Utah is marked by westerly trending highs marked A, B, and C that correspond with late Eocene, Oligocene and early Miocene igneous belts that are superimposed on the basement quiet zone. The crust is thin in western Utah and the Precambrian basement may have been heated enough to modify its original magnetic characteristics in contrast to the more normal Precambrian basement pattern in eastern Utah where the crust is thicker and cooler.

Gravity data—Regional Bouguer gravity and derivative gravity maps at a scale of 1:7,500,000 by Hildenbrand et al. (1982) and Simpson et al. (1986) have been published in color. A 1:1,000,000 simple Bouguer gravity map of Utah was compiled by Cook et al. (1975) from more detailed source maps of various scales. Utah's hingeline divides the state into an eastern part, including the topographically higher Uinta Mountains and the Colorado Plateau, that have lower Bouguer values, and a western part that is characterized by higher gravity. This reflects the seismic observation that the crust is thicker beneath eastern Utah.

According to Mabey et al. (1983), gravity data in western Utah reflect differences in density between less-dense valley fill in the basins and more-dense bedrock in the ranges. Gravity data, though inherently ambiguous, provide a quick and inexpensive way to estimate the distribution and thickness of late Cenozoic fill in Basin and Range valleys. Free air and isostatic anomalies indicate that the region is in approximate isostatic equilibrium.

Eaton et al. (1978) emphasized the gross bilateral symmetry of the Bouguer gravity and regional topography of the Great Basin. The low gravity and high topography of the Sierra Nevada on the west and the Wasatch Range-Colorado Plateau on the east enclose the high gravity and low elevations of the Lake Lahontan basin near Reno and the Bonneville basin in western Utah (Figure 114). The axis of symmetry trends slightly west of north and passes through the Ruby Mountains near Elko. Some students of the Great Basin believe that its bilateral symmetry, with a topographically higher and gravity lower center, reflects an upwelling convection current in the mantle that is causing extension in the Great Basin. Zuber and Parmentier (1986) noted a relationship between Bouguer gravity and regional topography with the width of basin-range tilt domains identified by Stewart (1980).

Locally, gravity techniques have been used to probe fault- and volcanic-related basins. Halliday and Cook (1980) noted that gravity lows are associated with calderas in the Marysvale volcanic field, and steep gravity gradients mark the locations of faults. Zoback (1983) used gravity data to define the geometry of crustal segmentation along the Wasatch fault zone, identifying transverse structural ridges in the gravity data that commonly correspond with salients (projections) along the mountain fronts.

Geothermal heat-flow data—Heat is generated in the mantle and crust of the earth from disseminated radioactive elements. This radiogenic heat powers the convective movement in the mantle and the tectonic motion of lithospheric plates, and rises to earth's surface where it dissipates. Surface heat flow is measured in boreholes by special instrumentation; heat flow measurements did not become common until the 1970's when interest in plate tectonics and geothermal power sources provided the incentive. A heat-flow map may be complex because it

FIGURE 113 — Aeromagnetic map of Utah and adjacent states. Darker shades are higher magnetic readings. Heavier line separates Utah into an eastern part dominated by readings from the Precambrian basement, and a western part showing belts of highs that correspond to middle Cenozoic volcanic areas. From Mabey et al. (1978). Refer to them for more detailed account of trends shown.

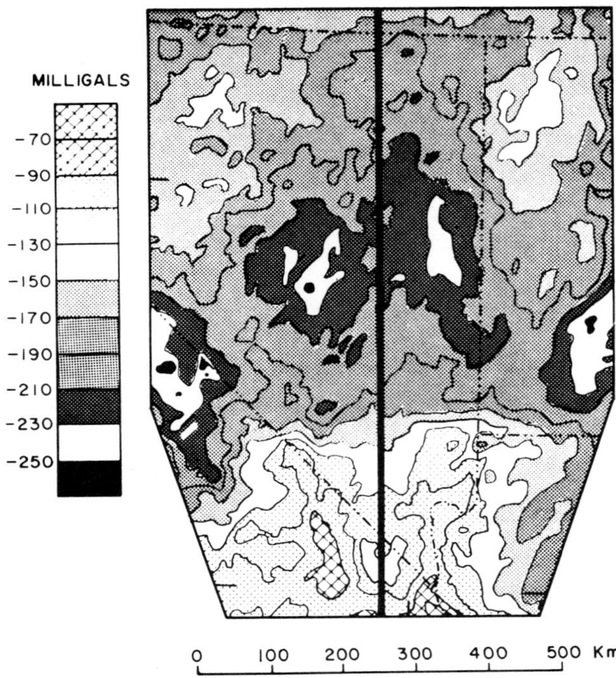

FIGURE 114 — Bilateral symmetry of the observed Bouguer gravity field of the Great Basin. From Eaton et al. (1978).

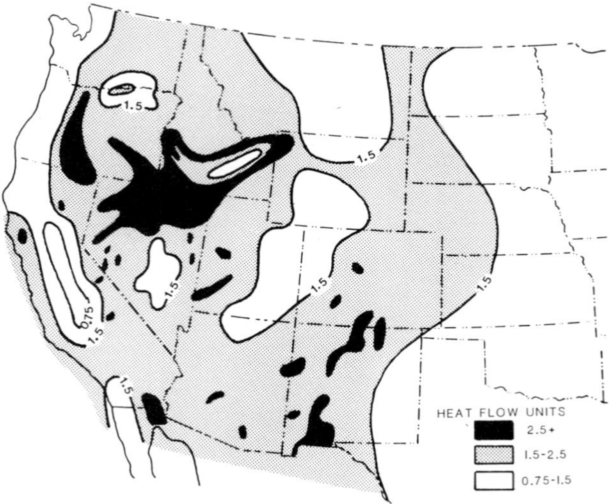

FIGURE 115 — Heat flow map of western U.S. showing bilateral symmetry over Great Basin. Compare with Figure 114. From Lachenbruch et al. (1985).

shows the combined effects of crust and mantle radioactivity, magmatic heat sources, regional hydrology, and structurally related heat conductivity contrasts, in addition to reflecting uneven coverage of heat-flow data points.

Enough heat-flow measurements have been obtained to show high heat flow values in the Great Basin and moderate heat flow in the Colorado Plateau (Figure 115). Blackwell (1978, p. 199) showed a belt of unusually high heat flow along the eastern margin of the Great Basin, following the trend of major fault zones (Hurricane, Wasatch) along Utah's hingeline. Maps of hot springs in Utah (Mundorff, 1970; Utah Geological and Mineral Survey staff, 1983) show that most of our major thermal springs lie in this belt. Utah's hottest geothermal spring, the Roosevelt Hot Spring near Milford, is clearly related to late Pleistocene rhyolitic intrusions in the Mineral Mountains (Ross, Nielson, and Moore, 1982). Mabey and Budding (1987) described the high-temperature geothermal resources of Utah; Budding and Bugden (1986) presented an annotated geothermal bibliography of the state.

Lachenbruch and Sass (1978) showed that heat flow measured at the surface, minus heat production from crustal radioactivity, is characteristically 50 to 100 percent greater in the Great Basin than in stable regions. They suggested that this extra heat is being transferred from the asthenosphere by convection in the lithosphere, stretching of rocks, and magmatic intrusion. They presented a model showing that as anomalous heat flow increases, so do rates of crustal extension in the Great Basin.

PLATE-TECTONIC PERSPECTIVE

The greatest modern advance in man's geologic understanding of planet Earth has been the realization that the Earth's interior is a radioactive heat-motor that keeps Earth's outer lithospheric plates constantly in ponderous motion. Tracking plate movements backwards through geologic time is possible because Earth's magnetic field leaves its imprint on iron-bearing rocks as they are deposited or emplaced. When igneous rock congeals along a mid-oceanic spreading ridge it locks the magnetic polarity signature of its cooling age into a record that has been mapped on the floors of all oceans (Figure 116). Oceanic crust is constantly being created along spreading ridges and, just as constantly, being consumed by subduction into the earth's interior along trenches. The maximum life of most ocean-floor rocks is about 200 million years, from birth at ridges to recyclement at trenches. So the magnetic stripes on present ocean floors record, with great fidelity, a partial record of the movement of continental and oceanic plates (and any attached continental crust) back to the beginning of Jurassic time.

To extend the record of plate movements further back in time beyond 200 m.y. we must turn to continental rocks. The magnetic field is not as widely, nor as uniformly, recorded in continental rocks as it is on the ocean floors. Nonetheless, using apparent north-polar wander plots (or more accurately, continental-wander plots relative to the pole) it is possible to trace the relative positions of continents back through Paleozoic time (Figure 16); some progress is being made in extending the record of continental wander into the Proterozoic, but the further back, the scantier and the more geologically overprinted the record, and the less accurate the results. Thus, as we view the phases of Utah's geologic history from a plate-tectonic perspective, the older phases are more poorly constrained than are younger ones.

Phase I—Metamorphic basement phase. It is not possible to reconstruct specifics of world paleogeography for Archean time, 2,500 million years ago and older. But we know that enough differentiation had taken place in the earth's primordial interior so that by about 3 billion years ago continental nuclei (cratons) that rose above the world ocean had begun to form and that sedimentary and igneous processes were operating that produced rocks similar in most respects to rocks formed throughout earth's later history. By late Archean time, 2,500 m.y. ago, northern Utah lay at the edge of the North American craton. What its course had been with respect to Earth's rotational pole during the vast Archean time interval is not known; because Earth's early heat flow was greater than now, the Archean crust was thinner and more mobile, and cratons probably moved around more. The reader interested in a world view of Archean development is referred to Cloud (1988).

Reconstruction of Early Proterozoic paleogeography is only slightly more satisfying. We see that the southern two-thirds of Utah gained its "basement" rocks during this time. These are now gneisses, schists, pegmatites and gneissic granites that were originally deposited as ocean floor basalts, island-arc volcanic and sedimentary rocks, plus some sandstones derived from weathering and erosion of the Archean continental nucleus. We have no idea as to the actual layout of world continents, spreading ridges or subduction zones at this time. We believe, however, that plate-tectonic processes were operating something like they do today because early Proterozoic rock compositions and structural relationships have comparable analogies in later plate-tectonic regimes that are well known. Thus, Condie (1987) suggested that early Proterozoic rocks represent additions to the Archean continental margin by collisional accretion of terranes, a plate-tectonic process that occurred frequently in later time in the interaction between the North American plate and plates of the Pacific basin. Condie noted that lithologic and geochemical characteristics of the Proterozoic accreted rocks suggest that they originally formed in oceanic arcs, continental-margin arcs, and back-arc basins. Early Proterozoic basement rocks are not exposed in western Utah or Nevada. However, they are believed to be present at depth westward to central Nevada, based on isotopic studies on Mesozoic and Cenozoic granites whose composition reflects the material that was melted to make them. Farmer and DePaolo (1983) determined the ratios of radiogenic to non-radiogenic neodymium and strontium isotypes in granite bodies between Salt Lake City and Reno; the ratios eastward from central Nevada are typical of granites derived from melted Precambrian continental crust, with little or no later input from mantle sources; westward from central Nevada no indication of Precambrian crustal basement can be found. Figure 117 shows their model of the pre-Jurassic crust of western Utah and Nevada.

Phase II—Miogeoclinal platform phase. About a billion years ago, in Late Proterozoic time, a new phase in Utah's geologic history was initiated that was characterized by deposition of thick packages of well-sorted sedimentary rocks. Late Proterozoic and Early Cambrian deposits are chiefly quartzitic sandstones; later Cambrian through Devonian rocks are chiefly shallow-marine carbonate rocks. Formation names of rocks deposited during this phase are shown on the correlation tables, Figures 15, 19, 26, and 31; thickness patterns are shown on Figures 14, 18, 24, 29, and 30. The Late Proterozoic thickness pattern differs from early Paleozoic patterns by having a basin of thick deposition, represented by the Uinta Mountain Group, that trends eastward across northeastern Utah; the Cambrian-Devonian miogeoclinal hingeline (Wasatch line of Kay, 1951) cuts across the west end of the Uinta Mountain trough and, during Cambrian and later time, the Uinta Mountain area behaved as part of the craton. In plate-tectonic parlance, the Late Proterozoic Uinta Mountain trough has been called an aulacogen, the term used for a narrow, elongate basin that extends into the craton from the miogeocline. It is thought to be a tectonic trough bounded by convergent normal faults.

The plate-tectonic explanation for the dramatic change from Phase I to Phase II is perhaps related to development of a north-south rift in central Nevada. Presumably the Precambrian crystalline basement shown on Figures 117 and 112 once extended farther to the west, but the western portion (as yet not identified as part of another continent) broke away and drifted off, leaving western North America a thinner and hotter raw trailing edge for a continental margin. Heat loss extending inland to the hingeline caused the western 100-mile-wide strip of residual North America to founder slowly; shallow-marine deposits accumulated across this sinking platform and formed the miogeocline we see today (Figures 117 and 112). Age of the rifting is in dispute; some prefer an early Late Proterozoic age; others argue for a late Late Proterozoic date. Burchfiel and Davis (1975) and Devlin and Bond (1988) suggested that two periods of rifting may be involved.

While the miogeocline developed, the North American continent moved slowly with respect to the poles, equator and other continents. Figure 28 shows that Utah drifted from a position slightly north of the equator in Cambrian time, to a position a bit south of the equator in Devonian time.

The mid-Paleozoic Antler orogeny that affected central

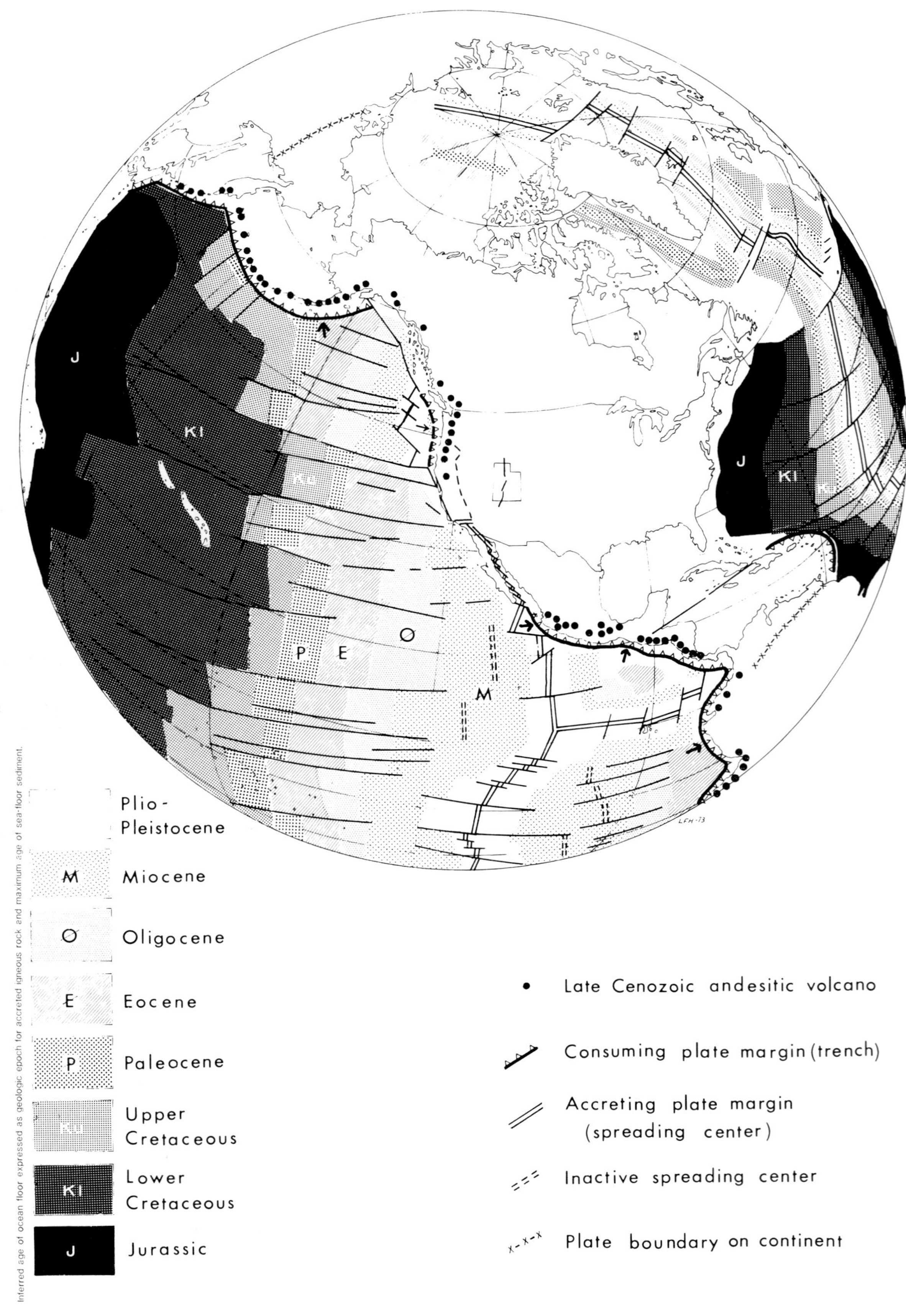

FIGURE 116 — Plate tectonic map of the western hemisphere showing Utah's position relative to spreading centers, subduction zones, and transform faults, with inferred age of sea floor rocks. See Drummond (1981).

Nevada is reflected in Utah in two ways. In the Stansbury Mountains of north-central Utah, the Stansbury uplift produced a regional unconformity and local coarse sedimentation. In western Utah the Pilot Shale is made of fine-grained detritus that represents the distal part of the Antler clastic wedge. Both the Antler and Stansbury disturbances have been ascribed to either accelerated plate convergence or to collision of western Nevada with a Pacific microplate.

Phase III—Oquirrh-Paradox Basin phase. Utah lies at the northwest end of a chain of linked basin-uplifts called the "Ancestral Rockies" that extend 1200 miles southeastward to Oklahoma. The Paradox Basin of the Four Corners area is genetically linked with the Uncompahgre paleo-highland, from which the basin received its arkosic sediments. The Oquirrh Basin of northwestern Utah (Figures 35, 39, and 43) is filled, for the most part, with fine-grained, far-traveled siliceous sand, derived from distant sources as well as the local source, the Northwest Utah High, that flanks it.

Uplift and depositional basin patterns of the Oquirrh-Paradox Basin phase in Utah's history are very much like those that formed in the Salton trough in the last 5 m.y. as a result of transform movement along the San Andreas fault (Lachenbruch, Sass, and Galanis, 1985). Stone and Stevens (1988) suggested that major southwesterly truncation of western North America occurred during Pennsylvanian time along a sinistral transform fault zone that paralleled the Ancestral Rockies trends. The coincidence in timing and direction of west coast truncation and Ancestral Rockies mid-continental rifting suggests that they are related. Plate-tectonic movements which caused the Ancestral Rockies uplifts and basins have not yet been unequivocally identified. The suggestion by Kluth and Coney (1981) and Kluth (1986) that the Ancestral Rockies were produced by continental collision between Africa, South America, and the southeast coast of North America is based on paleogeographic reconstructions (e.g. Babbach, Scotese, and Ziegler, 1980) that show Africa nuzzling up against North America in Pennsylvanian time. Other reconstructions (e.g. Smith, Hurley, and Briden, 1981, from which Figure 28 was derived) do not show Africa in this position. However, northward thrusting of this age formed the Ouachita Mountains fold and thrust belt in Oklahoma-Arkansas. The fault between the Uncompahgre Highland and the Paradox Basin is known to be a high-angle reverse fault; it may also have a lateral slip component. Post-Pennsylvanian salt anticlines in the Paradox Basin are aligned parallel to the Uncompahgre fault; Baars (1983) and Stevenson and Baars (1986) suggested that these and other alignments are controlled by basement fracture zones. Development of such deeply subsiding cratonic basins as the Oquirrh and the Paradox surely requires an extensional component. Whether this

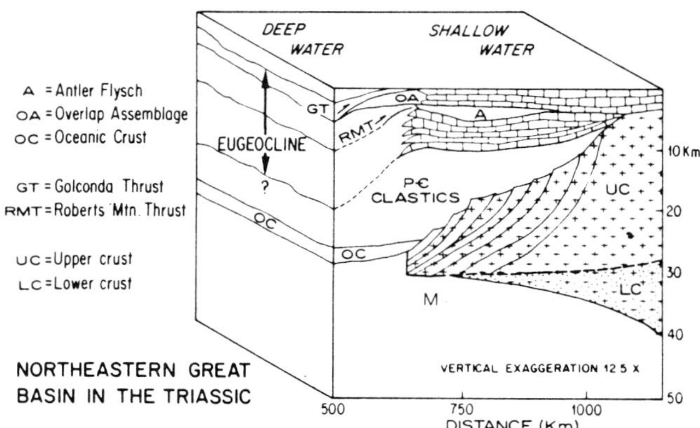

FIGURE 117 — Interpretive pre-Jurassic cross-section of the northern Great Basin. A, clastic rocks of the Antler orogeny; M, mantle. Compare with Figure 112. After Farmer and DePaolo (1983).

is tied to an important strike-slip displacement has not been documented to date.

Final plate-tectonic explanation of the Oquirrh-Paradox Basin phenomena awaits determination of two important factors: first, what large continent or continental fragment was in a position in southern Texas or the Gulf of Mexico to exert important extensional, compressive, or shear forces on North America at this time?; and second, is there an important strike-slip component to the faults that bound the Oquirrh and Paradox basins and other elements of the Ancestral Rockies system? Effects of the Sonoma orogeny that affected northwestern Nevada during late Permian and early Triassic time have not been identified in Utah.

Phase IV—Interior basin and Nevadan orogeny phase. This phase, lasting from the beginning of the Triassic into the early part of Cretaceous time, about 150 million years, is a catch-all of some different crustal behaviors in Utah; its interpretation is made the more difficult by the eradication of much of its record in western Utah by later events. Initially, Early Triassic shallow-marine deposits spread across the state, thickening westward in a fashion that mimics the miogeoclinal pattern of phase II. After a Middle Triassic hiatus, Late Triassic continental strata that contain a fair component of volcanic ash were deposited east of Utah's hingeline; how far west of the hingeline they may once have extended is not known. Early Jurassic sand deposits also thicken westward; in this case, the thickening is to the southwest rather than the northwest as in the case of Early Triassic deposits. Does

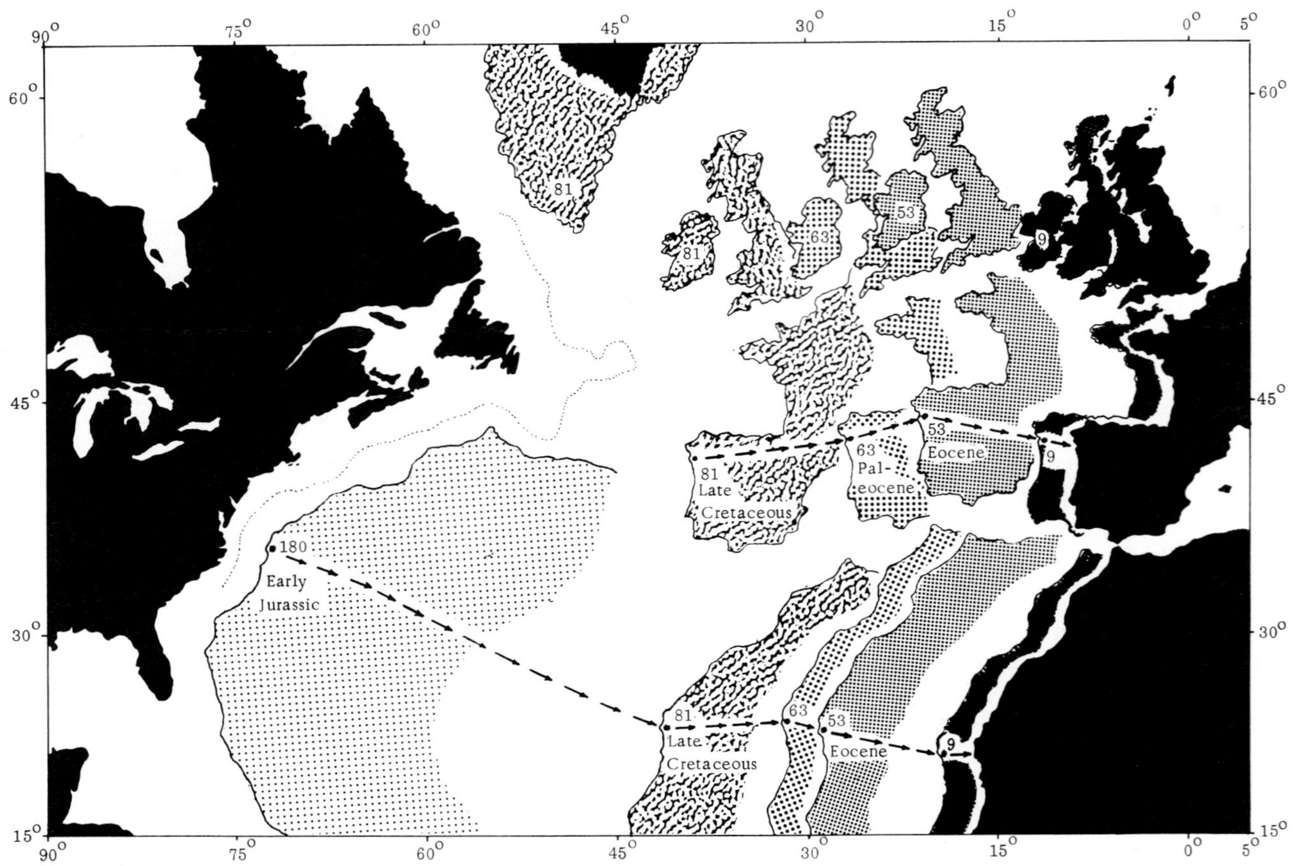

FIGURE 118 — Relative position of Europe and Africa with respect to North America showing changes in direction at 81, 63, 53, and 9 million years ago. After Pitman and Talwani (1972).

this mean that the miogeoclinal behavior may have been operative in western Utah since Late Proterozoic time, temporarily interferred with by local instabilities during phase III when local shelves and basins took the place of (or were superimposed on) the less complicated miogeoclinal pattern and partially resuming the old miogeoclinal pattern in early Mesozoic time? At about this time the Atlantic Ocean began to form as North America moved away from Africa (Figure 118).

Middle and Late Jurassic deposits (Figure 36) show a slightly different pattern. Again, some uncertainty exists about how far west Middle Jurassic deposits originally extended, but there is considerable certainty that in the 160–140 m.y. interval (Late Jurassic) no deposition was going on in northwestern Utah because of tectonism and plutonism related to the Nevadan orogeny. Nevadan orogeny is a term that has never been widely applied. Armstrong (1968a) showed it as overlapping in time with his newly named Sevier orogeny (Figure 119). Schweichert et al. (1984) discussed its effects in the Sierra Nevada. Their abstract notes "that the Nevadan orogeny involved underthrusting of island-arc rocks on the west and significant crustal shortening in the central and eastern belts (of the Sierra Nevada). These features suggest that the Nevadan orogeny resulted from collision of an island arc (western belt) with an andean-type arc (eastern belt) situated at the western edge of North America."

The North American continent gained much of what now constitutes western Nevada, California, Oregon, Washington, British Columbia, and Alaska by a process called terrane accretion, wherein island arcs or small fragments of continental or oceanic crust rafted along on Pacific Ocean plates that impinged against the west coast and became sutured onto North America. Jones et al. (1982) gave a popular account of the process; Silberling et al. (1987) presented a map showing the distribution of terranes accreted to the western U.S. The movement is, in part, collisional; another important component is lateral transfer of terrane material along strike-slip faults like the modern San Andreas fault. The accretion and strike-slip process has been going on for the past 200 m.y., having started at about the same time the North American plate began its rapid westward movement away from Africa, opening up the Atlantic Ocean. Utah, of course, lies too far inland to have been directly involved in the accretion process, but the effects of west coast plate interactions extended into Utah in the form of tectonic and plutonic orogenic events.

The events we are considering in Utah now lie more than 300 miles east of the Sierra Nevada. For simplicity, we can propose that Cretaceous compressional effects and Cenozoic extensional effects compensate each other so that the Jurassic distance across Nevada may have been about the same as at present. Miller et al. (1987) documented Late Jurassic plutonism and metamorphism in northwestern Utah and adjacent Nevada and Idaho. Allmendinger and Jordan (1982) traced a regional detachment of Jurassic age within the same area and speculated that the detachment (localized in Mississippian Chainman Shale) may be the oldest and structurally highest thrust of the Utah-Idaho-Wyoming thrust belt. Core complexes and two-mica granites that reflect Jurassic and perhaps Cretaceous events (Farmer and DePaolo, 1983) lie in this same area. Because the timing is about the same, and because the compressive action may be a distant effect of collision in the Sierra Nevada area, it seems useful to again extend the term Nevadan orogeny to apply to late Jurassic tectonism and plutonism in northwestern Utah. At this same time, Upper Jurassic Morrison Formation continental deposits, including many bentonitic ashfall layers, accumulated in a vast interior basin that extended from Montana to Arizona and from central Utah to central Colorado. Morrison deposits are generally only a few hundred feet thick in this area, and their wide extent bespeaks stability for this basin in the cratonic interior. At least part of the cobbles in Morrison conglomerates have been identified as having highland sources in western Utah and in southern Arizona.

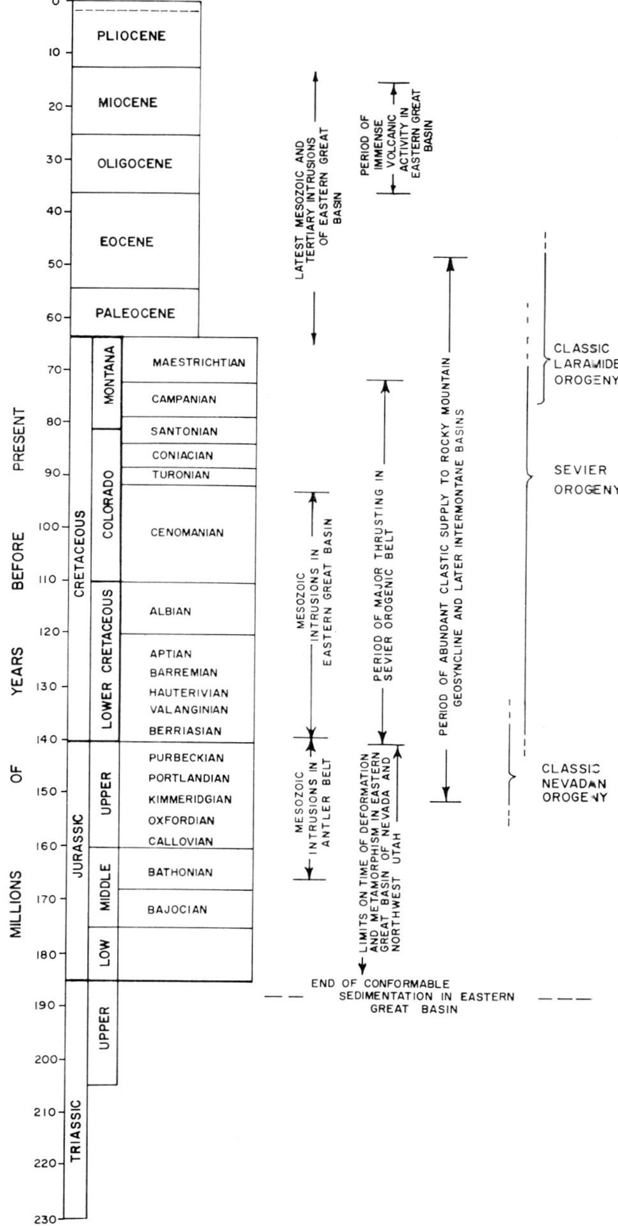

FIGURE 119 — Dating of orogenic events in Utah and Nevada. From Armstrong (1968a).

Steiner (1983) derived a polar wander track from Triassic and Jurassic redbeds in the Colorado Plateau. She noted only a small amount of polar wander from Late Triassic through Middle Jurassic time. But poles from the Morrison Formation indicate rapid motion for the North American plate in the Late Jurassic.

Phase V—Sevier compressional orogeny and foreland basin phase. Because of its profound effects, both structurally and depositionally, this phase of Utah's geologic history has been studied from many angles by many geologists. Cretaceous deposits in the foreland basin of eastern Utah form the chief time control for structural events in western Utah where thrusting and erosion were concurrent. Little plutonism of Cretaceous age has been identified in the Sevier orogenic hinterland of western Utah. Eastward transport of Paleozoic strata during Sevier thrusting was apparently a thin-skinned phenomenon with minor basement involvement. Royce et al. (1975) and Dixon (1982) summarized thrust relationships in the Utah-Idaho-Wyoming thrust belt showing that thrusts are progressively younger to the east. Lawton (1985) outlined the sequence of thrust events in central Utah. Subsidence of the clastic-filled Late Cretaceous foreland basin has been explained by Stockmal et al. (1986) as the flexural response of a lithosphere that was loaded by multiple thrust sheets.

Tracing separate thrusts in Utah's thrust belt becomes progressively more difficult the farther south you go because of increasing interference from later volcanic cover and basin-range faults and erosion. Actual thrusting probably extends no farther east than Cedar City, but broad synclines produced by compressive forces extend east to Kanab. From west to east these include the following broad, gently north-northeastward-plunging synclines: Shivwits syncline of the Beaver Dam Mountains, Pine Valley syncline, Kolob syncline, Paunsagunt syncline, and Straight Cliffs syncline. These broad downfolds adjoin one another along faults whose late Cenozoic movement has been down-to-the-west normal faulting. The westernmost faults, Gunlock and Hurricane, are associated with earlier narrow anticlinal folds that are probably of late Cretaceous age. The alignment of the axes of these broad synclines parallels the trend of the Sevier thrust belt.

There have been differing ideas as to when the Sevier orogeny started. Armstrong (1968a, 1974), who gave the orogeny its name, thought that it began possibly as early as latest Jurassic. But the major influx of western-sourced clastics into the foreland basin in eastern Utah started with the Cedar Mountain Formation about 100 m.y. ago; most current workers regard the Sevier orogeny as beginning then, in late Albian time, and continuing active until the late Campanian, about 75 m.y. ago, when the uppermost beds of the Mesaverde Group, the Price River Formation recorded the waning stages of deposition related to Sevier thrusting. Clastics in the overlying Maastrichtian-Paleocene North Horn Formation were derived from Laramide uplifts (Lawton, 1986).

The plate-tectonic setting of western North America during the Sevier orogeny has been compared to that of the modern Andes, with eastward subduction beneath the continent (Burchfiel and Davis, 1975; Hamilton, 1978). The Sierra Nevada batholith was emplaced 140–60 m.y. ago and other small Cretaceous plutons are scattered across Nevada almost to the Utah border. Jordan et al. (1983) described the varying relationship of structures in the eastern Andes with the angle of descent of the subducting plate on the west coast. They showed that tectonic segmentation of the Andes corresponds geographically with segmentation of the descending plate, which has some segments that are almost horizontal, and others with an eastward descent angle of nearly 30 degrees. Jordan et al. (1983, p. 355) compared structures in the Sevier orogenic belt in Utah to Andean structures that formed above a plate descending at 30 degree dip. In a recent effort to model the movement or traces of terranes across the north Pacific Basin, using the Hawaiian and Yellowstone hotspots as reference points, Debiche et al. (1987) noted that motion of plates relative to hotspots is well determined from 80 m.y. ago to the present, but that earlier Cretaceous pathways were only moderately well constrained, and that Jurassic motion is speculative. The basic difficulty is that, for those earlier plates, the oceanic crust record has been entirely consumed, and there are alternative models for the reconstruction of their relationships.

Phase VI—Laramide uplifts-Uinta Basin phase. The term "Laramide orogeny" was formerly used to include all Mesozoic deformation in Utah until Armstrong (1968a) restricted it by defining the Sevier orogeny to embrace the structural style where miogeoclinal sediments were deformed by low angle thrusts and folds in the Cretaceous hinterland of western Utah. Armstrong (1974) noted that the distinction between Sevier and Laramide structures is possible only in an east-west segment across the western U.S. between central Arizona and central Montana; northward into Canada and in southern Arizona and Mexico no such distinction can be made: there, Laramide is still the term that embraces all late Mesozoic-earliest Cenozoic deformation.

Uplift of the largest Laramide structure in Utah, the Uinta Mountains, is dated by sediments shed from it into the adjacent Uinta and Green River basins. The latest Cretaceous (Maastrichtian) North Horn Formation records the initial uplift, about 70 m.y. ago; Eocene deposits in the Uinta Basin record the end of the active rise.

Characteristically, Laramide uplifts are elongate asymmetrical anticlines bounded by a thrust or reverse fault on the steep side. COCORP deep-seismic data across the Wind River Mountains in Wyoming surprised geologists by showing that the Precambrian core of the range had been pushed about 15 miles out over adjoining basin sediments, and was obviously a compressive structure. Laramide-type deformation involved the whole crust, including the basement, in compression, buckling, and shear; it is a much less ductile type of deformation than is found in the Sevier hinterland, and the magnitude of crustal shortening is much less than either Sevier shortening or the late Mesozoic (Laramide-*sensu latu*) shortening that occurred at the same time in Canada and southern Arizona.

One of the most unusual aspects of the Laramide orogeny is how far inland it is from the active plate-margin. Several authors have proposed models to account for this. Jordan et al. (1983, p. 354) compared Laramide uplifts with late Cenozoic basement uplifts in the Andes in the region beneath which the descending subducting plate is nearly flat. Dickinson and Snyder (1978) had earlier proposed the same idea (Figure 120) and suggested that the gently descending slab of oceanic lithosphere scrapes along beneath the overriding continental lithospheric plate, within which local block uplifts, bounded by reverse faults, are produced. Magmatism is suppressed because the "fertile" asthenosphere is never penetrated by the descending slab. This inference is supported by the correlation between a prominent magmatic null in California and Nevada during the time Laramide uplifts were forming in the Rocky Mountain states. Chapin and Cather (1983) suggested that Laramide events were influenced not only by the shallow dip of the subducting Pacific plate, but also by changes in the direc-

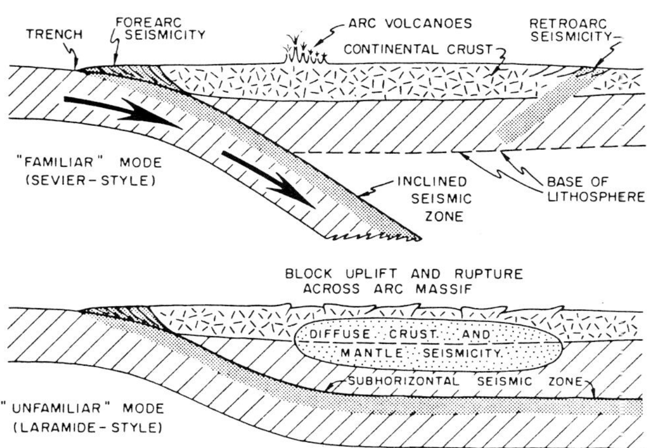

FIGURE 120 — Two modes of subduction that produce differing surface effects. From Dickinson and Snyder (1978).

tion and speed of the North American plate, as relationships in the opening North Atlantic basin changed. Hamilton (1981) put forth the idea that Laramide compressive deformation was caused by clockwise rotation of about 2 or 4 degrees of the Colorado Plateau region relative to the continental interior. This rotation absorbed a small part of the convergence between the North American plate and the Pacific plate that was being subducted beneath it.

It is, perhaps, easier to be satisfied with one of the above explanations for the Laramide uplifts than it is to visualize how the plate-tectonic mechanism is related to the associated basins like the Uinta Basin. Basin subsidence here has been explained (Hagen et al., 1985) by tectonic loading from the basement-involved uplifts. Other, less prominent, basins and uplifts are scattered across the Colorado Plateau portion of Utah. Smaller structures that are commonly thought of as Laramide include the San Rafael Swell, and the Monument and Circle Cliffs uplifts and their intervening basins. Although the Uinta Mountains are known to have a major thrust along their north flank, paralleled by lesser faults along the south flank, recent COCORP data show no fault associated with the northern portion of the San Rafael Swell (Allmendinger et al., 1986).

Phase VII—Explosive caldera and stratovolcano phase. Igneous activity moved into northern Utah about 40 m.y. ago (Figure 75); it had migrated southward from the Canadian border between 54 and 44 m.y. ago (Armstrong, 1978). It stayed in the northern half of the state initially, then in the period between 30 and 20 m.y. ago, it blossomed across the entire southern half of the state, extending eastward into Colorado and westward across Nevada. (Figures 79, 80; Steward and Carlson, 1978, plate 11-1). Given the fact that the general relationship along the west coast of North America was one of either subduction or transform movement (Figure 121), and given that in the present world there is a close association of active volcanism with consuming plate boundaries (Figure 75), it is not too surprising to find migrating igneous activity during phase VII as west coast plate interactions continually changed.

Two aspects of phase VII igneous activity stand out because they have no modern plate-tectonic counterpart: first, the unusual distance that igneous activity extended inland from the continental plate boundary in California; second, the incredible volume of ash-flow tuffs produced from explosive calderas during the maximum eruptive period between 30 and 20 m.y. ago.

Stewart and Carlson (1978), Stewart (1983), and Best et al. (1988) presented maps showing the belt of 30–20 m.y.

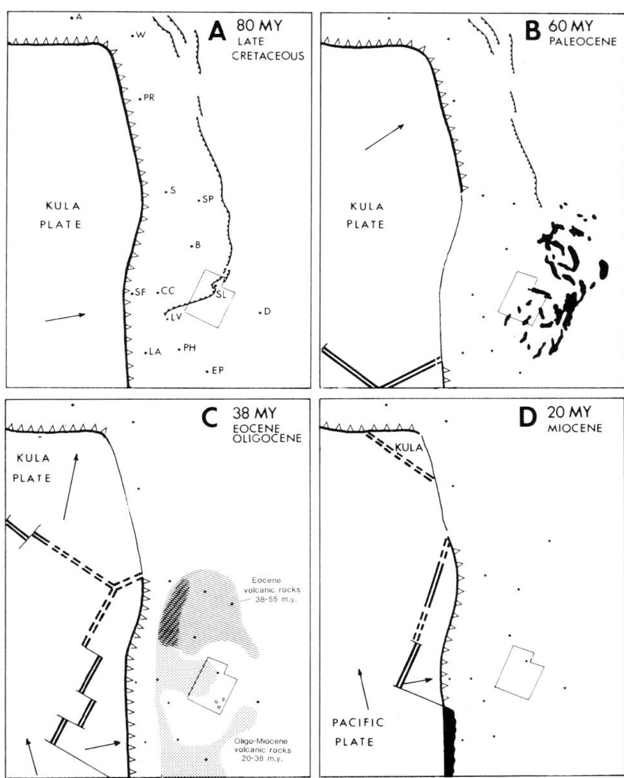

FIGURE 121 — Reconstructions of West Coast plate relations at 80, 60, 38, and 20 million years ago showing tectonic and volcanic activity of the same ages. After Atwater (1970) and Engebretson et al (1985).

igneous rock extending from the Sierra Nevada, across Nevada and Utah, into the San Juan Mountains of western Colorado. The distance is more than 800 miles (1300 km), nearly perpendicular to the west coast plate boundary. The modern Andes are often used as the prime example of structural and igneous activity related to a convergent plate boundary; nothing along the length of the Andes compares in width to this east-west zone of volcanism of late Oligocene and early Miocene age in the western U.S.

In the modern plate-tectonic world, where basalt flows and andesitic stratovolcanoes are commonplace, dacite and rhyolite ash-flow tuffs are unusual. Yet in the 30–20 m.y. period in this sector of the western U.S. they are the predominant rock type. Their volume, produced during the 10 million year span when calderas had their hey-day, was enormous; no modern counterpart of similar volume exists. They exceed, in volume, all of the igneous rocks formed in Utah, Nevada, and Colorado during all the rest of Cenozoic time; they were, in fact, an amazing outburst of east-west igneous activity.

What is the plate-tectonic explanation for this? At present we do not know. Christiansen and McKee (1978) reviewed plate-tectonic concepts that have been proposed for late Cenozoic volcanic and tectonic evolution of the Great Basin; many of their perceptions are pertinent to models of middle Cenozoic activity as well. Engebretson et al. (1985) acknowledged that as they traced west-coast plate relationships backwards through the Cenozoic specific interactions became less certain; they did not attempt to show the location of the subduction zone, nor did they attempt to restore the continental U.S. to its pre-basin-range configuration. Hence, they did not show specific plate relationships at the time of maximum activity 30–20 m.y. ago. Armstrong (1978, p. 278) expressed the situation as follows:

"Farther back in time than the Columbia event (13–16 m.y. ago), the tectonic situation remains obscure. I . . . am skeptical of all models that have been offered. There is still something missing—a lack of critical data or inadequate or false conceptions prevent the reconciliation of observed history with actualistic plate models. The salient features that are most puzzling are the broad extent of volcanic fields perpendicular to the continental margin and their irregular, but well documented, southward time transgression. The challenge of more work and thinking remains."

Phase VIII—Regional uplift, basin-range faulting, bimodal volcanism, and Lake Bonneville phase. The phenomena listed in the foregoing title are, along with the unlisted but implied phenomenon of regional extension (Figure 123), a by-product of an as-yet-unidentified source of high heat-flow; Figure 115 shows that the elevated western U.S., from the High Plains to the Sierra Nevada and Cascade Range, is an area where heat-flow values exceed the norm of 1.5 HFU. In the area of our particular concern, Utah, the Colorado Plateau is cooler than average, and the Great Basin hotter. To address the areas where most of the active tectonism has occurred within the last 15 m.y. we must direct our attention to the Great Basin. Eaton et al. (1978) persuasively promoted the significance of bilateral symmetry of geographic and gravitative features of the Great Basin (see Figure 114). One can see similar bilateral symmetry in the heat flow pattern across the Great Basin (Figure 115), as well as in the pattern of the COCORP reflection MOHO (Figure 112), and as illustrated in the distribution of late Cenozoic basalt: basaltic volcanism is concentrated on both the eastern and western sides of the southern margins of the Great Basin in Utah-Arizona, and in California-Nevada (Stewart and Carlson, 1978, plate 11-12). All of these features combine to show a continental crust that is generally thin under the Great Basin and even thinner along its margins where basalt from the mantle has erupted.

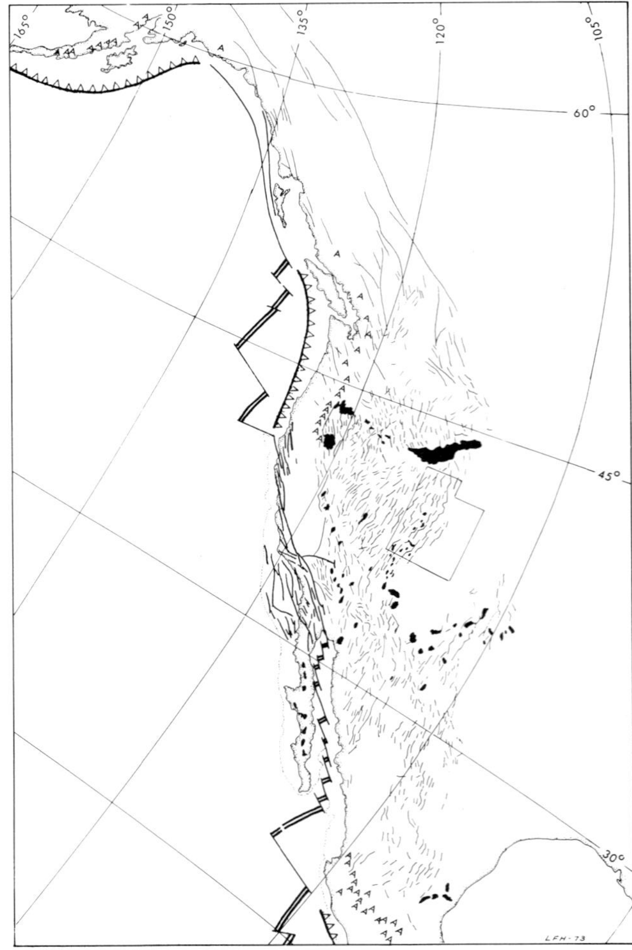

FIGURE 122 — Present-day West Coast plate relations, andesitic volcanoes (A), late Cenozoic basalts (black) and basin-range faults.

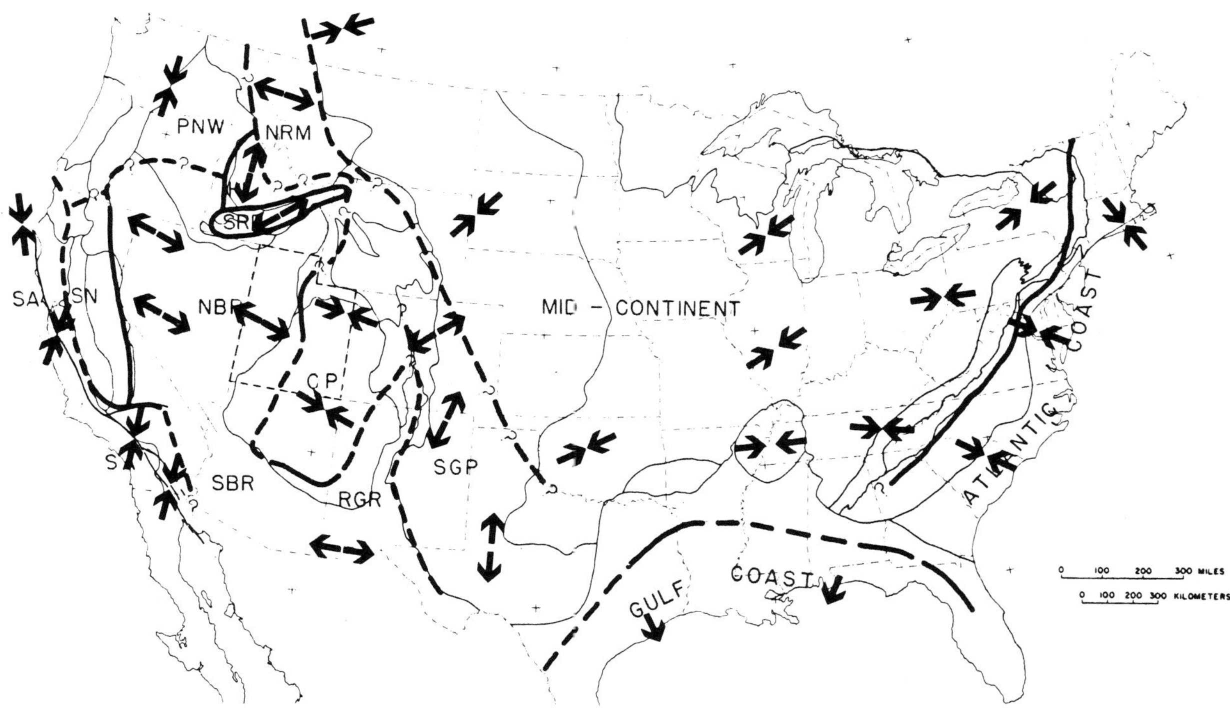

FIGURE 123 — Present day stress map showing extensional and compressional fields. From Zoback and Zoback (1980).

What sort of plate-tectonic model best explains the Late Cenozoic development of western North America to produce these features? Christiansen and McKee (1978, p. 300) reviewed various models relating to interactions of the North American plate with various plates on its Pacific margin, and models based on active upwelling of mantle beneath the Great Basin. They rejected the mantle upwelling models and suggested a model based on interaction of the North American plate with changing plate relationship in the Pacific basin. Their model has not gained wide acceptance because it is incompatible with the bilateral symmetry which requires a dynamic process, not a "one-sided" process. Later models have been tendered (Livaccari, 1979; Barrash and Venkatakriishnan, 1982), but in fact, no model has yet been proposed that has the support of the majority of geologists working on this complex problem. In plate-tectonic terms the western Cordillera of the United States is unique. Except for the Tibetan Plateau produced by the collision of India and Asia, no other area on Earth has such a broadly uplifted active area, particularly one that is apparently related to interaction between an oceanic plate and a continental plate (Figure 122). Many broad continent-continent collision mountain belts are active, but along the entire western margins of the American plates, both North and South, the Cordillera of the western U.S. is the broadest. Surely it is, among other properties, a manifestation of unusual heat transfer from the mantle; but the niceties of why this is located where we find it still eludes a general consensus.

FIGURE 124 — "Temples and Towers of the Virgin" at the east entrance to Zion National Park as seen from the south. Lithograph by W.H. Holmes in Dutton (1882).

ACKNOWLEDGMENTS

I thank many geologists who gave of their time and expertise to assist in bringing this book to its present state. Helpful friends with the U.S. Geological Survey include Bruce Bryant, Don Elston, Wally Hansen, Keith Ketner, Dave Lindsey, Marge MacLachlan, Dave Miller, "K" Molenaar, Hal Morris, Doug Nichols, Steve Oriel, Robert O'Sullivan, Fred Peterson, Barney Poole, Pete Rowley, Charlie Sandberg, Bill Sando, Betty Skipp, Jack Stewart, and Wit Witkind. Geologists with the Utah Geological and Mineral Survey who provided stratigraphic information and exchange of ideas are Genevieve Atwood, Hellmut Doelling, Don Mabey, Mary Siders, Doug Sprinkel, and Grant Willis. Petroleum and consulting geologists who gave me the benefit of their knowledge include Don Baars, Frank Royce, Tom Ryer, Joel Schneyer and John Welsh. Geologists who teach at universities across the country, but whose heart and research interests lie, at least partially, in Utah, and who provided useful input to this effort are Rick Allmendinger, Walt Arabasz, Dick Armstrong, Ron Blakey, Nick Christie-Blick, Dave Clark, Don Currey, Clyde Hardy, Peter Isaacson, Terry Jordan, Tim Lawton, Paul Link, Jack Oviatt, Lucian Platt, Scott Ritter, Richard Robison, Dave Rodgers, Rube Ross, Bert Rowell, Bob Smith, and Mac Weiss. Pete Palmer with the Geological Society of America also fits this group. My colleagues in the Department of Geology at Brigham Young University who have been particularly supportive of this project are Jim Baer, Myron Best, Eric Christiansen, Ken Hamblin, Bart Kowallis, Wade Miller and Keith Rigby. Stan Welch, a geologically-inclined botanist at BYU, provided an additional perspective. My daughter-in-law, Catherine, cheerfully and carefully lettered the interminable charts and correlation tables.

The initial impetus for this book came from teaching classes in Historical Geology and Field Geology during the last three decades at Brigham Young University. The logistical support of the university in producing this book is much appreciated.

Reader input for future revisions of this book would be much appreciated. If you note errors, or have suggestions for improvements in the presentation, please write to me at the address shown in the front of the book.

Lehi Hintze
Provo, Utah
May 1988

FIGURE 125 — West Temple, Zion National Park. Compare with Figure 124.

PHOTO CREDITS

James L. Baer—page i
W. Kenneth Hamblin—Figures 56, 65, 87, 148, 157, 161.
Janice Higgins—Figure 139.
Lehi F. Hintze—Figures 20, 22, 27, 52, 61, 63, 70, 71, 82, 83, 99, 126, 129, 138, 140, 141, 143, 144, 145, 146, 147, 150, 152, 154, 155, 156, 162, 163, 164.
Olympus Aerial Surveys Inc.—Figures 132, 134, 153.
Jack Rathbone—Figures 2, 8.
J. Keith Rigby Sr.—Figures 32, 47, 67.
John S. Shelton—Figures 5, 7, 44.
U.S. Department of Agriculture—Figures 100, 158.
U.S Bureau of Reclamation—Figures 6, 38, 45, 102, 137, 149, 165.
U.S. Forest Service—Figures 4, 57, 73, 84, 136, 142.
U.S. National Park Service—Figure 151.
Utah State Historical Society, Hal Rumel Collection—Figures 78, 90, 97, 104, 125, 127, 128, 130, 131, 133, 159.
Utah Travel Council—Figures 62, 77, 166.
Lynn Wakefield—Figures 41, 105, 135.
Woodward Clyde Associates—Figures 9, 94, 95.

REFERENCES CITED

Abbott, J.T., Best, M.G., and Morris, H.T., 1983, Geologic map of the Pine Grove-Blawn Mountain area, Beaver County, Utah: U.S. Geological Survey Miscellaneous Investigations Series Map I-1479.

Abbott, W.O., 1951, Cambrian diabase flow in central Utah: Compass of Sigma Gamma Epsilon, v. 19, no. 1, p. 5–9.

Adams, O.C., 1962, Geology of the Summer Ranch and North Promontory Mountains, Utah: unpublished master's thesis, Utah State University, Logan, 57 p.

Adamson, R.D., Hardy, C.T., and Williams, J.S., 1955, Tertiary rocks of Cache Valley, Utah and Idaho: Utah Geological Society Guidebook 10, p. 1–22.

Affleck, J., and Hunt, C.B., 1980, Magnetic anomalies and structural geology of stocks and laccoliths in the Henry Mountains, Utah: Utah Geological Association Publication 8, p. 107–112.

Ahlborn, R.C., 1977, Mesozoic-Cenozoic structural development of the Kern Mountains, eastern Nevada–western Utah: Brigham Young University Geology Studies v. 24, part 2, p. 117–131.

Akers, R.H., and Davis, L.E., 1985, Evidence for an unconformity based on conodonts in the lower Oquirrh Group, Samaria Mountain, southeastern Idaho: Geological Society of America Abstracts with Program, v. 17, p. 205.

Albers, S.H., 1975, Paleoenvironment of the Upper Triassic-Lower Jurassic Nugget Sandstone near Heber, Utah: University of Utah unpublished master's thesis, 88 p.

Alexander, D.W., 1978, Petrology and petrography of the Bridal Veil Limestone Member of the Oquirrh Formation at Cascade Mountain, Utah: Brigham Young University Geology Studies, v. 25, part 3, p. 11–26.

Allmendinger, R.W., 1983, Geologic map of the north Hansel Mountains, Idaho and Utah: U.S. Geological Survey Miscellaneous Field Studies Map MF-1643.

Allmendinger, R.W., and Jordan, T.E., 1981, Mesozoic evolution, hinterland of the Sevier orogenic belt: Geology, v. 9, p. 303–313, also Comment and Reply, v. 10, p. 442–444.

Allmendinger, R.W., and Jordan, T.E., 1984, Mesozoic structure of the Newfoundland Mountains, Utah: horizontal shortening and subsequent extension in the hinterland of the Sevier belt: Geological Society of America Bulletin, v. 95, p. 1280–1292.

Allmendinger, R.W., and Jordan, T.E., 1988, Geologic map and cross sections of the Newfoundland Mountains, northwest Utah: U.S. Geological Survey Miscellaneous Field Investigations Map (in press).

Allmendinger, R.W., Miller, D.M., and Jordan, T.E., 1984, Known and inferred Mesozoic deformation in the hinterland of the Sevier belt, northwest Utah: Utah Geological Association Publication 13, p. 21–34.

Allmendinger, R.W., and Platt, L.B., 1980, Stratigraphic variation and low-angle faulting in the North Hansel Mountains and Samaria Mountain, southern Idaho: Geological Society of America Memoir 157, p. 149–163.

Allmendinger, R.W. and six others, 1983, Cenozoic and Mesozoic structure of the eastern Basin and Range from COCORP seismic reflection data: Geology, v. 11, p. 532–536

Allmendinger, R.W., and six others, 1986, Phanerozoic tectonics of the Basin and Range-Colorado Plateau transition from COCORP data and geologic data: *in* Barazangi, M., and Brown L., eds., Reflection seismology—the continental crust: American Geo-physical Union Geodynamics Series, v. 14, p. 257–268.

Allmendinger, R.W., and five others, 1987a, Deep seismic reflection characteristics of the continental crust: Geology, v. 15, p. 304–310.

Allmendinger, R.W., and seven others, 1987b, Overview of the COCORP 40 degree north transect, western U.S.—the fabric of an orogenic belt: Geological Society of America Bulletin, v. 98, p. 308–319.

Andersen, D.W., and Picard, M.D., 1972, Stratigraphy of the Duchesne River Formation (Eocene-Oligocene?), northern Uinta Basin, northeastern Utah: Utah Geological and Mineral Survey Bulletin 97, 25 p.

Andersen, D.W., and Picard, M.D., 1974, Evolution of synorogenic clastic deposits in the intermontane Uinta Basin of Utah: *in* Dickenson, W.R., ed., Tectonics and Sedimentation, Society of Economic Paleontologists and Mineralogists Special Publication 22, p. 167–189.

Anderson, J.J., Itivari, T.A., and Rowley, P.D., 1987, Geologic map of the Little Creek Peak quadrangle, Iron and Garfield Counties, Utah: Utah Geological and Mineral Survey Map 104.

Anderson, J.J., and Rowley, P.D., 1975, Cenozoic stratigraphy of southwestern High Plateaus of Utah: Geological Society of America Special Paper 160, p. 1–52.

Anderson, J.J., and Rowley, P.D., 1987, Geologic map of the Panguitch NW quadrangle, Iron and Garfield Counties, Utah: Utah Geological and Mineral Survey Map 103.

Anderson, J.J., Rowley, P.D., Mehnert, H.H., Blackman, J.T., and Grant, T.C., 1986a, Geologic map of the Circleville Canyon area, southern Tushar Mountains and northern Markagunt Plateau, Beaver, Garfield, Iron, and Piute Counties, Utah: U.S. Geological Survey Miscellaneous Field Studies Map (in preparation).

Anderson, J.J., Rowley, P.D., Decatur, S.H., Machette, M.N., and Mehnert, H.H., 1986b, Geologic Map of the Nevershine Hollow area, eastern Black Mountains and southwestern Tushar Mountains, Beaver and Iron Counties, Utah: U.S. Geological Survey Miscellaneous Field Studies Map (in preparation).

Anderson, L.W., and Anderson, D.S., 1981, Weathering rinds on quartzarenite clasts as relative-age indicator and the glacial chronology of Mount Timpanogos, Wasatch Range, Utah: Artic and Alpine Research, v. 13, p. 25–31.

Anderson, L.W., and Miller, D.G., 1979, Quaternary fault map of Utah: Fugro Consulting Engineers and Geologists, distributed by Utah Geological and Mineral Survey, Salt Lake City.

Anderson, R.E., and Barnhard, T.P., 1986, Genetic relationship between faults and folds and determination of Laramide and neotectonic paleostress, western Colorado-transition zone, central Utah: Tectonics, v. 5, p. 335–357.

Anderson, R.E., and Barnhard, T.P., 1987, Neotectonic framework of the central Sevier Valley area, Utah, and its relationship to seismicity: U.S. Geological Survey Open-File Report 87-585, p. F1–134.

Anderson, R.E., Zoback, M.L., and Thompson, G.A., 1983, Implications of selected subsurface data on the structural form and evolution of some basins in the northern Basin and Range province, Nevada and Utah: Geological Society of America Bulletin, v. 94, p. 1055–1072.

Anderson, W.L., 1960, Stratigraphic section of the northern Silver Island Mountains: Utah Geological Society Guidebook 15, p. 114–115.

Arabasz, W.J., and Julander, D.R., 1986, Geometry of seismically active faults and crustal deformation within the Basin and Range-Colorado Plateau transition in Utah: Geological Society of America Special Paper 208, p. 43–74.

Arabasz, W.J., Pechmann, J.C., and Brown, E.D., 1987, Observational seismology and the evaluation of earthquake hazards and risk in the Wasatch Front area, Utah: U.S. Geological Survey Open-File Report 87-585, p. D1–39.

Armin, R.A., and Mayer, L., 1983, Subsidence analysis of the Cordilleran miogeocline-implications for timing of late Proterozoic rifting and amount of extension: Geology, v. 11, p. 702–705.

Armin, R.A., and Moore, W.J., 1981, Geology of the southeastern Stansbury Mountains and southern Onaqui Mountains, Tooele County, Utah: U.S. Geological Survey Open-File Report 81-0247.

Armstrong, F.C., 1969, Geologic map of the Soda Springs quadrangle, southeastern Idaho, U.S. Geological Survey Miscellaneous Geologic Investigations Map I-557.

Armstrong, R.L., 1968a, Sevier orogenic belt in Nevada and Utah: Geological Society of America Bulletin, v. 19, p. 429–58.

Armstrong, R.L., 1968b, Mantled gneiss domes in the Albion Range, southern Idaho: Geological Society of America Bulletin, v. 79, p. 1295–1314. See also v. 81, p. 909–910, 1970.

Armstrong, R.L., 1968c, The Cordilleran Miogeosyncline in Nevada and Utah: Utah Geological and Mineralogical Survey Bulletin, 78, 58 p.

Armstrong, R.L., 1969, K-Ar dating of laccolithic centers of the Colorado Plateau and vicinity: Geological Society of America Bulletin, v. 80, no. 10, p. 2081–86.

Armstrong, R.L., 1970a, Geochronology of Tertiary igneous rocks, eastern Basin and Range Province, western Utah eastern Nevada and vicinity, U.S.A.: Geochimica et Cosmochimica Acta, v. 34, p. 205–232.

Armstrong, R.L., 1970b, Mantled gneiss domes in the Albion Range, southern Idaho: a revision: Geological Society of America Bulletin,v. 81, no. 3, p. 909–10.

Armstrong, R.L., 1972, Low-angle (denudation) faults, hinterland of the Sevier Orogenic Belt, eastern Nevada and western Utah: Geological Society of America Bulletin, v. 83, p. 1729–54.

Armstrong, R.L., 1974, Magmatism, orogenic timing, and orogenic diachronism in the Cordillera from Mexico to Canada: Nature, v. 247, p. 348–351.

Armstrong, R.L., 1978, Cenozoic igneous history of the U.S. Cordillera from Lat 42 degrees to 49 degrees North: Geological Society of America Memoir 152, p. 265–282.

Armstrong, R.L., 1982, Cordilleran metamorphic core complexes, from Arizona to southern Canada: Annual Reviews Earth and Planetary Sciences, v. 10, p. 129–154.

Armstrong, R.L., Ekren, E.B., McKee, E.H., and Noble, D.C., 1969, Space-time relations of Cenozoic silicic volcanism in the Great Basin of the western United States: American Journal of Science, v. 267, p. 478–490.

Armstrong, R.L., and Higgins, R.E., 1973, K-Ar dating of the beginning of Tertiary volcanism in the Mojave Desert, California: Geological Society of America Bulletin, v. 84, p. 1095–1100.

Armstrong, R.L., Leeman, W.P., and Malde, H.E., 1975, K-Ar dating, Quaternary and Neogene volcanic rocks of the Snake River Plain, Idaho: American Journal of Science, v. 175, p. 225–251.

Armstrong, R.L., and Suppe, John, 1973, Potassium-argon geochronology of Mesozoic igneous rocks in Nevada, Utah, and southern California: Geological Society of America Bulletin, v. 84, p. 1375–92.

Arnow, Ted, and Mattick, R.E., 1968, Thickness of valley fill in the Jordan Valley east of the Great Salt Lake, Utah: U.S. Geological Survey Professional Paper 600-B, p. 79–82.

Ash, S.R., 1975, Chinle (Upper Triassic) flora of southeastern Utah: Four Corners Geological Society Guidebook 8, p. 143–147.

Atwater, Tanya, 1970, Implications of plate tectonics for the Cenozoic tectonic evolution of western North America: Geological Society of America Bulletin, v. 81, p. 3513–36.

Atwood, W.W., 1909, Glaciation of the Uinta and Wasatch Mountains: U.S. Geological Survey Professional Paper 61, 96 p.

Auby, W.L., 1986, Geology of the Levan 7½-minute quadrangle, Utah: Utah Geological and Mineral Survey Map (in preparation)

Averitt, Paul, 1962, Geology and coal resources of the Cedar Mountain quadrangle, Iron County, Utah: U.S. Geological Survey Professional Paper 389, 72 p.

Averitt, Paul, 1967, Geologic map of the Kanarraville quadrangle, Iron County, Utah: U.S. Geological Survey Geologic Quadrangle Map GQ-694.

Averitt, Paul, and Threet, R.L., 1973, Geologic map of the Cedar City quadrangle, Iron County, Utah: U.S. Geological Survey Geologic Quadrangle Map GQ-1120.

Axtell, D.C., 1967, Geology of the northern part of the Malad Range, Idaho: unpublished master's thesis, Utah State University, Logan, 71 p.

Baars, D.L., 1958, Cambrian stratigraphy of the Paradox Basin Region: Intermountain Association of Petroleum Geologists Guidebook 9, p. 93–101.

Baars, D.L., 1962, Permian System of Colorado Plateau: American Association of Petroleum Geologists Bulletin, v. 46, p. 149–218.

Baars, D.L., 1966, Pre-Pennsylvanian paleotectonics, Paradox Basin: American Association of Petroleum Geologists Bulletin, v. 50, no. 10, p. 2082–111.

Baars, D.L., 1972, Devonian System, *in* Geologic Atlas of the Rocky Mountain Region, p. 90–99. Rocky Mountain Association of Geologists, Denver.

Baars, D.L., 1973a, Permianland—the rocks of Monument Valley: Geology of the Canyons of the San Juan

River, Four Corners Geological Society River Guidebook, p. 27–34. Durango, Colorado.

Baars, D.L., 1973b, Rocks of the inner gorges—the Pennsylvanian System: Geology of the Canyons of the San Juan River, Four Corners Geological Society River Guidebook, p. 16–26. Durango, Colorado.

Baars D.L., 1975, Permian System of Canyonlands Country: Four Corners Geological Society Guidebook 8, p. 123–127.

Baars, D.L., 1979, The Permian System: Four Corners Geological Society Guidebook, 9th Field Conference, Permianland, p. 1–6.

Baars, D.L., 1983, The Colorado Plateau, a geologic history: University of New Mexico Press, 179 p., Albuquerque.

Baars, D.L., and Molenaar, C.M., 1971, Geology of Canyonlands and Cataract Canyon: Four Corners Geological Society Sixth Field Conference Guidebook, 99 p.

Baars, D.L., and Stevenson, G.M., 1979, Permian rocks of the San Juan Basin: Four Corners Geological Society Guidebook 9, Permianland, p. 59.

Bachman, M.E., 1959, Geology of the Water Hollow fault zone, Sevier and Sanpete Counties, Utah: unpublished master's thesis, Ohio State University, 95 p.

Baer, J.L., 1962, Geology of the Star Range, Beaver County, Utah: Brigham Young University Geology Studies, v. 9, part 2, p. 29–52. Summary in Utah Geological Publication 3, p. 33–38, 1973.

Baer, J.L., 1969, Paleoecology of cyclic sediments of the lower Green River Formation, central Utah: Brigham Young University Geology Studies, v. 16, part 1, p. 3–96.

Baer, J.L., 1976, Structural evolution of central Utah-Late Permian to Recent: Rocky Mountain Association of Geologists Symposium, p. 37–46.

Baer, J.L., 1985, Geologic summary of the Crawford Mountains, Rich County, Utah and Lincoln County, Wyoming: Utah Geological Association Publication 14, p. 5–14.

Bagshaw, L.H., 1977, Paleoecology of the lower Carmel Formation of the San Rafael Swell, Emery County, Utah: Brigham Young University Geology Studies, v. 24, part 2, p. 51–62.

Bagshaw, R.L., 1977, Foraminiferal abundance related to bentonitic ash beds in the Tununk Member of the Mancos Shale in southeastern Utah: Brigham Young University Geology Studies, v. 24, part 2, p. 33–49.

Baker, A.A., 1964a, Geology of the Aspen Grove quadrangle, Utah: U.S. Geological Survey Map GQ-239.

Baker, A.A., 1964b, Geology of the Orem quadrangle, Utah: U.S. Geological Survey Map GQ-241.

Baker, A.A., 1972a, Geologic map of the Bridal Veil Falls quadrangle, Utah: U.S. Geological Survey Map GQ-998.

Baker, A.A., 1972b, Geologic map of the northeast part of Spanish Fork Peak quadrangle, Utah: U.S. Geological Survey Open-File Report.

Baker, A.A., 1973, Geologic map of the Springville quadrangle, Utah: U.S. Geological Survey Map GQ-1103.

Baker, A.A., 1976, Geologic map of the west half of the Strawberry Valley quadrangle, Utah: U.S. Geological Survey Miscellaneous Investigations Map I-931.

Baker, A.A., and Crittenden, M.D. Jr., 1961, Geologic map of the Timpanogos Cave quadrangle, Utah: U.S. Geological Survey Map GQ-132.

Baker, C.H. Jr., 1970, Water resources of the Heber-Kamas-Park City area, north-central Utah: Utah Department of Natural Resources Technical Publication 27, 74 p.

Baker, C.H. Jr., 1974, Water resources of the Curlew Valley drainage basin, Utah and Idaho: Utah Department of Natural Resources Technical Publication 45, 83 p.

Bambach, R.K., Scotese, C.R., and Ziegler, A.M., 1980, Before Pangea—The geographies of the Paleozoic world: American Scientist, v. 68, p. 26–48.

Banks, R.L., 1986, Geology of the Fountain Green North 7½-minute quadrangle, Sanpete and Juab Counties, Utah: Utah Geological and Mineral Survey Map (in preparation).

Barrash, W., and Venkatakrishnan, R., 1982, Timing of late Cenozoic volcanic and tectonic events along the western margin of the North American Plate: Geological Society of America Bulletin, v. 93, p. 977–989.

Barton, M.D., 1987, Lithophile-element mineralization associated with Late Cretaceous two-mica granites in the Great Basin: Geology, v. 15, p. 337–340.

Baughman, R.L. 1959, Geology of the Musina graben area, Sevier and Sanpete Counties, Utah: unpublished master's thesis, Ohio State University, 110 p.

Bennett, V.C., and DePaulo, D.J., 1987, Proterozoic crustal history of the western U.S. as determined by neodymium isotopic mapping: Geological Society of America Bulletin, v. 99, p. 674–685.

Berg, R.R., 1981, Review of thrusting in the Wyoming foreland: University of Wyoming Contributions to Geology, v. 19, p. 93–104.

Berggren, W.A., Kent, D.V., Flynn J.J., Van Couvering, J.A., 1985, Cenozoic geochronology: Geological Society of America Bulletin v. 96, p. 1407–1418.

Berggren, W.A., and Van Couvering, J.A., 1979, Quaternary biogeography and biostratigraphy: *in* R.A. Robison and C. Teichert, eds., Treatise on Invertebrate Paleontology, Part A, p. A505–543.

Berry, Lea, 1987, Geology of the Porcupine Reservoir quadrangle, Cache County, Utah: Utah Geological and Mineral Survey Map (in preparation).

Berry, W.B.N., and Boucot, A.J., 1970, Correlation of the North American Silurian rocks: Geological Society of America Special Paper 102, 289 p.

Best M.G., 1987, Geologic map of the area between Hamlin Valley and Escalante Desert, Iron County, Utah: U.S. Geological Survey Map I-1774.

Best, M.G., 1988, Early Miocene change in direction of least principal stress, south western U.S.—conflicting inferences from dikes and metamorphic core-detachment fault terranes: Tectonics, v. 7, p. 249–259.

Best, M.G., Armstrong, R.L., Graustein, W.C., Embree, G.F., and Ahlborn, R.C., 1974, Mica granites of the Kern Mountains pluton, eastern White Pine County, Nevada—remobilized basement of the Cordilleran miogeosyncline?: Geological Society of America Bulletin, v. 85, p. 1277–1286.

Best, M.G., and Brimhall, W.H., 1974, Late Cenozoic alkalic basaltic magmas in the western Colorado Plateaus and the Basin and Range transition zone, U.S.A., and their bearing on mantle dynamics: Geological Society of America Bulletin v. 85, p. 1677–1690.

Best, M.G., Christiansen, E.H., Deino, A.L., Gromme, C.S., McKee, E.H., and Noble, D.C., 1988, Oligo-Miocene volcanism in the Great Basin of Nevada and Utah: New Mexico Bureau of Mines and Mineral Resources (in preparation).

Best, M.G., and Grant, S.K., 1987, Stratigraphy of the volcanic Oligocene Needles Range Group in southwestern Utah: U S. Geological Survey Professional Paper 1433-A, 28 p.

Best, M.G., Grant, S.K., Hintze, L.F., Cleary, J.G., Hutsinpillar, A., and Saunders, D.M., 1987, Geologic map of the Indian Peak (southern Needle) Range, Beaver and Iron Counties, Utah U.S. Geological Survey Map I-1795.

Best, M.G., and Hamblin, W.K., 1970, Implications of tectonism and volcanism in the western Grand Canyon: Utah Geological Society Guidebook 23, p. 75–79.

Best, M.G., and Hamblin, W.K., 1978, Origin of the northern Basin and Range province—implications from the geology of its eastern boundary: Geological Society of America Memoir 152, p. 313–340.

Best, M.G., Hintze, L.F., and Holmes, R.D., 1987, Geologic map of the southern Mountain Home and northern Indian Peak Ranges, Beaver County, Utah: U.S. Geological Survey Map I-1796.

Best, M.G., Lemmon, D.L., and Morris, H.T., 1988, Geologic map of the Shauntie Hills and Star Range, Beaver County, Utah: U.S. Geological Survey Map I-1904.

Best, M.G., McKee E.H., and Damon P.E., 1980, Space-time-composition patterns of late Cenozoic mafic volcanism, south-western Utah and adjoining areas: American Journal of Science, v. 280, p. 1035–1050.

Best, M.G., Mehnert, H.H., Keith, J.D., and Naeser, C.W., 1987, Miocene magmatism and tectonism in and near the southern Wah Wah Mountains, southwestern Utah: U.S. Geological Survey Professional Paper 1433-B, p. 31–47.

Best, M.G., Morris, H.T., Kopf, R.W., and Keith, J.D., 1987, Geologic map of the southern Pine Valley area, Beaver and Iron Counties, Utah, 1:48,000: U.S. Geological Survey Map I-1794.

Beus, S.S., 1963, Geology of the central Bluespring Hills, Utah-Idaho: unpublished Ph.D. thesis, University of California, Los Angeles, 233 p.

Beus, S.S., 1968, Paleozoic stratigraphy of Samaria Mountain, Idaho-Utah: American Association of Petroleum Geologists Bulletin, v. 52, p. 782–808.

Beus, S.S., 1969, Devonian stratigraphy in northwestern Arizona: Four Corners Geological Society Guidebook 5, p. 127–133.

Beus, S.S., 1984, Late Mississippian estuarine rock unit in the Grand Canyon, Arizona: Geological Society of American Abstracts with Programs, v. 16, p. 444.

Bick, K.F., 1966, Geology of the Deep Creek Mountains, Tooele and Juab Counties, utah: Utah Geological and Mineral Survey Bulletin 77, 120 p.

Bickford, M.E., and Cudzilo, T.F., 1975, U-Pb age of zircon from Vernal Mesa-type quartz monzonite, Unaweap Canyon, west-central Colorado: Geological Society of America Bulletin, v. 86, p. 1432–1434.

Biek, R.F., 1986, Geology of the Nephi 7½-minute quadrangle, Juab County, Utah: Utah Geological and Mineral Survey Map (in preparation).

Biggar, N.E., Harden, D.R., and Gillam, M.L., 1981, Quaternary deposits in the Paradox Basin: Rocky Mountain Association of Geologists Guidebook—Geology of the Paradox Basin, p. 129–145.

Biller, E.J., 1976, Stratigraphy and petroleum possibilities of lower Upper Devonian strata, southwestern Utah: U.S. Geological Survey Open-File Report 76-343.

Billingsley, G.H. Jr., 1979, Preliminary report of buried valleys at the Mississippian-Pennsylvanian boundary in western Grand Canyon; American Geological Institute Selected Guidebook Series No. 2, p. 115–117.

Billingsley, G.H. Jr., and Beus, S.S., 1985, The Surprise Canyon Formation—an upper Mississippian and lower Pennsylvanian rock unit in the Grand Canyon, Arizona: U.S. Geological Survey Bulletin 1605A, p. A27–A33.

Bissell, H.J. 1959, Silica in sediments of the upper Paleozoic of the Cordilleras area: *in* Silica in sediments—a symposium: Society of Economic Paleontologists and Mineralogists Special Publication 7, p. 150–185.

Bissell, H.J., 1959, Stratigraphy of the southern Oquirrh Mountains, Pennsylvanian System: Utah Geological Society Guidebook 14, p. 93–127.

Bissell, H.J., 1962, Pennsylvanian-Permian Oquirrh Basin of Utah: Brigham Young University Geology Studies, v. 9, part 1, p. 26–49.

Bissell, H.J., 1963, Lake Bonneville—geology of southern Utah Valley, Utah: U.S. Geological Survey Professional Paper 257-B, p. 101–130.

Bissell, H.J., 1973, Permian-Triassic boundary in the eastern Great Basin area: Canadian Society of Petroleum Geologists Memoir 2, p. 318–344.

Bissell, H.J., and Barker, H.K., 1977, Deep-water limestones of the Great Blue Formation (Mississippian) in the eastern part of the Cordilleran miogeosyncline in Utah: Society of Economic Paleontologists and Mineralogists Special Publication 25, p. 171–186.

Bissell, H.J., Rigby, J.K., and Proctor, P.D., 1959, Geology of the southern Oquirrh Mountains and Fivemile Pass—northern Boulter Mountains area, Tooele and Utah counties, Utah: Utah Geological Society Guidebook 14, p. 1–8.

Bjorklund, L.J., 1969, Reconnaissance of the groundwater resources of the upper Fremont River Valley, Wayne County, Utah: Utah Department of Natural Resources Technical Publication 22, 49 p.

Bjorklund, L.J., and McGreevy, 1971, Ground-water resources of Cache Valley, Utah and Idaho: Utah Department of Natural Resources Technical Publication 36, 66 p.

Bjorklund, L.J., and McGreevy, L.J., 1974 Groundwater resources of the lower Bear River drainage basin, Box Elder County, Utah: Utah Department of Natural Resources Technical Publication 44, 58 p.

Bjorklund, L.J., and Robinson, G.B. Jr., 1968, Groundwater resources of the Sevier River Basin between Yuba Dam and Leamington Canyon, Utah: U.S. Geological Survey Water Supply Paper 1848, 79 p.

Black, B.A., 1965, Nebo overthrust, southern Wasatch Mountains, Utah: Brigham Young University Geology Studies, v. 12, p.55–90.

Blackstone, D.L. Jr., 1977, Overthrust belt salient of the Cordilleran fold belt, western Wyoming-southeastern Idaho—northeastern Utah: Wyoming Geological Association Guidebook 29, p. 367–384.

Blackstone, D.L. Jr., 1983, Laramide compressional tectonics, southeastern Wyoming: University of Wyoming Contributions to Geology, v. 22, p. 1–38.
Blackwell, D.D., 1978, Heat flow and energy loss in the western U.S.: Geological Society of American Memoir 152, p. 175–208.
Blakey, R.C., 1974, Stratigraphic and depositional analysis of the Moenkopi Formation, southeastern Utah: Utah Geological and Mineral Survey Bulletin 104, 81 p.
Blakey, R.C., 1977, Petroliferous lithosomes in the Moenkopi Formation, southern Utah: Utah Geology, v. 4, no. 2, p. 67–84.
Blakey, R.C., 1980, Pennsylvanian and Early Permian paleogeography, southern Colorado Plateau and vicinity: Rocky Mountain Section, Society of Economic Paleontologists and Mineralogists, Symposium 1, Paleozoic Paleogeography of West-central U.S., p. 239–257.
Blakey, R.C., and Gubitosa, R., 1983, Late Triassic paleogeography and depositional history of the Chinle Formation, southern Utah and northern Arizona: *in* Reynolds, M.W., and Dolly, E.D., eds., Symposium on Mesozoic Paleogeography of West-central United States, Rocky Mountain Section of Society of Economic Paleontologists and Mineralogists, Denver, Colorado, p. 57–76.
Blakey, R.C., Peterson, Fred, Caputo, M.V., and Voorhees, B.J., 1983, Paleogeography of middle Jurassic continental shoreline and shallow marine sedimentation, southern Utah: *in* Reynolds, M.W., and Dolly, E.D., eds., Symposium on Mesozoic Paleogeography of West-central United States, Rocky Mountain Section of Society of Economic Geologists and Paleontologists, Denver, Colorado, p. 77–100.
Blank, H.R. Jr., 1959, Geology of the Bull Valley District, Washington County, Utah: University of Washington (Seattle) unpublished Ph.D. thesis, 176 p. map.
Blau, J.G., 1975, Geology of southern part of the James Peak quadrangle, Utah: unpublished master's thesis, Utah State University, Logan, 64 p.
Bloeser, Bonnie, 1985, *Melanocyrillum*, a new genus of structurally complex late Proterozoic microfossils from the Kwagunt Formation (Chuar Group), Grand Canyon, Arizona: Journal of Paleontology, May, 1985, p. 741–765.
Bohn, R.T., 1977, Subsurface correlation of Permian-Triassic strata in Lisbon Valley, Utah: Brigham Young University Geology Studies v. 24, part 2, p. 103–116.
Bolke, E.L., and Price, D., 1972, Hydrologic reconnaissance of the Blue Creek Valley area, Box Elder County, Utah: Utah Department of Natural Resources Technical Publication 37, 31 p.
Bond, G.G., and Komnitz, M.A., 1984, Construction of tectonic subsidence curves for the early Paleozoic miogeocline, southern Canadian Rocky Mountains—implications for subsidence mechanisms, age of breakup, and crustal thinning: Geological Society of America Bulletin, v. 95., p. 155–173.
Bordine, B.W., 1965, Paleoecologic implications of strontium, calcium, and magnesium in Jurassic rocks near Thistle, Utah: Brigham Young University Geology Studies, v. 12, p. 91–120.
Bortz, L.C., 1983, Hydrocarbons in the northern Basin and Range, Nevada and Utah: Geothermal Resources Council, Special Report 13, p. 179–213, Davis California 95617-1350.
Boucot, A.J., 1979, Silurian biogeography and biostratigraphy: *in* Treatise on Invertebrate Paleontology, Part A, R.A. Robison and C. Teichert, eds., p. A167–182.
Bowers, W.E., 1972, The Canaan Peak, Pine Hollow, and Wasatch Formations in the Table Cliff region, Garfield County, Utah: U.S. Geological Survey Bulletin 1331-B, 39 p.
Bowers, W.E., 1973a, Geologic map and coal resources of the Upper Valley quadrangle, Garfield County, Utah: U.S. Geological Survey Coal Investigations Map C-60.
Bowers, W.E., 1973b, Geologic map and coal resources of the Griffen Point quadrangle, Garfield County, Utah: U.S. Geological Survey Coal Investigations Map C-61.
Bowers, W.E., 1973c, Geology and coal resources of the Pine Lake quadrangle, Garfield County, Utah: U.S. Geological Survey Coal Investigations Map C-66.
Bowers, W.E., 1975, Geologic map and coal resources of the Henrieville quadrangle, Garfield and Kane Counties, Utah: U.S. Geological Survey Coal Investigations Map C-74.
Bowers, W.E., 1983; Geologic map and coal sections of the Butler Valley quadrangle, Kane County, Utah: U.S. Geological Survey Coal Investigations Map C-95.
Boyer, S.A., and Allison, M.L., 1987, Estimates of extension in the Basin and Range province: Geological Society of America, Abstracts with Programs, v. 19, p. 597.
Bradley, W.H., 1936, Geomorphology of the north flank of the Uinta Mountains: U.S. Geological Survey Professional Paper 185-I, 204 p.
Brady, M.J., 1965, Thrusting in the southern Wasatch Mountains, Utah: Brigham Young University Geology Studies, v. 12, p. 3–54.
Brady, M.J., and Koepnik, R.B., 1979, A Middle Cambrian platform-to-basin transition, House Range, west-central Utah: Brigham Young University Geology Studies, v. 26, part 1, p. 1–7.
Brady, M.J., and Rowell, A.J., 1976, An Upper Cambrian subtidal blanket carbonate, eastern Great Basin: Brigham Young University Geology Studies v. 23, part 2, p. 153–163.
Braithwaite, L.F., 1976, Graptolites from the Pogonip Group (lower Ordovician) of western Utah: Geological Society of America Special Paper 166, 106 p.
Bray, R.E., and Wilson, J.C., eds., 1975, Guidebook to the Bingham Mining District: Society of Economic Geologists, 155 p. Kennecott Copper Corporation, Salt Lake City.
Brenner, R.L., 1983, Late Jurassic tectonic setting and paleogeography of western interior, North America: *in* Reynolds, M.W., and Dolly, E.D., eds., Symposium on Mesozoic Paleogeography of West-central U.S., Rocky Mountain Section of Society of Economic Paleontologists and Mineralogists, Denver, Colorado, p. 119–132.
Bresler, S.L., 1981, Preliminary paleomagnetic poles and correlation of the Proterozoic Uinta Mountain Group, Utah and Colorado: Earth and Planetary Science Letters, v. 55, p. 53–64.
Briggs, D.E.G., and Robison, R.A., 1984, Exceptionally preserved nontrilobite arthropods and *Anomalocaris* from the Middle Cambrian of Utah: University of Kansas Paleontological Contributions Paper 111, 23 p.
Brimhall, W.H., and Merritt, L.B., 1981, Geology of Utah Lake—implications for resource management: Great Basin Naturalist Memoir 5, p. 24–42.
Britt, T.L., and Howard, E.L., 1982, Oil and gas potential of the central Utah hingeline-thrust belt region: Rocky Mountain Association of Geologists, Geologic Studies of the Cordilleran Thrust Belt, v. 2, Denver, Colorado, p. 475–505.
Bromfield, C.S., 1968, General geology of the Park City region, Utah: Utah Geological Society Guidebook 22, p. 10–29.
Bromfield, C.S., Baker, A.A., and Crittenden, M.D. Jr., 1970, Geologic map of the Heber quadrangle, Wasatch and Summit Counties, Utah: U.S. Geological Survey Geologic Quadrangle Map GQ-864.
Bromfield, C.S., and Crittenden, M.D. Jr., 1971, Geologic map of the Park City East quadrangle, Summit and Wasatch Counties, Utah: U.S. Geological Survey Geologic Quadrangle Map GQ-852.
Brown, R.P., and Cook, K.L., 1982, Regional gravity survey of the Sanpete-Sevier Valleys and adjacent areas in Utah: Utah Geological Association Publication 10, p. 121–135.
Bruhn, R.L., Picard, M.D., and Isby, J.S., 1986, Tectonics and sedimentology of Uinta Arch, western Uinta Mountains and Uinta Basin: American Association of Petroleum Geologists Memoir 41, p. 333–352.
Bryant, B., 1984, Reconnaissance geologic map of the Precambrian Farmington Canyon Complex and surrounding rocks in the Wasatch Mountains between Ogden and Bountiful, Utah: U.S. Geological Survey Miscellaneous Investigations Series Map I-1447.
Bryant, B., 1985, Structural ancestry of the Uinta Mountains: Utah Geological Association Publication 12, p. 115–120.
Bryant, B., 1988, Evolution and early Proterozoic history of the margin of the Archean continent in Utah: *in* Ernst, W.G., ed., Metamorphism and crustal evolution of the Western U.S.: Rubey Colloquium VII, Prentice Hall, N.Y., p. 432–445.
Bryant, B., and Nichols, D.J., 1985, Synorogenic conglomerates east of Salt Lake City—new data and interpretations: Geological Society of America Abstracts with Programs, v. 17, p. 210.
Bryant, B., and Nichols, D.J., 1988, Late Mesozoic and early Tertiary reactivation of an ancient crustal boundary along the Uinta trend and its interaction with the Sevier Orogenic Belt: *in* Perry, W.J., and Schmidt, C.J., eds., Interaction of the Rocky Mountain foreland and Cordilleran thrust belt: Geological Society of America (in press).
Budding, K.E., and Bugden, M.N., 1986, Annotated geothermal bibliography of Utah: Utah Geological and Mineral Survey Bulletin 121, 32 p.
Budge, D.R., 1966, Paleontology and stratigraphic significance of Late Ordovician and Silurian corals from the eastern Great Basin: unpublished Ph.D. dissertation, University of California, Berkeley, 572 p. 22 pl.
Budge, D.R., and Sheehan, P.M., 1980, Upper Ordovician through Middle Silurian of the eastern Great Basin, part 2, lithologic descriptions: Milwaukee Public Museum Contributions in Biology and Geology, no. 29, 80 p.
Bullock, K.C., 1973, Geology and iron ore deposits of Iron Springs District, Iron County, Utah: Brigham Young University Geology Studies, v. 20, part 2, p. 27–63.
Bullock, L.R., 1965, Paleoecology of the Twin Creek Limestone in the Thistle, Utah area: Brigham Young University Geology Studies, v. 12, p. 121–147.
Burchfiel, B.C., and Davis, G.A., 1975, Nature and controls of Cordilleran orogenesis, western United States—extensions of an earlier synthesis: American Journal of Science, v. 275A, p. 363–396.
Burton, S.M., 1973, Structural geology of the northern part of Clarkston Mountain, Malad Range, Utah-Idaho: unpublished master's thesis, Utah State University, Logan, 64 p.
Bushman, A.V., 1973, Pre-Needles Range silicic volcanism, Tunnel Spring Tuff (Oligocene), west-central Utah: Brigham Young University Geology Studies, v. 20, part 4, p. 159–190.
Buterbaugh, G.J., 1982, Petrology of the lower Middle Cambrian Langston Formation, north-central Utah and southeastern Idaho: unpublished master's thesis, Utah State University, Logan, 166 p.
Cadigan, R.A., 1967, Petrology of the Morrison Formation in the Colorado Plateau region: U.S. Geological Survey Professional Paper 556, 112 p.
Cadigan, R.A., 1967, Petrology of the Morrison Formation in the Colorado Plateau region: U.S. Geological Survey Professional Paper 556, 112 p.
Callaghan, E., and Parker, R.L., 1962, Geologic map of the Delano Peak quadrangle, Utah: U.S. Geological Survey Map GQ-153.
Campbell, D.R., 1978, Stratigraphy of pre-Needles Range Formation ash-flow tuffs in the northern Needle Range and southern Wah Wah Mountains, Beaver County, Utah: Brigham Young University Geology Studies, v. 25, part 3, p. 31–46.
Campbell, G.S., 1955, Weber pool of Rangely Field, Colorado Rocky Mountain Association of Geologists and Intermountain Association of Geologists Guidebook to the Geology of Northwest Colorado; p. 99–100, plate VI.
Campbell, J.A., 1979, Middle to late Cenozoic stratigraphy and structural development of the Canyon Range, central Utah: Utah Geology, v. 6, no. 1, p. 1–6.
Carpenter, C.H., Robinson, G.B. Jr., and Bjorklund, L.J., 1967, Ground-water conditions and geologic reconnaissance of the upper Sevier River basin, Utah: U.S. Geological Survey Water Supply Paper 1836, 91 p.
Carpenter, R.M., Pandolfi, J.M., and Sheehan, P.M., 1986, The Upper Ordovician carbonate ramp of the eastern Great Basin: Milwaukee Public Museum Contributions in Biology and Geology, no. 69, 92 p.
Carr, T.R., and Paull, R.K., 1983, Early Triassic stratigraphy and paleogeography of the Cordilleran miogeocline: *in* Reynolds, M.W., and Dolly, E.D., eds., Symposium on Mesozoic Paleogeography of West-central U.S., Rocky Mountain Section of Society of Economic Paleontologists and Mineralogists, Denver, Colorado, p. 39–55.
Carr, T.R., Paull, R.K., and Clark, D.L., 1984, Conodont paleoecology and biofacies analysis of the

Lower Triassic Thaynes Formation in the Cordilleran Miogeocline: Geological Society of America Special Paper 196, p. 283–293.
Carrara, P.E., 1980, Surficial geologic map of the Vernal 2-degree quadrangle, Colorado and Utah: U.S. Geological Survey Miscellaneous Investigations Series Map I-1204.
Carroll, R.E., 1987, Geology of the Standardville 7½-minute quadrangle, Carbon County, Utah: Brigham Young University Geology Studies v. 34, p. 1–31.
Carter, W.D., and Gualtieri, J.L., 1965, Geology and uranium-vanadium deposits of the La Sal quadrangle, San Juan County, Utah, and Montrose County, Colorado: U.S. Geological Survey Professional Paper 508, 79 p.
Case, J.E., Joesting, H.R., and Byerly, P.E., 1963, Regional geophysical investigations in the La Sal Mountains area, Utah and Colorado: U.S. Geological Survey Professional Paper 316-F, 116 p.
Case, R.W., and Cook, K.L., 1979, A gravity survey of the Sevier Lake are, Millard County, Utah: Utah Geology, v. 6, no. 1, p. 55–76.
Cashion, W.B., 1967a, Geologic map of the south flank of the Markagunt Plateau, northwest Kane County, Utah: U.S. Geological Survey Miscellaneous Geologic Investigations Map I-494.
Cashion, W.B., 1967b, Geology and fuel resources of the Green River Formation, southeastern Uinta Basin, Utah and Colorado: U.S. Geological Survey Professional Paper 548, 48 p.
Cashion, W.B., 1973, Geologic and structure map of the Grand Junction 2-degree quadrangle, Colorado and Utah: U.S. Geological Survey Miscellaneous Geologic Investigations Map I-736.
Cashion, W.B., and Donnell, J.R., 1974, Revision of nomenclature of the upper part of the Green River Formation of the Piceance Creek Basin, Colorado and the eastern Uinta Basin, Utah: U.S. Geological Survey Bulletin 1394-G, 9 p.
Cater, F.W., 1970, Geology of the Salt Anticline region in southwestern Colorado: U.S. Geological Survey Professional Paper 637, 80 p.
Chadwick, R.S., Barney, M.L., Beckstrand, D., Campbell, L., Carley, J.A., Jensen, E.H., McKinlay, C.R., Stock, S.S., and Stokes, H.A., 1975, Soil survey of Box Elder County, Utah, eastern part: U.S. Department of Agriculture Soil Conservation Service, Salt Lake City, 223 p.
Chamberlain, C.K., and Clark, D.L., 1973, Trace fossils and conodonts as evidence for deep-water deposits in the Oquirrh Basin of central Utah: Journal of Paleontology, v. 47, p. 663–682.
Chamberlain, R.L., 1980, Structure and stratigraphy of the Rex Peak quadrangle, Rich County, Utah: Brigham Young University Geology Studies, v. 27, part 3, p. 44–54.
Chamberlin, T.L., 1975, Stratigraphy of the Ordovician Ely Springs Dolomite in the southeastern Great Basin, Utah and Nevada: unpublished Ph.D. thesis, University of Illinois, Urbana, 201 p.
Chapin, C.E., and Cather, S.M., 1981, Eocene tectonics and sedimentation in the Colorado Plateau-Rocky Mountain area: Arizona Geological Society Digest, v. 14, p. 173–198; also in Rocky Mountain Association of Geologists 1983 Symposium, p. 33–56.
Cheevers, C.W., and Rawson, R.R., 1979, Facies analysis of the Kaibab Formation in northern Arizona, southern Utah and southern Nevada: Four Corners Geological Society Guidebook 9, Permianland, p. 105–113.
Cheney, T.M., Smart, R.A., Waring, R.G., and Warner, M.A., 1953, Stratigraphic sections of the Phosphoria Formation in Utah: U.S. Geological Survey Circular 306, 40 p.
Chidsey, T.C., 1978, Stratigraphy and attenuation faulting in the northern House Range, Utah: Utah Geology, v. 5, no. 2, p. 143–155.
Chidsey, T.C., Crook, S.R., and Link, P.K., 1985, Overthrusts and stratigraphy in the Wasatch, Bear River, and Crawford Ranges and Bear Lake Plateau, north-central Utah: Utah Geological Association Publication 14, p. 269–290.
Choquette, P.W., 1983, Platy algal reef mounds, Paradox Basin: American Association of Petroleum Geologists Memoir 33, p. 454–462.
Christenson, G.E., and Purcell, C., 1985, Correlation and age of Quaternary alluvial-fan sequences, Basin and Range province, southwestern U.S.: Geological Society of America Special Paper 203, p. 115–122.
Christiansen, E.H., Sheridan, M.F., and Burt, D.M., 1986, Geology and geochemistry of Cenozoic topaz rhyolites from the western U.S.: Geological Society of America Special Paper 205, 80 p.
Christiansen, R.L., and Lipman, P.W., 1972, Cenozoic volcanism and plate-tectonic evolution of the western United States: II Late Cenozoic: Philosophical Transactions of the Royal Society of London, v. 271, p. 249–284.
Christiansen, R.L., and McKee, E.H., 1978, Late Cenozoic volcanic and tectonic evolution of the Great Basin and Columbia intermontane regions: Geological Society of America Memoir 152, p. 283–311.
Christie-Blick, Nicholas, 1982a, Upper Precambrian (Eocambrian) Mineral Fork Tillite of Utah: a continental glacial and glaciomarine sequence: Discussion and reply: Geological Society of America Bulletin, v. 93, p. 184–187.
Christie-Blick, N. 1982b, Upper Proterozoic and Lower Cambrian rocks of the Sheeprock Mountains, Utah—regional correlation and significance: Geological Society of America Bulletin, v. 93, p. 735–750.
Christie-Blick, N., 1983, Structural geology of the southern Sheeprock Mountains, Utah—regional significance: Geological Society of America Memoir 157, p. 101–124.
Christie-Blick, N., 1985, Upper Proterozoic glacial-marine and subglacial deposits at Little Mountain, Utah: Brigham Young University Geology Studies, v. 32, part 1, p. 9–18.
Christie-Blick, N., Grotzinger, J.P., and von der Borch, C.C., 1988, Sequence stratigraphy in Proterozoic succession: Geology, v. 16, p. 100–104.
Clark, D.L., 1957, Marine Triassic stratigraphy in eastern Great Basin: American Association of Petroleum Geologists Bulletin, v. 41, p. 2192–2222.
Clark, D.L., 1979, Permian-Triassic boundary—Great Basin conodont perspective: Brigham Young University Geology Studies, v. 26, part 1, p. 85–90.
Clark, D.L., and Carr, T.R., 1984, Conodont biofacies and biostratigraphic schemes in western North America—a model: Geological Society of America Special Paper 196, p. 1–9.
Clark, D.L., Carr, T.R., Behnken, F.H., Wardlaw, B.R., and Collinson, J.W., 1979, Permian conodont biostratigraphy in the Great Basin: Brigham Young University Geology Studies, v. 26, part 3, p. 143–147.
Clark, D.L., Peterson, D.O., Stokes, W.L., Wardlaw, B., and Wilcox, J.D., 1977, Permian-Triassic sequence in northwest Utah: Geology, v. 5, p. 655–658.
Clark, D.L., and Rosser, S.V., 1976, Analysis of paleoecologic factors associated with the Triassic *Parachirognathus/Furnishius* conodont fauna in Utah and Nevada: Geological Association of Canada Special Paper 15, p. 295–311.
Clark, Donald L., 1986, Geology of the Juab 7½-minute quadrangle, Juab County, Utah: Utah Geological and Mineral Survey Map (in preparation).
Clark, J.M., and Fastovsky, D.E., 1986, Vertebrate biostratigraphy of the Glen Canyon Group in northern Arizona: *in*, The Beginning of the Age of Dinosaurs—Faunal Change across the Triassic-Jurassic Boundary, Kevin Padian, ed., Cambridge University Press, p. 285–301.
Clark, R.A., 1981, Stratigraphy, depositional environments, and carbonate petrology of the Toroweap and Kaibab formations, Lower Permian, Grand Canyon region, Arizona: unpublished Ph.D. thesis, New York State University, Binghamton, 454 p.
Clark, R.S., 1953, Structure and stratigraphy of Government Ridge Utah: unpublished master's thesis, Brigham Young University 75 p.
Clem, K.M., 1985, Oil and gas production summary of the Uinta Basin: Utah Geological Association Publication 12, p. 159–167.
Clem, K.M., and Brown, K.W., 1984, Petroleum resources of the Paradox Basin: Utah Geological and Mineral Survey Bulletin 119, 162 p.
Clem, K., and Brown, K.W., 1985, Summary of oil and gas activity in northeastern Utah and southeastern Idaho: Utah Geological Association Publication 14, p. 157–166.
Cloud, Preston, 1988, Oasis in space—Earth history from the beginning: W.W. Norton & Co., New York, 508 p.
Cluff, L.S., Brogan, G.E., and Glass, C.E., 1970, Wasatch Fault, northern portion: Woodward-Clyde & Associates, Oakland, California. Report prepared for Utah Geological and Mineral Survey, Salt Lake City, 1:24,000 scale maps of area from Jordan Narrows to Brigham City.
Cluff, L.S., Brogan, G.E., and Glass, C.E., 1973, Wasatch Fault, southern portion: Woodward-Lundgren & Associates, Oakland, California. Report prepared for Utah Geological and Mineral Survey, Salt Lake City. 1:24,000 scale maps from Alpine to Fayette.
Cluff, L.S., Glass, C.E., and Brogan, G.E., 1974, Investigation and evaluation of the Wasatch fault north of Brigham City,and Cache Valley faults, Utah and Idaho: Woodward-Lundgren & Associates, Oakland, California. Report prepared for U.S. Geological Survey and Utah Geological and Mineral Survey. 1:24,000 scale maps.
Cobban, W.A., 1976, Ammonite record from the Mancos Shale of the Castle Valley-Price-Woodside area, east-central Utah: Brigham Young University Geology Studies, v. 22, part 3, p. 117–126.
Cobban, W.A., and Hook, S.C., 1984, Mid-Cretaceous molluscan biostratigraphy and paleogeography of southwestern part western interior United States: Geological Association of Canada Special Paper 27, p. 257–271.
Cocks, L.R.M., Holland, C.H., Rickards, R.B., and Strachan, I., 1971, A correlation of Silurian rocks in the British Isles: Geological Society of London Journal, v. 127, p. 103–136.
Cohee, G.V., Glaessner, M.F., and Hedberg, H.D., eds., 1978, Contributions to the geologic time scale: American Association of Petroleum Geologists, Studies in Geology, no. 6, 388 p. Includes 30 papers on geochronological dating by authorities on each geologic period.
Cohenour, R.E., 1959, Sheeprock Mountains, Precambrian and Paleozoic stratigraphy, igneous rocks, structure, geomorphology and economic geology: Utah Geological and Mineral Survey Bulletin 63, 201 p.
Collinson, J.W., and Hasenmueller, W.A., 1978, Early Triassic paleogeography and biostratigraphy of the Cordilleran miogeosyncline: Pacific Section, Society of Economic Paleontologists and Mineralogists, Symposium 2, Mesozoic Paleogeography, p. 175–187.
Collinson, J.W., Kendall, C.G.S., and Marcantel, J.B., 1976, Permian-Triassic boundary in eastern Nevada and west-central Utah: Geological Society of America Bulletin, v. 87, p. 821–824.
Colman, S.M., 1983, Influence of the Onion Creek salt diapir on the late Cenozoic history of Fisher Valley southeastern Utah: Geology, v. 11, p. 240–243.
Colman, S.M., and Pierce, K.L., 1977, Summary table of Quaternary dating methods: U.S. Geological Survey Map MF-904.
Compton, R.R., 1972, Geologic map of the Yost quadrangle, Box Elder County, Utah, and Cassia County, Idaho: U.S. Geological Survey Miscellaneous Investigation Map I-672.
Compton, R.R., 1975, Geologic map of the Park Valley quadrangle, Box Elder County, Utah, and Cassia County, Idaho: U.S. Geological Survey Miscellaneous Investigation Map I-873.
Compton, R.R., 1980, Fabrics and strains in quartzites of a metamorphic core complex, Raft River Mountains, Utah: Geological Society of America Memoir 153, p. 385–398.
Compton, R.R., 1983, Displaced Miocene rocks on the west flank of the Raft River-Grouse Creek core complex, Utah: Geological Society of America Memoir 157, p. 271–280.
Compton, R.R., Todd, V.R., Zartman, R.E., and Naeser, C.W., 1977, Oligocene and Miocene metamorphism, folding, and low-angle faulting in northwestern Utah: Geological Society of America Bulletin, v. 88, p. 1237–1250. See also Discussion and Reply in v. 90, p. 305–309.
Condie, K.C., 1964, Crystallization Po2 of syenite porphyry from Navajo Mountain, southern Utah: Geological Society of America Bulletin, v. 74, p. 359–362.
Condie, K.C., 1987, Early Proterozoic arc terranes and continental accretion in the southwestern U.S.: Geological Society of America Abstracts with Programs, v. 19, p. 625.

Conner, D.C., and Covlin, R.J., 1977, Development geology of Pineview Field, Summit County, Utah: Wyoming Geological Association 29th Annual Field Conference Guidebook, p. 639–650.

Coogan, J.C., and Yonkee, W.A., 1985, Salt detachments in the Jurassic Preuss redbeds within the Meade and Crawford thrust systems, Idaho and Wyoming: Utah Geological Association Publication 14, p. 75–82.

Cook, H.E., and Taylor, M.E., 1977, Comparison of continental slope and shelf environments in the Upper Cambrian and lowest Ordovician of Nevada: *in* Deep-Water Carbonate Environments, H.E. Cook and Paul Enos, eds., Society of Economic Paleontologists and Mineralogists, Special Publication 25, p. 51–82.

Cook, K.L., and Berg, J.W.Jr., 1961, Regional gravity survey along the central and southern Wasatch Front, Utah: U.S. Geological Survey Professional Paper 316-E, p. 75–89.

Cook, K.L., Berg, J.W.Jr., Johnson, W.W., and Novotuny, R.T., 1966, Some Cenozoic structural basins in the Great Salt Lake area, Utah, indicated by regional gravity surveys: Utah Geological Society Guidebook 20, p. 57–75.

Cook, K.L., Gray, E.F., Iverson, R.M., and Strohmeier, M.T., 1980, Bottom gravity meter regional survey of the Great Salt Lake, Utah: Utah Geological and Mineral Survey Bulletin 116, p. 125–143.

Cooper, J.E. Jr., 1956, Petrography of the Moroni Formation, southern Cedar Hills, Utah: unpublished master's thesis, Ohio State University, 86 p.

Corbett, M. 1978, Geologic map of the northern Portneuf Range, Idaho: U.S. Geological Survey Open-File Report OF 78-1018. Scale 1:48,000.

Costain, J.K., 1960, Geology of the Gilson Mountains and vicinity Juab County, Utah: unpublished Ph.D. thesis, University of Utah, 139 p.

Cotter, E., 1975, The role of deltas in the evolution of the Ferron Sandstone and its coals, Castle Valley, Utah: Brigham Young University Geology Studies v. 22, part 3, p. 15–41.

Couch, E.L., Kent, H.C., and Knepp, R.A., 1985, Central and Southern Rockies Region—Correlation of stratigraphic units of North America: American Association of Petroleum Geologists, COSUNA Project Chart, Tulsa.

Covington, H.R., 1983, Structural evolution of the Raft River Basin, Idaho: Geological Society of America Memoir 157, p. 229–237.

Craig, L.C., 1972, Mississippian System: *in* Geologic Atlas of the Rocky Mountain Region. Rocky Mountain Association of Geologists, Denver, p. 100–110.

Craig, L.C., 1981, Lower Cretaceous rocks, southwestern Colorado and southeastern Utah: Rocky Mountain Association of Geologists Guidebook to Geology of the Paradox Basin, p. 195–200.

Craig, L.C., and Dickey, D.D., 1956, Jurassic strata of southeastern Utah and southwestern Colorado: Intermountain Association of Petroleum Geologists Guidebook 7, p. 93–104.

Craig, L.C., and Varnes, K.L., 1979, History of the Mississippian System—an interpretive summary: U.S. Geological Survey Professional Paper 1010, p. 371–406.

Crawford, K.A., 1979, Echo Canyon and Evanston conglomerates, north-central Utah thrust belt: unpublished master's thesis, University of Wisconsin-Madison, 143 p.

Crecraft, H.R., Nash, W.P., and Evans, S.H. Jr., 1981, Late Cenozoic volcanism at Twin Peaks, Utah, geology and petrology: Journal of Geophysical Research, v. 86, p. 10303–10320.

Crittenden, M.D. Jr., 1959, Mississippian stratigraphy of the central Wasatch and western Uinta Mountains, Utah: Intermountain Association of Petroleum Geologists Guidebook 10, p. 63–74.

Crittenden, M.D. Jr., 1963a, Emendation of the Kelvin Formation and Morrison (?) Formation near Salt Lake City, Utah: U.S. Geological Survey Professional Paper 475-B, p. 95–98.

Crittenden, M.D. Jr., 1963b, New data on the isostatic deformation of Lake Bonneville: U.S. Geological Survey Professional Paper 454-E, 31 p.

Crittenden, M.D. Jr., 1972a, Geologic map of the Browns Hole quadrangle, Utah: U.S. Geological Survey Geologic Quadrangle Map GQ-968.

Crittenden, M.D. Jr., 1972b, Willard thrust and the Cache allochthon, Utah: Geological Society of America Bulletin, v. 83, p. 2871–2880.

Crittenden, M.D. Jr., 1974, Regional extent and age of thrusts near Rockport reservoir and relation to possible exploration targets in northern Utah: American Association of Petroleum Geologists Bulletin, v. 58, p. 2428–2435.

Crittenden, M.D. Jr., 1976, Stratigraphic and structural setting of the Cottonwood area, Utah: Rocky Mountain Association of Geologists, Symposium on Geology of the Cordilleran Hingeline, p. 363–379.

Crittenden, M.D. Jr., Christie-Blick, N., and Link, P.K., 1983, Evidence for two pulses of glaciation during the late Proterozoic in northern Utah and southeastern Idaho: Geological Society of America Bulletin, v. 94, p. 437–450.

Crittenden, M.D. Jr., and Peterman, Z.E., 1975, Provisional Rb/Sr age of the Precambrian Uinta Mountain Group, northeastern Utah: Utah Geology, v. 2, no. 1, p. 75–77.

Crittenden, M.D. Jr., Schaeffer, F.E., Trimble, D.E., and Woodward, L.A., 1971, Nomenclature and correlation of some Upper Precambrian and basal Cambrian sequences in western Utah and southeastern Idaho: Geological Society of America Bulletin v. 82, p. 581–602.

Crittenden, M.D. Jr., and Sorensen, M.L., 1980, The Facer Formation, a new early Proterozoic unit in northern Utah: U.S. Geological Survey Bulletin 1482-F, 28 p.

Crittenden, M.D. Jr., and Sorensen, M.L., 1985, Geologic map of the North Ogden quadrangle and part of the Ogden and Plain City quadrangles, Box Elder and Weber Counties, Utah: U.S. Geological Survey Miscellaneous Investigation Map I-1606.

Crittenden, M.D. Jr., Wallace, C.A., and Sheridan, M.J., 1967, Mineral resources of the High Uintas Primitive Area, Utah: U.S. Geological Survey Bulletin 1230-I, 27 p.

Crittenden, M.D. Jr., and Wallace, C.A., 1973, Possible equivalents of the Belt Supergroup in Utah: Belt Symposium, v. 1, p. 116–138, Idaho Bureau of Mines and Geology, Moscow.

Crook, S.R., Link, P.K., and Chidsey, T.C., 1985, Structure and stratigraphy of the Paris and Meade thrust plates and transition to the Basin and Range Province-Bear River, Preuss, and Bannock Ranges, southeastern Idaho: Utah Geological Association Publication 14, p. 291–313.

Crosby, G.W., 1959, Geology of the south Pavant Range, Millard and Sevier Counties, Utah: Brigham Young University Geology Studies, v. 6, no. 3, 59 p.

Crosby, G.W., 1976, Tectonic evolution in Utah's miogeosyncline shelf boundary zone: Rocky Mountain Association of Geologists Symposium, p. 27–36.

Cross, A.T., Maxfield, E.B., Cotter, E., and Cross, C.C., 1975, Field guide and road log to the western Book Cliffs, Castle Valley and parts of the Wasatch Plateau: Brigham Young University Geology Studies, v. 22, part 2, 132 p.

Crowell, J.C., 1982, Continental glaciation through geologic times: U.S. National Academy of Science Press, Studies in Geophysics—"Climate in Earth History" Washington D.C., p. 77–82.

Cullins, H.L., 1971, Geologic map of the Rangely quadrangle, Rio Blanco County, Colorado: U.S. Geological Survey Geologic Quadrangle Map GQ-903.

Cunningham, C.G., and Steven, T.A., 1979, Mount Belknap and Red Hills calderas and associated rocks, Marysvale volcanic field, west-central Utah: U.S. Geological Survey Bulletin 1468, 34 p.

Cunningham, C.G., Steven, T.A., Rowley, P.D., Glassgold, L.B., and Anderson, J.J., 1983, Geologic map of the Tushar Mountains and adjoining areas, Marysvale volcanic field, Utah: U.S. Geological Survey Miscellaneous Investigations Map I-1430A.

Currey, D.R., 1980, Coastal geomorphology of Great Salt Lake and vicinity: Utah Geological and Mineral Survey Bulletin 116, p. 69–82.

Currey, D.R., 1982, Lake Bonneville—selected features of relevance to neotectonic analysis: U.S. Geological Survey Open File Report 82-1070, 30 p.

Currey, D.R., Atwood G., and Mabey, D.R., 1983, Major levels of Great Salt lake and Lake Bonneville: Utah Geological and Mineral Survey Map 73.

Currey, D.R., and Oviatt, C.G., 1985, Durations, average rates, and probable causes of Lake Bonneville expansions, still-stands, and contractions during the last deep-lake cycle 32,000 to 10,000 years ago: Geographical Journal of Korea, v. 10, p. 1085–1099.

Currey, D.R., Oviatt, C.G., and Czarnomski, J.E., 1984, Late Quaternary geology of Lake Bonneville and Lake Waring: Utah Geological Association Publication 13, p. 227–238.

Currey, D.R., Oviatt, C.G., and Plyler, G.B., 1983, Lake Bonneville stratigraphy, gemorphology, and isostatic deformation in west-central Utah: Utah Geological and Mineral Survey Special Studies 62, p. 63–82.

Dalton, R.O. Jr., 1972, Stratigraphy of the Bass Formation: unpublished master's thesis, Northern Arizona University (Flagstaff), 78 p.

Dane, C.H., 1954, Stratigraphic and facies relationships of upper part of Green River Formation and lower part of Uinta Formation in Duchesne, Uintah, and Wasatch Counties, Utah: American Association of Petroleum Geologists Bulletin, v. 38, p. 405–425.

Dane, C.H., 1955, Stratigraphic and facies relationships of the upper part of the Green River Formation and the lower part of the Uinta Formation in Duchesne, Uintah, and Wasatch Counties, Utah: U.S. Geological Survey Oil and Gas Investigation Chart 52 (2 sheets).

Danzl, R.B., 1985, Depositional and diagenetic environment of the Salt Lake Group at Oneida Narrows, southeastern Idaho: unpublished master's thesis, Idaho State University, Pocatello, 177 p. Preliminary summary in Northwest Geology v. 11, p. 22–30, 1982.

Davidson, E.S., 1967, Geology of the Circle Cliffs area, Garfield and Kane Counties, Utah: U.S. Geological Survey Bulletin 1229, 140 p.

Davies, S.F., 1980, Geology of the Grayback Hills, north-central Tooele County, Utah: unpublished master's thesis, University of Utah, 162 p.

Davis, B.L., 1959, Petrology and petrography of the igneous rocks of the Stansbury Mountains, Tooele County, Utah: Brigham Young University Geology Studies, v. 6, no. 2, 56 p.

Davis, D.A., and Cook, K.L., 1984, Gravity survey of Utah and Goshen valleys and adjacent areas, Utah: Geological Society of America Abstracts with Programs, v. 16, p. 219.

Davis, F.D., 1983a, Geologic map of the central Wasatch Front, Utah: Utah Geological and Mineral Survey Map 54-A.

Davis, F.D., 1983b, Geologic map of the southern Wasatch Front, Utah: Utah Geological and Mineral Survey Map 55-A.

Davis, L.E., and Webster, G.D., 1987, Conodont biostratigraphy of the Lake Point Limestone and recognition of the Mississippian-Pennsylvanian boundary and a potential Morrowan stratotype: Geological Society of America Abstracts with Programs, v. 19, p. 269. (Abstract only), complete paper in Compte Rendu, 11th International Congress of Carboniferous Stratigraphy and Geology, Beijing (in press).

Davis, L.J., 1984, Beryllium deposits in the Spor Mountain area, Juab County, Utah: Utah Geological Association Publication 13, p. 173–183.

Davis, R.L., 1983, Geology of the Dog Valley-Red Ridge area, southern Pavant Mountains, Millard County, Utah: Brigham Young University Geology Studies, v. 30, part 1, p. 19–36.

Dawson, J.C., 1970, Sedimentology and stratigraphy of the Morrison Formation (Upper Jurassic) in northwestern Colorado and northeastern Utah: unpublished Ph.D. thesis, University of Wisconsin, 142 p.

Dean, J.S., 1981, Carbonate petrology and depositional environments of the Sinbad Limestone Member of the Moenkopi Formation in the Teasdale Dome Area, Wayne and Garfield Counties, Utah: Brigham Young University Geology Studies, v. 28, part 3, p. 19–47.

Debiche, M.G., Cox A., and Engebretson, D.C., 1987, Motion of allochthonous terranes across the North Pacific basin: Geological Society of America Special Paper 207, 49 p.

Decourten, F.L., 1985, A specimen of *Ornithomimus velox* from the terminal Cretaceous Kaiparowits Formation of southern Utah: Journal of Paleontology, v. 59, p. 1091–1099.

DeGraff, J.V., 1976, Quaternary geomorphic features of the Bear River Range, north-central Utah: unpublished master's thesis, Utah State University, Logan, 215 p.

Delaney, P.T., Pollard, D.D., Ziony, J.I. and McKee, E.H., 1986, Field relations between dikes and joints—emplacement processes and paleostress analysis: Journal of Geophysical Research, v. 91, p. 4920–4938.

Demars, L.C., 1956, Geology of the northern part of Dry Mountain, southern Wasatch Moutains, Utah: Brigham Young University Geology Studies, v. 3, no. 2, 49 p.

Dennis, P.E., Maxey, G.B., and Thomas, H.E., 1946, Ground-water in Pavant Valley, Millard County, Utah: Utah Department of Natural Resources Technical Publication, no. 3, 96 p.

Deputy, E.J., 1984, Petrology of the Middle Cambrian Ute Formation, north-central Utah and southeastern Idaho; unpublished master's thesis, Utah State University, Logan, 124 p.

Derr, M.E., 1974, Sedimentary structure and depositional environment of paleochannels in the Jurassic Morrison Formation near Green River, Utah: Brigham Young University Geology Studies, v. 21, part 3, p. 3–39.

Devlin, W.J., 1988, Initiation of the early Paleozoic miogeocline—evidence from the uppermost Proterozoic-Lower Cambrian Hamill Group of southeastern British Columbia: Canadian Journal of Earth Science, v. 25, p. 1–19.

Dickinson, W.R., 1976, Sedimentary basins developed during evolution of Mesozoic-Cenozoic arc-trench system in western North America: Canadian Journal of Earth Science, v. 13, p. 1268–1287.

Dickinson, W.R., and Snyder, W.S., 1978, Plate tectonics of the Larmide orogeny: Geological Society of America Memoir 151, p. 355–366.

Dixon, J.S., 1982, Regional structural synthesis, Wyoming salient of western overthrust belt: American Association of Petroleum Geologists, v. 66, p. 1560–1580.

Dockal, J.A., 1980, Petrology and sedimentary facies of Redwall Limestone (Mississippian) of Uinta Mountains, Utah and Colorado: unpublished Ph.D. thesis, University of Iowa, 440 p.

Doelling, H.H., 1964, Geology of the northern Lakeside Mountains and the Grassy Mountains and vicinity, Tooele and Box Elder Counties, Utah: unpublished Ph.D. thesis, University of Utah, 354 p.

Doelling, H.H., 1972, Central Utah coal fields, Sevier-Sanpete, Wasatch Plateau, Book Cliffs, and Emery: Utah Geological and Mineral Survey Monograph 3, 570 p.

Doelling, H.H., 1975, Geology and mineral resources of Garfield County, Utah: Utah Geological and Mineral Survey Bulletin 107, 175 p.

Doelling, H.H., 1980, Geology and mineral resources of Box Elder County, Utah: Utah Geological and Mineral Survey Bulletin 115, 251 p.

Doelling, H.H., 1983, Observations on Paradox Basin salt anticlines: Grand Junction Geological Society Guidebook—Northern Paradox Basin and Uncompahgre Uplift; p. 81–90.

Doelling, H.H., 1985, Geologic map of Arches National Park and vicinity, Grand County, Utah: Utah Geological and Mineral Survey Map 74.

Doelling, H.H., and Davis, F.D., 1978, Coal drilling, Johns Valley, Garfield County, Utah: Utah Geology, v. 5, no. 2, p. 113–124.

Doelling, H.H., and Graham, R.L., 1972a, Southwestern Utah coal fields—Alton, Kaiparowits Plateau and Kolob-Harmony: Utah Geological and Mineral Survey Monograph 1, 333 p.

Doelling, H.H., and Graham, R.L., 1972b, Eastern and northern Utah coal fields—Vernal, Henry Mountains, Sego, La Sal-San Juan, Tabby Mountain, Coalville, Henry's Fork, Goose Creek and Lost Creek: Utah Geological and Mineral Survey Monograph 2, 409 p.

Dommer, M.L., 1980, Geology of the Drum Mountains, Millard and Juab Counties, Utah: Brigham Young University Geology Studies, v. 27, part 3, p. 55–72.

Donnell, J.R., 1969, Paleocene and Lower Eocene units in the southern part of the Piceance Creek Basin, Colorado: U.S. Geological Survey Bulletin 1274-M, 18 p.

Douglass, R.C., Moore, W.J., and Huddle, J.W., 1974, Stratigraphy and microfauna of the Oquirrh Group in the western Traverse Mountains and northern Lake Mountains, Utah: U.S. Geological Survey Journal of Research, v. 2, no. 1, p. 97–104.

Dover, J.H., 1985, Geologic map of Mount Naomi roadless area, Cache County, Utah and Franklin County, Idaho: U.S. Geological Survey Miscellaneous Field Studies Map MF-1566-B.

Driese, S.G., 1983, Diagenetic aspects of Morgan Formation (Pennsylvanian) shelf carbonates, northern Utah and Colorado: Brigham Young University Geology Studies, v. 30, part 1, p. 1–18.

Driese, S.G., 1985, Interdune pond carbonates, Weber Sandstone (Pennsylvanian-Permian), northern Utah and Colorado: Journal of Sedimentary Petrology, v. 55, nol 2, p. 187–195.

Driese, S.G., Carr, T.R., and Clar, D.L., 1984, Quantitative analysis of Pennsylvanian shallow-water conodont biofacies, Utah and Colorado: Geological Society of America Special Paper 196, p. 233–250.

Drummond, K.J., 1981, Plate-tectonic map of the Circum-Pacific region, northeast quadrant: American Association of Petroleum Geologists, Tulsa.

Drummond, K.J., 1983, Geologic map of the Circum-Pacific region, northeast quadrant: American Association of Petroleum Geologists, Tulsa.

Drummond, K.J., 1984, Geodynamic map of the Circum-Pacific region, northeast quadrant: American Association of Petroleum Geologists, Tulsa.

Dutro, J.T. Jr., 1979, Stop 10, Dry Lake section, Utah, *in* Carboniferous of the Northern Rocky Mountains: Field Trip No. 15, Ninth International Congress of Carboniferous Stratigraphy and Geology: American Geological Institute Selected Guidebook Series, No. 3, p. 50–51.

Dutro, J.T. Jr., Gordon, M. Jr., and Huddle, J.W., 1979, Paleontological zonation of the Mississippian System: U.S. Geological Survey Professional Paper 1010, p. 407–429.

Dutton, C.E., 1882, Atlas to accompany the monograph on the Tertiary history of the Grand Canon district: U.S. Geological Survey Monograph 2, 23 folio sheets.

Eardley, A.J., 1944, Geology of the north-central Wasatch Mountains, Utah: Geological Society of America Bulletin, v. 55, p. 819–894.

Eardley, A.J., 1963, Oil seeps at Rozel Point: Utah Geological and Mineralogical Survey Special Studies 5, 26 p.

Eardley, A.J., Shuey, R.T., Gvostdetsky, V., Nash, W.P., Picard, M.D., Grey, D.C., and Kukla, G.J., 1973, Lake cycles in the Bonneville Basin, Utah: Geological Society of America Bulletin, v. 84, p. 211–216.

Earll, F.N., 1957, Geology of the central Mineral Range, Beaver County, Utah: unpublished Ph.D. thesis, University of Utah, 112 p.

Eaton, J.G., 1987, Biostratigraphic framework for Late Cretaceous non-marine sequence, Kaiparowits Plateau, southern Utah: Geological Society of America Abstracts with Programs, v. 19, p. 650–651. Extracted from University of Colorado (Boulder) Ph.D. dissertation, 1987.

Eaton, J.G., and six others, 1987, Stratigraphy, correlation and tectonic setting of Late Cretaceous rocks in the Kaiparowits and Black Mesa basins: Arizona Bureau of Geology and Mineral Technology Special Paper 5, p. 113–125.

Eddington, P.K., Smith, R.B., and Renggli, C., 1987, Kinematics of Basin and Range intraplate extension: *in* Continental Extensional Tectonics, Geological Society (of London) Special Publication 28, p. 371–392.

Edson, D.J., Scholl, M.R.Jr., and Zabriskie, W.E., 1954, Clear Creek Gas Field, central Utah: Intermountain Association of Geologists Guidebook 5, p. 89–93.

Eliason, J.F., 1969, Hyrum and Beirdneau formations of north-central Utah and southeastern Idaho: unpublished master's thesis, Utah State University, Logan, 92 p.

Ellingson, J.A., 1973 The Mule Ear Diatreme: Geology of the Canyons of the San Juan River, Four Corners Geological Society River Guidebook, p. 43–50, Durango, Colorado.

Elston, D.P., 1979, Late Precambrian Sixtymile Formation and orogeny at top of Grand Canyon Supergroup, northern Arizona: U.S. Geological Survey Professional Paper 1092, 20 p.

Elston, D.P., and McKee, E.H., 1982, Age and correlation of the late Proterozoic Grand Canyon disturbance, northern Arizona: Geological Society of America Bulletin, v. 93, p. 681–699.

Elston, D.P., and Scott, G.R., 1976, Unconformity at the Cardenas-Nankoweap contact (Precambrian), Grand Canyon Supergroup, northern Arizona: Geological Society of America Bulletin, v. 87, p. 1763–1772.

Engebretson, D.C., Cox, A., and Gordon, R.G., 1985, Relative motions between oceanic and continental plates in the Pacific Basin: Geological Society of America Special Paper 206, 59 p.

England, P. and Jackson, J. 1987, Migration of the seismic—aseismic transition during uniform and nonuniform extension of the continental lithosphere: Geology, v. 15, p. 291–294.

Erickson, M.P., 1963, Volcanic geology of western Juab County, Utah: Utah Geological Society Guidebook 17, p. 23–35.

Ethington, R.L, and Clark, D.L., 1981, Lower and Middle Ordovician conodonts from the Ibex area, western Millard County, Utah: Brigham Young University Geology Studies, v. 28, part 2, 155 p., 14 plates.

Evans, S.H. Jr., and Steven, T.A., 1982, Rhyolites in the Gillies Hill-Woodtick Hill area, Beaver County, Utah: Geological Society of America Bulletin, v. 93, p. 1131–1141.

Evernden, J.F., Savage, D.E., Curtis, G.H., and James, G.T., 1964, Potassium-argon dates and the Cenozoic mammalian chronology of North America: American Journal of Science v. 262, p. 145–198.

Eyer, J.A., 1969, Gannett Group of western Wyoming and southeastern Idaho: American Association of Petroleum Geologists Bulletin, v. 53, p. 1368–1390.

Fagadau, S.P. 1949, An investigation of the Flagstaff Limestone between Manti and Willow Creek Canyons in the Wasatch Plateau, central Utah: unpublished master's thesis, Ohio State University, 80 p.

Farmer, G.L., and DePaolo, D.J., 1983, Origin of Mesozoic and Tertiary granite in the western U.S. and implications for pre-Mesozoic crustal structure: Journal of Geophysical Research, v. 88, p. 3379–3401.

Felix, C., 1956, Geology of the eastern part of the Raft River Range: Utah Geological Society Guidebook 11, p. 76–97.

Felt, V.L., 1981, Geology of the Antelope Peak area of the southern San Francisco Mountains, Beaver County, Utah: Brigham Young University Geology Studies, v. 28, part 3, p. 53–65.

Fisher, D.J., Erdmann, C.E., and Reeside, J.B., 1960, Cretaceous and Tertiary formations of the Book Cliffs, Carbon, Emery, and Grand Counties, Utah and Garfield and Mesa Counties, Colorado: U.S. Geological Survey Professional Paper 332, 80 p.

Fisher, M.J., and Dunay, R.E., 1984, Palynology of the Petrified Forest Member of the Chinle Formation (Upper Triassic) Arizona, U.S.A.: Pollen et Spores, v. 26, p. 241–284.

Flint, R.F., and Denny, C.S., 1958, Quaternary geology of Boulder Mountain, Aquarius Plateau, Utah: U.S. Geological Survey Bulletin 1061-D, 164 p.

Flores, R.M., and Marley, W.E.III, 1979 Physical stratigraphy and coal correlation of the Blackhawk Formation and the Star Point Sandstone near Emery, Utah: U.S. Geological Survey Miscellaneous Field Studies Map MF-1068.

Flores, R.M., Marley, W.E., Sanchez, J.D., Blanchard, L.F., and Muldoon, W.J., 1982, Coal correlations and depositional environments of Cretaceous Blackhawk Formation and Star Point Sandstone, Wasatch Plateau, Utah: Utah Geological and Mineral Survey Bulletin 118, p. 70–75. Also Geological Society of America Bulletin, v. 95, p. 540–550 (1984).

Flower, R.H., and Gordon, M. Jr., 1959, More Mississippian belemnites: Journal of Paleontology, v. 33, p. 809–842.

Fograsher, A.C., 1956, Stratigraphy of the Green River and Crazy Hollow Formations of part of the Cedar Hills, central Utah: unpublished master's thesis, Ohio State University, 88 p.

Ford, T.D., and Breed, W.J., 1973, Late Precambrian Chuar Group, Grand Canyon, Arizona: Geological Society of America Bulletin, v. 84, p. 1243–1260.

Forester, R.M., and Bradbury, J.P., 1981, Paleoenvironmental implications of ostracodes and diatoms from selected samples in Pliocene and Pleistocene lacustrine sediments in the Beaver basin, Utah: U.S. Geological Survey Open-File Report 81-390, 44 p.

Fouch, T.D., 1975, Lithofacies and related hydrocarbon accumulations in Tertiary strata of the western and central Uinta Basin, Utah: Rocky Mountain Association of Geologists Symposium on Deep Drilling Frontiers, p. 163–173.

Fouch, T.D., 1976, Revision of the lower part of the Tertiary System in the central and western Uinta Basin, Utah: U.S. Geological Survey Bulletin 1405-C, 7 p.

Fouch, T.D., Hanley, J.H., and Forester, R.M., 1979, Preliminary correlation of Cretaceous and Paleogene lacustrine and related non-marine sedimentary and volcanic rocks in parts of the eastern Great Basin of Nevada and Utah: Rocky Mountain Association of Geologists Basin and Range Symposium, p. 305–312.

Fouch, T.D., Lawton, T.F., Nichols, D.J., Cashion, W.B., and Cobban, W.A., 1982, Chart showing preliminary correlation of major Albian to middle Eocene rock units from the Sanpete Valley in central Utah to the Book Cliffs in eastern Utah: Utah Geological Association Publication 10, p. 267–272.

Fouch, T.D., Lawton, T.F., Nichols, D.J., Cashion, W.B., and Cobban, W.A., 1983, Patterns and timing of synorogenic sedimentation in Upper Cretaceous rocks of central and northeast Utah: *in* Reynolds, M.W., and Dolly, E.D., eds., Symposium on Mesozoic Paleogeography of West-central U.S., Rocky Mountain Section of Society of Economic Paleontologists and Mineralogist, Denver, Colorado, p. 305–336.

Foutz, D.R., 1960, Geology of the Wash Canyon area, southern Wasatch Mountains, Utah: Brigham Young University Geology Studies, v. 7, no. 2, 37 p.

Fowkes, E.J., 1964, Pegmatites of Granite Peak Mountain, Tooele County, Utah: Brigham Young University Geology Studies, v. 11, p. 97–127.

Frahme, C.W., and Vaughn, E.B., 1983, Paleozoic geology and seismic stratigraphy of the northern Uncompahgre front, Grand County, Utah: Rocky Mountain Association of Geologists Symposium on Rocky Mountain Foreland Basins and Uplifts, Denver, p. 201–211.

Frey, R.W., and Howard, J.D., 1985, Trace fossils from the Panther Member, Star Point Formation, Coal Creek Canyon, Utah: Journal of Paleontology, v. 59, p. 370–404.

Friz, D.R., 1980, Paleocurrent directions in Late Triassic Kayenta Formation, Capitol Reef National Park, Utah: Utah Geological Association Publication 8, p. 123–127.

Furnish, W.M., 1973, Permian stage names: Canadian Society of Petroleum Geologists Memoir 2, p. 522–548. A symposium volume on The Permian and Triassic Systems and their mutual boundary, eds. A. Logan and L.V. Hills, Calgary.

Gable, D.J., and Hatton, T., 1983, Maps of vertical crustal movements in the conterminous U.S. over the last 10 million years: U.S. Geological Survey Map I-1315.

Galloway, C.L., 1970, Structural geology of eastern part of the Smithfield quadrangle, Utah: unpublished master's thesis, Utah State University, Logan, 115 p.

Galyardt, G.L., and Rush, F.E., 1981, Geologic map of the Crater Springs Known Geothermal Resources area and vicinity, Juab and Millard Counties, Utah: U.S. Geological Survey Miscellaneous Investigations Series Map I-1297.

Gans, P.B., 1987, An open-system two-layer crustal stretching model for the eastern Great Basin: Tectonics, v. 6, p. 1–12.

Gans, P.B., Clark, D.H., Miller, E.L., Wright, J.E., and Sutter, J.F., 1986, Structural development of the Kern Mountains and northern Snake Range: Geological Society of America Abstracts with Programs, v. 18, p. 108.

Gans, P.B., and Miller, E.L., 1983, Style of mid-Tertiary extension in east-central Nevada: Utah Geological and Mineral Survey Special Studies 59, p. 107–139.

Gans, P.B., Miller, E.L., McCarthy, J., and Ouldcott, M.L., 1985, Tertiary extensional faulting and evolving ductile-brittle transition zones in the northern Snake Range and vicinity—new insights from seismic data: Geology, v. 13, p. 189–193.

Gardiner, L.L., 1974, Environmental analysis of the Upper Cambrian Nounan Formation, Bear River Range and Wellsville Mountain, north-central Utah: unpublished master's thesis, Utah State Univeristy, Logan, 121 p.

Garrand, L.J., 1985, Phosphate in the Uinta Basin, northeast Utah; Utah Geological Association Publication 12, p. 263–270.

Gartner, A.E., 1986, Geometry, emplacement history, petrography, and chemistry of a basaltic intrusive complex, San Rafael and Capitol Reef areas, Utah: U.S. Geological Survey Open-File Report 86-81, 66 p.

Garvin, R.F., 1969, Stratigraphy and economic significance, Currant Creek Formation, northwest Uinta Basin, Utah: Utah Geological and Mineral Survey Special Studies 27, 53 p.

Gates, J.S., 1982, Hydrogeology of the Gunnison-Fairview-Nephi area, central Utah: Utah Geological Association Publication 10, p. 151–161.

Gates, J.S., 1984, Hydrogeology of northwestern Utah and adjacent parts of Idaho and Nevada: Utah Geological Association Publication 13, p. 239–248.

Gates, J.S., 1987, Ground-water in the Great Basin part of the Basin and Range Province, western Utah: Utah Geological Association Publication 16, p. 75–90.

Gates, J.S., and Kruer, S.A., 1981, Hydrologic reconnaissance of the southern Great Salt Lake Desert and summary of the hydrology of west-central Utah: Utah Department of Natural Resources Technical Publication No. 71, 45 p.

Gates, J.S., Steiger, J.I., and Green, R.T., 1984, Ground-water reconnaissance of the central Weber River area, Morgan and Summit Counties, Utah: Utah Department of Natural Resources Technical Publication 77, 59 p.

George, S.E., 1985, Geology of the Fillmore and Kanosh quadrangles, Millard County, Utah: Brigham Young University Geology Studies v. 32, part 1, p. 39–62.

Gilbert, G.K., 1890, Lake Bonneville: U.S. Geological Survey Monograph 1, 438 p.

Gill, J.R., and Hail, W.J. Jr., 1975, Stratigraphic sections across Upper Cretaceous Mancos Shale-Mesaverde Group boundary, eastern Utah and western Colorado: U.S. Geological Survey Oil and Gas Investigations Chart OC-68.

Gilland, J.K., 1979, Paleoenvironment of a carbonate lens in the lower Navajo Sandstone near Moab, Utah: Utah Geology, v. 6, no. 1, p. 29–38.

Gilliland, W.N., 1951, Geology of the Gunnison quadrangle, Utah: University of Nebraska Studies, new series no. 8, 101 p.

Gilluly, J., 1929, Geology and oil and gas prospects of part of the San Rafael Swell, Utah: U.S. Geological Survey Bulletin 806-C, p. 69–130.

Gilluly, J., 1932, Geology and ore deposits of the Stockton and Fairfield quadrangles, Utah: U.S. Geological Survey Professional Paper 173, 171 p.

Girdley, W.A., 1974, Kaibab Limestone and associated strata, Circle Cliffs, Utah: Utah Geology, v. 1, no. 1, p. 5–20.

Glick, L.L., and Miller, D.M., 1986, Geologic map of the Lucin 4 NW quadrangle, Box Elder County, Utah: Utah Geological and Mineral Survey Map 93.

Glick, L.L., and Miller, D.M., 1987a, Geologic map of the Pigeon Mountain quadrangle, Box Elder County, Utah: Utah Geological and Mineral Survey Map 94.

Glick, L.L., and Miller, 1987b, Geologic map of the Jackson quadrangle, Box Elder County, Utah: Utah Geological and Mineral Survey Map 95.

Goode, H.D., 1973a, Preliminary geologic map of the Bald Knoll quadrangle, Utah: U.S. Geological Survey Miscellaneous Field Studies Map MF-520.

Goode, H.D., 1973b, Preliminary geologic map of the Skutumpah Creek quadrangle, Utah: U.S. Geological Survey Miscellaneous Field Studies map MF-521.

Gordon, M. Jr., 1984, Biostratigraphy of the Chainman Shale: Western Geological Excusions sponsored by the Geological Society of America and the Department of Geology, University of Nevada-Reno, Volume 1, p. 74–77. Editor Joseph Lintz, Jr., Reno.

Gori, P.L., and Hays, W.W., eds., 1987, Assessment of regional earthquake hazards and risk along the Wasatch Front, Utah: U.S. Geological Survey Open-File Report 87-585, two volumes, 1294 p.

Gosney, T.C., 1982, Conodont biostratigraphy of the Pinyon Peak Limestone and the Fitchville Formation, Late Devonian-Early Mississippian, North Salt Lake City, Utah: Brigham Young University Geology Studies v. 29, part 2, p. 27–39.

Granath, V.C.H., Papike, J.J., and Labotka, T.C., 1983, The Notch Peak contact metmorphic aureole, Utah—petrology of the Big Horse Limestone Member of the Orr Formation: Geological Society of America Bulletin, v. 94, p. 889–906.

Grant, S.K., 1978, Stratigraphic relations of the Escalante Desert Formation near Lund, Utah: Brigham Young University Geology Studies, v. 25, part 3, p. 27–30.

Gray, W.E., 1975, Structural geology of the southern part of Clarkston Mountain, Malad Range, Utah: unpublished master's thesis, Utah State University, Logan, 58 p.

Greenhalgh, B.R., 1980, Fitchville Formation: a study of the biostratigraphy and depositional environments in west-central Utah County, Utah: Brigham Young University Geology Studies, v. 27, part 1, p. 9–29.

Greenman, E.R., 1982, Petrology and mineralogy of Tertiary volcanic rocks in the vicinity of the Rozel Hills and Black Mountain, Box Elder County, Utah: unpublished master's thesis, Utah State University, Logan, 75 p.

Gregory, H.E., 1948, Geology and geography of central Kane County, Utah: Geological Society of America v. 59, p. 211–248.

Gregory, H.E., 1950a, Geology and geography of the Zion Park region, Utah and Arizona: U.S. Geological Survey Professional Paper 220, 200 p.

Gregory, H.E., 1950b, Geology of eastern Iron County, Utah: Utah Geological and Mineral Survey Bulletin 37, 153 p.

Gregory, H.E., 1951, Geology and geography of the Paunsaugunt region, Utah: U.S. Geological Survey Professional Paper 226, 116 p.

Grier, S., 1983, Tertiary stratigraphy and geologic history of the Sacramento pass area, Nevada: Utah Geological and Mineral Survey Special Studies 59, p. 139–144.

Gries, R., 1983, North-south compression of Rocky Mountain foreland structures: Rocky Mountain Association of Geologists Symposium on Rocky Mountain Foreland Basins and Uplifts, Denver, p. 9–32.

Grogger, P.K., 1974, Glaciation of the High Uinta primitive area, Utah: with emphasis on the northern slopes: unpublished Ph.D. thesis, University of Utah: 209 p.

Gunther, L.F., and Gunther, V.G., 1981, Some Middle Cambrian fossils of Utah: Brigham Young University Geology Studies, v. 28, part 1, 81 p.

Gustafson, V.O., 1981, Petroleum geology of the Devonian and Mississippian rocks of the Four Corners region: Rocky Mountain Association of Geologists Guidebook—Geology of the Paradox Basin, p. 101–109.

Gustason, E.R., 1987, Depostional history of the mid-Cretaceous Dakota Formation and early foreland basin evolution, southern Utah: Geological Society of America Abstracts with Programs, v. 19, p. 687. See also Kauffman et al., 1987; and Eaton et al., 1987.

Gutschick, R.A., and Rodriquez, J., 1979, Biostratigraphy of the Pilot Shale, (Devonian-Mississippian) and contemporaneous strata in Utah, Nevada, and Montana: Brigham Young University Geology Studies, v. 26, part 1, p. 37–63.

Gutschick, R.G., and Sandberg, C.A., 1983, Mississippian continental margins of the conterminous U.S.: Society of Economic Paleontologists and Mineralogists Special Publication 33, p. 79–96.

Gwynn, J.W., ed., 1980, Great Salt Lake: Utah Geological and Mineral Survey Bulletin 116, 400 p. Review papers by many authors covering many aspects of Great Salt Lake.

Hackman, R.J., and Wyant, D.G., 1973, Geology, structure, and uranium deposits of the Escalante quadrangle, Utah and Arizona: U.S. Geological Survey Miscellaneous Geological Investigations Map I-744.

Hagen, E.S., Schuster, M.W., and Furlong, K.P., 1985, Tectonic loading and subsidence of intermontane basins, Wyoming foreland province: Geology, v. 13, p. 585–592.

Hale, L.A., 1960, Frontier Formation-Coalville, Utah and nearby areas of Wyoming and Colorado: Wyoming Geological Association Guidebook 15, p. 137–146.

Hale, L.A., and Van de Graaff, F.R., 1964, Cretaceous stratigraphy and facies patterns, northeastern Utah and adjacent areas: Intermountain Association of Petroleum Geologists Guidebook 13, p. 115–138.

Hall, C.D., 1981, The Tintic Quartzite in Rock Canyon, Utah County, Utah: Brigham Young University Geology Studies, v. 28, part 3, p. 67–79.

Hallam, Anthony, 1982, Jurassic climate: *in* Climate in Earth History, National Academy Press, Washington D.C., p. 159–163.

Halliday, M.E., and Cook, K.L., 1980, Regional gravity survey, northern Marysvale volcanic field, south-central Utah: Geological Society of America Bulletin, part 1, v. 91, p. 502–508.

Hamblin, W.H., 1965, Origin of "reverse drag" on the downthrown side of normal faults: Geological Society of America Bulletin, v. 76, p. 1145–1164.

Hamblin, W.K., 1970, Late Cenozoic basalt flows of the western Grand Canyon: Utah Geological Society Guidebook 23, p. 21–37.

Hamblin, W.K., 1976, Patterns of displacement along the Wasatch fault: Geology, v. 4, p. 619–622.

Hamblin, W.K., 1984, Direction of absolute movement along the boundary faults of the Basin and Range—Colorado Plateau margin: Geology, v. 12, p. 116–119.

Hamblin, W.K., 1987, Late Cenozoic volcanism in the St. George basin, Utah: Geological Society of America Centennial Field Guide v.2, Rocky Mountain Section, p. 291–294.

Hamblin, W.K., Damon, P.E., and Bull, W.B., 1981, Estimates of vertical crustal strain rates along the western margins of the Colorado Plateau: Geology, v. 9, p. 293–298.

Hamilton, Warren, 1978, Mesozoic tectonics of the western U.S.: *in* Howell, D.G., and McDougall, A.K., eds., Mesozoic Paleogeography of the Western U.S., Society of Economic Paleontologists and Mineralogists, Pacific Section, Los Angeles, p. 33–70.

Hamilton, Warren, 1981, Plate-tectonic mechanism of Laramide deformation: University of Wyoming Contributions to Geology, v. 19, p. 87–92.

Hamilton, Warren, 1982, Structural evolution of the Big Maria Mountains, northeastern Riverside County, southeastern California: *in* Frost, E.G., and Martin, D.L., eds., Mesozoic-Cenozoic tectonic evolution of the Colorado River region, California, Arizona, and Nevada: San Diego, California, Cordilleran Publishers, p. 1–28.

Hamilton, W.L., 1978, Geological map of Zion National Park, Utah, 1:31680: Zion Natural History Association, Springdale, Utah.

Hannum, Cheryl, 1980, Sandstone and conglomerate breccia pipes and dikes of the Kodachrome Basin area, Kane County, Utah: Brigham Young University Geology Studies, v. 27, part 1, p. 31–50.

Hansen, G.H., and Scoville, H.C., 1955, Drilling records of oil and gas in Utah: Utah Geological and Mineralogical Survey Bulletin 50, 110 p.

Hansen, W.R., 1965, Geology of the Flaming Gorge area, Utah-Colorado-Wyoming: U.S. Geological Survey Professional Paper 490, 196 p.

Hansen, W.R., 1984, Post-Laramide tectonic history of the eastern Uinta Mountains, Utah, Colorado, and Wyoming: The Mountain Geologist, v. 21, no. 1, p. 5–29.

Hansen, W.R., Rowley, P.D., and Carrara, P.E., 1983, Geologic map of Dinosaur National Monument and vicinity, Utah and Colorado: U.S. Geological Survey Miscellaneous Investigations Series Map I-1407.

Hanson, A.M., 1949, Geology of the southern Malad Range and vicinity in northern Utah: unpublished Ph.D. thesis, University of Wisconsin, 128 p.

Harden, D.R., Biggar, N.E., and Gillam, M.L., 1985, Quaternary deposits and soils in and around Spanish Valley, Utah: Geological Society of America Special Paper 203, p. 43–64.

Hardy, C.T., 1962, Mesozoic and Cenozoic stratigraphy of north-central Utah: Brigham Young University Geology Studies, v. 9, part 1, p. 50–64.

Hardy, C.T., and Muessig, S., 1952, Glaciation and drainage changes in the Fish Lake Plateau, Utah: Geological Society of America Bulletin, v. 60, p. 1109–1116.

Harland, W.B., Cox, A.V., Llewellyn, P.G., Smith, A.G., and Walters, R., 1982, A geologic time scale: Cambridge Earth Science Series, Cambridge University Press, Great Britain, 131 p.

Harmala, J.C., 1982, Conodont biostratigraphy of some Mississippian rocks in northeastern Nevada and northwestern Utah: unpublished master's thesis, Arizona State University, Tempe, 271 p.

Harper, G.D., and Link, P.K., 1986, Geochemistry of Upper Proterozoic rift-related volcanics, northern Utah and southeastern Idaho: Geology, v. 14, p. 864–867.

Harris, D.R., 1980, Exhumed paleochannels in the Lower Cretaceous Cedar Mountain Formation near Green River, Utah: Brigham Young University Geology Studies v. 27, part 1, p. 51–66.

Harris, H.D., 1959, Late Mesozoic positive area in western Utah: American Association of Petroleum Geologists Bulletin, v. 43, p. 2636–2652.

Harrison, J.E., and Peterman, Z.E., 1984, Introduction to correlation of Precambrian rock sequences: U.S. Geological Survey Professional Paper 1241-A, 7 p.

Hauge, T.A., and 10 others, 1987, Crustal structure of western Nevada from COCORP deep seismic-reflection data: Geological Society of America Bulletin, v. 98, p. 320–329.

Hausel, W.D., and Nash, W.P., 1977, Petrology of Tertiary and Quaternary volcanic rocks, Washington County, southwestern Utah: Geological Society of America Bulletin, v. 88, p. 1831–1842.

Hauser, E.C., and nine others, 1987a, Crustal structure of eastern Nevada from COCORP deep seismic reflection data: Geological Society of America Bulletin, v. 99, p. 833–844.

Hauser, E.C., and six others, 1987b, COCORP Arizona transect—strong crustal reflections and offset Moho beneath the transition zone: Geology, v. 15, p. 1103–1106.

Hawks, R. L. Jr., 1980, Stratigraphy and structure of the Cedar Hills, Sanpete County, Utah: Brigham Young University Geology Studies v. 27, part 1, p. 67–80.

Hawley, C.C., Robeck, R.C., and Dyer, H.B., 1968, Geology, altered rocks and ore deposits of the San Rafael Swell, Emery County, Utah: U.S. Geological Survey Bulletin 1239, 115 p.

Hay, H.W. Jr., 1982, Petrology of the Middle Cambrian Blacksmith Formation, north-central Utah: unpublished master's thesis, Utah State University, Logan, 157 p.

Haymond, D.E., 1981, Geology of the Longlick and White Mountain area, southern San Francisco Mountains, Beaver County, Utah: Brigham Young University Geology Studies, v. 28, part 3, p. 81–100.

Haynes, D.D., Vogel, J.D., Wyant, D.G., 1972, Geology, structure, and uranium deposits of the Cortez quadrangle, Colorado and Utah: U.S. Geological Survey Map I-629.

Haynie, A.V. Jr., 1957, Worm Creek Quartzite Member of the St. Charles Formation: unpublished master's thesis, Utah State University, Logan, 39 p.

Hays, J.D., 1960, A study of the South Flat and related formations of central Utah—Part I, Petrology; Part II, Palynology: unpublished master's thesis, Ohio State University, 146 p.

Hedge, C.E., Houston, R.S., Tweto, O.L., Peterman, Z.E., Harrison, J.E., and Reid, R.R., 1986, The Precambrian of the Rocky Mountain Region: U.S. Geological Survey Professional Paper 1241-D, 17 p.

Hedge, C.E., Peterman, Z.E., Case, J.E., and Obradovich, J.D., 1958, Precambrian geochronology of the northwestern Uncompahgre Plateau, Utah and Colorado: U.S. Geological Survey Professional Paper 600-C, p. 91–96.

Hedge, C.E., Stacey, J.S., and Bryant, Bruce, 1983, Geochronology of the Farmington Canyon Complex, Wasatch Mountains, Utah: Geological Society of America Memoir 157, p. 37–44.

Heller, P.L., Bowdler, S.S., Chambers, H.P., Coogan, J.C., Hagen, E.S., Shuster, M.W., Winslow, N.S., and Lawton, T.F., 1986, Time of initial thrusting in the Sevier orogenic belt, Idaho-Wyoming and Utah: Geology, v. 14, p. 388–391.

Hely, A.G., Mower, R.W., and Harr, C.A., 1971, Water resources of Salt Lake County, Utah: Utah Department of Natural Resources Technical Publication 31, 239 p.

Henderson, G.V., 1964, Geology of the north half of the Soldier Summit quadrangle, Utah: Intermountain Association of Petroleum Geologists Guidebook 13, p. 157–168.

Herr, R.G., and Picard, M.D., 1981, Petrography of Upper Cambrian Lodore Formation, northeast Utah and northwest Colorado: University of Wyoming Contributions to Geology, v. 20, no. 1, p. 1–21.

Herr, R.G., Picard, M.D., and Evans, S.H. Jr., 1982, Age and depth of burial, Cambrian Lodore Formation, northeastern Utah and northwestern Colorado: University of Wyoming Contributions to Geology, v. 21, no. 2, p. 115–121.

Heylmun, E.B., Cohenour, R.E., and Kayser, R.B., 1965, Drilling records for oil and gas in Utah: Utah Geological and Mineral Survey Bulletin 74, 518 p.

Heyman, O.G., 1983, Distribution and structural geometry of faults and fields along the northwestern Uncompahgre uplift, western Colorado and eastern Utah: Grand Junction Geological Society Guidebook to the Northern Paradox Basin and Uncompahgre Uplift, p. 45–57.

Higgins, J.M., 1982, Geology of the Champlin Peak Quadrangle, Juab and Millard Counties, Utah: Brigham Young University Geology Studies, v. 29, part 2, p. 40–58.

High, L.R. Jr., and Picard, M.D., 1975, Sedimentary cycles in the Nugget Sandstone, northeastern Utah: Utah Geology, v. 2, no. 2, p. 117–124.

Hildebrand, R.T., and Newman, K.R., 1985, Miocene sedimentation in the Goose Creek basin, south-central Idaho, northeastern Nevada, and northwestern Utah: *in* Cenozoic Paleogeography of the West-central United States, Rocky Mountain Paleogeography Symposium 3, Society of Economic Geologists and Paleontologists, Denver, p. 55–70.

Hildenbrand, T.G., Simpson, R.W., Godson, R.H., and Kane, M.F., 1982, Digital colored residual and regional Bouquer gravity maps of the conterminous U.S. with cut-off wavelengths of 250 km and 1000 km: U.S. Geological Survey Map GP-953A.

Hill, R.B., 1982, Depositional environments of the Upper Cretaceous Ferron Sandstone south of Notom, Wayne County, Utah: Brigham Young University Geology Studies, v. 29, part 2, p. 59–84.

Hintze, L.F., 1951, Lower Ordovician detailed stratigraphic sections of western Utah: Utah Geological and Mineralogical Survey Bulletin 39, 99 p.

Hintze, L.F., 1952, Lower Ordovician trilobites from western Utah and eastern Nevada: Utah Geological Mineralogical Survey Bulletin 48, 249 p.

Hintze, L.F., 1962, Precambrian and lower Paleozoic rocks of north-central Utah: Brigham Young University Geology Studies v. 9, part 1, p. 8–16.

Hintze, L.F., 1971, Wasatch Fault zone east of Provo, Utah: Utah Geological Association Publication 1, p. F1–F10.

Hintze, L.F., 1973, Lower and Middle Ordovician stratigraphic sections in the Ibex area, Millard County, Utah: Brigham Young University Geology Studies, v. 20, part 4, p. 3–36.

Hintze, L.F., 1974a, Preliminary geologic map of The Barn quadrangle, Millard County, Utah: U.S. Geological Survey Miscellaneous Field Studies Map MF-633.

Hintze, L.F., 1974b, Preliminary geologic map of the Conger Mountain quadrangle, Millard County, Utah: U.S. Geological Survey Miscellaneous Field Studies Map MF-634.

Hintze, L.F., 1974c, Preliminary geologic map of the Crystal Peak quadrangle, Millard County, Utah: U.S. Geological Survey Miscellaneous Field Studies Map MF-635.

Hintze, L.F., 1974d, Preliminary geologic map of the Wah Wah Summit quadrangle, Millard and Beaver Counties, Utah: U.S. Geological Survey Miscellaneous Field Studies Map MF-637.

Hintze, L.F., 1978a, Geologic map of the Y-Mountain area east of Provo, Utah: Brigham Young University Geology Studies, Special Publication 5.

Hintze, L.F., 1978b, Sevier orogenic attenuation faulting in the Fish Springs and House ranges, western Utah: Brigham Young University Geology Studies, v. 25, part 1, p. 11–24.

Hintze, L.F., 1979, Preliminary zonations of Lower Ordovician of western Utah by various taxa: Brigham Young University Geology Studies, v. 26, part 2, p. 13–19.

Hintze, L.F., 1980a, Preliminary geologic map of the Fish Springs NW and Fish Springs SW quadrangles, Juab and Tooele Counties, Utah: U.S. Geological Survey Miscellaneous Field Studies Map MF-1148.

Hintze, L.F., 1980b, Preliminary geologic map of the Sand Pass NW quadrangle, Juab County, Utah: U.S. Geological Survey Miscellaneous Field Studies Map MF-1149.

Hintze, L.F., 1980c, Preliminary geologic map of Fish Springs NE and Fish Springs SE quadrangles, Juab and Tooele Counties, Utah: U.S. Geological Survey Miscellaneous Field Studies Map MF-1147.

Hintze, L.F., 1981a, Preliminary geologic map of the Marjum Pass and Swasey Peak SW quadrangles, Millard County, Utah: U.S. Geological Survey Miscellaneous Field Studies Map MF-1332.

Hintze, L.F., 1981b, Preliminary geologic map of the Swasey Peak and Swasey Peak NW quadrangles, Millard County, Utah: U.S. Geological Survey Miscellaneous Field Studies Map MF-1333.

Hintze, 1981c, Preliminary geologic map of the Whirlwind Valley NW and Whirlwind Valley SW quadrangles, Millard County, Utah: U.S. Geological Survey Miscellaneous Field Studies Map MF-1335.

Hintze, L.F., 1982, Ibexian Series (Lower Ordovician) type section, western Utah, U.S.A.: International Union of Geological Sciences Publication No. 12, p. 7–10.

Hintze, L.F., 1984, Geology of the Cricket Mountains, Millard County, Utah: U.S. Geological Survey Open-File Report 84-683, 14 p.

Hintze, L.F., 1985a, Geologic map of the Shivwits and West Mountain Peak quadrangles, Washington County, Utah: U.S. Geological Survey Open-File Report 85-119, 19 p., maps.

Hintze, L.F., 1985b, Geologic map of the Castle Cliff and Jarvis Peak quadrangles, Washington County, Utah: U.S. Geological Survey Open-File Report 85-120, 19 p. maps.

Hintze, L.F., 1985c, Great Basin Region—Correlation of stratigraphic units of North America: American Association of Petroleum Geologists, COSUNA Project Chart, Tulsa.

Hintze, L.F., 1986a, Geologic map of the Mormon Gap and Tweedy Wash quadrangles, Millard County, Utah, and Lincoln and White Pine Counties, Nevada: U.S. Geological Survey Miscellaneous Field Studies Map MF-1872.

Hintze, L.F., 1986b, Stratigraphy and structure of the Beaver Dam Mountains, southwestern Utah: Utah Geological Association Publication 15, p. 1–36.

Hintze, L.F., 1986c, CUSMAP—Minerals appraisal: Utah Geological and Mineral Survey Notes, v. 20, no. 4, p. 3–7.

Hintze, L.F., 1987, Exceptionally fossiliferous Lower Ordovician strata in the Ibex area, western Millard County, Utah: Geological Society of America Centennial Field Guide, Volume 2, p. 261–264, Boulder, Colorado.

Hintze, L.F., and Anderson, R.D., 1985, Geologic map of the Motoqua and Gunlock quadrangles, Washington County, Utah: U.S. Geological Survey Open-File Report (in preparation).

Hintze, L.F., and Best, M.G., 1987, Geologic map of the Mountain Home Pass and Miller Wash quadrangles, Millard and Beaver Counties, Utah, and Lincoln County, Nevada: U.S. Geological Survey Miscellaneous Field Studies Map MF-1950.

Hintze, L.F., Lemmon, D.M., and Morris, H.T., 1984, Geologic map of the Frisco Peak quadrangle, Millard and Beaver Counties, Utah: U.S. Geological Survey Miscellaneous Investigations Map I-1573.

Hintze, L.F., Miller, J.F., and Taylor, M.E., 1980, Upper Cambrian-Lower Ordovician Notch Peak Formation in western Utah: U.S. Geological Survey Open-File Report 80-776, 67 p., also 1987, U.S. Geological Survey Professional Paper 1393.

Hintze, L.F., and Palmer, A.R., 1976, Upper Cambrian Orr Formation, its subdivisions and correlatives in western Utah: U.S. Geological Survey Bulletin 1405-G, 25 p.

Hintze, L.F., and Robison, R.A., 1975, Middle Cambrian stratigraphy of the House, Wah Wah, and adjacent ranges in western Utah: Geological Society of American Bulletin, v. 86, p. 881–891.

Hintze, L.F. and Robison, R.A., 1987, The House Range, western Utah—Cambrian mecca: Geological Society of America Centennial Field Guide, Volume 2, p. 257–269, Boulder Colorado.

Hintze, L.F., Taylor, M.E., and Miller, J.F., 1988, Upper Cambrian-Lower Ordovician Notch Peak Formation in western Utah: U.S. Geological Survey Professional Paper 1393.

Hite, R.J., and Cater, F.W., 1972, Pennsylvanian rocks and salt anticlines, Paradox Basin, Utah and Colorado, *in* Geologic Atlas of the Rocky Mountain Region, p. 133–137. Rocky Mountain Association of Geologists, Denver.

Hodgkinson, K.A., 1961, Permian stratigraphy of northeastern Nevada and northwestern Utah: Brigham Young University Geology Studies, v. 8, p. 167–196.

Hogg, N.C., 1972, Shoshonitic lavas in west-central Utah: Brigham Young University Studies, v. 19, part 2, p. 133–184.

Holladay, J.C., 1984, Geology of the northern Canyon Ridge, Millard and Juab Counties, Utah: Brigham Young University Geology Studies, v. 31, part 1, p. 1–28.

Holmes, C.N., Page, B.M., and Averitt, Paul, 1948, Geology of the bituminous sandstone deposits near Sunnyside, Carbon County, Utah: U.S. Geological Survey Oil and Gas Investigation Preliminary Map 86. Summary in Intermountain Association of Petroleum Geologists Guidebook 7, p. 171–177, (1956).

Holmes, W.F., 1984, Ground-water hydrology and projected effects of ground-water withdrawals in the Sevier Desert, Utah: Utah Department of Natural Resources Technical Publication 79, 43 p.

Holmes, R.D., 1979, Thermoluminescence dating of Quaternary basalts—continental basalts from the eastern margin of the Basin and Range Province, Utah and northern Arizona: Brigham Young University Geology Studies, v. 26, part 2, p. 51–65.

Hood, J.W., 1972, Hydrologic reconnaissance of the Promontory Mountains area, Box Elder County, Utah: Utah Department of Natural Resources Technical Publication 38, 35 p.

Hood, J.W., 1976, Characteristics of aquifers in the northern Uinta Basin area, Utah and Colorado: Utah Department of Natural Resources Technical Publication 53, 63 p.

Hood, J.W. and Fields, F.K., 1978, Water resources of the northern Uinta Basin area, Utah and Colorado, with special emphasis on ground-water supply: Utah Department of Natural Resources Technical Publication 62, 66 p.

Hood, J.W. and Patterson, D.J., 1984, Bedrock aquifers in the northern San Rafael Swell area, Utah, with special emphasis on the Navajo Sandstone: Utah Department of Natural Resources Technical Publication 78, 117 p.

Hood, J.W., Price, Don, and Waddell, K.M., 1969, Hydrologic reconnaissance of Rush Valley, Tooele County, Utah: Utah Department of Natural Resources Technical Publication 23, 58 p.

Hood, J.W. and Rush, F.E., 1966, Water resources appraisal of the Snake Valley area, Utah and Nevada: Utah Department of Natural Resources Technical Publication 14, 60 p.

Hood, J.W. and Waddell, K.M., 1968, Hydrologic reconnaissance of Skull Valley, Tooele County, Utah: Utah Department of Natural Resources Technical Publication 18, 54 p.

Hoover, J.D., 1974, Periodic Quaternary volcanism in the Black Rock Desert, Utah: Brigham Young University Geology Studies, v. 21, part 1, p. 3–72.

Hornbacher, D., 1985, Geology and structure of Kodachrome Basin State Reserve and vicinity, Kane and Garfield Counties, Utah: unpublished master's thesis, Loma Linda University, Riverside, California, 179 p.

Hose, R.K., 1966, Devonian stratigraphy of the Confusion Range, west-central Utah: U.S. Geological Survey Professional Paper 550-B, p. 36–41.

Hose, R.K., 1977, Structural geology of the Confusion Range, west-central Utah: U.S. Geological Survey Professional Paper 971, 9 p.

Hose, R.K. and Repenning, C.A., 1959, Stratigraphy of Pennsylvanian, Permian, and Lower Triassic rocks of Confusion Range, west-central Utah: American Association of Petroleum Geologists Bulletin, v. 43, p. 2167–2196.

Huddle, J.W. and McCann, F.T., 1947, Geologic map of the Duchesne River area, Wasatch and Duchesne Counties, Utah: U.S. Geological Survey Oil and Gas Investigations Preliminary Map 75.

Huddle, J.W., Mapel, W.J., and McCann, 1951, Geology of the Moon Lake area, Duchesne County, Utah: U.S. Geological Survey Oil and Gas Investigations Map OM-115.

Huff, L.C. and Lesure, F.G., 1965, Geology and uranium deposits of the Montezuma Canyon area, San Juan County, Utah: U.S. Geological Survey Bulletin 1190, 102 p.

Hunt, C.B., 1956, Cenozoic geology of the Colorado Plateau: U.S. Geological Survey Professional Paper 279, 99 p.

Hunt, C.B., 1958, Structural and igneous geology of the La Sal Mountains, Utah: U.S. Geological Survey Professional Paper 294-I, p. 305–364.

Hunt, C.B., 1983, Development of the La Sal and other laccolithic mountains on the Colorado Plateau: Grand Junction (Colorado) Geological Society Guidebook to Northern Paradox Basin and Uncompahgre Uplift, p. 29–32.

Hunt, C.B., Averitt, P., and Miller, R.L., 1953, Geology and geography of the Henry Mountains region, Utah; U.S. Geological Survey Professional Paper 228, 234 p. Republished with minor comments in Utah Geological Association Publication 8, p. 25–106, 1980.

Hunt, C.B., Varnes, H.D., and Thomas, H.E., 1953, Lake Bonneville—geology of northern Utah Valley, Utah: U.S. Geological Survey Professional Paper 257-A, 99 p.

Hunt, R.E., 1954, South Flat Formation, New Upper Cretaceous formation of central Utah: American Association of Petroleum Geologists Bulletin, v. 38, no. 1 p. 118–128.

Huntoon, P.W., 1981, Grand Canyon monoclines—vertical uplift or horizontal compression?: University of Wyoming Contributions to Geology, v. 19, p. 127–134.

Huntoon, P.W., 1982, The Meander anticline, Canyonlands, Utah—an unloading structure resulting from horizontal gliding on salt: Geological Society of America Bulletin, v. 93, p. 941–950.

Huntoon, P.W., Billingsley, G.H. Jr., Breed, W.J., Sears, J.W., Ford, T.D., Clark, M.D., Babcock, R.S., and Brown E.H., 1976, Geologic map of the Grand Canyon National Park, Arizona: Museum of Northern Arizona, Flagstaff.

Huntoon, P.W., Billingsley, G.H. Jr., and Breed, W.J., 1982, Geologic map of Canyonlands National Park and vicinity, Utah: Canyonlands Natural History Association, Moab, Utah.

Huntoon, P.W., Breed, W.J., and Billingsley, G.H., 1986, Geologic map of Capitol Reef National Park and vicinity, Utah; Utah Geological and Mineral Survey Map 87.

Hurst, Carolyn, 1982, Detailed gravity survey delineating buried strike-slip faults in the Crawford Mountain portion of the Utah-Idaho-Wyoming overthrust belt: Brigham Young University Geology Studies, v. 29, part 2, p. 85–102.

Imlay, R.W., 1967, Twin Creek Limestone (Jurassic) in the western interior of the United States: U.S. Geological Survey Professional Paper 540, 105 p.

Imlay, R.W., 1980, Jurassic paleobiogeography of the conterminous United States in its continental setting: U.S. Geological Survey Professional Paper 1062, 134 p.

Irwin, C.D., 1971, Stratigraphic analysis of upper Permian and lower Triassic strata in southern Utah: American Association of Petroleum Geologists Bulletin, v. 55, p. 1976, 2007.

Irwin, C.D., 1976, Permian and lower Triassic reservoir rocks of central Utah: Rocky Mountain Association of Geologists, Symposium of Geology of Cordilleran Hingeline, p. 193–202, Denver.

Isaacson, P.E., Measures, E.A., and Siegmann, S.A., 1985, Depositional paleoenvironments and paleogeography of the Mississippian Lodgepole and Monroe Canyon Formations and equivalents, southeastern Idaho: Utah Geological Association Publication 14, p. 15–28.

Isby, J.S., and Picard, M.D., 1983, Currant Creek Formation: record of tectonism in Sevier-Laramide orogenic belt, north-central Utah: University of Wyoming Contributions to Geology, v. 22, no.2, p. 91–108.

Isby, J.S., and Picard, M.D., 1985, Depositional setting of Upper Cretaceous-Lower Tertiary Current Creek Formation, north-central Utah: Utah Geological Association Publication 12, p. 39–49.

Izett, G.A., 1981, Volcanic ash beds—recorders of upper Cenozoic silicic pyroclastic volcanism in the western U.S.: Journal of Geophysical Research, v. 86, p. 10200–10222.

Izett, G.A., and Wilcox, R.E., 1982, Map showing localities and inferred distributions of the Huckleberry Ridge, Mesa Falls, and Lava Creek ash beds (Pearlette Family ash beds) of Pliocene and Pleistocene age in the western U.S. and southern Canada: U.S. Geological Survey Map I-1325.

Jaanusson, V. 1979, Ordovician biogeography and biostratigraphy: *in* R.A. Robison and C. Teichert, eds., Treatise on Invertebrate Paleontology, Part A, p. A136–166.

Jackson, A.L., 1983, Humates and their development at Harley Dome, Utah: Grand Junction Geological Society Guidebook to Northern Paradox Basin and Uncompahgre Plateau, p. 17–19.

Jackson, D. 1985, American gilsonite-mining solid hydrocarbon: Utah Geological Association Publication 12, p. 257–281.

Jacobson, S.R., and Nichols, D.J., 1982, Palynological dating of syntectonic units in the Utah-Wyoming thrust belt: the Evanston Formation, Echo Canyon Conglomerate, and Little Muddy Creek Conglomerate: *in* R.B. Powers, ed., Rocky Mountain Association of Geologists, Symposium, Geologic Studies of the Cordilleran thrust belt, p. 735–750.

Jado, A.R.I., 1978, Stratigraphy and petrology of the Park City Formation in northwestern Colorado and northeastern Utah: unpublished Ph.D. thesis, Colorado School of Mines, 263 p.

James, B.H., 1980, Paleoenvironments of the lower Triassic Thaynes Formation near Diamond Fork in Spanish Fork Canyon, Utah County, Utah: Brigham Young University Geology Studies, v. 27, part 1, p. 81–100.

James, L.P., 1984, The Queen of Sheba gold and the Boston-Deep Creek silver mines, Goshute Indian Reservation, Juab County, Utah: Utah Geological Association Publication 13, p. 193–200.

Jarrett, R.D., and Malde, H.E., 1987, Paleodischarge of the late Pleistocene Bonneville flood, Snake River, Idaho, computed from new evidence: Geological Society of America Bulletin, v. 99, p. 127–134.

Jefferson, W.S., 1982, Structural and stratigraphic relations of upper Cretaceous to lower Tertiary orogenic sediments in the Cedar Hills, Utah: Utah Geological Association Publication 10, p. 65–80.

Jensen, M.E., 1986, Tertiary geologic history of the Slate Jack Canyon quadrangle, Juab and Utah Counties, Utah: Brigham Young University Geology Studies, v. 33, no. 1, p. 1–19.

Jensen, N.R., 1986, Geology of the Fairview 7½-minute quadrangle, Sanpete County, Utah: Brigham Young University Geology Studies, (in press).

Jenson, John, 1986, Stratigraphy and facies analysis of the upper Kaibab and lower Moenkopi formations in southwest Washington County, Utah: Brigham Young University Geology Studies, v. 33, p. 21–43.

Johnson, B.T., 1984, Depositional environment of the Iron Springs Formation, Gunlock, Utah: Brigham Young University Geology Studies, v. 31, part 1, p. 29–46.

Johnson, J.G., and Murphy, M.A., 1984, Time-rock model for Siluro-Devonian continental shelf, western United States: Geological Society of America Bulletin, v. 95, p. 1349–1359.

Johnson, J.G., and Sandberg, C.A., 1977, Lower and Middle Devonian continental-shelf rocks of the western U.S.: *in* Murphy, M.A. Berry, W.B.N., and Sandberg, C.A., eds., Western North America—Devonian, California University Riverside Campus Museum Contribution 4, p. 121–143.

Johnson, J.L., 1978, Stratigraphy of the coal-bearing Blackhawk Formation on North Horn Mountain, Wasatch Plateau, Utah: Utah Geology, v. 5, no. 1, p. 57–77.

Johnson, R.C., 1984, New names for units in the lower part of the Green River Formation, Piceance Basin, Colorado: U.S. Geological Survey Bulletin 1529 I, 20 p.

Johnson, R.C., 1985, Early Cenozoic history of the Uinta and Piceance Creek Basins, Utah and Colorado, with special reference to the development of Eocene Lake Uinta: *in* Cenozoic Paleogeography of west-central United States, Rocky Mountain Symposium 3, Society of Economic Geologists and Paleontologists, Denver, p. 247–276.

Johnson, R.C., and May, Fred, 1980, A study of the Cretaceous-Tertiary unconformity in the Piceance Creek Basin, Colorado—The underlying Ohio Creek Formation (Upper Cretaceous) redefined as a member of the Hunter Canyon or Mesaverde Formation: U.S. Geological Survey Bulletin 1482-B, 27 p.

Johnson, R.C., May, Fred, Hansley, P.L., Pitman, J.K., and Fouch, T.C., 1980, Petrography, X-ray mineralogy, and palynology of a measured section of the upper Cretaceous Mesaverde Group in Hunter Canyon, western Colorado: U.S. Geological Survey Oil and Gas Investigations Chart OC-91.

Johnson, S.R., and Vaninetti, G.E., 1982, Multiple barrier island and deltaic progradational sequences in Upper Cretaceous coal-bearing strata northern Kaiparowits Plateau, Utah: Utah Geological and Mineral Survey Bulletin 118, p. 62–69.

Jones, D.L., Cox, A., Coney, P., and Beck, M., 1982, The growth of western North America: Scientific American, November 1982, p. 70–84.

Jordan, T.E., 1979, Lithofacies of the upper Pennsylvanian and lower Permian western Oquirrh Group, northwest Utah: Utah Geology, v. 6, no. 2, p. 41–56.

Jordan, T.E., 1981, Enigmatic deep-water depositional mechanisms, upper part of the Oquirrh Group, Utah: Journal of Sedimentary Petrology, v. 51, p. 879–894.

Jordan, T.E., 1983, Structural geometry and sequence, Bovine Mountain, northwestern Utah: Geological Society of America Memoir 157, p. 215–228.

Jordan, T.E., 1985a, Geologic map of the Bulls Pass quadrangle, Box Elder County, Utah: U.S. Geological Survey Miscellaneous Field Studies Map MF-1491.

Jordan, T.E., 1985b, Tectonic setting and petrography of Jurassic foreland basin sandstones, Idaho-Wyoming-Utah: Utah Geological Association Publication 14, p. 201–214.

Jordan, T.E., and Allmendinger, R.W., 1979, Upper Permian and Lower Triassic stratigraphy of the Stansbury Mountains, Utah: Utah Geology, v. 6, no. 2, p. 69–74.

Jordan, T.E., and Douglass, R.C., 1980, Paleogeography and structural development of the late Pennsylvanian to early Permian Oquirrh Basin, northwestern Utah: *in* Paleozoic Paleogeography of the west-central United States, Rocky Mountain Section, Society of Economic Paleontologists and Mineralogists, Denver, p. 217–238.

Jordan, T.E., Isacks, B.L., Allmendinger, R.W., Brewer, J.A., Ramos, V.A., and Ando, C.J., 1983, Andean tectonics related to geometry of subducted Nazca plate: Geological Society of America Bulletin, v. 94, p. 341–361.

Kanter, L.R., Dyer, R. and Dohmen, T.E., 1981, Laramide crustal shortening in the northern Wyoming province: University of Wyoming Contributions to Geology, v. 19, p. 135–142.

Karklins, O.L., 1986 Chesterian (Late Mississippian) bryozoans from the upper Chainman Shale and the lowermost Ely Limestone of western Utah: Paleontological Society Memoir 17, published with the 1986 Journal of Paleontology, 48 p.

Kauffman, E.G., 1979, Cretaceous biogeography and biostratigraphy: *in* R.A. Robison and C. Teichert eds.,Treatise on Invertebrate Paleontology, Part A, p. A418–A487.

Kauffman, E.G., 1984, Paleobiogeography and evolutionary response dynamic in the Cretaceous western interior seaway of North America: Geological Association of Canada Special Paper 27, p. 273–306.

Kauffman, E.G., Sageman, B.B., Gustason, E.R., and Elder, W.P., 1987, Field trip guidebook: High-resolution event stratigraphy, Greenhorn cyclothem (Cretaceous) western interior basin of Colorado and Utah: Geological Society of America, Rocky Mountain Section Regional Meeting, Boulder, Colorado, May, 1987, 198 p.

Kay, J.L., 1957, Eocene vertebrates of the Uinta Basin, Utah: Intermountain Association of Petroleum Geologists Guidebook 8, p. 110–115.

Kay, Marshall, 1951, North American geosynclines: Geological Society of America Memoir 48, 143 p.

Keigwin, L., and Keller, G., 1984, Middle Miocene cooling from equatorial Pacific DSDP Site 77B: Geology, v. 12, p. 16–19.

Kelley, V.C., and Clinton, N.J., 1960, Fracture systems and tectonic elements of the Colorado Plateau: University of New Mexico Publications in Geology, no. 6, 104 p.

Kellogg, H.E., 1977, Geology and petroleum of the Mancos B Formation, Douglas Creek Arch area, Colorado and Utah: Rocky Mountain Association of Geologists Symposium, p. 167–179.

Kennedy, R.R., 1960, Geology between Pine (Bullion) Creek and Tenmile Creek, eastern Tushar Range, Piute County, Utah: Brigham Young University Geology Studies, v. 7, no. 4, 58 p.

Kent, D.V., and Gradstein, F.M., 1985, A Cretaceous and Jurassic geochronology: Geological Society of America Bulletin, v. 96, p. 1419–1427.

Kepper, J.C., 1972, Paleoenvironmental patterns in middle to lower Upper Cambrian interval in eastern Great Basin: American Association of Petroleum Geologists Bulletin, v. 56, no. 3, p. 503–527.

Kepper, J.C., 1976, Stratigraphic relationships and depositional facies in a portion of the Middle Cambrian of the Basin and Range Province: Brigham Young University Geology Studies, v. 23, pt. 2, p. 75–91.

Kepper, J.C., 1981, Sedimentology of a Middle Cambrian outer shelf margin with evidence for syndepositional faulting, eastern California and western Nevada: Journal of Sedimentary Petrology, v. 51, p. 807–821.

Kerns, R.L. Jr., 1987, Review of the petroleum activity of the Utah portion of the Great Basin: Utah Geological Association Publication 16.

Kerr, P.F., Brophy, G.P., Dahl, H.M., Green, J., and Woolard, L.E., 1957, Marysvale, Utah, uranium area: Geological Society of America Special Paper 64, 212 p.

Ketner, K.B., 1968, Origin of Ordovician quartzite in the Cordilleran miogeosyncline: U.S. Geological Survey Professional Paper 600-B, p. 169–177.

Kinney, D.M., 1951, Geology of the Uinta River and Brush Creek-Diamond Mountain areas, Duchesne and Uinta Counties, Utah: U.S. Geological Survey Oil and Gas Investigations Map OM-123.

Kiser, L.C., 1976, Stratigraphy and petroleum potential of the Permian Kaibab Beta Member, east central Utah: Rocky Mountain Association of Geologists, 1976 Symposium, p. 161–172.

Kistler, R.W., 1983, Isotope geochemistry of plutons in the northern Great Basin: Geothermal Resources Council Special Report No. 13, p. 3–8. Davis, CA, 95617-1350.

Kitzmiller, J.M. II, 1982, Preliminary geologic map of the Joes Valley Reservoir quadrangle, Emery and Sanpete Counties, Utah: Utah Geological and Mineral Survey Map 67.

Klemperer, S.L., Hauge, T.A., Hauser, E.C., Oliver, J.E., and Potter, C.J., 1986, The Moho in the northern Basin and Range province, Nevada, along the COCORP 40 degree north seismic-reflection transect: Geologic Society of America Bulletin, v. 97, p. 603–618.

Kluth, C.F., 1986, Plate tectonics of the Ancestral Rocky Mountains: American Association of Petroleum Geologists Memoir 41, p. 353–369.

Kluth, C.F., and Coney, P.J., 1981, Plate tectonics of the Ancestral Rocky Mountains: Geology, v. 9, p. 10–15.

Knoll, A.H., Blick, Nicholas, and Awramik, S.M., 1981, Stratigraphic and ecologic implications of Late Precambrian microfossils from Utah: American Journal of Science, v. 281, p. 247–263.

Knowles, S.P., 1985, Geology of the Scofield 7½-minute quadrangle, Carbon, Emery, and Sanpete Counties, Utah: Brigham Young University Geology Studies, v. 32, part 1, p. 85–100.

Koch, W.J., 1976, Lower Triassic facies in the vicinity of the Cordilleran hingeline—western Wyoming, southeastern Idaho, and Utah: Rocky Mountain Association of Geologists 1976 Symposium, Denver, p. 203–217.

Kocurek, G., 1981, Erg reconstruction-the Entrada sandstone of northern Utah and Colorado: Paleogeography, Palaeo-climatology and Palaeoecology, v. 36, nos. 1–2, p. 125–153.

Kocurek, G., and Dott, R.H., Jr., 1983, Jurassic paleogeography and paleoclimate of the central and southern Rocky Mountains region: *in* Reynolds, M.W., and Dolly, E.D., eds., Symposium on Mesozoic Paleogeography of West-central U.S., Rocky Mountain section of Society of Economic Paleontologists and Mineralogists, Denver, Colorado, p. 101–116.

Koelmel, M.H., 1986, Post-Mississippian paleotectonic, stratigraphic, and diagenetic history of the Weber Sandstone in the Rangely field area, Colorado: American Association of Petroleum Geologists Memoir 41, p. 371–396.

Koepnick, R.B., 1976, Depositional history of the Upper Dresbachian-Lower Franconian (Upper Cambrian) Pterocephaliid biomere from west-central Utah: Brigham Young University Geology Studies, v. 23, part 2, p. 123–138.

Konopka, E.H., and Dott, R.H. Jr., 1982, Stratigraphy and sedimentology, lower part of the Butterfield Peaks Formation (middle Pennsylvanian), Oquirrh Group, at Mt. Timponogos, Utah: Utah Geological Association Publication 10, p. 215–234.

Kopaska-Merkel, D.C., 1983, Paleontology and depositional environments of the Whirlwind Formation (Middle Cambrian), west-central Utah: unpublished Ph.D. thesis, Kansas University, 215 p.

Kornze, L.D., 1984, Geology of the Mercur Gold Mine: Utah Geological Association Publication 13, p. 201–214.

Kowallis, B.J., and Heaton, Julie, 1984, Fission-track stratigraphy of Upper Jurassic and Lower Cretaceous bentonites from the Morrison and Cedar Mountain formations of central Utah: Geological Society of America Abstracts with Programs, v. 16, no. 6, p. 565.

Kowallis, B.J., and Heaton, J.S., 1987, Fisson-track dating of bentonites and bentonitic mudstones from the Morrison Formation in central Utah: Geology, v. 15, p. 1138–1142.

Kummel, B., 1954, Triassic stratigraphy of southeastern Idaho and adjacent areas: U.S. Geological Survey Professional Paper 254-H, p. 165–194.

Kummel, B., 1979, Triassic biogeography and biostratigraphy: *in* R.A. Robison and C. Teichert (eds.), Treatise on Invertebrate Paleontology, Part A, Geological Society of America, p. A351–A389.

Lachenbruch, A.H., and Sass, J.H., Models of an extending lithosphere and heat flow in the Basin and Range province: Geological Society of America Memoir 152, p. 209–250.

Lambert, D.J., 1976, A detailed stratigraphic study of initial deposition of Tertiary lacustrine sediments near Mills, Utah: Brigham Young University Geology Studies, v. 23, part 3, p. 9–36.

Lambert, R.E., 1984, Shnabkaib Member of the Moenkopi Formation-depositional environment and stratigraphy near Virgin, Washington County, Utah: Brigham Young University Geology Studies, v. 31, part 1, p. 47–66.

Larsen, N.W., 1960, Geology and ground-water resources of northern Cedar Valley, Utah County, Utah: Brigham Young University Geology Studies, v. 7, no. 1, 42 p.

Larson, E.E., Patterson, P.E., Curtis, G., Drake, R., and Mutschler, F.E., 1985, Petrologic, paleomagnetic, and structural evidence of a Paleozoic rift system in Oklahoma, New Mexico, Colorado, and Utah: Geological Society of America Bulletin, v. 96, p. 1364–1372.

Larson, J.A., and Clark, D.L., 1979, Lower Permian (Sakmarian) portion of the Oquirrh Formation, Utah: Brigham Young University Geology Studies, v. 26, part 3, p. 135–142.

Lautenschlager, H.K., 1952, Geology of the central part of the Pavant Range, Utah: unpublished Ph.D. thesis, Ohio State University, 188 p.

Law, B.E., 1979, Coal deposits of the Emery coal zone, Henry Mountains coal field, Utah: U.S. Geological Survey Miscellaneous Field Studies Map MF-1082A and 1082B.

Law, B.E., 1980, Tectonic and sedimentological controls of coal bed depositional patterns in Upper Cretaceous Emery Sandstone, Henry Mountains coal field, Utah: Utah Geological Association Publication 8, Henry Mountains Symposium, p. 323–336.

Lawrence, J.C., 1965, Stratigraphy of Dakota and Tropic formations of Cretaceous age in southern Utah: Utah Geological Society Guidebook 19, p. 71–91.

Lawton, T.F., 1982, Lithofacies correlations within the upper Cretaceous Indianola Group, central Utah: Utah Geological Association Publication 10, p. 199–213.

Lawton, T.F., 1983, Late Cretaceous fluvial systems and the age of foreland uplifts in central Utah: Rocky Mountain Association of Geologists Symposium on Rocky Mountain Foreland Basins and Uplifts, Denver, p. 181–200.

Lawton, T.F., 1985, Style and timing of frontal structures, thrust belt, central Utah: American Association of Petroleum Geologists Bulletin, v. 69, p. 1145–1159.

Lawton, T.F., 1986, Fluvial systems of the Upper Cretaceous Mesaverde Group and Paleocene North Horn Formation, central Utah—a record of transition from thin-skinned to thick-skinned deformation in the foreland region: American Association of Petroleum Geologists Memoir 41, p. 423–442.

Lawyer, G.F., 1972, Sedimentary features and paleoenvironments of the Dakota Sandstone near Hanksville, Utah: Brigham Young University Geology Studies, v. 19, part 2, p. 89–120.

Leatham, W.B., 1985, Ages of the Fish Haven and lowermost Laketown dolomites in the Bear River Range, Utah: Utah Geological Association Publication 14, p. 29–38. See also Geological Society of America abstracts with Programs, v. 19, p. 313.

Lee, W.T., 1908, Water resources of Beaver Valley, Utah: U.S. Geological Survey Water Supply Paper 217, 57 p.

Leedom, S.H., 1974, Little Drum Mountains, an early Tertiary shoshonitic volcanic center in Millard County, Utah: Brigham Young University Geology Studies, v. 21, part 1, p. 73–108.

Leeman, W.P., 1974, Late Cenozoic alkali-rich basalt from the western Grand Canyon area, Utah and Arizon—isotopic composition of strontium: Geological Society of America Bulletin, v. 85, p. 1691–1696.

Leeman, W.P. 1982, Tectonic and magmatic significance of strontium isotopic variations in Cenozoic volcanic rocks from the western U.S.: Geological Society of America Bulletin, v. 93, p. 487–503.

Lelek, J.J., 1982, Anschutz Ranch East field, northeast Utah and southwest Wyoming: Rocky Mountain Association of Geologists Geologic Studies of the Cordilleran Thrust Belt, p. 619–631.

Lemmon, D.M., and Morris, H.T., 1979, Preliminary geologic map of the Frisco quadrangle, Beaver County, Utah: U.S. Geological Survey Open-File Report 79-724.

Lemmon, D.M., and Morris, H.T., 1984, Geologic map of the Beaver Lake Mountains quadrangle, Beaver and Millard Counties, Utah: U.S. Geological Survey Miscellaneous Investigations Map I-1572.

Lemmon, D.M., and Morris, H.T., 1979, Preliminary geologic map of the Milford quadrangle, Beaver County, Utah: U.S. Geological Survey Open-File Report 79-1471.

Lessard, R.H., 1973, Micropaleontology and paleoecology of the Tununk Member of the Mancos Shale: Utah Geological and Mineral Survey Special Studies 45, 28 p.

Lewis, R.Q., and Trimble, D.E., 1959, Geology and uranium deposits of Monument Valley, San Juan County, Utah: U.S. Geological Survey Bulletin 1087-D, 131 p. 105–131.

Liese, H.C., 1957, Geology of the northern Mineral Range, Millard and Beaver Counties, Utah: unpublished master's thesis, Univeristy of Utah, 88 p.

Lindsey, D.A., 1979, Geologic map and cross-sections of Tertiary rocks in the Thomas Range and northern Drum Mountains, Juab County, Utah: U.S. Geological Survey Miscellaneous Investigations Series Map I-1176.

Lindsey, D.A., 1982, Tertiary volcanic rocks and uranium in the Thomas Range and northern Drum Mountains, Juab County, Utah: U.S. Geological Survey Professional Paper 1221, 71 p.

Lindsey, D.A., 1982, Volcanism and uranium mineralization at Spor Mountain, Utah: American Association of Petroleum Geologists Studies in Geology 13, p. 89–98; Philip Goodell, ed., Symposium on Uranium in Volcanic and Volcaniclastic Rocks, Tulsa.

Lindsey, D.A., Glanzman, R.K., Naeser, C.W., and Nichols, D.J., 1981, Upper Oligocene evaporites in basin fill of Sevier Desert region, western Utah: American Association of Petroleum Geologists, v. 65, p. 251–260.

Lindsey, K.A., 1982, Upper Proterozoic and Lower Cambrian Brigham Group, Oneida Narrows, southeastern Idaho: Northwest Geology, v. 11, p. 13–21.

Lines, G.C., 1979, Hydrology and surface morphology of the Bonneville Salt Flats and Pilot Valley playa, Utah: U.S. Geological Survey Water-Supply Paper 2057, 107 p.

Link, P.K., 1982a, Structural geology of the Oxford and Malad Summit quadrangles, Bannock Range, southeastern Idaho: Rocky Mountain Association of Geologists, Geologic studies of the Cordilleran Thrust Belt, v. 2, p. 851–858.

Link, P.K., 1982b, Idaho-Wyoming thrust belt, regional geology: Northwest Geology, v. 11, p. 1–12.

Link, P.K., 1983, Glacial and tectonically influenced sedimentation in the upper Proterozoic Pocatello Formation, southeastern Idaho: Geological Society of America Memoir 157, p. 165–181.

Link, P.K., 1984, Comment on "Subsidence analysis of the Cordilleran miogeocline": Geology, v. 12, p. 699.

Link, P.K., 1988, Outline and references for Proterozoic Z rocks of the United States Cordillera: *in* Geological Society of America DNAG volume on Precambrian of the United States, (in press).

Link, P.K., Crook, S.R., and Chidsey, T.C., 1985, Hinterland structure, Paleozoic stratigraphy and duplexes of the Willard thrust system-Bannock, Wellsville, and Wasatch Ranges, southeastern Idaho and northern Utah: Utah Geological Association Publication 14, p. 315–328.

Link, P.K., and LeFebre, G.B., 1983, Upper Proterozoic diamictites and volcanic rocks of the Pocatello Formation and correlative units, southeastern Idaho and northern Utah: Utah Geological and Mineral Survey Special Studies 60, p. 1–32.

Link, P.K., LeFebre, G.B., Pogue, K.R., and Burgel, W.D., 1985, Structural geology between the Putnam thrust and the Snake River Plain, southeastern Idaho; Utah Geological Association Publication 14, p. 97–118.

Lipman, P.W., Prostka, H.J., and Christiansen, R.L., 1971, Evolving subduction zones in the western United States, as interpreted from igneous rocks: Science, v. 174, p. 821–5.

Lipman, P.W., Rowley, P.D., Mehnert, H.H., Evans, S.H. Jr., Nash, W.P., and Brown, F.H., 1978, Pleistocene rhyolite of the Mineral Mountains, Utah—geothermal and archeological significance: U.S. Geological Survey Journal of Research, v. 6, no. 1, p. 133–147.

Little, T.A., 1987, Stratigraphy and structure of metamorphosed upper Paleozoic rocks near Mountain City, Nevada: Geological Society of America Bulletin, v. 98, p. 1–17.

Little, W.W., 1988, Geomorphology of Lake Bonneville deltas: Brigham Young University Geology Studies v. 35 (in press).

Livaccari, R.F., 1979, Late Cenozoic tectonic evolution of the western U.S.: Geology, v. 7, p. 72–75.

Livaccari, R.F., and Gallo, D.G., 1986, Plate tectonic processes of late Cenozoic extension in the Basin and Range province of the western U.S.—transtension to free-face tectonics: Geological Society of America Special Paper 208.

Livingston, V.E. Jr., 1955, Sedimentation and stratigraphy of the Humbug Formation in central Utah: Brigham Young University Geology Studies, v. 2, no. 6, 60 p.

Lochman-Balk, C., 1971, Cambrian of the craton of the United States: *in* C.H. Holland, ed., Cambrian of the New World, Wiley-Interscience, New York, p. 79–167.

Lochman-Balk, C., 1972, Cambrian System, *in* Geologic Atlas of the Rocky Mountain Region, p. 60–75. Rocky Mountain Association of Geologists, Denver.

Lochman-Balk, C., 1976, Cambrian section of the central Wasatch Mountains: Rocky Mountain Association of Geologists, Symposium on Geology of the Cordilleran Hingeline, p. 103–108.

Logan, A., and Hills, L.V., eds., 1973, The Permian and Triassic Systems and their Mutual Boundary: Canadian Society of Petroleum Geologists Memoir 2, Calgary.

Lohman, S.W., 1965a, Geology and artesian water supply, Grand Junction area, Colorado: U.S. Geological Survey Professional Paper 451, 149 p.

Lohman, S.W., 1965b, The geologic story of Colorado National Monument: Colorado and Black Canyon Natural History Association, 56 p.

Lohman, S.W., 1975, The geologic story of Arches National Park: U.S. Geological Survey Bulletin 1393, 113 p.

Lohmann, K.C., Lower Dresbachian (Upper Cambrian) platform-to-basin transition in eastern Nevada and western Utah—an evaluation through lithologic cycle correlation: Brigham Young University Geology Studies, v. 23, part 2, p. 111–122.

Lohrengel, C.F., 1969, Palynology of the Kaiparowits Formation, Garfield County, Utah: Brigham Young University Geology Studies, v. 16, part 3, p. 61–180.

Loleit, A.J., 1963, Cambrian stratigraphic problems of the Four Corners area: Four Corners Geological Society Guidebook 4, p. 21–30.

Loope, D.B., 1984, Eolian origin of upper Paleozoic sandstones, southeastern Utah: Journal of Sedimentary Petrology, v. 55, no. 2, p. 563–580.

Loring, A.K., 1976, Distribution in time and space of late Phanerozoic normal faulting in Nevada and Utah: Utah Geology, v. 3, no. 2, p. 97–109.

Loucks, G.G., 1975, The search for Pineview Field, Summit County, Utah: Rocky Mountain Association of Geologists Symposium, p. 255–264.

Lowrey, R.O., 1976, Paleoenvironment of the Carmel Formation at Sheep Creek Gap, Daggett County, Utah: Brigham Young University Geology Studies, v. 23, part 1, p. 173–203.

Lucas, P.T., and Drexler, J.M., 1975, Altamont-Bluebell: a major fractured and overpressured stratigraphic trap, Uinta Basin Utah: Rocky Mountain Association of Geologists Symposium on Deep Drilling Frontiers, p. 265–273.

Lucchitta, Ivo, and Hendricks, J.D., 1983, Characteristics, depositional environment, and tectonic interpretations of the Proterozoic Cardenas Lavas, eastern Grand Canyon, Arizona: Geology, v. 11, p. 177–181.

Luedke, R.B., and Smith, R.L., 1978, Map showing distribution, composition, and age of late Cenozoic volcanic centers in Colorado, Utah, and southwestern Wyoming: U.S. Geological Survey Miscellaneous Investigations Series Map I-1091-B.

Luedke, R.G., and Smith, R.L., 1984, Map showing distribution, composition and age of late Cenozoic volcanic centers in the western conterminous U.S.: U.S. Geological Survey Map I-1523.

Lupe, R., 1979, Stratigraphic sections of the Upper Triassic Chinle Formation, San Rafael Swell to the Moab area, Utah: U.S. Geological Survey Oil and Gas Investigations Chart OC-89.

Lupe, R., 1983, Stratigraphic sections of subsurface Jurassic rocks in the San Juan Basin, New Mexico, Colorado, Utah, and Arizona: U.S. Geological Survey Oil and Gas Investigations Chart OC-118.

Lupe, R., 1984, Stratigraphic sections of upper Triassic Chinle Formation in the Capitol Reef, Circle Cliffs, and Monument Uplift areas, southeast Utah: U.S. Geological Survey Oil and Gas Investigations Chart OC-125.

Mabey, D.R., 1987, Subsurface geology along the Wasatch Front: U.S. Geological Survey Open-File Report 87-585, p. C1–39.

Mabey, D.R., and Budding, K.E., 1987, High-temperature geothermal resources of Utah: Utah Geological and Mineral Survey Bulletin 123, 64 p.

Mabey, D.R., Oliver, H.W., and Hildenbrand, T.G., 1983, Regional gravity and magnetic anomalies in the northern Basin and Range Province: Geothermal Resources Council, Special Report No. 13, p. 307–314.

Mabey, D.R., Zietz, I., Eaton, G.P., and Kleinkopf, M.D., 1978, Regional magnetic patterns in part of the Cordillera in the western U.S.: Geological Society of America Memoir 152, p. 93–106.

Machette, M.N., 1982, Guidebook to the late Cenozoic geology of the Beaver Basin, south-central Utah: U.S. Geological Survey Open-File Report 82-850, 42 p.

Machette, M.N., 1983, Geologic map of the southwest quarter of the Beaver quadrangle, Beaver County, Utah: U.S. Geological Survey Miscellaneous Investigations Maps I-1444.

Machette, M.N., 1985, Late Cenozoic geology of the Beaver Basin, southwestern Utah: Brigham Young University Geology Studies, v. 32, part 1, p. 19–37.

Machette, M.N., 1985, Calcic soils of the southwestern U.S.: Geological Society of America Special Paper 203, p. 1–21.

Machette, M.N., Personius, S.F., and Nelson, A.R., 1987, Quaternary geology along the Wasatch Fault Zone—segmentation recent investigations, and preliminary conclusions: U.S. Geological Survey Open-File Report 87-585, p. A1–72.

Machette, M.N. and Steven, T.A., 1983, Geologic map of the northwest quarter of the Beaver quadrangle, Beaver County, Utah: U.S. Geological Survey Miscellaneous Investigations Map I-1445.

Machette, M.N., Steven, T.A., Cunningham, C.G., and Anderson, J.J., 1984, Geologic map of the Beaver Quadrangle, Beaver and Piute Counties, Utah: U.S. Geological Survey Miscellaneous Investigations Map I-1520.

Mack, G.H. and Rasmussen, K.A., 1984, Alluvial-fan sedimentation of the Cutler Formation (Permo-Pennsylvanian) near Gateway, Colorado: Geological Society of America Bulletin v. 95, p. 109–116.

Mackin, J.H., Nelson, W.H., and Rowley, P.D., 1976, Geologic map of the Cedar City NW quadrangle, Iron County, Utah: U.S. Geological Survey Geologic Quadrangle Map GQ-1295.

Mackin, J.H. and Rowley, P.D., 1975, Geologic map of the Avon SE quadrangle, Iron County, Utah: U.S. Geological Survey Geologic Quadrangle Map GQ-1294.

Mackin, J.H., and Rowley, P.D., 1976, Geologic map of the Three Peaks quadrangle, Iron County, Utah: U.S. Geological Survey Geologic Quadrangle Map GQ-1297.

MacLachlan, M.E., 1972, Triassic System: Geologic Atlas of the Rocky Mountain Region, Rocky Mountain Association of Geologists, Denver, p. 166–176.

MacLachlan, M.E., 1985, Chart showing the major stratigraphic units used in southeastern Idaho and northern Utah: Utah Geological Association Publication 14, p. 39–46.

Madsen, D.B., and Currey, D.R., 1979, Late Quaternary glacial and vegetation changes, Little Cottonwood Canyon area, Wasatch Mountains, Utah: Quaternary Research, v. 12, p. 254–270.

Maione, S.J., 1971, Stratigraphy of the Frontier Sandstone Member of the Mancos Shale (Upper Cretaceous) on the south flank of the eastern Uinta Mountains, Utah and Colorado: Wyoming Geological Association Earth Science Bulletin, September, 1971, p. 27–58.

Malde, H.E., 1960, Evidence in the Snake River Plain, Idaho, of a catastrophic flood from Pleistocene Lake Bonneville: U.S. Geological Survey professional Paper 400B, p. 255–297.

Malde, H.E., 1968, The catastrophic late Pleistocene Bonneville Flood in the Snake River Plain, Idaho: U.S. Geological Survey Professional Paper 596, 52 p.

Mallory, W.W., 1972, Regional synthesis of the Pennsylvanian System, *in* Geologic Atlas of the Rocky Mountain Region, p. 111–127. Rocky Mountain Association of Geologists, Denver.

Mallory, W.W., 1975, Pennsylvanian System—middle and southern Rocky Mountains, northern Colorado Plateau, and eastern Great Basin region: U.S. Geological Survey Professional Paper 853, p. 265–278.

Mallory, W.W., 1979, Mississippian System—central Rocky Mountains and northern Colorado Plateau: U.S. Geological Survey Professional Paper 1010, p. 209–220.

Mann, D.C., 1974, Clastic Laramide sediments of the Wasatch Hinterland, northeastern Utah: unpublished master's thesis, University of Utah, 112 p.

Marcantel, E.L. and Weiss, M.P., 1968, Colton Formation (Eocene-fluviatile) and associated lacustrine beds, Gunnison Plateau, central Utah: Ohio Journal of Science v. 68/1, p. 40–49.

Marine, I.W., 1962, Water supply possibilities at Capitol Reef National Monument, Utah: U.S. Geological Survey Water Supply Paper 1475-G, p. 201–208.

Marzolf, J.E. and Dunne, G.C., 1983, Evolution of early Mesozoic tectonostratigraphic environments—southwestern Colorado Plateau to southern Inyo Mountains: Utah Geological and Mineral Survey Special Studies 60 p. 33–39.

Matthews, V. III, 1986, A case for brittle deformation of the basement during the Laramide revolution in the Rocky Mountain foreland province: The Mountain Geologist, v. 23, p. 1–5.

Mattox, S., 1987, Geology of the Hells Kitchen Canyon SE quadrangle, Sanpete County, Utah: Utah Geological and Mineral Survey Map 98.

Mauger, R.L., 1977, K-Ar ages of biotite from tuffs in Eocene rocks of the Green River, Washakie, and Uinta Basins, Utah, Wyoming, and Colorado: University of Wyoming Contributions to Geology, v. 15, no. 1, p. 17–41.

Maughan, E.K., 1979, Petroleum source rock evaluation of the Permian Park City Group in the northeastern Great Basin, Utah Nevada and Idaho: Rocky Mountain Association of Geologists Basin and Range Symposium, p. 523–530.

Maughan, E.K. and Perry, W.J., Jr., 1986, Lineaments and their tectonic implications in the Rocky Mountains and adjacent plains region: American Association of Petroleum Geologists Memoir 41, p. 41–54.

Maurer, R.E., 1970, Geology of the Cedar Mountains, Tooele County, Utah: unpublished Ph.D. thesis, University of Utah, 184 p.

Maw, G.G., 1968, Lake Bonneville history in the Cutler Dam quadrangle, Cache and Box Elder Counties, Utah: unpublished master's thesis, Utah State University, Logan, 67 p.

Maxey, G.B., 1958, Lower and Middle Cambrian stratigraphy in northern Utah and southeastern Idaho: Geological Society of America Bulletin, v. 69, p. 647–88.

Maxfield, E.B., 1957, Sedimentation and stratigraphy of the Morrowan Series in central Utah: Brigham Young University Geology Studies v. 4, no. 1, 46 p.

Maxfield, E.B., 1976, Foraminifera from the Mancos Shale of east-central Utah: Brigham Young University Geology Studies, v. 23, part 3, p. 67–162.

Maxson, J.H., 1961, Geologic map of the Bright Angel quadrangle, Grand Canyon National Park, Arizona: Grand Canyon National History Association.

McCalpin, J., Robison, R.M., and Garr, J.D., 1987, Neotectonics of the Hansel Valley-Pocatello Valley corridor northern Utah and southern Idaho: U.S. Geological Survey Open-File Report 87-585, p. G1–G44.

McClellan, P.H., 1981, Nonmarine carbonates of Neogene Lake Idaho in Utah: American Association of Petroleum Geologists Bulletin, v. 65, p. 955–956.

McCollum, L.B., 1987, Depositional changes within a Cambrian outer passive margin sequence, northern Great Basin: Geological Society of America Abstracts with Programs, v. 19, p. 320.

McCollum, L.B., and McCollum, M.B., 1984, Comparison of a Cambrian medial shelf sequence with an outer shelf margin sequence, northern Great Basin: Utah Geological Association Publication 13, p. 35–44.

McCoy, W.D., 1977, A reinterpretation of certain aspects of the late Quaternary glacial history of Little Cottonwood Canyon, Wasatch Mountains, Utah: unpublished master's thesis, University of Utah, Salt Lake City, 84 p.

McCoy, W.D., 1987, Quaternary aminostratigraphy of the Bonneville Basin, western U.S.: Geological Society of America Bulletin, v. 99, p. 99–112.

McDermott, J.G., 1986, Geology of the Chriss Canyon 7½ minute quadrangle, Utah: Utah Geological and Mineral Survey Map (in preparation).

McDonald, R.E., 1976, Tertiary tectonics and sedimentary rocks along the transition—Basin and Range province to plateau and thrust belt province, Utah: Rocky Mountains Association of Geologists Symposium, p. 281–317.

McDowell, R.R., 1982, Depositional environments of the Middle Cambrian Lodore, Flathead, and lower Gros Ventre formations: unpublished master's thesis, University of Kansas, Lawrence, 360 p.

McGill, G.E. and Stromquist, A.W., 1979, Grabens of Canyonlands National Park, Utah—geometry, mechanics and kinematics: Journal of Geophysical Research: v. 84, p. 4547.

McGookey, D.P., 1958, Geology of the northern portion of the Fish Lake Plateau, Sevier County, Utah: unpublished Ph.D. thesis, Ohio State University.

McGookey, D.P., 1972, Cretaceous System: *in* Geologic Atlas of the Rocky Mountain Region: Rocky Mountain Association of Geologists, Denver, p. 190–228.

McKee, E.D., 1934, The Coconino Sandstone—its history and origin: Carnegie Institution of Washington Publication 440, p. 77–115.

McKee, E.D., 1945, Cambrian history of the Grand Canyon region: Carnegie Institution of Washington Publication 563, 232 p.

McKee, E.D., 1954, Stratigraphy and history of the Moenkopi Formation of Triassic age: Geological Society of America Memoir 61, 133 p.

McKee, E.D., 1982, The Supai Group of Grand Canyon: U.S. Geological Survey Professional Paper 1173, 504 p.

McKee, E.D. and Gutschick, R.C., 1969, History of the Redwall Limestone of northern Arizona: Geological Society of America Memoir 114, 726 p.

McKee, E.D., Oriel, S.S., Ketner, K.B., MacLachlan, M.E., Goldsmith, J.W., MacLachlan, J.C., and Mudge, M.R., 1959, Paleotectonic maps of the Triassic System: U.S. Geological Survey Map 1-300.

McKee, E.D. and Oriel, S.S., 1967, Paleotectonic maps of the Permian System: U.S. Geological Survey Miscellaneous Geologic Investigations Map I-450, 164 p., map folio. See also U.S. Geological Survey Professional Paper 515, for text to accompany these maps.

McKee, E.H., 1971, Tertiary igneous chronology of the Great Basin of western United States: Geological Society of America Bulletin v. 82, p. 3497–3502.

Medaris, L.G., Jr., Byers, C.W., Mickelson, D.M., and Shanks, W.C., *eds*., 1983 Proterozoic geology—selected papers from an international Proterozoic symposium: Geological Society of America Memoir 161, 315 p.

Mehnert, H.N., Rowley, P.D., and Lipman, P.W., 1978, K-Ar ages and geothermal implications of young rhyolites in west-central Utah: Isochron/West, no. 21, p. 3–7.

Meibos, L.C., 1983, Structure and Stratigraphy of the Nephi NW (Sugarloaf) quadrangle, Juab County, Utah: Brigham Young University Geology Studies v. 30, part 1, p. 37–58.

Melton, R.A., 1972, Paleoecology and paleoenvironment of the upper Honaker Trail Formation near Moab, Utah: Brigham Young University Geology Studies v. 19, part 2, p. 45–88.

Mercier, J.M., 1982, Rocky Mountain Coal Field Trip, second day road log—northern Wasatch coal field of central Utah: Utah Geological and Mineral Survey Bulletin 118, p. 301–317.

Merewether, E.A. and Cobban, W.A., 1986, Biostratigraphic units and tectonism in the mid-Cretaceous foreland of Wyoming, Colorado, and adjoining areas: American Association of Petroleum Geologists Memoir 41, p. 443–468.

Merrill, R.C., 1972, Geology of the Mill Fork area, Utah: Brigham Young University Geology Studies, v. 19, part 1, p. 65–88.

Metter, R.E., 1955, Geology of part of the southern Wasatch Mountains, Utah: unpublished Ph.D. thesis, Ohio State University, 243 p.

Michaels, Roger, 1986, Geology of the Scipio Pass quadrangle, Utah: Brigham Young University Geology Studies (in preparation).

Migliaccio, R.R., 1958, Middle Cambrian trilobites from the Ophir Shale of central Utah: Compass of Sigma Gamma Epsilon, v. 35, no. 4, p. 298–301.

Millard, A.W. Jr., 1983, Geology of the southwestern quarter of the Scipio North (15-minute) quadrangle, Millard and Juab Counties, Utah: Brigham Young University Geology Studies, v. 30, part 1, p. 59–81.

Millen, T.M., 1982, Stratigraphy and petrology of the Green River Formation (Eocene), Gunnison Plateau, central Utah: unpublished master's thesis, Northern Illinois University, DeKalb, 216 p.

Miller, D.M., 1978, Deformation associated with Big Bertha Dome, Albion Mountains, Idaho: unpublished Ph.D. thesis, University of California, Los Angeles, 225 p.

Miller, D.M., 1980, Structural geology of the northern Albion Mountains, south-central Idaho: Geological Society of America Memoir 153, p. 399–423.

Miller, D.M., 1983, Allochthonous quartzite sequence in the Albion Mountains, Idaho, and proposed Proterozoic Z and Cambrian correlatives in the Pilot Range, Utah and Nevada: Geological Society of America Memoir 157, p. 191–213.

Miller, D.M., 1984, Sedimentary and igneous rocks of the Pilot Range and vicinity, Utah and Nevada: Utah Geological Association Publication 13, p. 45–63.

Miller, D.M., 1985, Geologic map of the Lucin quadrangle, Box Elder County, Utah: Utah Geological and Mineral Survey Map 78.

Miller, D.M. and Lush, A.P., 1981, Preliminary geologic map of the Pilot Peak and adjacent quadrangles, Elko County, Nevada, and Box Elder County, Utah: U.S. Geological Survey Open-File Report 81-658.

Miller, D.M., Lush, A.P., and Schneyer, J.D., 1982, Preliminary geologic map of Patterson Pass and Crater Island NW quadrangles, Box Elder County, Utah, and Elko County, Nevada: U.S. Geological Survey Open-File Report 82-834.

Miller, D.M., Armstrong, R.L., Compton, R.R., and Todd, V.R., 1983, Geology of the Albion-Raft River-Grouse Creek Mountains area, northwestern Utah and southern Idaho: Utah Geological and Mineral Survey Special Studies 59, p. 1–62.

Miller, D.M. and Schneyer, J.D., 1985, Geologic map of the Tecoma quadrangle, Box Elder County, Utah and Elko County, Nevada: Utah Geological and Mineral Survey Map 77.

Miller, D.M. and Glick, L.L., 1986, Geologic map of the Lemay Island quadrangle, Box Elder County, Utah: Utah Geological and Mineral Survey Map 91.

Miller, D.M., Hillhouse, W.C., Zartman, R.E., and Lanphere, M.A., 1987, Geochronology of intrusive and metamorphic rocks in the Pilot Range, Utah and Nevada, and comparison with regional patterns: Geological Society of America Bulletin, v. 99, p. 866–879.

Miller, E.L., Gans, P.B., and Garing, J., 1983, The Snake Range decollement—an exhumed mid-Tertiary ductile-brittle transition: Tectonics, v. 2, p. 239–263.

Miller, E.L., Gans, P.B., Wright, J.E., and Sutter, J.F., 1986, Metamorphic history of the east-central Basin and Range Province: tectonic setting and relation to magmatism: Geological Society of America Abstracts with Programs, v. 18, p. 158.

Miller, J.F., 1969, Conodont fauna of the Notch Peak Limestone (Cambro-Ordovician), House Range, Utah: Journal of Paleontology v. 43, no. 2, p. 413–39.

Miller, J.F., 1984, Cambrian and Ordovician conodont evolution, biofacies, and provincialism: Geological Society of America Special Paper 196, p. 43–68.

Miller, S.T., Martindale, S.G., and Fedewa, W.T., 1984, Permian stratigraphy of the Leach Mountains, Elko County, Nevada: Utah Geological Association Publication 13, p. 65–78.

Misch, Peter, and Hazzard, J.C., 1962, Stratigraphy and metamorphism of Late Precambrian rocks in central northeastern Nevada and adjacent Utah: American Association of Petroleum Geologists Bulletin, v. 46, p. 289–343.

Mitchell, G.C., 1985, Permian-Triassic stratigraphy of the northwest Paradox Basin area, Emery, Garfield, and Wayne Counties, Utah: The Mountain Geologist, v. 22, no. 4, p. 149–166.

Mitchell, G.C. and McDonald, R.E., 1987, Subsurface Tertiary strata, depositional model, and hydrocarbon potential of the Sevier Desert basin, west-cenral Utah: Utah Geological Association Publication 16.

Moir, G.J., 1974, Depositional environments and stratigraphy of the Cretaceous rocks, southwestern Utah: unpublished Ph.D. thesis, University of California at Los Angeles, 316 p.

Molenaar, C.M., 1975, Some notes on Upper Cretaceous stratigraphy of the Paradox Basin: Four Corners Geological Society Guidebook, 8th Field Conference, Canyonlands, p. 191–192.

Molenaar, C.M., 1983, Major depositional cycles and regional correlatives of Upper Cretaceous rocks, southern Colorado Plateau and adjacent areas: *in* Reynolds, M.W., and Dolly, E.D., eds., Symposium on Mesozoic Paleogeography of West-central U. S., Rocky Mountain Section of Society of Economic Paleontologists and Mineralogists, Denver, Colorado, p. 201–224.

Montgomery, S.B. and Everitt, B.L., 1984, Summary of the hydrogeology of Park Valley, Utah: Utah Geological Association Publication 13, p. 249–259.

Moore, W.J., 1973, Preliminary geologic map of western Traverse Mountains and northern Lake Mountain, Salt Lake and Utah Counties, Utah: U.S. Geological Survey Miscellaneous Field Studies Map MF-490.

Moore, W.J., 1973, Summary of radiometric ages of igneous rocks in the Oquirrh Mountains, north-central Utah: Economic Geology v. 68, p. 97–101. Also U.S. Geological Survey Professional Paper 629-B.

Moore, W.J. and Sorensen, M.L., 1979, Geologic map of the Tooele 1 x 2 degree quadrangle, Utah: U.S. Geological Survey Miscellaneous Investigations Map I-1132.

Morris, H.T., 1964a, Geology of the Eureka quadrangle, Utah and Juab Counties, Utah: U.S. Geological Survey Bulletin 1142-K, 29 p.

Morris, H.T., 1964b, Geology of the Tintic Junction quadrangle, Tooele, Juab, and Utah Counties, Utah: U.S. Geological Survey Bulletin 1142-L, 23 p.

Morris, H.T., 1975, Geologic map and sections of the Tintic Mountain quadrangle and adjacent part of the McIntyre quadrangle, Juab and Utah Counties, Utah: U.S. Geological Survey Miscellaneous Investigations Series Map I-833.

Morris, H.T., 1977, Geologic map and sections of the Furner Ridge quadrangle, Juab County, Utah: U.S. Geological Survey Miscellaneous Investigations Series Map I-1045.

Morris, H.T., 1983, Interrelations of thrust and transcurrent faults in the central Sevier orogenic belt near Leamington, Utah: Geological Society of America Memoir 157, p. 75–81.

Morris, H.T., Best, M.G., Kopf, R.W., and Keith, J.D., 1982: Preliminary geologic map and cross-sections of the Tetons quadrangle, Beaver and Iron Counties, Utah: U.S. Geological Survey Open-File Report 82-778.

Morris, H.T., Douglass, R.C., and Kopf, R.W., 1977, Stratigraphy and microfaunas of the Oquirrh Group in the southern East Tintic Mountains, Utah: U.S. Geological Survey Professional Paper 1025, 22 p.

Morris, H.T., and Lovering, T.S., 1961, Stratigraphy of the East Tintic Mountains, Utah: U.S. Geological Survey Professional Paper 361, 145 p.

Morris, H.T., and Lovering, T.S., 1979, General geology and mines of the East Tintic mining district, Utah and Juab Counties, Utah: U.S. Geological Survey Professional Paper 1024, 203 p.

Morris, S.K., 1980, Geology and ore deposits of Mineral Mountain, Washington County, Utah: Brigham Young University Geology Studies, v. 27, part 2, p. 85–102.

Morrison, R.B., 1965, New evidence on Lake Bonneville stratigraphy and history from southern Promontory Point, Utah: U.S. Geological Survey Professional Paper 525-C, p. 110–119.

Morton, L.B., 1984, Geology of the Mount Ellen quadrangle, Henry Mountains, Garfield County, Utah: Brigham Young University Geology Studies, v. 31, part 1, p. 67–95. Also Utah Geological and Mineral Survey Map 90 (1986).

Mount, J.F., Gevirtzman, D.A., and Signor, P.W.III, 1983, Precambrian-Cambrian transition problem in western North America—Part 1, Tommotian fauna in the southwestern Great Basin and its implications for the base of the Cambrian System: Geology, v. 11, p. 224–226.

Moussa, M.T., 1965, Geology of the Soldier Summit quadrangle, Utah: unpublished Ph.D. thesis, University of Utah, 129 p.

Moussa, M.T., 1969, Green River Formation (Eocene) in the Soldier Summit area, Utah: Geological Society of America Bulletin, v. 80, p. 1737–1748.

Mower, R.W., 1982, Hydrology of the Beryl-Enterprise area, Escalante Desert, Utah, with emphasis on ground-water: Utah Department of Natural Resources Technical Publication 73, 56 p.

Mower, R.W., and Cordova, R.M., 1974, Water resources of the Milford area, Utah, with emphasis on groundwater: Utah Department of Natural Resources Technical Publication 43, 99 p. Summary in Utah Geological Association Publication 3 (1973), p. 63–72.

Moyle, R.W., 1959, Stratigraphy of the southern Oquirrh Mountains, Manning Canyon Shale: Utah Geological Society Guidebook 14, p. 59–92.

Muessig, S.J., 1951, Geology of a part of Long Ridge, Utah: unpublished Ph.D. dissertation, Ohio State University, 213 p.

Mullens, T.E., 1960, Geology of the Clay Hills area, San Juan County, Utah: U.S. Geological Survey Bulletin 1087-H, p. 259–336.

Mullens, T.E., 1969, Geologic map of the Causey Dam quadrangle, Weber County, Utah: U.S. Geological Survey Geologic Quadrangle Map GQ-790.

Mullens, T.E., 1971, Reconnaissance study of the Wasatch, Evanston, and Echo Canyon Formations in part of northern Utah: U.S. Geological Survey Bulletin 1311-D, p. 1–31.

Mullens, T.E., and Izett, G.A., 1963, Geology of the Paradise quadrangle, Utah: U.S. Geological Survey Geologic Quadrangle Map GQ-185. Also U.S. Geological Survey Bulletin 1181-S, 32 p., 1964.

Mullens, T.E., and Cole, T.H., 1972, Geologic map of the northeast quarter of the Morgan 15-minute quadrangle, Morgan and Weber Counties, Utah: U.S. Geological Survey Mineral Investigations Field Studies Map MF-304.

Mullens, T.E., and Laraway, W.H., 1964, Geology of the Devils Slide quadrangle, Morgan and Summit Counties, Utah: U.S. Geological Survey Mineral Investigations Field Studies Map MF-290.

Mullens, T.E., and Laraway, W.H., 1973, Geologic map of the Morgan 7½-minute quadrangle, Morgan County, Utah: U.S. Geological Survey Mineral Investigations Field Studies Map MF-318.

Mundorff, J.C., 1970, Major thermal springs of Utah: Utah Geological and Mineral Survey Water-Resources Bulletin 13, 60 p.

Munger, R.D., Greene, J., Peace, F.S., and Liming, J.A., 1965, Pre-Pennsylvanian stratigraphy of the

Kaiparowits region, south-central Utah and north-central Arizona: Utah Geological Society Guidebook 19, p. 13–29.

Murphy, B.E., 1983, Geology of the Limekiln Knoll quadrangle, north-central Utah: unpublished master's thesis, Bryn Mawr College, 52 p. Also to be published by Utah Geological and Mineral Survey Map (in press).

Murphy, C.M., 1975, Stratigraphy and metamorphic petrography of a late Precambrian diamictite, Deep Creek Mountains, Utah: unpublished master's thesis, University of Nebraska (Lincoln), 127 p.

Murphy, D.R., 1954, Fauna of the Morrowan rocks of central Utah: Brigham Young University Geology Studies, v. 1, no. 3, 64 p.

Naeser, C.W., Bryant, Bruce, Crittenden, M.D. Jr., and Sorensen, M.L., 1983, Fission-track ages of apatite in the Wasatch Mountains, Utah, an uplift study: Geological Society of America Memoir 157, p. 29–36.

Nash, W.P., and Smith R.P., 1977, Pleistocene volcanic ash deposits in Utah; Utah Geology, v. 4, no. 1, p. 35–42.

Nations, D., Wilt, J.C., and Hevly, R.H., 1985, Cenozoic paleogeography of the west-central United States, Rocky Mountain Paleogeography Symposium 3, Society of Economic Geologists and Paleontologists, Denver, p. 335–355.

Nelson, M.E., 1971, Stratigraphy and paleontology of Norwood Tuff and Fowkes Formation, northeastern Utah and southwestern Wyoming: unpublished Ph.D. dissertation, University of Utah, Salt Lake City, 181 p.

Nelson, M.E., 1972, Age and stratigraphic relation of the Fowkes Formation and Norwood Tuff, southwestern Wyoming and northeastern Utah: Geological Society of America Abstracts with Programs, v. 4, no. 6, p. 397–398.

Nelson, R.B., 1966, Structural development of northernmost Snake Range, Kern Mountians, and Deep Creek Range, Nevada and Utah: American Association of Petroleum Geologists Bulletin, v. 50, p. 921–951.

Nethercott, M.A., 1985, Geologic map and coal resources of the Deadman Canyon 7½-minute quadrangle, Carbon County, Utah: Utah Geological and Mineral Survey Map 76; also Brigham Young University Geology Studies v. 33, part 1, p. 45–85.

Newman, G.J., 1980, Conodonts and biostratigraphy of the lower Mississippian in western Utah and eastern Nevada: Brigham Young University Geology Studies, v. 27, part 2, p. 103–121.

Newman, K.R., 1974, Palynomorph zones in early Tertiary formations of the Piceance Creek and Uinta Basins, Colorado and Utah: Rocky Mountain Association of Geologists Guidebook 25, p. 47–55.

Nichols, D.J., and Jacobsen, S.R., 1982a, Palynostratigraphic framework for the Cretaceous (Albian-Maestrichtian) of the overthrust belt of Utah and Wyoming: Palynology, v. 6, p. 119–147.

Nichols, D.J., and Jacobsen, S.R., 1982b, Cretaceous biostratigraphy in the Wyoming thrust belt: The Mountain Geologist, v. 19, no. 3, p. 73–78.

Nichols, D.J., Jacobsen, S.R., and Tschudy, R.H., 1982, Cretaceous palynomorph biozones for the central and northern Rocky Mountain region of the United States: *in* Powers, R.B., ed.,Rocky Mountain Association of Geologists, Symposium on Geologic Studies of the Cordilleran thrust belt, p. 721–733.

Nichols, D.J., and Warner, M.A., 1978, Palynology, age, and correlation of the Wanship Formation and their implications for the tectonic history of northeastern Utah: Geology, v. 6, p. 430–433.

Nielson, D.L., Evans, S.H., and Sibbett, B.S., 1986, Magmatic, structural, and hydrothermal evolution of the Mineral Mountains intrusive complex, Utah: Geological Society of America Bulletin, v. 97, p. 765–777.

Nielson, D.R., 1988, Depositional environments and petrology of the Jurassic Carmel Formation in the Beaver Dam Mountains, southwestern Utah: Brigham Young University Geology Studies (in press).

Nielson, R.L., 1977, Geomorphic evolution of the Crater Hill volcanic field of Zion National Park: Brigham Young University Geology Studies, v. 24, part 1, p. 55–70.

Nielson, R.L., 1981, Depositional environment of the Toroweap and Kaibab formations of southwestern Utah: unpublished Ph.D. thesis, University of Utah, Salt Lake City, 495 p.

Nielson, R.L., and Johnson, J.L., 1979, The Timpoweap Member of the Moenkopi Formation, Timpoweap Canyon, Utah: Utah Geology, v. 6, no. 1, p. 17–25.

Nielson, W., and Erickson, A.J., 1980, Soil survey of Navajo Indian Reservation, San Juan County, Utah: U.S. Department of Agriculture, Soil Conservation Service, 119 p.

Nolan, T.B., 1935, The Gold Hill mining district, Utah: U.S. Geological Survey Professional Paper 177, 172 p.

Norris, A.W., 1979, Devonian in the western hemisphere: *in* Robison, R.A., and Teichert, C., eds., Treatise on Invertebrate Paleontology, Part A., p. A218–253.

Norris, G.E., 1981, Conodonts and biostratigraphy of Mississippian Brazer Dolomite, Crawford Mountains, Utah: Journal of Paleontology v. 55, p. 1270–1283.

Oaks, R.Q., James, W.C., Francis, G.G., and Schulingkamp, W.J. III, 1977, Summary of Middle Ordovician stratigraphy and tectonics, northern Utah, southern and central Idaho: Wyoming Geological Association Guidebook 29, p. 101–118.

Oberhansley, G.C., 1980, Geology of the Fairview Lakes quadrangle, Sanpete County, Utah: Brigham Young University Geology Studies, v. 27, part 3, p. 73–95. Also Utah Geological and Mineral Survey Map 56.

Oliveira, M.E., 1975, Geology of the Fish Springs Mining District, Fish Springs Range, Utah: Brigham Young University Geology Studies, v. 22, part 1, p. 69–104.

Olmore, S.D., 1972, Pillow lavas of the Chinle Formation in the southern Wah Wah Mountains, Beaver County, Utah: Geological Society of America Abstracts with Programs, v. 4, p. 400.

Olsen, P.E., and Galton, P.M., 1977, Triassic-Jurassic tetrapod extinctions—are they real?: Science, v. 197, p. 983–986.

Olson, R.H., 1960, Geology of the Promontory Range, Box Elder County, Utah: unpublished Ph.D. thesis, University of Utah Preliminary summary available in Utah Geological Society Guidebook, 11, p. 41–75.

Oriel, S.S., 1968, Preliminary geologic map of the SW quarter of the Bancroft quadrangle, Bannock and Caribou Counties, Idaho: U.S. Geological Survey Miscellaneous Field Studies Map MF-299.

Oriel, S.S., and Armstrong, F.C., 1971, Uppermost Precambrian and lowest Cambrian rocks in southeastern Idaho: U.S. Geological Survey Professional Paper 394, 52 p.

Oriel, S.S., and Armstrong, F.C., 1986, Tectonic development of the Idaho-Wyoming thrust belt—author's commentary: American Association of Petroleum Geologists Memoir 41, p. 267–279.

Oriel, S.S., and Platt, L.B., 1980, Geologic map of the Preston 1 x 2 degree quadrangle, southeastern Idaho and western Wyoming: U.S. Geological Survey Miscellaneous Investigations Map I-1127.

Oriel, S.S., and Tracey, J.I.Jr., 1970, Uppermost Cretaceous and Tertiary stratigraphy of Fossil Basin, southwestern Wyoming: U.S. Geological Survey Professional Paper 635, 53 p.

Osmond, J.C., 1954, Dolomites in Silurian and Devonian of east-central Nevada: American Association of Petroleum Geologists Bulletin, v. 38, no. 9, p. 1911–1956.

Osmond, J.C., 1962, Stratigraphy of Devonian Sevy Dolomite in Utah and Nevada: American Association of Petroleum Geologists Bulletin, v. 46, no. 11, p. 2033–2056.

Osterwald, F.W., and Maberry, J.O., 1974, Engineering geologic map of the Woodside quadrangle, Emery and Carbon Counties, Utah: U.S. Geological Survey Miscellanous Investigations Series Map I-798.

Osterwald, F.W., Maberry, J.O., and Dunrud, C.R., 1981, Bedrock, surficial, and economic geology of the Sunnyside coal-mining district, Carbon and Emery Counties, Utah: U S. Geological Survey Professional Paper 1166, 68 p.

O'Sullivan, R.B., 1965, Geology of the Cedar Mesa-Boundary Butte area, San Juan County, Utah: U.S. Geological Survey Bulletin 1186, 128 p.

O'Sullivan, R.B., 1970, The upper part of the Upper Triassic Chinle Formation and related rocks, southeastern Utah and adjacent areas: U.S. Geological Survey Professional Paper 644-E, 21 p.

O'Sullivan, R.B., 1980a, Stratigraphic sections of middle Jurassic San Rafael Group from Wilson Arch to Bluff in southeastern Utah: U.S. Geological Survey Oil and Gas Investigations Chart OC-102.

O'Sullivan, R.B., 1980b, Stratigraphic sections of Middle Jurassic San Rafael Group and related rocks from the Green River to the Moab area in east-central Utah: U.S. Geological Survey Miscellaneous Field Studies Map MF-1247.

O'Sullivan, R.B., 1981a, Stratigraphic sections of some Jurassic rocks from near Moab, to Slick Rock, Colorado: U.S. Geological Survey Oil and Gas Investigations Chart OC-107.

O'Sullivan, R.B., 1981b, Stratigraphic sections of Middle Jurassic Entrada Sandstone and related rocks from Salt Valley to Dewey Bridge in east-central Utah: U.S. Geological Survey Oil and Gas Investigations Chart OC-113.

O'Sullivan, R.B., 1984, The base of the Upper Jurassic Morrison Formation in east-central Utah: U.S. Geological Survey Bulletin, no. 1561, 20 p.

O'Sullivan, R.B., and Craig, L.C., 1973, Jurassic rocks of northeast Arizona and adjacent areas: New Mexico Geological Society Guidebook 24, p. 79–85.

O'Sullivan, R.B., and Green, M.W., 1973, Triassic rocks of northeast Arizona and adjacent areas: New Mexico Geological Society Guidebook 24, p. 72–78.

O'Sullivan, R.B., and Maberry, J.D., 1975, Marine trace fossils in the upper Jurassic Bluff Sandstone, Southeastern Utah: U.S. Geological Survey Bulletin 1395-I, 15 p.

O'Sullivan, R.B., and MacLachlan, M.E., 1975, Triassic rocks of the Moab-White Canyon area, southeastern Utah: Four Corners Geological Society Guidebook 8, p. 129–141.

O'Sullivan, R.B., and Pierce, F.W., 1983, Stratigraphic diagram of Middle Jurassic San Rafael Group and associated formations from the San Rafael Swell to Bluff in southeastern Utah: U.S. Geological Survey Oil and Gas Investigations Chart OC-119.

O'Sullivan, R.B., and Pipiringos, G.IV., 1983, Stratigraphic sections of Middle Jurassic Entrada Sandstone and related rocks from Dewey Bridge, Utah, to Bridgeport, Colorado: U.S. Geological Survey Oil and Gas Investigations Chart OC-122.

Orgill, J.R., 1971, The Permian-Triassic unconformity and its relationship to the Moenkopi, Kaibab, and White Rim formations in and near the San Rafael Swell, Utah: Brigham Young University Geology Studies, v. 18, part 3, p. 131–179.

Ott, V.D., 1980, Geology of the Woodruff Narrows quadrangle, Utah-Wyoming: Brigham Young University Geology Studies, v. 27, part 2, p. 67–84.

Ott, V.D., Potter, E.C., and Kreckel, K.R., 1985, South Crawford Mountain prospect, Rich County, Utah—a case history: Utah Geological Association Publication 14, p. 193–200.

Otto, E.P., and Picard, M.D., 1975, Stratigraphy and oil and gas potential of Entrada Sandstone (Jurassic), northeastern Utah: Rocky Mountain Association of Geologists Symposium, p. 129–139.

Oviatt, C.G., 1984, Lake Bonneville stratigraphy at the Old River Bed and Leamington, Utah; unpublished Ph.D. dissertation, University of Utah, Salt Lake City, 122 p.

Oviatt, C.G., 1985, Preliminary notes on the Paleozoic stratigraphy and structural geology of the Honeyville quadrangle, northern Wellsville Mountain Utah: Utah Geological Association Publication 14, p. 47–54.

Oviatt, C.G., 1986a, Geologic map of the Honeyville quadrangle, Utah: Utah Geological and Mineral Survey Map 88.

Oviatt, C.G., 1986b, Geologic map of the Cutler Dam quadrangle, Box Elder and Cache Counties, Utah: Utah Geological and Mineral Survey Map 91.

Oviatt, C.G., 1987, Lake Bonneville stratigraphy at the Old River Bed, Utah: American Journal of Science, v. 287, p. 383–398.

Oviatt, C.G., McCoy, W.D., and Reider, R.G., 1987, Evidence for shallow early or middle Wisconsin-age lake in the Bonneville Basin, Utah: Quaternary Research, v. 27, p. 248–262.

Paddock, R.E., 1956, Geology of the Newfoundland Mountains, Box Elder County, Utah: unpublished master's thesis, University of Utah, 101 p.

Pakiser, L.C., 1985, Seismic exploration of the crust and upper mantle of the Basin and Range Province: Geological Society of America, DNAG Special Volume 1,

Geologists and Ideas, Drake, E.T., and Jordan, W.M., eds., p. 453–469.
Palmer, A.R., 1971, Cambrian of the Great Basin and adjacent areas, western United States: *in* Holland, C.H., ed., Cambrian of the New World, Wiley-Interscience, p. 1–78.
Palmer, A.R., 1979, Cambrian biogeography and biostratigraphy: *in* Robison, R.A., and Teichert, C., eds., Treatise on Invertebrate Paleontology, Part A, p. A119–A135.
Palmer, A.R., 1981, Subdivision of the Sauk Sequence: Short papers for the Second International Symposium on the Cambrian System: U.S. Geological Survey Open-File Report 81-743, p. 160–162.
Palmer, A.R., 1983, The decade of North American geology 1983 geologic time scale: Geology, v. 11, p. 503–504.
Palmer, A.R., and Hintze, L.F., 1988, Middle Cambrian to Lower Ordovician rocks: *in* Burchfiel, B.C., Zoback, M.L., and Lipman P.W., eds., The Cordilleran Orogen—Conterminous U.S.: Geological Society of America, Decade of North American Geology (DNAG) Series, v. G3, in press, Boulder, Colorado.
Pampeyan, E.H., 1986, Geologic map of the Lynndyl 30 x 60 minute quadrangle, Tooele, Juab, Utah, and Millard Counties, Utah: U.S. Geological Survey Miscellaneous Geological Investigations Map I-(in press).
Papson, R.P., 1977, Road log along Pleistocene Lake Bonneville flood path; *in* Greeley, Ronald and Kings, J.S., eds., Volcanism of the Eastern Snake River Plain, Idaho, National Aeronautics and Space Administration publication CR-154621, p. 272–293.
Parker, J.M., 1981, Lisbon field area, San Juan County, Utah: Rocky Mountain Association of Geologists Guidebook—Geology of the Paradox Basin, p. 89–100.
Parker, J.W., and Roberts, J.W., 1966, Regional Devonian and Mississippian stratigraphy, central Colorado Plateau region: American Association of Petroleum Geologists Bulletin, v. 50, p. 2404–2433.
Parker, L.R., 1975, The paleoecology of the fluvial coal-forming swamps and associated floodplain environments in the Blackhawk Formation of central Utah: Brigham Young University Geology Studies, v. 22, part 3, p. 99–116.
Parry, W.T., and Bruhn, R.L., 1987, Fluid inclusion evidence for minimum 11 km vertical offset on the Wasatch Fault, Utah: Geology, v. 15, p. 67–70.
Patton, T.L., and Lent, R.L., 1980, Current hydrocarbon exploration activity in the Great Salt Lake: Utah Geological and Mineral Survey Bulletin 116, p. 115–124.
Paull, R.K., and Paull, R.A., 1982, Permian-Triassic unconformity in the Terrace Mountains, northwestern Utah: Geology, v. 10, p. 582–587.
Paull, R.K., Paull, R.A., and Anderson, A.L., 1985, Conodont biostratigraphy and depositional history of the lower Triassic Dinwoody Formation in the Meade Plate, southeastern Idaho: Utah Geological Association Publication 14, p. 55–66.
Peacock, C.H., 1953, Geology of the Government Hill area, a part of Long Ridge, Utah: unpublished master's thesis, Brigham Young University, 96 p.
Perry, L.I., 1976, Gold Springs mining district, Iron County, Utah: Utah Geology, v. 3, no. 1, p. 23–50.
Petersen, C.A., 1974, Summary of stratigraphy in the Mineral Range, Beaver and Millard Counties, Utah: Utah Geology, v. 1, no. 1, p. 45–50.
Petersen, C.A., 1975, Geology of the Roosevelt Hot Springs area, Beaver, County, Utah: Utah Geology, v. 2., no. 2, p. 109–116.
Petersen, H.N., 1953, Structure and Paleozoic stratigraphy of the Currant Creek area near Goshen, Utah: unpublished master's thesis, Brigham Young University, 61 p.
Petersen, L.M., and Roylance, M.M., 1982, Stratigraphy and depositional environments of the Upper Jurassic Morrison Formation near Capitol Reef National Park, Utah: Brigham Young University Geology Studies, v. 29, part 2. p. 1–12.
Petersen, S.M., and Pack, R.T., 1982, Paleoenvironments of the Upper Jurassic Summerville Formation near Capitol Reef National Park, Utah: Brigham Young University Geology Studies, v. 29, part 2, p. 13–25.
Peterson, A.R., 1976, Paleoenvironments of the Colton Formation, Colton, Utah: Brigham Young University Geology Studies, v. 23, part 1, p. 3–35.
Peterson, D.L., 1974, Bouguer gravity map of part of the northern Lake Bonneville Basin, Utah and Idaho: U.S. Geological Survey Miscellaneous Field Studies, Map MF-627.
Peterson, D.O., 1953, Structure and stratigraphy of the Little Valley area, Long Ridge, Utah: unpublished master's thesis, Brigham Young University, 96 p.
Peterson, Fred, 1969, Four new members of the Upper Cretaceous Straight Cliffs Formation in the southeastern Kaiparowits region, Kane County, Utah: U.S. Geological Survey Bulletin 1274-J, 28 p.
Peterson, Fred, 1973, Geologic map of the southwest quarter of the Gunsight Butte Quadrangle, Kane and San Juan Counties, Utah and Coconino County, Arizona: U.S. Geological Survey Mineral Investigations Field Study Map MF-306.
Peterson, Fred, 1975, Geologic map of the Sooner Bench quadrangle, Kane County, Utah: U.S. Geological Survey Miscellaneous Investigations Series Map I-874.
Peterson, Fred, 1980, Geologic map and coal deposits of the Big Hollow Wash Quadrangle, Kane County, Utah: U.S. Geological Survey Coal Investigations Map C-84.
Peterson, Fred, 1980, Sedimentology of the uranium-bearing Salt Wash Member and Tidwell unit of the Morrison Formation in the Henry and Kaiparowits Basins, Utah: Utah Geological Association Publication 8, Henry Mountains Symposium, p. 305–322.
Peterson, Fred, 1984, Fluvial sedimentation on a quivering craton—influence of slight crustal movements on fluvial processes, Upper Jurassic Morrison Formation, western Colorado Plateau: Sedimentary Geology, v. 38, p. 21–49.
Peterson, Fred, 1986, Jurassic paleotectonics in the west-central part of the Colorado Plateau, Utah and Arizona; *in*, Peterson, J.A., ed., Paleotectonics and sedimentation in the Rocky Mountain region, United States: American Association of Petroleum Geologists Memoir 41, p. 563–596.
Peterson, Fred, 1987, The search for source areas of Morrison (Upper Jurassic) clastics on the Colorado Plateau: Geological Society of America Abstracts with Programs, v. 19, p. 804.
Peterson, Fred, 1988, in press, Stratigraphy and nomenclature of Middle and Upper Jurassic rocks, western Colorado Plateau, Utah and Arizona: U.S. Geological Survey Bulletin 1633-B.
Peterson, Fred, and Barnum, B.E., 1973a, Geologic map and coal resources of the northeast quarter of the Cummings Mesa quadrangle, Kane County, Utah: U.S. Geological Survey Coal Investigations Map C-63.
Peterson, Fred, and Barnum, B.E., 1973b, Geologic map and coal resources of the northwest quarter of the Cummings Mesa Quadrangle, Kane County, Utah: U.S. Geological Survey Coal Investigations Map C-64.
Peterson, Fred and Barnum, B.E., 1973c, Geologic map of the southeast quarter of the Cummings Mesa Quadrangle, Kane and San Juan Counties, Utah and Coconino County, Arizona: U.S. Geological Survey Miscellaneous Investigations Map I-758.
Peterson, Fred, and Barnum, B.E., 1973d, Geologic map of the southwest quarter of the Cummings Mesa Quadrangle, Kane and San Juan Counties, Utah, and Coconino County, Arizona: U.S. Geological Survey Miscellaneous Investigations Map I-759.
Peterson, Fred, and Pipiringos, G.N., 1979, Stratigraphic relations of the Navajo Sandstone to Middle Jurassic formations, southern Utah and northern Arizona: U.S. Geological Survey Professional Paper 1035-B, 43 p.
Peterson, Fred, Ryder, R.T., and Law, B.E., 1980, Stratigraphy, sedimentology, and regional relationships of the Cretaceous system in the Henry Mountains region, Utah: Utah Geological Association Publication 8, Henry Mountains Symposium, p. 151–170.
Peterson, Fred, and Turner-Peterson, C.E., 1987, The Morrison Formation of the Colorado Plateau—recent advances in sedimentology, stratigraphy, and paleotectonics: Hunteria, v. 2, no. 1,p. 1–18. University Museum, Boulder, Colorado.
Peterson, J.A., 1972, Jurassic System: Geologic Atlas of the Rocky Mountain Region, Rocky Mountain Association of Geologists, Denver, p. 177–189.
Peterson, J.A., 1977, Paleozoic shelf-margins and marginal basins, western Rocky Mountains—Great Basin, United States: Wyoming Geological Association 29th Annual Field Conference Guidebook, p. 135–153.
Peterson, J.A., 1980, Permian paleogeography and sedimentary provinces, west-central U.S.: Symposium 1, Paleozoic Paleogeography of the west-central U.S., Rocky Mountain Section, Society of Economic Paleontologists and Mineralogists, p. 271–292. Denver, Colorado.
Peterson, J.A., and Smith, D.L., 1986, Rocky Mountain paleogeography through geologic time: American Association of Petroleum Geologists Memoir 41, p. 3–19.
Picard, M.D., 1975, Facies, petrography and petroleum potential of Nugget Sandstone (Jurassic), southwestern Wyoming and northeastern Utah: Rocky Mountain Association of Geologists Symposium, p. 109–127.
Picard, M.D., 1977, Petrography and stratigraphy of productive beds in the Morgan Formation, Church Buttes Unit No. 19, southwest Wyoming: Wyoming Geological Association 29th Annual Field Conference Guidebook, p. 179–196.
Picard, M.D., and High, L.R. Jr., 1972, Paleoenvironmental reconstructions in an area of rapid facies change, Parachute Creek Member of Green River Formation (Eocene), Uinta Basin, Utah: Geological Society of America Bulletin, v. 83, p. 2689–2708.
Picard, M.D., Thompson, W.D., and Williamson, C.R., 1973, Petrology, geochemistry, and stratigraphy of black shale facies of Green River Formation (Eocene), Uinta Basin, Utah: Utah Geological and Mineral Survey Bulletin 100, 55 p.
Picha, F., and Gibson, R.I., 1985, Cordilleran hingeline—Late Precambrian rifted margin of the North American craton and its impact on the depostional and structural history, Nevada and Utah: Geology, v. 13, p. 465–468.
Piekarski, Lee, 1980, Relative age determination of Quaternary fault scarps along the southern Wasatch, Fish Springs, and House ranges, Utah: Brigham Young University Geology Studies, v. 27, part 2, p. 123–129.
Pierce, C.R., 1974, Geology of the southern part of the Little Drum Mountains, Utah: Brigham Young University Geology Studies, v. 21, part 1, p. 109–130.
Pierce, K.L., 1986, Dating methods: *in* National Academy Press, Studies in geophysics, Active Tectonics, p. 196–214.
Pinnell, M.L., 1972, Geology of the Thistle quadrangle, Utah: Brigham Young University Geology Studies, v. 19, part 1, p. 89–130.
Pinney, R.I., 1965, Preliminary survey of Mississippian biostratigraphy (conodonts) in the Oquirrh Basin of central Utah: unpublished Ph.D. thesis, University of Wisconsin (Madison), 199 p.
Pipiringos, G.N., 1979, Preliminary geologic map of the Agency Draw NE quadrangle, Uintah County, Utah: U.S. Geological Survey Miscellaneous Field Studies Map MF-1078.
Pipiringos, G.N., and Imlay, R.W., 1979, Lithology and subdivisions of the Jurassic Stump Formation in southeastern Idaho and adjoining areas: U.S. Geological Survey Professional Paper 1035-C, 25 p.
Pipiringos, G.N., and O'Sullivan, R.B., 1975, Chert pebble unconformity at the top of the Navajo Sandstone in southeastern Utah: Four Corners Geological Society Guidebook, 8th Field Conference, Canyonlands, p. 149–156.
Pipiringos, G.N., and O'Sullivan, R.B., 1978, Principal unconformities in Triassic and Jurassic rocks, western Interior United States—a preliminary survey: U.S. Geological Survey Professional Paper 1035-A, 29 p.
Pitman, J.K., Fouch, T.D., and Goldhaber, M.B., 1982, Depositional setting and diagenetic evolution of some Tertiary unconventional reservoir rocks, Uinta Basin, Utah: American Association of Petroleum Geologists Bulletin, v. 66, p. 1581–1596.
Pitman, W.C. III, and Talwani, Manik, 1972, Sea-floor spreading in the North Atlantic: Geological Society of America Bulletin, v. 83, p. 619–646.
Platt, L.B., 1977, Geologic map of the Ireland Springs-Samaria area, southeastern Idaho and northern Utah: U.S. Geological Survey Miscellaneous Field Studies Map MF-890.
Podrebarac, T.J., 1976, Trace fossils of the Brigham Quartzite in northern Utah and southeastern Idaho;

unpublished master's thesis, University of Utah, 158 p.

Poole, F.G., 1961, Stream directions in Triassic rocks of the Colorado Plateau: U.S. Geological Survey Professional Paper 424-C, p. 139–141.

Poole, F.G., 1962, Wind direction in late Paleozoic to middle Mesozoic time on the Colorado Plateau: U.S. Geological Survey Professional Paper 450-D, p. 147–151.

Poole, F.G., and Stewart, J.H., 1964, Chinle Formation and Glen Canyon Sandstone in northeastern Utah and northwestern Colorado: U.S. Geological Survey Professional Paper 501-D, p. 30–39.

Poole, F.G., Baars, D.L., Drewes, Harald, Hayes, P.T., Ketner, K.B., McKee, E.D., Teichert, Curt, and Williams, J.S., 1967, Devonian of the southwestern U.S.: Alberta Society of Petroleum Geologists, International Symposium on the Devonian System, v. 1, p. 879–912.

Poole, F.G., and Sandberg, C.A., 1988, Upper Ordovician—Cincinnatian rocks, Lower Silurian to Middle Devonian rocks: *in* Burchfiel, B.C., Zoback, M.L., and Lipman, P.W., eds., The Cordilleran Orogen—Conterminous U.S.: Geological Society of America, Decade of North American Geology (DNAG) Series, v. G3, in press, Boulder, Colorado.

Prince, D., 1963, Mississippian coal cyclothems in the Manning Canyon Shale of central Utah: Brigham Young University Geology Studies, v. 10, p. 83–103.

Prammani, P., 1957, Geology of the east-central part of the Malad Range, Idaho: unpublished master's thesis, Utah State University, Logan, 67 p.

Rascoe, B., Jr., and Baars, D.L., 1972, Permian System, *in* Geologic Atlas of the Rocky Mountain Region, p. 143–165. Rocky Mountain Association of Geologists, Denver.

Rawson, R.R., 1957, Geology of the southern part of the Spanish Fork Peak quadrangle, Utah: Brigham Young University Geology Studies, v. 4, no. 2, 33 p.

Rawson, R.R., and Turner-Peterson, C.E., 1979, Marine-carbonate, sabkha, and eolian facies transitions within the Permian Toroweap Formation, northern Arizona: Four Corners Geological Society Guidebook 9, Permianland, p. 87–104.

Rees, M.N., 1986, A fault-controlled trough through a carbonate platform: the Middle Cambrian House Range embayment: Geological Society of America Bulletin, v. 97, p. 1054–1069.

Rees, M.N., Brady, M. J., and Rowell, A. J., 1976, Depositional environments of the Upper Cambrian Johns Wash Limestone (House Range, Utah): Journal of Sedimentary Petrology, v. 46, p. 38–47.

Reinemund, J.A., and Terman, M.J., 1982, Plate-tectonic map of the Circum-Pacific region, Pacific Basin sheet: American Association of Petroleum Geologists, Tulsa.

Rehrig, W.A., 1986, Processes of regional Tertiary extension in the western Cordillera—insights from the metamorphic core complexes: Geological Society of America Special Paper 208, p. 97–121.

Reynolds, R.L., Hudson, M.R., Fishman, N.S., and Campbell, J.A., 1985, Paleomagnetic and petrologic evidence bearing on the age and origin of uranium deposits in the Permian Cutler Formation, Lisbon Valley, Utah: Geological Society of America Bulletin, v. 96 p.719–730.

Richmond, G.M., 1962, Quarternary stratigraphy of the La Sal Mountains, Utah: U.S. Geological Survey Professional Paper 324, 134 p.

Richmond, G.M., 1964, Glaciation of Little Cottonwood and Bells Canyons, Wasatch Mountains, Utah: U.S. Geological Survey, Professional Paper 454.-D, 41 p.

Rigby, J.K., 1958, Geology of the Stansbury Mountains, Tooele County, Utah: Utah Geological Society Guidebook 13, p. 1–134.

Rigby, J.K., 1959, Upper Devonian unconformity in central Utah: Geological Society of America Bulletin, v. 70, p. 207–218.

Rigby, J.K., 1959, Stratigraphy of the southern Oquirrh Mountains, lower Paleozoic succession: Utah Geological Society Guidebook 14, p. 9–31.

Rigby, J.K., and Clark, D.L., 1962, Devonian and Mississippian Systems in central Utah: Brigham Young University Geology Studies, v. 9. part 1, p. 17–25.

Rigby, J.K., and Hintze, L.F., 1977, Early Middle Ordovician corals from western Utah: Utah Geology, v. 4, no. 2., p. 105–111.

Rigo, R.J., 1968, Middle and Upper Cambrian stratigraphy in the autochthon and allochthon of northern Utah: Brigham Young University Geology Studies v. 15, part 1, p. 31–66.

Ritter, S.M., 1986, Taxonomic revision and phylogeny of post-Early Permian crisis *bisselli-whitei* Zone conodonts with comments on Late Paleozoic diversity: Geologica et Palaeontologica v. 20, p. 139–165, Marburg, Germany.

Ritter, S.M., 1987, Biofacies-based refinement of Early Permian conodont biostragraphy, in central and western USA: in, Austin, R.L., ed., Conodonts: Investigative Techniques and Applications, p. 382–403. Ellis Horwood, Ltd., publisher, Great Britain.

Ritzma, H.R., 1971, Faulting on the north flank of the Uinta Mountains, Utah and Colorado: Wyoming Geological Association Guidebook 23, p. 145–150.

Ritzma, H.R., 1974, Dating of igneous dike, eastern Uinta Mountains: Utah Geology, v. 1, no. 1, p. 95.

Robinson, G.B. Jr., 1971, Ground-water hydrology of the San Pitch River drainage basin, Sanpete County, Utah: U.S. Geological Survey Water Supply Paper 1896, 80 p.

Robison, R.A., 1960 Lower and Middle Cambrian stratigraphy of the eastern Great Basin: Intermountain Association of Petroleum Geologists Guidebook 11, p. 43–52.

Robison, R.A., 1962, Late Middle Cambrian faunas from the Wheeler and Marjum Formations of western Utah. unpublished Ph.D. dissertation, University of Texas, 304 p.

Robison, R.A., 1964a, Late Middle Cambrian faunas from western Utah: Journal of Paleontology, v. 38, no. 3, p. 510–566.

Robison, R.A., 1964b, Upper Middle Cambrian stratigraphy of western Utah: Geological Society of America Bulletin, v. 75, p. 995–1010.

Robison, R.A., 1971, Additional Middle Cambrian trilobites from the Wheeler Shale of Utah: Journal of Paleontology, v. 45, no. 5, p. 796–804.

Robison, R.A., 1976, Middle Cambrian trilobite biostratigraphy of the Great Basin: Brigham Young University Geology Studies, v. 23, part 2, p. 93–109.

Robison, R.A., 1982, Some Middle Cambrian agnostoid trilobites from western North America: Journal of Paleontology, v. 56, p. 132–160.

Robison, R.A., 1984, Cambrian agnostida of North America and Greenland. Part 1, Ptychagnostidae: University of Kansas Paleontological Contributions, Paper 109, 59 p.

Robison, R.A., and Hintze, L.F., 1972, An Early Cambrian trilobite faunule from Utah: Brigham Young University Geology Studies, v. 19, part 1, p. 3–15.

Robison, R.A., and Palmer, A.R., 1968, Revision of Cambrian stratigraphy, Silver Island Mountains, Utah: American Association of Petroleum Geologists Bulletin, v. 52, p. 167–171.

Robison, R.A., Rosova, A.V., Rowell, A.J., and Fletcher, T.P., 1977, Cambrian boundaries and divisions: Lethaia, v. 10, p. 257–262.

Robison, S.F., 1986, Paleocene (Puercan-Torrejonian) mammalian faunas of the North Horn Formation, central Utah: Brigham Young University Geology Studies, v. 33, no. 1, p. 87–133.

Rodgers, D.W., 19984 Stratigraphy, correlation and depositional environments of Upper Proterozoic and Lower Cambrian rocks of the southern Deep Creek Range, Utah: Utah Geological Association Publications 13, p. 79–91.

Rodgers, D.W., 1985, Tectonic evolution of the southern Deep Creek Range, Nevada-Utah: Geological Society of America Abstracts with Programs, v. 17, p. 405.

Rodgers, D.W., and Sutter, J.F., 1986, Thermal history of the Deep Creek Range, Nevada-Utah, and implications for Cordilleran tectonics: Geological Society of America Abstracts with Programs, v. 18, p. 177.

Roehler, H.W., 1972a Geologic map of the Red Creek Ranch quadrangle, Wyoming, Utah, and Colorado: U.S. Geological Survey Geologic Quadrangle Map GQ-1001.

Roehler, H.W., 1972b, Geologic maps of the Brushy Point and Razorback Ridge quadrangles, Rio Blanco and Garfield Counties, Colorado: U.S. Geological Survey Geologic Quadrangle Maps GQ-1018 and GQ-1019.

Roehler, H.W., 1986, McCourt Sandstone Tongue and Glades coal bed of the Rock Springs Formation, Wyoming and Utah: Geological Society of America Special Paper 210, p. 141–154.

Rogers, J.C., 1984, Depositional environments and paleoecology of two quarry sites in the Middle Cambrian Marjum and Wheeler Formations, House Range, Utah: Brigham Young University Geology Studies v. 31, p. 97–115.

Rose, P.R., 1976, Mississippian carbonate shelf margins, western United States: U.S. Geological Survey Journal of Research v. 4, no. 2, p. 449–466. Also printed in Rocky Mountain Association of Geologists 1976 Hingeline Symposium, p. 135–161 also in Wyoming Geological Association Guidebook 29, p. 155–72.

Ross, C.A., 1979, Carboniferous biogeography and biostratigraphy: *in* Robison, R.A. and Teichert, C., eds., Treatise on Invertebrate Paleontology, Part A, p. A254–290.

Ross, C.A., and Ross, J.P., 1979, Permian biogeography and biostratigraphy: *in* Robison, R.A. and Teichert, C., eds., Treatise on Invertebrate Paleontology, Part A, p. A291–A350.

Ross, H.P., Nielson, D.L., and Moore, J.N., 1982, Roosevelt Hot Springs geothermal system, Utah—case study: American Association of Petroleum Geologists Bulletin, v. 66, p. 879–902.

Ross, R.J. Jr., 1951, Stratigraphy of the Garden City Formation in northeastern Utah and its trilobite faunas: Yale University Peabody Museum Bulletin 6, 161 p.

Ross, R.J. Jr., 1976, Ordovician sedimentation in the western U.S.A.: *in* Bassett, M.G., ed., The Ordovician System, Proceedings, Palaeontological Association Symposium, Birmingham, 1974, University of Wales Press and National Museum of Wales, Cardiff, p. 73–105. Also in Rocky Mountain Association of Geologists 1976 Symposium, p. 109–133.

Ross, R.J. Jr., 1975, Early Paleozoic trilobites, sedimentary facies, lithospheric plates, and ocean currents: Fossils and Strata, no. 4, p. 307–329, Oslo.

Ross, R.J. Jr., 1977, Ordovician paleogeography of the western United States: *in* Stewart, J.H., Stevens, C.H., and Fritsche, A.E., eds., Paleozoic paleogeography of the western United States, Pacific Coast Paleogeography Symposium 1, Society of Economic Paleontologists and Mineralogists, Los Angeles, p. 19–38.

Ross, R.J. Jr., and others, 1982, The Ordovician System in the United States: International Union of Geological Sciences Publication No. 12, 73 p., Ottawa, Canada, and Paris, France.

Ross, R.J. Jr., James, N.P., Hintze, L.F., and Poole, F.G., 1987, Architecture and evolution of a Whiterockian, early middle Ordovician, carbonate platform, Basin Ranges of western U.S.A.: *in* Controls on Carbonate Platform and Basin Development, Society of Economic Paleontologists and Mineralogists Symposium, Los Angeles, (in press).

Ross, R.J. Jr., Hintze, L.F., and Poole, F.G., 1988, Middle Ordovician, Whiterockian and Mohawkian rocks: *in* Burchfiel, B.C., Zoback, M.L., and Lipman, P.W., eds., The Cordilleran Orogen—Conterminous U.S.: Geological Society of America, Decade of North American Geology (DNAG) Series, v. G3, in press, Boulder, Colorado.

Ross, R.J. Jr., and Naeser, C.W., 1984, The Ordovician time scale—new refinements: *in* Bruton, D.L., (ed), Aspects of the Ordovician System, Palaeontological Contributions from the University of Oslo, no. 295, Universitets Forlaget, p. 5–10.

Rowell, A.J., Rees, M.N., and Suczek, C.A., 1979, Margin of the North American continent during Late Cambrian time: American Journal of Science, v. 279, p. 1–18.

Rowell, A.J., Robison, R.A., and Strickland, D.K., 1982, Aspects of Cambrian agnostoid phylogeny and chronocorrelation: Journal of Paleontology , v. 56, p. 161–182.

Rowley, P.D., 1975, Geologic map of the Enoch NE quadrangle, Iron County, Utah: U.S. Geological Survey Geologic Quadrangle Map GQ-1301.

Rowley, P.Q., 1976, Geologic map of the Enoch NW quadrangle, Iron County, Utah: U.S. Geologic Survey Geologic Quadrangle Map GQ-1302.

Rowley, P.D., Anderson, J.J., Williams, P.L., and Fleck, R.J., 1978, Age of structural differentiation between the Colorado Plateaus and Basin and Range provinces in southwestern Utah: Geology, v. 6, p. 51–55.

Rowley, P.D., Cunningham, C.G., Steven, T.A., Mehnert, H.H., and Naeser, C.W., 1988a, Geologic map of the Antelope Range quadrangle, Sevier and Piute Counties, Utah; Utah Geological and Mineral Survey Map 106.

Rowley, P.D., Cunningham, C.G., Steven, T.A., Mehnert, H.H., and Naeser, C.W., 1988b, Geologic map of the Marysvale quadrangle, Piute County, Utah: Utah Geological and Mineral Survey Map 105.

Rowley, P.D., Hansen, W.R., Tweto, O., and Carrara, P.E., 1985, Geologic map of the Vernal 1 x 2 degree quadrangle, Colorado, Utah, and Wyoming: U.S. Geological Survey Map I-1526.

Rowley, P.D., Hereford, R., and Williams, V.S., 1986, Geologic map of the Adams Head-Johns Valley area, southern Sevier Plateau, Garfield County, Utah: U.S. Geological Survey Miscellaneous Field Studies Map I-(in press).

Rowley, P.D., Kinnery, D.M., and Hansen, W.R., 1979, Geologic map of the Dinosaur Quarry quadrangle, Uintah County, Utah: U.S. Geological Survey Geologic Quadrangle Map GQ-1513.

Rowley, P.D., Steven, T.A., Anderson, J.J., and Cunningham, C.G., 1979, Cenozoic stratigraphic and structural framework of southwestern Utah: U.S. Geological Survey Professional Paper 1149, 22 p.

Rowley, P.D., Steven, T.A., and Kaplan, A.M., 1981, Geologic map of the Monroe NE quadrangle, Sevier County, Utah: U.S. Geological Survey Miscellaneous Field Studies Map MF-1330.

Rowley, P.D., Steven, T.A., and Mehnert, H.H., 1981b, Origin and structural implications of upper Miocene rhyolites in Kingston Canyon, Piute County, Utah: Geological Society of America Bulletin, v. 92, part 1, p. 590–602.

Rowley, P.D., and Threet, R.L., 1976, Geologic map of the Enoch quadrangle, Iron County, Utah: U.S. Geological Survey Geologic Quadrangle Map GQ-1296.

Rowley, P.D., Williams, P.L., and Kaplan, A.M., 1986a, Geologic map of the Greenwich quadrangle, Piute County, Utah: U.S. Geological Survey Geologic Quadrangle Map GQ-1589.

Rowley, P.D., Williams, P.L., and Kaplan, A.M., 1986b, Geologic map of the Koosharem quadrangle, Sevier and Piute Counties, Utah: U.S. Geological Survey Geologic Quadrangle Map GQ-1590.

Royse, F. Jr., Warner, M.A., and Reese, D.L., 1975, Thrust belt structural geometry and related stratigraphic problems, Wyoming, Idaho, northern Utah: Rocky Mountain Association of Geologists Symposium, p. 41–54.

Rubey, W.W., 1973, New Cretaceous Formations in the western Wyoming thrust belt: U.S. Geological Survey Bulletin 1372-I, 35 p.

Rubey, W.W., Oriel, S.S., and Tracey, J.I. Jr., 1975, Geology of the Sage and Kemmerer 15-minute quadrangles, Lincoln County, Wyoming: U.S. Geological Survey Professional paper 855, 18 p.

Rubin, D.M., and Hunter, R.E., 1987, Field guide to sedimentary structures in the Navajo and Entrada sandstones in southern Utah and northern Arizona: Arizona Bureau of Geology and Mineral Technology, Geological Survey Branch, Special Paper 5, p. 126–139.

Runyon, D.M., 1977, Structure, stratigraphy and tectonic history of the Indianola quadrangle, central Utah: Brigham Young University Geology Studies, v. 24 part 2, p. 63–82.

Russon, M.P., 1987, Geology, depositional environments, and coal resources of the Helper 7½-minute quadrangle, Carbon County, Utah: Brigham Young University Geology Studies v. 34, p. 131–168.

Ryder, R.T., Fouch, T.D., and Elison, J.H., 1976, Early Tertiary sedimentation in the western Uinta Basin, Utah: Geological Society of America Bulletin, v. 87, p. 496–512.

Ryer, T.A., 1976, Cretaceous stratigraphy of the Coalville and Rockport areas, Utah: Utah Geology, v. 3, no. 2, p. 71–83, also summarized in Geological Society of America Bulletin, v. 88, p. 177–188, (1977).

Ryer, T.A., 1981, Cross-section of the Ferron Sandstone Member of the Mancos Shale in the Emery coal field, Emery and Sevier Counties, central Utah: U.S. Geological Survey Miscellaneous Field Studies Map MF-1357.

Ryer, T.A., 1981, Deltaic coals of Ferron Sandstone Member of Mancos Shale—predictive model for Cretaceous coal-bearing strata of western interior: American Association of Petroleum Geologists Bulletin, v. 65, p. 2323–2340.

Ryer, T.A., 1982, Possible eustatic control on the location of Utah Cretaceous coal fields: Utah Geological and Mineral Survey Bulletin 118, p. 274–300. Also see Field trip on Ferron Sandstone same volume, p. 274–300.

Ryer, T.A., 1983, Transgressive-regressive cycles and the occurrence of coal in some Upper Cretaceous strata of Utah: Geology , v. 11, p. 207–210.

Ryer, T.A., and McPhillips, M., 1983, Early Late Cretaceous paleogeography of east-central Utah: Rocky Mountain section, Society of Economic Paleontologists and Mineralogists, Mesozoic Paleogeography Symposium, p. 253–272.

Ryer, T.A., and Lovekin, J.R., 1986, The Upper Cretaceous Vernal delta of Utah—depositional or paleotectonic feature?: American Association of Petroleum Geologists Memoir 41, p. 497–510.

Sacks, P.E., and Platt, L.B., 1985, Depth and timing of decollement extension, southern Portneuf Range, southeastern Idaho: Utah Geological Association Publication 14, p. 119–128.

Sadlick, Walter, 1965, Biostratigraphy of the Chainman Formation, eastern Nevada and western Utah: unpublished Ph.D. thesis, University of Utah, 227 p.

St. Aubin-Hietpas, L.A., 1983, Carbonate petrology and paleoecology of Permo-Carboniferous rocks, Mountain Home Range, Millard County, Utah: Brigham Young University Geology Studies, v. 30, pt. 1, p. 113–143.

Saleeby, J.B., 1986, Centennial Continent/Ocean Transect #10, C-2 central California offshore to Colorado Plateau: Geological Society of America DNAG Continent-ocean Transect C-2.

Sanchez, J.D., and Brown, T.L., 1983, Stratigraphic framework and coal resources of the Upper Cretaceous Blackhawk Formation in the Muddy Creek and Nelson Mountain areas of the Wasatch Plateau Coal Field, Emery, Sevier, and Sanpete Counties, Utah: U.S. Geological Survey Coal Investigations Map C-94A.

Sandberg, C.A., 1976, Conodont biofacies of late Devonian *Polygnathus styriacus* zone in western U.S.: Geological Association of Canada Special Paper 15, p. 171–186.

Sandberg, C.A., and Dreesen, R., 1984, Late Devonian icriodontid biofacies models and alternate shallow-water conodont zonation: Geological Society of America Special Paper 196, p. 143–178.

Sandberg, C.A., and Gutschick, R.C., 1979 Guide to conodont biostratigraphy of Upper Devonian and Mississippian rocks along the Wasatch Front and Cordilleran Hingeline, Utah: Brigham Young University Geology Studies, v. 26, part 3, p. 107–133.

Sandberg, C.A., and Gutschick, R.C., 1980, Sedimentation and biostratigraphy of Osagean and Meramecian starved basin and foreslope, western United States: Rocky Mountain Section, Society of Economic Geologists and Paleontologists, Symposium 1, Paleozoic Paleogeography of west-central United States, p. 129–148.

Sandberg, C.A., Gutschick, R.C., Johnson, J.G., Poole, F.G., and Sando, W.J., 1982, Middle Devonian to Late Mississippian geologic history of the Overthrust Belt region, western U.S.: Rocky Mountain Association of Geologists, Geologic Studies of the Cordilleran Thrust Belt, v. 2, p. 691–719.

Sandberg, C.A., and Gutschick, R.C., 1984, Distribution, microfauna, and source-rock potential of Mississippian Delle Phosphatic Member of Woodman Formation and equivalents, Utah and adjacent States, *in* Woodward, Jane, Meissner, F.F., and Clayton, J.L., eds., Hydrocarbon source rocks of the Greater Rocky Mountain region: Denver, Colorado, Rocky Mountain Association of Geologists, p. 135–178.

Sandberg, C.A., and Poole, F.G., 1977, Conodont biostratigraphy and depositional complexes of upper Devonian cratonic-platform and continental-shelf rocks in the western United States, *in* Murphy, M.A., and others eds., Western North America-Devonian: California University Riverside Campus Museum Contribution 4, p. 144–122.

Sandberg, C.A., Gutschick, R.C., Johnson, J.G., Poole, F.G., and Sando, W.J., 1982, Middle Devonian to Late Mississippian geologic history of the overthrust belt region, western U.S.: *in* Powers, R.B., ed., Rocky Mountain Association of Geologists, Geologic studies of the Cordilleran Thrust Belt, vol. II, Denver, p. 691–719.

Sandberg, C.A., Poole, F.G., and Gutschick, R.C., 1980, Devonian and Mississippian stratigraphy and conodont zonation of Pilot and Chainman Shales, Confusion Range, Utah: Rocky Mountain Section, Society of Economic Paleontologists and Mineralogists, Paleozoic Paleogeography Symposium no. 1, Denver, p. 71–79.

Sandborn, A.F., 1977, Possible future petroleum of Uinta and Piceance basins and vicinity, northeast Utah and northwest Colorado: Rocky Mountain Association of Geologists Symposium, p. 151–166.

Sanderson, I.D., 1974, Sedimentary structures and their environmental significance in the Navajo Sandstone, San Rafael Swell, Utah: Brigham Young University Geology Studies, v. 21, part 1, p. 215–246.

Sanderson, I.D., 1984, Mount Watson Formation, an interpreted braided-fluvial deposit in the Uinta Mountain Group (Upper Precambrian), Utah: Mountain Geologist, v. 21, no. 4, p. 157–164.

Sanderson, I.D., and Wiley, M.T., 1986, The Jesse Ewing Canyon Formation, an interpreted alluvial fan deposit in the basal Uinta Mountain Group (Middle Proterozoic), Utah: American Association of Petroleum Geologists Bulletin (in press).

Sando, W.J., and Bamber, E.W., 1985, Coral zonation of the Mississippian System in the western interior province of North America: U.S. Geological Survey Professional Paper 1334, 61 p.

Sando, W.J., Dutro, J.T. Jr., and Gere, W.T., 1959, Brazer Dolomite (Mississippian), Randolph quadrangle, northeast Utah: American Association of Petroleum Geologists, v. 43, p. 2741–2769.

Sando, W.J., Dutro, J.T. Jr., Sandberg, C.A., and Mamet, B.L., 1976, Revision of Mississippian stratigraphy, eastern Idaho and northeastern Utah: U.S. Geological Survey Journal of Research v. 4, p. 467–479.

Sanford, R.F., 1982, Preliminary model of regional Mesozoic groundwater flow and uranium deposition in the Colorado Plateau: Geology, v.10, p. 348–352.

Sass, J.H., Lachenbruch, A.H., Munroe, R.J., Greene, G.W., and Moses, T.H. Jr., 1971, Heat flow in the western United States: Journal of Geophysicists Research, v. 76, p. 6376–6413.

Scarbrough, B.E., 1984, Structure and petrology of Tertiary volcanic rocks in parts of Toms Cabin Spring and Lucin NW quadrangles, Box Elder County, Utah, unpublished master's thesis, Utah State University, Logan, 77 p.

Schaeffer, F.E., 1960, Stratigraphy of the Silver Island Mountains: Utah Geological Society Guidebook 15, p. 15–113.

Schaffer, A.R., 1984, Paleoecology and geomorphology of Holocene deposits in north-central Utah: Geological Society of America Abstracts with Programs, v. 16, p. 645.

Schell, E.M., 1969, Summary of the geology of the Sheep Creek Canyon geological area and vicinity, Daggett County, Utah: Intermountain Association Geology Guidebook 16, p. 143–52.

Scheu, S.R., 1985, Geophysical interpretations east of Preston, Idaho, based on gravity and magnetic data: Utah Geological Association Publication 14, p. 261–268.

Schirmer, T.W., 1985, Basement thrusting in north-central Utah—a model for the development of the northern Utah highland: Utah Geological Association Publication 14, p. 129–144.

Schneider, M.C., 1964, Geology of the Pavant Mountains west of Richfield, Sevier county, Utah: Brigham Young University Geology Studies, v. 11, p. 129–139.

Schneider, M.C., 1967, Early Tertiary continental sediments of central and south-central Utah: Brigham Young University Geology Studies, v. 14, p. 143–194.

Schneyer, J.D., 1984, Geology of the Leppy Hills area, southern Silver Island Mountains, near Wendover, Utah: unpublished master's thesis, University of Texas, Austin, 56 p. Published in modified form in Utah Geological Association Publication 13, p. 93–116.

Schurer, V.C., 1979, The Kirkman-Diamond Creek Formation—a chaotic terrane in the Oquirrh Mountains, Utah: Geological Society of America Abstracts with Programs, v. 11, p. 301.

Schwans, P., 1988, Depositional patterns, controls and facies in the alluvial foreland of the overthrust belt—

effects of tectonism and eustasy on sedimentation, central Utah: Geological Society of America Special Paper (in press).

Schweickert, R.A., Bogen, N.L., Girty, G.H., Hanson, R.E., and Merguerian, C., 1984, Timing and structural expression of the Nevadian orogeny, Sierra Nevada, California: Geological Society of America Bulletin, v. 95, p. 967–979.

Scott, R.A., 1982, Aspects of the palynology of the Chinle Formation (Upper Triassic), Colorado Plateau, Arizona, Utah, and New Mexico: U.S. Geological Survey Open-File Report 82-0937, 19 p.

Scott, W.E., McCoy, W.D., Shroba, R.R., and Rubin, M., 1983, Reinterpretation of the exposed record of the last two cycles of Lake Bonneville, western U.S.: Quaternary Research, v. 20, p. 261–285.

Scott, W.E., Pierce, K.L., Bradbury, J.P., and Forester, R.M., 1982, Revised Quaternary stratigraphy and chronology in the American Falls area, southeastern Idaho: Idaho Bureau of Mines and Geology Bulletin 26, p. 581–595.

Sears, J.W., Graff, P.J., and Holden, G.S., 1982, Tectonic evolution of lower Proterozoic rocks, Uinta Mountains, Utah and Colorado: Geological Society of America Bulletin, v. 93, p. 990–997.

Sheehan, P.M., 1971, Silurian brachiopoda, community ecology, and stratigraphic geology in western Utah and eastern Nevada, with a section on Later Ordovician stratigraphy: unpublished Ph.D. dissertation, University of California, Berkeley, 537 p.

Sheehan, P.M., 1979, Silurian continental margin in northern Nevada and northwestern Utah: University of Wyoming Contributions to Geology, v. 17, no. 1, p. 25–35.

Sheehan, P.M., 1980a, Paleogeography and marine communities of the Silurian carbonate shelf in Utah and Nevada: *in* Fouch, T.D., and Magathan, E.R., eds., Paleozoic Paleogeography Symposium 1, Rocky Mountain section, Society of Economic Paleontologists and Mineralogists, Denver, p. 19–37.

Sheehan, P.M., 1980b, Brachiopods of the Tony Grove Lake Member of the Laketown Dolomite: Milwaukee Public Museum Contributions in Biology and Geology no. 30, 23 p.

Sheehan, P.M., 1982a, Late Llandovery and Wenlock brachiopods of the eastern Great Basin: Milwaukee Public Museum Contributions in Biology and Geology no. 50, 83 p.

Sheehan, P.M., 1982b, Brachiopod macroevolution at the Ordovician-Silurian boundary: Third North American Paleontological Convention, Proceedings, v. 2, p. 477–481.

Sheehan, P.M., 1986, Late Ordovician and Silurian carbonate-platform margin near Bovine and Lion mountains, northeastern Utah: Milwaukee Public Museum Contributions in Biology and Geology no. 70, 16 p.

Sheldon, R.P., Cressman, E.R., Cheney, T.M., and McKelvey, V.E., 1967, Permian—Middle Rocky Mountains and northeastern Great Basin: U.S. Geological Survey Professional Paper 515, p. 157–174.

Shoemaker, E.M., and Newman, W.L., 1959, Moenkopi Formation in salt anticline region, Colorado and Utah: American Association of Petroleum Geologists Bulletin, v. 43, p. 1835–1851.

Shuey, R.T., Schellinger, D.K., Johnson, E.H., and Alley, L.B., 1973, Aeromagnetics and the transition between the Colorado Plateau and Basin Range provinces: Geology, v. 1, p. 107–110.

Sibbett, B.S., and Nielson, D.L., 1980, Geology of the central Mineral Mountains, Beaver County, Utah: Earth Science Laboratory, University of Utah Research Institute Report ESL-33, DOE/ET/28392-40, 42. p.

Siders, M.A., 1985a, Geology of the Pinon Point quadrangle, Iron County, Utah: Utah Geological and Mineral Survey Map 84.

Siders, M.A., 1985b, Geology of the Beryl Junction quadrangle, Iron County, Utah: Utah Geological and Mineral Survey Map 85.

Silberling, N.J., Jones, D.L., Blake, M.C. Jr., and Howell, D.G., 1987, Lithotectonic terrane map of the western conterminous U.S.: U.S. Geological Survey Map MF-1874C.

Simpson, R.W., Jachens, R.C., Saltus, R.W., and Blakely, R.J., 1986, Isostatic residual gravity, topographic, and first-vertical-derivative gravity maps of the conterminous United States: U.S. Geological Survey Map GP-975.

Sims, P.K., 1979, Precambrian subdivided: Geotimes, December, 1979, p. 15.

Sirrine, G.K., 1953, Geology of Warm Springs Mountain near Goshen, Utah: unpublished master's thesis, Brigham Young University, 83 p.

Skipp, B., 1979, Mississippian-Great Basin region: U.S. Geological Survey Professional Paper 1010, p. 273–328.

Sloan, R.E., 1987, Paleocene and latest Cretaceous mammal ages, biozones magnetozones, rates of sedimentation, and evolution: Geological Society of America Special Paper 209, p. 165–200.

Smith, A.G., Hurley, A.M. and Briden, J.C., 1981, Phanerozoic paleocontinental world maps: Cambridge Earth Science Series, Cambridge University Press, 102 p.

Smith, Curtis, 1983, Geology depositional environments, and coal resources of the Mt. Pennell 2 NW Quadrangle, Garfield County, Utah: Brigham Young University Geology Studies, v. 30, part 1, p. 145–169.

Smith, C.V., 1965, Geology of the North Canyon area, southern Wasatch Mountains, Utah: Brigham Young University Geology Studies, v. 3, no. 7, 32 p.

Smith, J.D., 1986, Depositional environments of the Tertiary Colton and basal Green River Formations in Emma Park, Utah: Brigham Young University Geology Studies, v. 33, no. 1, p. 134–174.

Smith, J.F., Jr., 1982, Geologic map of the Strevell 15-minute quadrangle Cassia County, Idaho: U.S. Geological Survey Miscellaneous Investigation Series Map I-1403.

Smith, J.F., Jr., 1983, Paleozoic rocks in the Black Pine Mountains, Cassia County, Idaho: U.S. Geological Survey Bulletin 1536, 36 p.

Smith, J.F., Huff, L.C., Hinrichs, E.N., and Luedke, R.G., 1963, Geology of the Capitol Reef area, Wayne and Garfield Counties, Utah: U.S. Geological Survey Professional Paper 363, 102 p.

Smith, J.T., and Cook, K.L., 1985, Geologic interpretation of gravity anomalies of northeastern Utah: Utah Geological Association Publication 12, p. 121–147.

Smith, L.E., Hosford, G.F., Sears, R.S., Sprouse, D.P., and Stewart, M.D., 1952, Stratigraphic sections of the Phosphoria Formation in Utah 1947–48: U.S. Geological Survey Circular 211, 48 p.

Smith L.S., 1976, Paleoenvironments of the upper Entrada Sandstone and the Curtis Formation on the west flank of the San Rafael Swell, Emery County, Utah: Brigham Young University Geology Studies, v. 23, part 1, p. 113–171.

Smith, R.B., and Bruhn, R.L., 1984, Intraplate extensional tectonics of the eastern Basin-Range—inferences on structural style from seismic reflection data, regional tectonics, and thermal-mechanical models of brittle-ductile deformation: Journal of Geophysical Research, v. 89, p. 5733–5762.

Smith, R.B., Nagy, W.C., Julander, K.A., Viveiros, J.J., Barker, C.A., and Gants, D.G., 1988, Geophysical and tectonic framework of the eastern Basin and Range-Colorado Plateau-Rocky Mountain transition: Geophysical Framework of the Continental U.S. Geological Society of America Memoir (in press).

Smith, R.B., and Sbar, M.L., 1974, Contemporary tectonics and seismicity of the western U.S., with emphasis on the Intermountain Seismic Belt: Geological Society of America Bulletin, v. 85, p. 1205–1218.

Smith, R.P., and Nash, W.P., 1976, Chemical correlation of volcanic ash deposits in the Salt Lake Group, Utah, Idaho, and Nevada: Journal of Sedimentary Petrology, v. 46, no. 4, p. 930–939.

Smith, T.L., 1957, Geology of the Antimony Canyon area, Garfield and Piute Counties, Utah: unpublished masters thesis, University of Utah, 45 p.

Snelling, N.J., 1985, An interim time-scale: in, The Chronology of the Geological Record, Geological Society of London Memoir 10, p. 161–165.

Snoke, A.W., and Howard, K.A., 1984, Geology of the Ruby Mountains-East Humboldt Range, Nevada—a Cordilleran metamorphic core complex: Geological Society of America, Annual Meeting, Reno, Nevada, Western Geologic Excursions, v. 4, p. 260–302.

Snoke, A.W., and Lush, A.P., 1984, Polyphase Mesozoic-Cenozoic deformation history of the northern Ruby Mountains-East Humboldt Range, Nevada: Geological Society of America, Annual Meeting, Reno, Nevada, Western Geologic Excursions, v. 4, p. 232–260.

Snyder, W.S., Dickinson, W.R., and Silberman, M.L., 1976, Tectonic implications of space-time patterns of Cenozoic magmatism in western United States: Earth and Planetary Science Letters, v. 32, p. 91–106.

Solien, M.A., 1979, Conodont biostratigraphy of the Lower Triassic Thaynes Formation, Utah: Journal of Paleontology v. 53, p. 276–306.

Solien, M.A., Morgan, W.A., and Clark, D.L., 1979, Structure and stratigraphy of a Lower Triassic conodont locality, Salt Lake City, Utah: Brigham Young University Geology Studies, v. 26, part 3, p. 165–177.

Sorensen, M.L., 1982, Geologic map of the Stansbury Roadless Area, Tooele County, Utah: U.S. Geological Survey Miscellaneous Field Studies Map MF-1353-A.

Sorensen, M.L., and Crittenden, M.D. Jr., 1972, Geologic map of part of the Wasatch Range near North Ogden, Utah: U.S. Geological Survey Miscellaneous Field Studies Map MF-428.

Sorensen, M.L., and Crittenden, M.D. Jr., 1976, Preliminary geologic map of the Mantua quadrangle and part of the Willard quadrangle, Box Elder, Weber, and Cache Counties, Utah: U.S. Geological Survey Miscellaneous Field Studies Map MF-720.

Sorensen, M.L., and Crittenden, M.D., Jr., 1976b, Type locality of Walcott's Brigham Formation, Box Elder County, Utah: Utah Geology, v. 3, no. 2, p. 117–121.

Sorensen, M.L., and Crittenden, M.D., Jr., 1979, Geologic map of the Huntsville quadrangle, Weber and Cache Counties, Utah: U.S. Geological Survey Geologic Quadrangle Map GQ-1503.

Sorensen, M.L., Korzeb, S.L., and Neubert, J.T., 1983, Mineral resource potential map of the Birdseye, Nephi, and Santaquin roadless areas, Juab and Utah Counties, Utah: U.S. Geological Survey Miscellaneous Field Studies Map 1574.

Sperry, S.W., 1980, Flagstaff Formation—depositional environment and paleoecology of clastic deposits near Salina, Utah: Brigham young University Geology Studies, v. 27, part 2, p. 153–173.

Spieker, E.M., 1931, The Wasatch Plateau coal field, Utah: U.S. Geological Survey Bulletin 819, 201 p.

Spieker, E.M., 1949, Transition between the Colorado Plateaus and the Great Basin in central Utah: Utah Geological Society Guidebook 4, 106 p.

Spieker, E.M., and Baker, A.A., 1928, Geology and coal resources of the Salina Canyon district, Sevier County, Utah: U.S. Geological Survey Bulletin 796C, p. 125–170.

Spinosa, Claude, Harrison, R.L., and O'Dell, I., 1977, Unique Carboniferous (Visean-Namurian) ammonoid succession in Chainman Shale, Confusion Range, Utah: Geological Society of America Abstracts with Programs, v. 9, p. 766.

Spreng, W.C., 1979, Upper Devonian and Lower Mississippian strata on the flanks of the western Uinta Mountains, Utah: Brigham Young University Geology Studies v. 26, part 2, p. 67–79.

Sprinkel, D.A., 1976, Structural geology of Cutler Dam quadrangle and northern part of Honeyville quadrangle, Utah: unpublished master's thesis, Utah State University, Logan, 68 p.

Sprinkel, D.A., 1982, Twin Creek-Arapien shale relations in central Utah: Utah Geological Association Publication 10, p. 169–180.

Staatz, M.H., 1972, Geologic map of the Dugway Proving Ground SW quadrangle, Tooele County, Utah: U.S. Geological Survey Geologic Quadrangle Map GQ-992.

Staatz, M.H., and Carr, W.J., 1964, Geology and mineral deposits of the Thomas and Dugway Ranges, Juab and Tooele Counties, Utah: U.S. Geological Survey Professional Paper 415, 188 p.

Stacey, J.S., and Hedlund, D.C., 1983, Lead-isotopic compositions of diverse igneous rocks and ore deposits from southwestern New Mexico and their implications for early Proterozoic crustal evolution in the western United States: Geological Society of America Bulletin, v. 94, p. 43–57.

Stacey, J.S., and Zartman, R.E., 1978, A lead and strontium isotopic study of igneous rocks and ores from the Gold Hill mining district, Utah: Utah Geology, v. 5, no. 1, p. 1–15.

Standlee, L.A., 1982, Structure and stratigraphy of Jurassic rocks in central Utah—their influence on tectonic development of the Cordilleran foreland thrust belt:

Rocky Mountain Association of Geologists, Geologic Studies of the Cordilleran Thrust Belt, v. 1, p. 357–382.
Stanley, K.O., and Collinson, J.W., 1979, Depositional history of Paleocene-lower Eocene Flagstaff Limestone and coeval rocks, central Utah: American Association of Petroleum Geologists Bulletin, v. 63, p. 311–323.
Stanton, R.G., 1976, Paleoenvironments of the Summerville Formation on the west side of the San Rafael Swell, Emery County, Utah: Brigham Young University Geology Studies, v. 23, part 1, p. 37–73.
Staub, A.M., 1975, Geology of the Picture Rock Hills quadrangle, southwestern Keg Mountains, Juab County, Utah: unpublished master's thesis, University of Utah, 87 p.
Steed, D.A., 1980, Geology of the Virgin River Gorge, Arizona: Brigham Young University Geology Studies, v. 27, part 3, p. 96–115.
Steed, R.H., 1954, Geology of Circle Cliffs Anticline: Intermountain Association of Petroleum Geologists Guidebook 5, p. 99–102.
Steiner, M.B., 1983, Mesozoic apparent polar wander and plate motions of North America: *in* Reynolds, M.W. and Dolly, E.D., eds., Symposium on Mesozoic Paleogeography of west-central U.S., Rocky Mountain Section, Society of Economic Paleontologists and Mineralogists, Denver, Colorado, p. 1–11.
Stephens, E.V., 1973, Geologic map and coal resources of the Wide Hollow Reservoir Quadrangle, Garfield County, Utah: U.S. Geological Survey Coal Investigations Map C-55.
Stephens, J.C., 1974a, Hydrologic reconnaissance of the northern Great Salt Lake Desert and summary hydrologic reconnaissance of northwestern Utah: Utah Department of Natural Resources Technical Publication 42, 55 p.
Stephens, J.C., 1974b, Hydrologic reconnaissance of the Wah Wah Valley drainage basin, Millard and Beaver Counties, Utah: Utah Department of Natural Resources Technical Publication 47, 45 p.
Stephens, J.C., 1976, Hydrologic reconnaissance of the Pine Valley drainage basin, Millard, Beaver, and Iron Counties, Utah: Utah Department of Natural Resources Technical Publication 51, 31 p.
Stephens, J.C., 1977, Hydrologic reconnaissance of the Tule Valley drainage basin, Juab and Millard Counties, Utah: Utah Department of Natural Resources Technical Publication 56, 29 p.
Stephens, J.C., and Sumision, C.T., 1978, Hydrologic reconnaissance of the Dugway Valley-Government Creek area, west-central Utah: Utah Department of Natural Resources Technical Publication 59, 33 p.
Stephens, J.C., and Wood, J.W., 1973, Hydrologic reconnaissance of Pilot Valley, Utah and Nevada: Utah Department of Natural Resources Technical Publication 41, 33 p.
Steven, T.A., 1981, Three Creeks Caldera, southern Pavant Range, Utah: Brigham Young University Geology Studies, v. 28, part 3, p. 1–7.
Steven, T.A., 1984, Igneous activity and related ore deposits in the western and southern Tushar Mountains, Marysvale volcanic field, west-central Utah: U.S. Geological Survey Professional Paper 1299 A-B.
Steven, T.A., Cunningham, C.G., Naeser, C.W., and Mehnert, H.H., 1979, Revised stratigraphy and radiometric ages of volcanic rocks and mineral deposits in the Marysvale area, west-central Utah: U.S. Geological Survey Bulletin 1469, 40 p.
Steven, T.A., and Morris, H.T., 1983, Geologic map of the Cove Fort quadrangle, west-central Utah: U.S. Geological Survey Miscellaneous Investigations Series Map I-1481.
Steven, T.A., Rowley, P.D., and Cunningham, C.G., 1984, Calderas of the Marysvale volcanic field, west-central Utah: Journal of Geophysical Research, v. 89, p. 8751–8764.
Stevens, C.H., 1977, Permian depositional provinces and tectonics, western U.S.: *in* Symposium 1, Paleozoic Paleogeography of the Western U.S., Pacific Section, Society of Economic Paleontologists and Mineralogists, p. 113–135, Los Angeles.
Stevens, C.H., 1979, Lower Permian of the central Cordilleran miogeosyncline: Geological Society of America Bulletin, v. 90, p. 381–455.
Stevens, C.H., 1981, Evaluation of the Wells fault, northeastern Nevada and northwestern Utah: Geology, v. 9, p. 534–537.
Stevens, C.H., and Armin, R.A., 1983, Microfacies of the Middle Pennsylvanian part of the Oquirrh Group, central Utah: Geological Society of America Memoir 157, p. 83–100.
Stevens, C.H., and Stone, P., 1988, Restoration of the late Paleozoic continental margin of western U.S.: Geological Society of America Abstracts with Programs, v. 20, no. 3, p. 235.
Stevenson, G.M., and Baars, D.L., 1986, The Paradox—a pull-apart basin of Pennsylvanian age: American Association of Petroleum Geologists Memoir 41, p. 513–539.
Stevenson, G.M., and Beus, S.S., 1982, Stratigraphy and depositional setting of the upper Precambrian Dox Formation in Grand Canyon: Geological Society of America Bulletin, v. 93, p. 163–173.
Stewart, J.H., 1969, Major Upper Triassic lithogenetic sequences in Colorado Plateau region: American Association of Petroleum Geologists Bulletin, v. 53, no. 9, p. 1866–1879.
Stewart, J.H., 1971, Basin and Range structure: a system of horsts and grabens produced by deep-seated extension: Geological Society of America Bulletin, v. 82, p. 1019–1044.
Stewart, J.H., 1972, Initial deposits in the Cordilleran Geosyncline: evidence of a Late Precambrian (850 m.y.) continental separation: Geological Society of America Bulletin, v. 83, p. 1345–1360.
Stewart, J.H., 1976, Late Precambrian evolution of North America—plate-tectonic implications: Geology, v. 4, p. 11–15.
Stewart, J.H., 1980, Regional tilt patterns of late Cenozoic basin-range fault blocks, western U.S.: Geological Society of America Bulletin, v. 91, part 1, p. 460–464.
Stewart, J.H., 1983, Cenozoic structure and tectonics of the northern Basin and Range province, California, Nevada, and Utah: Geothermal Resources Council, Special Report no. 13, p. 25–40. Davis, California, 95617-1350.
Stewart, J.H., Anderson, T.H., Haxel, G.B., Silver, L.T., and Wright, J.E., 1986, Late Triassic paleogeography of the southern Cordillera—the problem of a source for voluminous volcanic detritus in the Chinle Formation of the Colorado Plateau region: Geology, v. 14, p. 567–570. See also forum discussion and reply Geology, v. 15, p. 578–579.
Stewart, J.H., Moore, W.J., and Zietz, I., 1977, East-west patterns of Cenozoic igneous rocks, aeromagnetic anomalies, and mineral deposits, Nevada and Utah: Geological Society of America, v. 88, p. 67–77.
Stewart, J.H., Poole, F.G., and Wilson, R.F., 1972a, Stratigraphy and origin of the Chinle Formation and related Upper Triassic strata in the Colorado Plateau region: U.S. Geological Survey Professional Paper 690, 336 p.
Stewart, J.H., Poole, F.G., and Wilson, R.F., 1972b, Stratigraphy and origin of the Triassic Moenkopi Formation and related strata in the Colorado Plateau region: U.S. Geological Survey Professional Paper 691, 195 p.
Stewart, J.H. and Suczek, C.A., 1977, Cambrian and latest Precambrian paleogeography and tectonics in the western United States: *in* Stewart, J.H., Stevens, C.H., and Fritsche, A.E., eds., Paleozoic paleogeography of the western United States, Pacific Coast Section, Society of Economic Paleontologists and Mineralogists, Los Angeles, p. 1–17.
Stifel, P.B., 1964, Geology of the Terrace and Hogup Mountains, Box Elder County, Utah: unpublished Ph.D. thesis, University of Utah, 173 p.
Stockmal, G.S., Beumont, C., and Boutilier, R., 1986, Geodynamic models of convergent margin tectonics—transition from rifted margin to overthrust belt and consequences for foreland-basin development: American Association of Petroleum Geologists Bulletin, v. 77, p. 181–190.
Stokes, W.L., 1952, Uranium-vanadium deposits of the Thompson area, Grand County, Utah: Utah Geological and Mineral Survey Bulletin 46, 48 p. Summarized in Utah Geological Society Guidebook 9 (1954) p. 78–94.
Stokes, W.L., 1959, Jurassic rocks of the Wasatch Range and vicinity: Intermountain Association of Petroleum Geologists Guidebook 10, p. 109–114.
Stokes, W.L., 1976, What is the Wasatch Line?: Rocky Mountain Association of Geologists Symposium, p. 11–25.
Stokes, W.L., 1980, Stratigraphic interpretations of Triassic and Jurassic beds, Henry Mountains area: Utah Geological Association Publication 8, p. 113–122.
Stokes, W.L., 1982, Geologic comparisons and contrasts, Paradox and Arapien basins: Utah Geological Association Publication 10, p. 1–11.
Stokes, W.L., and Arnold, D.E., 1958, Northern Stansbury Range and the Stansbury Formation: Utah Geological Society Guidebook, 13, p. 135–149.
Stokes, W.L., and Cohenour, R.E., 1956, Geologic atlas of Emery County, Utah: Utah Geological and Mineral Survey Bulletin, 52, 92 p.
Stokes, W.L., and Madsen, J.H. Jr., 1979, Environmental significance of pterosaur tracks in the Navajo Sandstone, Grand County, Utah: Brigham Young University Geology Studies, v. 26, part 2, p. 21–26.
Stolle, J.M., 1978, Stratigraphy of the lower Tertiary and upper Cretaceous (?) continental strata in the Canyon Range, Juab County, Utah: Brigham Young University Geology Studies, v. 25, part 3, p. 117–139.
Stone, P., and Stevens, C.H., 1988, Pennsylvanian and Early Permian paleogeography of east-central California—implications for the shape of the continental margin and timing of continental truncation: Geology, v. 16, p. 330–333.
Stover, C.W., Reagor, B.G., and Algermissen, S.T., 1986, Seismicity map of the state of Utah: U.S. Geological Survey Map MF-1856.
Stuart-Alexander, D.E., Shoemaker, E.M., and Moore, H.J., 1972, Geologic map of the Mule Ear Diatreme, San Juan County, Utah: U.S. Geological Survey Miscellaneous Geologic Investigations Map I-674.
Sullivan, K.R., 1987, Isotopic ages of igneous intrusions in southeastern Utah—evidence for a mid-Cenozoic Reno-San Juan magmatic zone: Brigham Young University Geology Studies (in preparation), also abstract in Geological Society of America Abstracts with Programs, v. 19, p. 337.
Surdam, R.C., and Stanley, K.O., 1979, Lacustrine sedimentation during the culminating phase of Eocene Lake Gosiute, Wyoming (Green River Formation): Geological Society of America Bulletin, v. 90, p. 93–110.
Surdam, R.C., and Stanley, K.O., 1980, Effects of changes in drainage-basin boundaries on sedimentation in Eocene lakes Gosiute and Uinta of Wyoming, Utah, and Colorado: Geology, v. 8, p. 135–139.
Swan, F.H., Schwartz, D.P., Cluff, L.S., Hanson, K.L., and Knuepfer, P.L., 1981, Study of earthquake recurrence intervals on the Wasatch Fault at the Hobble Creek site, Utah: U.S. Geological Survey Open-File Report 81-229.
Swanson, R.W., Carswell, L.D., Sheldon, R.P., and Cheney, T.M., 1956, Stratigraphic sections of the Phosphoria Formation, 1953: U.S. Geological Survey Circular 375, 30 p.
Swayze, G.A., and Holden, G.S., 1984, Metamorphic development of the Proterozoic Red Creek orogen, Uinta Mountains, northeastern Utah: Geological Society of America Abstracts with Programs, v. 16, p. 672.
Sweet, W.C., 1984, Graphic correlation of upper Middle and Upper Ordovician rocks, North American Midcontinent Province: *in*, Bruton, D.L., ed., Aspects of the Ordovician System, Paleontological Contributions from the University of Oslo, no. 295, Universtetsforlaget, p. 23–35.
Swensen, A.J., 1975, Geologic map of the Bingham District: Society of Economic Geologists, Guidebook to the Bingham Mining District by Exploration Services Department of Kennecott Copper Corporation, scale 1:24,000.
Swenson, J.L. Jr., Erickson, D.T., Donaldson, K.M., and Shiozaki, J.I., 1970, Soil Survey of Carbon-Emery area, Utah: U.S. Department of Agriculture, Soil Conservation Service, Salt Lake City, Utah 78 p.
Swenson, J.L. Jr., Archer, W.M., Donaldson, K.M., Shiozaki, J.J., Broderick, J.H., and Woodward, L., 1972, Soil survey of Utah County, central part: U.S. Department of Agriculture Soil Conservation Service, Salt Lake City, 161 p.
Swenson, J.L. Jr., Beckstrand, D., Erickson, D.T., McKinley, C., Shiozaki, J.J., and Tew, R., 1981, Soil survey of Sanpete Valley area, Utah: U.S. Department of Agriculture, Salt Lake City, 179 p.
Swetland, P.J., Clayton, J.L., and Sable, E.G., 1978, Petroleum source-bed potential of Mississippian-Pennsylvanian rocks in part of Montana, Idaho, Utah

and Colorado: The Mountain Geologist, v. 14, no. 3 p. 79–87.

Szabo, E., and Wenger, S.A., 1975, Stratigraphy and tectogenesis of the Paradox Basin: Four Corners Geological Society Guidebook 8, p. 193–210.

Tapp, S.C., 1963, Kanab Creek unit area: Intermountain Association of Petroleum Geologists Guidebook 12, p. 193–198.

Taylor, D.W., 1981, Carbonate petrology and depositional environments of the limestone member of the Carmel Formation, near Carmel Junction, Kane County, Utah: Brigham Young University Geology Studies, v. 28, part 3, p. 117–133.

Taylor, M.E., Landing, Ed, and Gillett, S.L., 1981, The Cambrian-Ordovician transition in the Bear River Range, Utah-Idaho: a preliminary evaluation: U.S. Geological Survey Open-File Report 81-743, p. 222–227.

Taylor, M.E., and Miller, J.F., 1981, Upper Cambrian and Lower Ordovician stratigraphy and biostratigraphy, southern House Range, Utah: *in* Cambrian stratigraphy and paleontology of the Great Basin and vicinity, western U.S.—Guidebook for Field Trip 1, Second International Symposium on the Cambrian System, Denver, Colorado.

Taylor, M.E., and Repetski, J.E., 1985, Early Ordovician eustatic sea-level changes in northern Utah and southeastern Idaho: Utah Geological Association Publication 14, p. 237–248.

Terrell, F.M., 1972, Lateral facies and paleoecology of Permian Elephant Canyon Formation, Grand County, Utah: Brigham Young University Geology Studies, v. 19, part 2, p. 3–44.

Thaden, R.E., Trites, A.F., and Finnell, T.L., 1964, Geology and ore deposits of the White Canyon area, San Juan and Garfield Counties, Utah: U.S. Geological Survey Bulletin 1125, 166 p.

Thomas, G.E., 1960, The South Flat and related formations in the northern part of the Gunnison Plateau, Utah: unpublished master's thesis, Ohio State University, 137 p.

Thomas, J.B., 1985, Stratigraphy, sedimentology, and petrology of the early Tertiary Claron Formation in the Red Canyon area of south-central High Plateaus, Utah: unpublished master's thesis, Kent State University, Ohio, 135 p.

Thomas, W.D., 1976, Dikes of the Clear Creek area, Wasatch Plateau, Utah: unpublished master's thesis, Utah State University, Logan, 73 p.

Thompson, A.E., and Stokes, W.L., 1970, Stratigraphy of the San Rafael Group, southwest and south-central Utah: Utah Geological and Mineral Survey Bulletin 87, 53 p.

Thompson, R.S., 1987, Radiocarbon dating of Pleistocene lake sediments in the Great Basin by accelerator mass spectrometry: Geological Society of America Abstracts with Programs, v. 19, p. 868.

Thomson, K.C., 1973, Mineral deposits of the Deep Creek Mountains, Tooele and Juab Counties, Utah: Utah Geological and Mineral Survey Bulletin 99, 76 p.

Thomson, K.C., and Perry, L.I., 1975, Reconnaissance study of the Stateline mining district, Iron County, Utah: Utah Geology, v.2, no. 1, p. 27–48.

Tidwell, W.D., Thayn, G.F., and Roth, J.L., 1976, Cretaceous and early Tertiary floras of the Intermountain area, a survey: Brigham Young University Geology Studies, v. 22, part 3, p. 77–96.

Tidwell, W.D., Medlyn, D.A., and Simper, A.D., 1974, Flora of the Manning Canyon Shale, part II: Lepidodendrales: Brigham Young University Geology Studies, v. 21, part 3, p. 119–146.

Tingey, D., 1986, Dikes in the northern Wasatch Plateau: Brigham Young University Geology Studies (in preparation).

Todd, V.R., 1973, Structure and petrology of metamorphosed rocks in central Grouse Creek Mountains, Box Elder County, Utah: unpublished Ph.D. dissertation, Stanford University, 373 p., Dissertation Abstracts, 1973, v. 34, p. 1155-B.

Todd, V.R., 1980, Structure and petrology of a Tertiary gneiss complex in northwestern Utah: Geological Society of America Memoir 153, p. 349–384.

Todd, V.R., 1983, Late Miocene displacement of pre-Tertiary and Tertiary rocks in the Matlin Mountains, northwestern Utah: Geological Society of America Memoir 157, p. 239–270.

Tomida, Y. and Butler, R.F., 1980, Dragonian mammals and Paleogene magnetic polarity stratigraphy of the North Horn Formation, central Utah: American Journal of Science, v. 280, p. 787–811.

Tooker, E.W., 1983, Variations in structural style and correlation of thrust plates in the Sevier foreland thrust belt, Great Salt Lake area, Utah: Geological Society of America Memoir 157, p. 61–73.

Tooker, E.W., and Roberts, R.J., 1970, Upper Paleozoic rocks in the Oquirrh Mountains and Bingham mining district, Utah: U.S. Geological Survey Professional Paper 629-A, 76 p.

Tooker, E.W., and Roberts, R.J., 1971, Structures related to thrust faults in the Stansbury Mountains, Utah: U.S. Geological Survey Professional Paper 750-B, p. 1–12.

Tooker, E.W., and Roberts, R.J., 1971, Geologic map of the Garfield quadrangle, Salt Lake and Tooele Counties, Utah: U.S. Geological Survey Geologic Quadrangle Map GQ-922. See also GQ 923 and 924.

Traver, W.M., 1949, Investigation of Coyote Creek antimony deposits, Garfield County, Utah: U.S. Bureau of Mines Report of Investigation 4470.

Trexler, D.W., 1966, The stratigraphy and structure of the Coalville area, northeastern Utah: Colorado School of Mines Contributions, no. 2, 69 p.

Trimble, D.E., 1976, Geology of the Michaud and Pocatello quadrangles, Bannock and Power Counties, Idaho: U.S. Geological Survey Bulletin 1400.

Trimble, L.M., and Doelling, H.H., 1978, Geology and uranium-vanadium deposits of the San Rafael River mining area, Emery County, Utah: Utah Geological and Mineral Survey Bulletin 113, 122 p.

Tschudy, R.H., Tschudy, B.D., and Craig, L.C., 1984, Palynological evaluation of Cedar Mountain and Burro Canyon formations, Colorado Plateau: U.S. Geological Survey Professional Paper 1281, 24 p.

Tucker, L.M., 1954, Geology of the Scipio quadrangle, Utah: unpublished Ph.D. dissertation, Ohio State University, 360 p.

Tweto, O. 1977, Nomenclature of Precambrian rocks in Colorado, U.S. Geological Survey Bulletin 1422-D, 22 p.

Tyler, N., and Ethridge, F.G., 1983, Depositional setting of the Salt Wash Member of the Morrison Formation, southwest Colorado: Journal of Sedimentary Petrology, v. 53, p. 67–82.

Tynan, M.C., 1980, Conodont biostratigraphy of the Mississippian Chainman Formation, western Millard County, Utah: Journal of Paleontology, v. 54, p. 1282–1309.

U.S. Department of Agriculture—Soil Conservation Service, 1959, Soil Survey of East Millard area, Utah, 101 p., maps.

U.S. Department of Agriculture—Soil conservation Service, 1960, Soil Survey of Beryl-Enterprise area, Utah, 75 p., maps.

U.S. Department of Agriculture—Soil Conservation Service, 1962, Soil Survey of San Juan area, Utah, 49 p., maps.

U.S. Department of Agriculture—Soil Conservation Service, 1968, Soil Survey of Davis-Weber area, Utah; 149 p., maps.

U.S. Department of Agriculture—Soil Conservation Service, 1970, Soil Survey of Carbon-Emery area, Utah, 78 p., maps.

U.S. Department of Agriculture—Soil Conservation Service, 1972, Soil Survey of Utah County, Utah, central part, 161 p., maps.

U.S. Department of Agriculture—Soil Conservation Service, 1974, Soil Survey of Salt Lake area, Utah, 132 p., maps.

U.S. Department of Agriculture—Soil Conservation Service, 1975, Soil Survey of Box Elder County, Utah, Eastern Part, 223 p., maps.

U.S. Department of Agriculture—Soil Conservation Service, 1976, Soil Survey of Beaver-Cove Fort area, Utah: 138 p., maps.

U.S. Department of Agriculture—Soil Conservation Service, 1976, Soil Survey of Heber Valley area, Utah, 124 p., maps.

U.S. Department of Agriculture—Soil Conservation Service, 1977, Soil Survey and Interpretation-Parleys Park portion of Summit County, Utah, 302 p., maps.

U.S. Department of Agriculture—Soil Conservation Service, 1977, Soil Survey of Washington County area, Utah, 140 p., maps.

U.S. Department of Agriculture—Soil Conservation Service, 1977, Soil Survey of Delta area, Utah; 77 p., maps.

U.S. Department of Agriculture—Soil Conservation Service, 1980, Soil Survey of Navajo Indian Reservation, San Juan County, Utah, 119 p., maps.

U.S. Department of Agriculture—Soil Conservation Service, 1980, Soil Survey of Morgan area, Utah, 300 p., maps.

U.S. Department of Agriculture—Soil Conservation Service, 1981, Soil Survey of Sanpete Valley area, Utah; 179 p., maps.

U.S. Department of Agriculture—Soil Conservation Service, 1982, Soil Survey of Rich County, Utah, 273 p., maps.

U.S. Department of Agriculture—Soil Conservation Service, 1984, Soil Survey of Fairfield-Nephi area, Utah, 361 p., maps.

Untermann, G.E., and Untermann, B.R., 1954, Geology of Dinosaur National Monument and vicinity, Utah-Colorado: Utah Geological and Mineral Survey Bulletin 42, 221 p.

Uresk, J., 1979, Sedimentary environment of the Cretaceous Ferron Sandstone near Caineville, Utah: Brigham Young University Geology Studies, v. 26, part 2, p. 81–100.

Utah Geological and Mineral Survey Staff, 1983, Energy resources map of Utah: Utah Geological and Mineral Survey Map 68.

Uygur, K., and Picard, MD., 1985, Mesozoic stratigraphy of Uinta Basin, northeast Utah: Utah Geological Association Publication 12, p. 21–37.

Valenti, G.L., 1980, Structural geology of the Laketown quadrangle, Rich County, Utah: unpublished M.S. thesis University of Wyoming, Laramie, 97 p. Map published separately as Utah Geological and Mineral Survey Map 58, 1982, Salt Lake City.

Valenti, G.L., 1982, Structural geology of the Laketown quadrangle and petroleum exploration in the Bear Lake area, Rich County, Utah: Rocky Mountain Association of Geologists Geologic Studies of the Cordilleran Thrust Belt, p. 859–868.

Van de Graaff, F.R., 1972, Fluvial-deltaic facies of the Castlegate Sandstone (Cretaceous), east-central Utah: Journal of Sedimentary Petrology, v. 42, no. 3 p. 558–571.

VanDorston, P.L., 1970, Environmental analysis of Swan Peak Formation in Bear River Range, north-central Utah and southeastern Idaho: American Association of Petroleum Geologists Bulletin, v. 54, p. 1140–1154.

Van Horn, Richard, 1981, Geologic map of pre-Quaternary rocks of the Salt Lake City North quadrangle, Davis and Salt Lake Counties, Utah: U.S. Geological Survey Miscellaneous Investigations Series Map I-1330.

Van Horn, Richard, 1982, Surficial geologic map of the Salt Lake City North quadrangle, Davis and Salt Lake Counties, Utah: U.S. Geological Survey Miscellaneous Investigations Series Map I-1404.

Van Horn, R., and Crittenden, M.D. Jr., 1987, Map showing surficial units and bedrock geology of the Fort Douglas quadrangle and parts of the Mountain Dell and Salt Lake City North quadrangles, Davis, Salt Lake, and Morgan counties, Utah: U.S. Geological Survey Miscellaneous Investigations Map I-1762.

Van Kooten, G.K., 1987, Structure and hydrocarbon potential beneath the Iron Springs laccolith, southwestern Utah: Geological Society of America Abstracts with Programs, v. 19, also complete paper in press American Association of Petroleum Geologists.

Varnes, D.J., and Van Horn, R., 1984, Surficial geologic map of the Oak City area, Millard County, Utah: U.S. Geological Survey Open-File Report 84-115.

Veevers, J.J., and Powell, C.M., 1987, Late Paleozoic glacial episodes in Gondwanaland reflected in transgressive-regressive depositional sequences in Euramerica: Geological Society of America Bulletin, v. 98, p. 475–487.

VerPloeg, A.J., and DeBruin, R.H., 1982, The search for oil and gas in Idaho-Wyoming-Utah salient of the overthrust belt: Wyoming Geological Survey Report of Investigations 21, 108 p.

Villien, A., and Kligfield, R.M., 1986, Thrusting and synorogenic sedimentation in central Utah: American Association of Petroleum Geologists Memoir 41, p. 281–307.

vonBitter, P.H., Sandberg, C.A., and Orchard, M.J., 1986, Phylogeny, speciation, and paleoecology of the early Carboniferous (Mississippian) conodont genus *Mesognathus*: Royal Ontario Museum Life Sciences Contribution 143, 115 p.

Von Tish, D.B., Allmendinger, R.W., and Sharp, J.W., 1985, History of Cenozoic extension in central Sevier Desert, west-central Utah, from COCORP seismic reflection data: American Association of Petroleum Geologists Bulletin, v. 69, p. 1077–1087.

Vorce, C.L., 1979, Sedimentary petrology and stratigraphic relationships of the North Horn Formation and the Flagstaff Limestone at Long Ridge, central Utah: unpublished master's thesis, Ohio State University, 90 p.

Wach, P.H., 1967, Geology of the west-central part of the Malad Range, Idaho: unpublished master's thesis, Utah State University, Logan, 79 p.

Waite, R.H., 1956, Upper Silurian Brachiopoda from the Great Basin: Journal of Paleontology, v. 30, no. 1, p. 15–18.

Waldrop, H.A., Sutton, R.L., Peterson, F., and Horton, G.W., 1967, Preliminary geologic maps and coal deposits of the Nipple Butte and Gunsight Butte quadrangles, Kane County, Utah: Utah Geological and Mineral Survey Maps 24 A-G.

Walker, C.S., 1983, Stratigraphy and depositional environment of the Upper Proterozoic and Cambrian Brigham Group near Mink Creek, Idaho: unpublished master's thesis, Idaho State University, Pocatello, 55 p.

Wallace, C.A., and Crittenden, M.D. Jr., 1969, The stratigraphy, depositional environment, and correlation of the Precambrian Uinta Mountain Group, western Uinta Mountains, Utah: Intermountain Association of Geologists Guidebook 16, p. 127–142.

Wallace, R.E., 1984, Patterns and timing of Late Quaternary faulting in the Great Basin province and relation to some regional tectonic features: Journal of Geophysical Research, v. 89, p. 5763–5769.

Walton, P.T., 1954, Wasatch Plateau gas fields, Utah: Intermountain Association of Petroleum Geologists Guidebook 5, p. 79–85.

Wang, Y.F., 1970, Geological and geophysical studies of the Gilson Mountains and vicinity, Juab County, Utah: unpublished Ph.D. thesis, University of Utah, 126 p.

Wardlaw, B.R., and Collinson, J.W., 1978, Stratigraphic relations of Park City Group (Permian) in eastern Nevada and western Utah: American Association of Petroleum Geologists Bulletin, v. 62, p. 1171–1184.

Wardlaw, B.R., Collinson, J.W., and Maughan, E.K., 1979, Stratigraphy of Park City Group equivalents (Permian) in southern Idaho, northeastern Nevada, and northwestern Utah: U.S. Geological Survey Professional Paper 1163-C, p. 9–16.

Warner, M.M., 1966, Sedimentational analysis of the Duchesne River Formation, Uinta Basin, Utah: Geological Society of America Bulletin, v. 77, p. 945–958.

Wassenburg, G.J., and Lanphere, M.A., 1965, Age determination in the Precambrian of Arizona and Nevada: Geological Society of America Bulletin, v. 76, p. 735–758.

Weaver, C.L., 1980, Geology of the Blue Mountain quadrangle, Beaver and Iron Counties, Utah: Brigham Young University Geology Studies, v. 27, part 3, p. 116–132.

Webb, G.W., 1958, Middle Ordovician stratigraphy in eastern Nevada and western Utah: American Association of Petroleum Geologists Bulletin, v. 42, no. 10, p. 2335–2377.

Webster, G.D., 1984, Conodont zonations near the Mississippian-Pennsylvanian boundary in the eastern Great Basin: *in* Lintz, Joseph, ed., Western Geological Excursions, v. 1, Mackay School of Mines, University of Nevada, Reno, p. 4–6, 78–81.

Webster, G.D., Gordon, Mackensie, Langenheim, R.L., and Henry, T.W., 1984, Road logs for the Mississippian-Pennsylvanian boundary in the eastern Great Basin: *in* Lintz, Joseph, ed., Western Geological Excursions, v. 1, Mackay School of Mines, Reno, p. 1–86.

Weimer, R.J., 1986, Relationship of unconformities, tectonics, and sea level changes in the Cretaceous of the western Interior, U.S.: American Association of Petroleum Geologists Memoir 41, p. 397–422.

Weiss, M.P., 1969, Oncolites, paleoecology, and Laramide tectonics, central Utah: American Association of Petroleum Geologists Bulletin, v. 53, p. 1105–1120.

Weiss, M.P., 1982, Relation of the Crazy Hollow Formation to the Green River Formation, central Utah: Utah Geological Association Publication 10, p. 285–289.

Wells, L.F., 1954, Stratigraphic correlations of central and south-central Utah: Intermountain Association of Petroleum Geologists Guidebook 5, p. 15–17, charts.

Wells, M.L., 1988, Extensional kinematics in the cover rocks of a metamorphic core complex, Black Pine Mountains, southern Idaho: unpublished master's thesis, Cornell University, Ithaca, N.Y.; see also Wells, M.L., and Allmendinger, R.W., 1987, Geological Society of America Abstracts with Programs, v.19, p. 886.

Wells, R.B., 1963, Orthoquartzites of the Oquirrh Formation: Brigham Young University Geology Studies, v. 10, p. 51–81.

Welsh, J.E., 1973, Geology of the Beaver Lake Mountains, Beaver County, Utah: Utah Geological Association Publication 3, p. 49–53.

Welsh, J.E., 1976, Relationships of Pennsylvanian-Permian stratigraphy to the late Mesozoic thrust belt in the eastern Great Basin, Utah and Nevada: Rocky Mountain Association of Geologists, Symposium on Geology of the Cordilleran Hingeline, p. 153–160.

Welsh, J.E., 1983, Differences in the Oquirrh Group and the Granger Mountain Group between the Oquirrh Mountains and Wallsburg Ridge, Wasatch Mountains, Utah: Geological Society of America Abstracts with Programs, v. 15, p. 290.

Welsh, J.E., and Bissell, H.J., 1979, The Mississippian and Pennsylvanian (Carboniferous) Systems in the United States: U.S. Geological Survey Professional Paper 1110-Y, 35 p.

Welsh, J.E., James, A.E., and Smith, W.H., 1961, Kennecott Copper Corporation stratigraphic correlation chart, Bingham Mining District and adjacent areas: Utah Geological Society Guidebook 16, plate 5.

Welsh, J.E., Stokes, W.L., and Wardlaw, B.R., 1979, Regional stratigraphic relationships of the Permian "Kaibab" or Black Box Dolomite of the Emery High, Central Utah: Four Corners Geological Society Guidebook, v. 9, p. 143–149.

Wender, L.E., and Nash, W.P., 1979, Petrology of Oligocene and early Miocene calc-alkalic volcanism in the Marysvale area, Utah—Summary: Geological Society of America Bulletin, v. 90, p. 2–3.

Wengerd, S.A., 1958, Pennsylvanian stratigraphy, southwest shelf, Paradox Basin: Intermountain Association of Petroleum Geologists Guidebook 9, p. 109–134.

Wengerd, S.A., 1962, Pennsylvanian sedimentation in Paradox Basin, Four Corners, Region: American Association of Petroleum Geologists Symposium, Pennsylvanian System in U.S., p. 264–330.

Wengerd, S.A., and Matheny, M.L., 1958, Pennsylvanian System of Four Corners region: American Association of Petroleum Geologists Bulletin, v. 42, p. 2048–2106.

Wernicke, B., 1981, Low-angle normal faults in the Basin and Range Province—nappe tectonics in an extending orogen: Nature, v. 291, p. 645–648.

Wernicke, B., Spencer, J.E., Burchfiel, B.C., and Guth, P.L., 1982, Magnitude of crustal extension in the southern Great Basin: Geology, v. 10, p. 499–502; also Comment by Frank Royce and Reply by authors, *in* Geology, v. 11, p.495–497, 1983.

West, Judy, and Lewis, Helen, 1982, Structure and palinspastic reconstruction of the Absaroka thrust, Anschutz Ranch area, Utah and Wyoming: Rocky Mountain Association of Geologists Geologic Studies of the Cordilleran Thrust Belt, p. 633–639.

Wheeler, R.F., 1980, Geology of the Sewing Machine Pass quadrangle, central Wah Wah Range, Beaver County, Utah: Brigham Young University Geology Studies, v. 27, part 2, p. 175–191.

Wheeler, R.L., and Krystinik, K.B., 1987, Persistent and nonpersistent segmentation of the Wasatch Fault zone, Utah—statistical analysis for evaluation of seismic hazard: U.S. Geological Survey Open-File Report 87-585, p. B1–124.

Whelan, J., and Bowdler, J., 1979, Geology of the Antelope Springs quadrangle, Millard County, Utah: Utah Geology, v. 6, no. 2, p. 81–85.

Whitaker, R.M., 1975, Upper Pennsylvanian and Permian strata of northeast Utah and northwest Colorado: Rocky Mountain Association of Geologists 1975 Symposium, p. 75–85.

White, B.O., 1953, Geology of the West Mountain and northern portion of Long Ridge, Utah County, Utah: unpublished master's thesis, Brigham Young University.

White, David, 1929, Flora of the Hermit Shale, Grand Canyon, Arizona; Carnegie Institution of Washington Publication 405, 221 p.

White, M.A., and Jacobson, M.I., 1983, Structures associated with the southwest margin of the ancestral Uncompahgre uplift: Grand Junction Geological Society Guidebook to the Northern Paradox Basin and Uncompahgre Uplift, p. 33–39.

White, W.W., 1973, Paleontology and depositional environments of the Cambrian Wheeler Formation, Drum Mountains, west-central Utah: unpublished master's thesis, University of Utah, Salt Lake City, 134 p.

Whitlock, W.W., 1984, Geology of the Steele Butte quadrangle, Garfield County, Utah: Brigham Young University Geology Studies, v. 31, part 1, p. 141–165.

Whitney, J.W., 1981, Surficial geologic map of the Grand Junction 1 x 2 degree quadrangle, Colorado and Utah: U.S. Geological Survey Miscellaneous Investigations Series Map I-1289.

Wickwire, D.W., Davis, L.E., and Webster, G.D., 1985, Conodont biostratigraphy of a Lodgepole Limestone equivalent at Samaria Mountain, southeastern Idaho: Utah Geological Association Publication 14, p. 67–74.

Willard, P.D., 1972, Tertiary igneous rocks of northeastern Cache Valley, Idaho: unpublished master's thesis, Utah State University, Logan, 53 p.

Williams, J.S., 1962, Lake Bonneville: Geology of southern Cache Valley, Utah: U.S. Geological Survey Professional Paper 257-C, p. 131–152.

Williams, J.S., 1971, The Beirdneau and Hyrum formations of north-central Utah: Smithsonian Contributions to Paleobiology, no. 3, p. 219–229.

Williams, J.S., and Taylor, M.E., 1964, Lower Devonian Water Canyon Formation of northern Utah: Wyoming University Contributions to Geology, v. 3, no. 2, p. 38–53.

Williams, N.C., 1953, Late Precambrian and early Paleozoic geology of the western Uinta Mountains, Utah: American Association of Petroleum Geologists Bulletin, v. 37, p. 2734–2742.

Williams, N.C., 1957, Cambrian stratigraphy of the south flank of the Uinta Mountains: Intermountain Association of Petroleum Geologists Guidebook 8, p. 53–55.

Williams, P.L., 1964, Geology, structure, and uranium deposits of the Moab 2-degree quadrangle, Colorado and Utah: U.S. Geological Survey Miscellaneous Investigations Series Map I-360.

Williams, P.L., 1972, Map showing types of bedrock and surficial deposits, Salina 2-degree quadrangle, Utah: U.S. Geological Survey Miscellaneous Geologic Investigations Map I-591-H.

Williams, P.L., Covington, H.R., and Pierce, K.L., 1982, Cenozoic stratigraphy and tectonic evolution of the Raft River Basin, Idaho; Idaho Bureau of Mines and Geology Bulletin 26, p. 491–504.

Williams, P.L., and Hackman, R.J., 1971, Geology, structure and uranium deposits of the Salina 2-degree quadrangle, Utah: U.S. Geological Survey Miscellaneous Geologic Investigations Map I-591.

Willis, G.C., 1985, Geology of the Sego Canyon SW quadrangle, Grand County, Utah: Utah Geological and Mineral Survey map 89. Also Brigham Young University Geology Studies, v. 33, part 1, p. 175–208.

Willis, G.C., 1986, Geologic map of the Salina quadrangle, Sevier County, Utah: Utah Geological and Mineral Survey map 83, 20 p.

Willis, G.C., 1988, Geologic map of the Aurora quadrangle, Sevier County, Utah: Utah Geological and Mineral Survey Map (in press).

Willis, G.C., and Kowallis, B.J., 1988, Newly recognized Cedar Mountain Formation in Salina Canyon, Sevier County, Utah: Brigham Young University Geology Studies, v. 35, (in press).

Wilson, R.F., 1965, Triassic and Jurassic strata of southwestern Utah: Utah Geological Society Guidebook 19, p. 31–46.

Wilson, R.F., 1967, Whitmore Point, a new member of the Moenave Formation in Utah and Arizona: Plateau, v. 40, no. 1, p. 29–40.

Wilson, R.F., and Stewart, J.H., 1967, Correlation of Upper Triassic and Triassic (?) formations between southwestern Utah and southern Nevada: U.S. Geological Survey Bulletin 1244-D, 20 p.

Winter, M.B., 1985, Geology and petrology of the Cub River Diabase, a late Pliocene differentiated mafic intrusion, northeastern Cache Valley, Idaho: Utah Geological Association Publication 14, p. 215–226.

Winterfeld, G.F., 1982, Mammalian paleontology of the Fort Union Formation (Paleocene), eastern Rocky Springs Uplift, Sweetwater County, Wyoming: University of Wyoming Contributions to Geology, v. 21, no. 1, p. 73–112.

Witkind, I.J., 1964, Geology of the Abajo Mountains area, San Juan County, Utah: U.S. Geological Survey Professional Paper 453, 110 p.

Witkind, I.J., 1979, Reconnaissance geologic map of the Wellington quadrangle, Carbon County, Utah: U.S. Geological Survey Miscellaneous Investigations Map I-1178.

Witkind, I.J., 1980, Reconnaissance geologic map of the Mounds quadrangle, Carbon and Emery Counties, Utah: U.S. Geological Survey Miscellaneous Investigations Map I-1202.

Witkind, I.J., 1981, Reconnaissance geologic map of the Redmond quadrangle, Sanpete and Sevier Counties, Utah: U.S. Geological Survey Miscellaneous Investigations Series Map I-1304-A.

Witkind, I.J., 1983, Overthrusts and salt diapirs, central Utah: Geological Society of America Memoir 157, p. 45–60.

Witkind, I.J., 1987, Implications of deformation along the east flank of the Charleston-Nebo thrust plate, central Utah: U.S. Geological Survey Professional Paper 1170-F, 29 p.

Witkind, I.J., 1988, Potential geologic hazards near the Thistle landslide, Utah County, Utah: Association of Engineering Geologists Bulletin, v. 25, p. 83–94.

Witkind, I.J., and Hardy, C.T., 1983, Arapien Shale of central Utah—a dilemma in stratigraphic nomenclature: U.S. Geological Survey Bulletin 1537-A, p. 5–20.

Witkind, I.J., and Marvin, R.F., 1988, Significance of new potassium-argon ages from the Goldens Ranch and Moroni formations, Sanpete-Sevier Valley area, central Utah: U.S. Geological Survey (in progress).

Witkind, I.J., and Page, W.R., 1983, Geological map of the Thistle area, Utah County, Utah: Utah Geological and Mineral Survey Map 69.

Witkind, I.J., and Page, W.R., 1984, Origin and significance of the Wasatch and Valley Mountains monoclines, Sanpete-Sevier Valley area, central Utah: Mountain Geologist, v. 21, no. 4, p. 143–156.

Witkind, I.J., Standlee, L.A., and Maley, K.F., 1986, Age and correlation of Cretaceous rocks previously assigned to the Morrison (?) Formation, Sanpete-Sevier Valley area, central Utah: U.S. Geological Survey Bulletin 1584, 9 p.

Witkind, I.J., Weiss, M.P., and Brown, T.P., 1987, Geologic map of the Manti 30' x 60' quadrangle, Carbon, Emery, Juab, Sanpete and Sevier Counties, Utah: U.S. Geological Survey Miscellaneous Investigations Map I-1631.

Woodburne, M.O., 1987, A prospectus of the North American Mammal ages: *in* Cenozoic Mammals of North America, University of California Press, Riverside, (in press).

Woodland, R.B., 1958, Stratigraphic significance of Mississippian endothyroid foraminifera in central Utah: Journal of Paleontology, v. 32, p. 791–814.

Woodward, L.A., 1965, Late Precambrian stratigraphy of northern Deep Creek Range, Utah: American Association of Petroleum Geologists Bulletin, v. 49, p. 310–316.

Woodward, L.A., 1968, Lower Cambrian and upper Precambrian strata of Beaver Mountains, Utah: American Association of Petroleum Geologists Bulletin, v. 52, p. 1279–1290.

Woodward, L.A., 1972, Upper Precambrian stratigraphy of central Utah: Utah Geological Association Publication 2, p. 1–5.

Woodward, L.A., 1976, Stratigraphy of younger Precambrian rocks along the Cordilleran hingeline, Utah and southern Idaho: Rocky Mountain Association of Geologists, 1976 Symposium, p. 83–90.

Woodward, N.B., 1987, Geological applicability of critical-wedge thrust-belt models: Geological Society of America Bulletin, v. 99, p. 827–832.

Wright, J.C., and Dickey, D.D., 1963, Block diagram of the San Rafael Group and underlying strata in Utah and part of Colorado: U.S. Geological Survey Oil and Gas Investigations Chart OC-63.

Wright, J.C., and Dickey, D.D., 1978, North-south cross-sections of the Jurassic San Rafael Group in Utah and western Colorado: U.S. Geological Survey Open-File Report 78-965.

Wright, J.C., Dickey, D.D., and Snyder, R.P., 1979, Stratigraphic sections of Jurassic San Rafael Group and adjacent rocks in Wayne County, Utah: U.S. Geological Survey Open-File Report 79-1126, 106 p.

Wright, R.E., 1961, Stratigraphic and tectonic interpretation of Oquirrh Formation, Stansbury Mountains, Utah: Brigham Young University Geology Studies, v. 8, p. 147–166.

Yancey, T.E., Ishibashi, G.D., and Bingman, P.T., 1980, Carboniferous and Permian stratigraphy of the Sublett Range, south-central Idaho: *in* Fouch, T.D., and Magathan, E.R., eds., Paleozoic stratigraphy of west-central United States, Symposium 1, Rocky Mountain Section of Society of Economic Paleontologists and Mineralogists, Denver, p. 259–269.

Yingling, V.L., and Heller, P.L., 1987, Timing and record of foreland basin development in central Utah during the initiation of the Sevier Orogeny: Geological Society of America Abstracts with Programs, v. 19, p. 907.

Young, G.E., 1976, Geology of Billies Mountain quadrangle, Utah County, Utah: Brigham Young University Geology Studies, v. 23, part 1, p. 205–280.

Young, J.C., 1955, Geology of the southern Lakeside Mountains, Utah: Utah Geological and Mineral Survey Bulletin 56, 110 p.

Young, R.A., and Hartman, J.H., 1984, Early Eocene fluviolacustrine sediments near Grand Canyon, Arizona; evidence for Laramide drainage across northern Arizona into southern Utah: Geological Society of America Abstracts with Programs, v. 16, p. 703.

Young, R.A., and Carpenter, C.H., 1965, Groundwater conditions and storage in the central Sevier Valley, Utah: U.S. Geological Survey Water Supply Paper 1787, 95 p.

Young, R.G., 1955, Sedimentary facies and intertonguing in the Upper Cretaceous of the Book Cliffs, Utah-Colorado: Geological Society of America Bulletin, v. 66, p. 177–202.

Young, R.G., 1966, Stratigraphy of coal-bearing rocks of Book Cliffs, Utah-Colorado: Utah Geological and Mineral Survey Bulletin 80, p. 7–22.

Young, R.G., 1975, Genesis of western Book Cliffs, coal: Brigham Young University Geology Studies, v. 22, part 3, p. 3–14.

Young, R.G., 1975, Lower Cretaceous rocks of northwestern Colorado and northeastern Utah: Rocky Mountain Association of Geologists Symposium, p. 141–147.

Young, R.G., 1983a, Petroleum in northeastern Grand County, Utah: Grand Junction Geological Society Guidebook to Northern Paradox Basin and Uncompahgre Uplift, p. 1–7.

Young, R.G., 1983b, Book Cliffs coal field, western Colorado: Grand Junction Geological Society Guidebook to Northern Paradox Basin and Uncompahgre Uplift, p. 9–15.

Youngs, R.R., Swan, F.H., Power, M.S., Schwartz, D.P., and Green R.K., 1987, Probabalistic analysis of earthquake ground-shaking hazard along the Wasatch Front, Utah: U.S. Geological Survey Open-File Report 87-585, p. M1–M110.

Zabriskie, W.E., 1970, Petrology and petrography of Permian carbonate rocks, Arcturus Basin, Nevada and Utah: Brigham Young University Geology Studies, v. 17, part 2, p. 83–160.

Zawiskie, J., Chapman, D., and Alley, R., 1982, Depositional history of the Paleocene-Eocene Colton Formation, north-central Utah: Utah Geological Association Publication 10, p. 273–284.

Zelazek, D.P., 1981, Petrology of Middle Cambrian Blacksmith Formation, southeastern Idaho and northernmost Utah: unpublished master's thesis, Utah State University, Logan, 141 p.

Zeller, H.D., 1973a, Geologic map and coal resources of the Caracass Canyon quadrangle, Garfield and Kane Counties, Utah: U.S. Geological Survey Coal Investigations Map C-56.

Zeller, H.D., 1973b, Geologic map and coal resources of theCanaan Creek quadrangle, Garfield County, Utah: U.S. Geological Survey Coal Investigations Map C-57.

Zeller, H.D., 1973c, Geologic map and coal resources of the Death Ridge quadrangle, Garfield and Kane Counties, Utah: U.S. Geological Survey Coal Investigations Map C-58.

Zeller, H.D., 1973d, Geologic map and coal resources of the Dave Canyon quadrangle, Garfield County, Utah: U.S. Geological Survey Coal Investigations Map C-59.

Zeller, H.D., 1978, Geologic map and coal resources of the Collet Top quadrangle, Kane County, Utah: U.S. Geological Survey Coal Investigations Map C-80.

Zeller, H.D., and Stephens, E.V., 1973, Geologic map and coal resources of the Seep Flat Quadrangle, Garfield and Kane Counties, Utah: U.S. Geological Survey Coal Investigations Map C-65.

Zietz, I., 1982, Composite magnetic anomaly map of the United States: U.S. Geological Survey Map GP-954A.

Zietz, I., Shuey, R., and Kirby, J.R. Jr., 1976, Aeromagnetic map of Utah: U.S. Geological Survey Map GP-907.

Ziegler, A.M., Scotese, C.R., McKerrow, W.S., Johnson, M.E., and Bambach, R.K., 1979, Paleozoic paleogeography: Annual Review of Earth and Planetary Sciences, v. 7, p. 473–502. See also Paleobiogeography and Climatology, *in* Niklas, K.J., ed., Paleobotany, Paleoecology, and Evolution, New York, Praeger, P. 231–266; also see Cambrian world paleogeography, biogeography, and Climatology (abstract only), U.S. Geological survey Open-File Report 81–743, p. 252.

Zoback, M.L., 1978, Basin and Range rifting in northern Nevada—clues from a mid-Miocene rift and its subsequent offsets: Geology, v. 6, p. 111–116.

Zoback, M.L., 1983, Structure and Cenozoic tectonism along the Wasatch fault zone, Utah: Geological Society of America Memoir 157, p. 3–27.

Zoback, M.L., 1987, Superimposed late Cenozoic, Mesozoic, and possible Proterozoic deformation along the Wasatch Fault zone in central Utah: U.S. Geological Survey Open-File Report 87-585, p. E1–43.

Zoback, M.L., Anderson, R.E., and Thompson, G.A., 1981, Cainozoic evolution of the state of stress and style of tectonism of the Basin and Range province of the western United States: Philosophical Transactions of the Royal Society of London, v. 300, p. 407–434.

Zoback, M.L., and Anderson, R.E., 1983, Style of Basin-Range faulting as inferred from seismic reflection data in the Great Basin, Nevada and Utah: Geothermal Resources Countil, Special Report no. 13, p. 363–381, Davis, California, 95617-1350.

Zoback, M.L., and Zoback, M.D., 1980, State of stress in the conterminous U.S.: Journal of Geophysical Research, v. 85, p. 6113–6156.

Zuber, M.T. and Parmentier, E.M., 1986, Extension of continental lithosphere—a model for two scales of Basin and Range deformation: Journal of Geophysical Research, v. 91, p. 4826–4838.

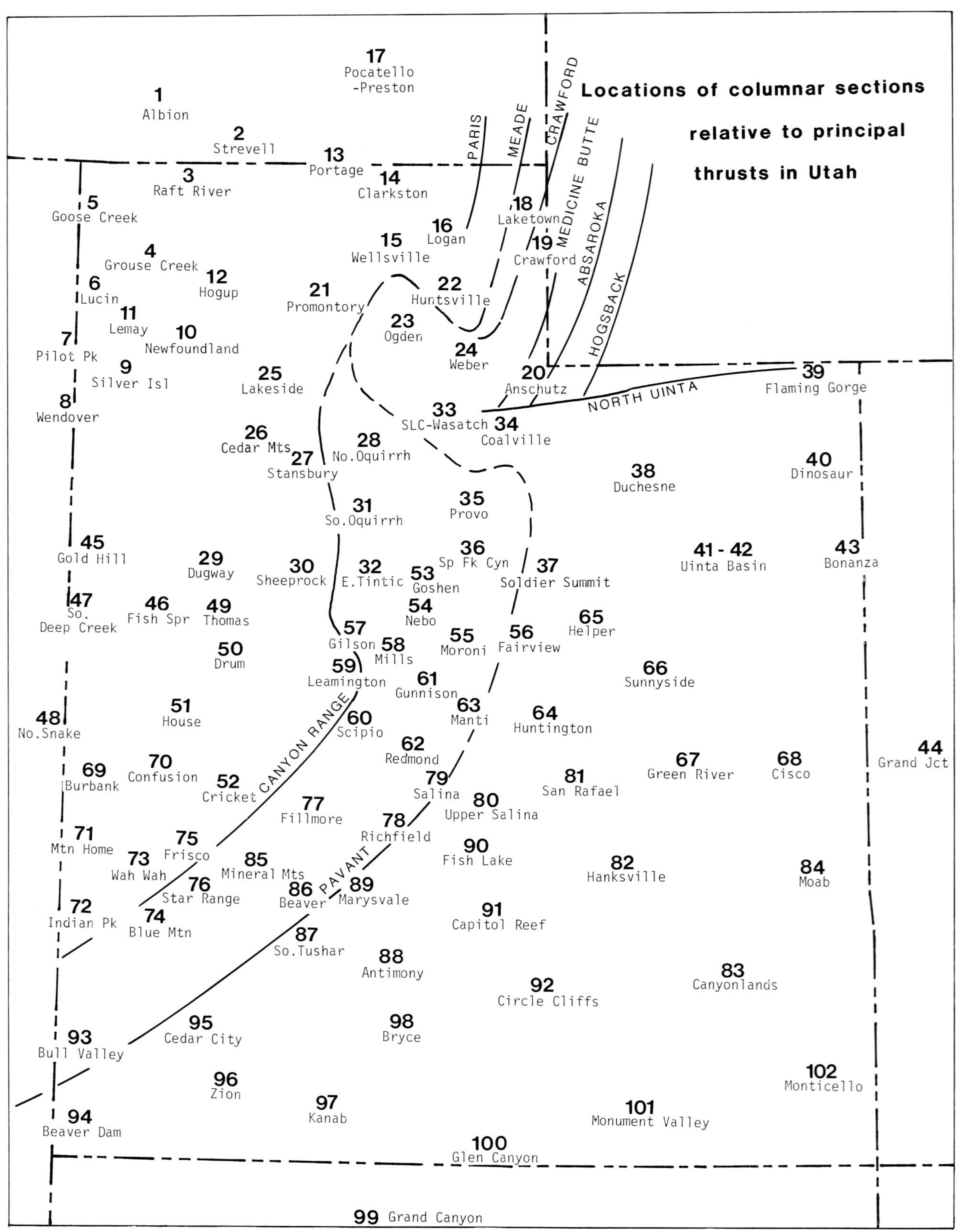

Locations of stratigraphic charts that follow. Trends of leading edges of Cretaceous thrusts are shown in gray. Pre-Cretaceous rocks west of the thrust belt were moved eastward during the Sevier orogeny.

CHART EXPLANATION

Condensing stratigraphic sections from the original sources to the charts has necessarily required eliminating much detail; citations at the bottom of each chart will lead those requiring more data to the source literature. In all instances I have attempted to assign stratigraphic units to their proper time according to current consensus, which may vary from that of some of the original age designations.

In order to give a visual impression of the relative thickness and bedding characteristics of the various units, each chart has its own schematic scale; because of the great range of rock thicknesses in the various parts of Utah, it was not feasible to use a uniform reduction scale for the stratigraphic thicknesses on all charts.

References listed on the charts and elsewhere in this book are not exhaustive; stress has been placed on recent stratigraphic literature from which the interested reader may pursue older reports.

Chart

1 ALBION AREA, IDAHO

Age	Unit	FEET	Notes
Q	Alluvium, Colluvium	0-100	
Q	Glacial deposits, Albion Mts	0-50	
Q	Snake River Basalt Group	0-200	0.5 m.y.
Q	Raft Formation	0-300	
PLIO	Salt Lake Fm: Upper basin fill	0-300	
PLIO / MIOCENE	Salt Lake Fm: Upper tuffaceous mbr	0-4000	7 m.y. Cedar Knoll Tuff
MIOCENE	Salt Lake Fm: Basalt of N. Cotterel Mt	0-100	
MIOCENE	Salt Lake Fm: Jim Sage Volc Mbr: Upper rhyolite flows & domes	0-1500	7-9 m.y.; 9-11 m.y.
MIOCENE	Salt Lake Fm: Jim Sage Volc Mbr: Middle unit	100-300	vitrophyre breccia
MIOCENE	Salt Lake Fm: Jim Sage Volc Mbr: Lower rhyolite flows	0-1500	
MIOCENE	Salt Lake Fm: Lower tuffaceous mbr	600-1500	
Φ	Almo granitic pluton		28 m.y. ductile strain
K	CRETACEOUS DEFORMATION 70-80 M.Y. MADE NW-TRENDING FOLDS		
J	JURASSIC REGIONAL METAMORPHISM 160 M.Y. WITH NE-TRENDING FOLDS		
	ALL PRE-CENOZOIC ROCKS ARE IN FAULT-BOUNDED TECTONIC SHEETS THAT COVER DOMED ARCHEAN ROCKS		
₮	Dinwoody Formation	slice	
P	Rex Chert Mbr, Phosphoria Fm	slice	
P	Oquirrh Group: Sandy limestone	0-500	
P	Oquirrh Group: West Canyon Ls	0-200	
M	Manning Canyon Phyllite	0-200	black
	3000-4000 FEET OF SIL-DEV-MISS CARBONATE STRATA MISSING BY FAULTING HERE		
ORDO	Metamorphosed Fish Haven Dol	0-50	
ORDO	Metamorphosed Eureka Qtzt	0-50	Kanosh Shale
ORDO	Metamorphosed Pogonip Group	0-1000 ?	
	3000-5000 FEET OF CAMBRIAN CARBONATE ROCKS FOUND IN OTHER RANGES IN THIS REGION ARE NOT RECOGNIZED IN THE ALBION AREA		
CAMB ?	Raft River Mountains Sequence, PARAUTOCHTHON: Schist of Mahogany Peaks or Quartzite of Thompson Flat	0-300	
PROTEROZOIC ?	Raft River Mountains Sequence, PARAUTOCHTHON: Qtzt of Clarks Basin	0-200	
PROTEROZOIC ?	Raft River Mountains Sequence, PARAUTOCHTHON: Schist of Stevens Spring	100-500	
PROTEROZOIC ?	Raft River Mountains Sequence, PARAUTOCHTHON: Quartzite of Yost	0-100	
	THRUST FAULT COMMONLY HERE		
PROTEROZOIC ?	Raft River Mountains Sequence, AUTOCHTHON: Schist of Upper Narrows	200-600	
PROTEROZOIC ?	Raft River Mountains Sequence, AUTOCHTHON: Elba Quartzite	200-700	
	MAJOR UNCONFORMITY		
ARCH N	Raft River Mountains Sequence, AUTOCHTHON: Green Creek Complex, 2500 m.y. K-Ar	gneiss, biotite schist, amphibolite	

Mt. Harrison sequence
Harrison Summit Quartzite 2500 ft
Schist of Willow Creek 1600 ft
Daley Creek Qtzt 7000-8500 ft

Note: Correlation between Mt Harrison sequence and other pre-Mesozoic rocks is uncertain. In addition, there are other tectonic sequences of smaller extent whose original identity is not known

MAP WITH TEXT - Armstrong, 1968; Miller, 1978, 1980, 1983.
GUIDEBOOK TO PRE-CENOZOIC - Miller, Armstrong, Compton, & Todd, 1983;
CENOZOIC - Williams, Covington, and Pierce, 1982.

Chart

2 STREVELL - BLACK PINE MTS

Age	Unit	FEET	Notes
Q	Alluvium & colluvium	0-200	
Q	Basalt, tuff, gravel & ss	0-200	
TERT	Salt Lake Formation and Curlew Valley fill undivided	1000-5000	8-10 m.y. FT on ash
	CENOZOIC ROCKS LIE UNCONFORMABLY ON PALEOZOIC STRATA PALEOZOIC ROCKS HEATED 200-400 DEG. CENT. BETWEEN 80-110 M.Y.		
PERMIAN	Limestone, silicified ls & chert	400-500	Smith, 1982, shows small altered dikes & sills in Mississippian here R. W. Allmendinger, 1985 letter, believes that they pre-date the thermal event
PERMIAN	Dark-gray ls of Plate II Unit A & Unit B of Plate I (northern sequence)	2400-4700	*Ellisonia*, *Neogondolella*
PERMIAN	Beds west of Dry Canyon, Plate I of northern sequence	9000	*Parafusulina*, *Schwagerina*, *Sweetognathus*
	WEST DRY CANYON FAULT SEPARATES SEQUENCES		
PENNSYLVANIAN	Oquirrh Group: Sandstone & siltstone (southern sequence upper plate)	9000	*Triticites*; *Caninia*, *Multithecopora*; *Beedeina*, *Fusulinella*
PENNSYLVANIAN	Oquirrh Group: Limestone member, Limestone & dolomite mbr, Ls, ss, & qtzt member (southern sequence, middle plate)	*600-7500 *strain within this member locally exceed 150 percent layer-parallel extension	*Fusulinella*, *Neognathodus*; *Idiognathodus*, *Ozarkodina*; corals & bryozoans; *Rhachistognathus*, *Adetognathus*; FAULTED
MISSISSIPPIAN	Manning Canyon Argillite	6500?	unusual reported thickness probably the result of tectonism; tight isoclinal folds mapped locally by Smith, 1982.
DEV	Jefferson (Hyrum) Dolomite	900	*Amphipora*; white dolomite
Є - O - S	Metasedimentary rocks-marble, dolomite, quartzite & schist	unknown	subsurface only

MAP WITH TEXT - Smith, 1981, 1983. CENOZOIC - Baker, 1974; Covington, 1983. NOTE - In adjacent Sublett Range, Yancey et al, 1980, record a total of at least 13,000 feet of Oquirrh Group strata plus 2600 feet of post-Oquirrh Permian rocks including very thick Rex Chert, and overlain by Triassic Dinwoody Formation. Wells, 1988.

Chart

3 RAFT RIVER RANGE

Age	Unit	Feet	Remarks
Q	Alluvium, colluvium, landslides	0-100	
Q	Bonneville & glacial deposits	0-50	6,600 yr Mazama ash
P	Basalt flows E of Peplin Mts	0-100	3.2 m.y.
MIOCENE	Dacite flows near Standrod	0-100	9 m.y ?
MIOCENE	Tuffaceous mudstone sandstone, conglomerate and buried valley-filling sedimentary and igneous deposits	several thousand	Oligo-Miocene ductile strain imposed on all pre-Pennsylvanian units.
CENOZOIC DEPOSTIS OVERLAP OLDER ROCKS UNCONFORMABLY			
PENN	Oquirrh Group: Calcareous sandstone	0-1000	Metamorphic grade decreases eastward in the range
PENN	Oquirrh Group: Gray sandy limestone		*Pseudofusulinella*
PENN	Oquirrh Group: Dolomite marble		*Triticites*
M	Manning Canyon or Chainman Phyllite	0-300	
3000-4000 FEET OF SILURIAN, DEVONIAN, & MISSISSIPPIAN STRATA, REGIONALLY PRESENT, ARE TECTONICALLY MISSING IN THIS RANGE			
ORDO	Metamorphosed Fish Haven Dolo	0-600	Lower Paleozoic & Proterozoic strata have been greatly thinned tectonically. Stretched-grain ratios range up to 1:7:20
ORDO	Metamorphosed Eureka Qtzt	10-320	
ORDO	Metamorphosed Pogonip Group	0-500	
3000-5000 FEET OF CAMBRIAN CARBONATE ROCKS MISSING HERE			
€ ?	Schist of Mahogany Peaks	50-300	garnet & staurolite
€ ?	Quartzite of Clarks Basin	0-400	
PROTEROZOIC ?	Metamorphosed granite prophyry intrudes schist of Stevens Spring		
PROTEROZOIC ?	Schist of Stevens Spring	0-600	graphite-bearing quartz-muscovite schist
PROTEROZOIC ?	Quartzite of Yost	0-400	
THRUST FAULT COMMONLY LIES NEAR THIS HORIZON			
PROTEROZOIC ?	Schist of Upper Narrows	0-1500	muscovite schist; biotitic & feldspathic gneiss & schist
PROTEROZOIC ?	Elba Quartzite	50-1500	green-mica quartzite; feldspathic quartzite; white-mica schist
			MAJOR UNCONFORMITY
ARCHEAN	Metamorphosed granite (adamellite of Compton)	intrusive	2.5 b.y. Rb-Sr whole rock
ARCHEAN	Metamorphosed pegmatite	intrusive	
ARCHEAN	Metamorphosed mafic rocks	0-200	hornblende schist
ARCHEAN	Older schist	300-1000	relict features show parent rocks were shale, mudstone, siltstone & felsdpathic sandstone

MAP WITH TEXT - Compton, 1972, 1975; Doelling, 1980; Felix, 1956; QUATERNARY - Nash & Smith, 1977; Montgomery & Everitt, 1984; Gates, 1984; Currey et al, 1984; ROAD LOG - Miller et al. 1983: STRUCTURE - Compton, 1980.

Chart

4 GROUSE CREEK & MATLIN MTS

Age	Unit	Feet	Remarks
Q	Alluvium & colluvium	0-100	
Q	Lake Bonneville deposits	0-200	
PLI	Basalt flows E of Matlin Mts	0-100	3.2 m.y.
PLI / MIOCENE	Salt Lake Formation or Group: Poorly sorted conglomerate and silicic vitric tuff with some volcanic flows as noted below:	3000-6000	Note: Rocks already metamorphosed by Jurassic-age heating and doming were displaced to the east by denudational faulting that accompanied volcanism some time after 11.1 to 10.5 m.y. ago See Todd, 1983.
MIOCENE	Rhyolite flows & domes near Etna		8 m.y.
MIOCENE	Dacite welded tuff		9 m.y.
MIOCENE	Rhyolitic volcanism		11.7 m.y.
MIOCENE	Dacite flows & domes near Lucin		13 m.y.
MIOCENE	Basaltic volcanism		14.4 m.y.
POST-OLIGOCENE ROCKS & TECTONIC DEPOSITS ARE UNCONFORMABLE ON OLDER ROCKS			
Φ	Granite in central Grouse Ck Mt (Adamellite of Todd)	two stocks	24.9 m.y. Rb-Sr
E	Granodiorite at Immigrant Pass	pluton	38.2 m.y. Rb-Sr
PALEOZOIC & MESOZOIC STRATA ARE ALLOCHTHONOUS & ATTENUATED.			
TR	Thaynes Formation	0-1200	Only the Oquirrh Group on Bovine Mt has been preserved close to its depositional thickness Contacts between all the other Paleozoic units are probably low-angle faults.
PERMIAN	Gerster Limestone	525	
PERMIAN	Black chert-Rex ?	sliver	
PERMIAN	Plympton ? Formation	200 ?	
PERMIAN	Kaibab Limestone	100 ?	
PERMIAN	Pequop Formation	100 ?	
PERMIAN	Oquirrh Group (1000 feet in thrust slices, 14,000 feet at Bovine Mt shown here): Unit 6	2600	thin-bedded sandy & silty limestone; *Pseudofusulina, Schwag.*
PERMIAN	Oquirrh Group: Unit 5	2000	laminated ss with interbedded cg
PERMIAN / PENN	Oquirrh Group: Units 3 & 4	4600	Sandstone, siltstone, & sandy limestone; *Triticites*
PENN	Oquirrh Group: Unit 2	1600	*Triticites* thin-bedded silty ls
PENN	Oquirrh Group: Unit 1	3200	limestone with minor ss, chert, & cg beds
M	Metamorphosed Chainman and Diamond Peak Fm	0-800	
D	Guilmette Formation	0-1500	
D	Simonson Dolomite	0-500	
S	Laketown Dolomite	0-1100	
ORDO	Fish Haven Dolomite	160*	*thickness shown is from unmetamorphosed Bovine Mt Section. Unit is also present in metamorphosed and attenuated sections in central Grouse Creek Mts.
ORDO	Eureka Quartzite	360*	
ORDO	Crystal Peak & Swan Pk Fms	150	
ORDO	Garden City Formation or Pogonip Group	400*	
ROCKS BELOW ARE AUTOCHTHONOUS			
€ ?	Schist of Mahogany Peaks	0-10	
€ ?	Quartzite of Clarks Basin	up to 600	All autochthonous rocks were greatly attenuated by Jurassic metamorphism. Their original thickness is not known.
PR ?	Schist of Stevens Spring	up to 500	
PR ?	Elba Quartzite	up to 100	
AR	Granite	intrusive	
AR	gneiss & schist	basement	

MAP WITH TEXT - Todd, 1973, 1980, 1983; Compton, 1983; Scarbrough, 1984; Doelling, 1980; **GROUNDWATER -Stephens** & Wood, 1973; TECTONICS - Compton et al, 1977; Allmendinger et al, 1984; **OQUIRRH** - Jordan, 1983; **ROAD** LOG - Miller et al, 1983.

Chart
5 GOOSE CREEK - RHYOLITE MTS

Period	Group	Unit	FEET	Remarks
Q		Alluvium & colluvium	0-100	
		Lake Bonneville deposits	0-200	
P		Gravel	0-200	
MIOCENE	Salt Lake Fm	Rhyolite flow	0-100	8.4 m.y. (Idavada Volcs)
		Vitric tuff member	1000+	*Merychippus* tooth; 12.8 m.y. rhyolite (Jarbridge Rhyolite of Hilderbrand & Newman)
		Green facies and Conglomerate of Cotton Creek	4000-5000	-green facies; ostracodes; gastropods; -cg of Cotton Creek
		Red cg & sed. breccia	250-300	mudflow deposits
		Limestone member	2000-2500	yellow, thin-bedded gastropods, ostracodes pelecypods
		Basalt of Mill Cr	50-60	16.4 m.y.
		Older Member	unknown	cg & ash
ANGULAR UNCONFORMITY				
TRIASSIC		Thaynes Formation	2000-3000	conodonts; *Meekoceras*
		Dinwoody Formation	2400	pelecypods; lingulid brachiopods
PERMIAN		Gerster Limestone	1200	*Kuvelousia*; *Spiriferina pulchra*
		Murdock Mountain Formation	1700	Rex Chert equiv
		Meade Pk Tongue, Phosphoria	400	*Pseudogastrioceras*; *Neogondolella*
		Grandeur Formation	3000	*Composita*; *Aviculopinna*
		Unnamed formation	1500	siltstone and very fine-grained ss
		Pequop Formation	4000	thin-bedded, platy; *Parafusulina*; *Schwagerina*; thin-bedded ss & slts
		Oquirrh Group	0-500 ?	Wolfcampian fusulinids
M		Diamond Peak Fm	400	
INTERVAL CONCEALED OR FAULTED				
D		Guilmette Formation	200+	stromatoporoids
		Simonson Dolomite	600+	

MAP WITH TEXT - Doelling, 1980; and summary by Doelling of thesis mapping in progress by A. R. Young. MIOCENE - Hildebrand & Newman, 1985.

Chart
6 LUCIN AREA

Period	Group	Unit	FEET	Remarks
Q		Alluvial, eolian & lake deposits	0-100	
		Lake Bonneville deposits	0-200	
UNCONFORMITY				
PLIO		Diabase dikes & sills	intrusive	Snake River Group ?
MIOCENE	Salt Lake Formation or Group	Basin-filling fanglomerate and lacustrine deposits including cg, ss, slts, ls and vitric tuff. Interbedded with volcanic units as shown below:	more than 3000	
		Rhyolite of Rhyolite Butte		8.8 m.y. 300 ft thick
		Basalt flows & dikes		10 m.y.
		Welded rhyolite tuff		12 m.y. 30-60 ft tk
		Rhyolite flow		
		Vitric tuff		100-2000 ft thick
POST-OLIGOCENE ROCKS ARE UNCONFORMABLE ON ALL OLDER ROCKS				
EO - Φ		Older rhyolitic tuff	up to 2000	37 m.y.
		McGinty Monzogranite	intrusive stock	37 m.y.
PALEOZOIC STRATA HERE WERE SUBJECTED TO MESOZOIC HEATING AND TECTONISM PRIOR TO TERTIARY INTRUSION				
PERMIAN		Chert and dolomite	980	highly fractured and silicified; Leonardian conodonts
		Pequop ? Formation	920*	*Parafusulina*; *Schwagerina*
MISS		Chainman Shale and Diamond Peak Fm	600*	
		Tripon Pass Limestone	80*-350	conodonts
DEVONIAN		Guilmette Formation	970*	conodonts
		Simonson Dolomite	700-900	*Amphipora*; conodonts
SIL		Thick-bedded dolomite	680*-1010	Sevy Dolomite and Laketown Dolo equiv conodonts
ORDOVICIAN		Ely Springs Dolomite	110*-420	small horn corals
		Eureka Quartzite	260*-520	*thinning probably caused by faulting
		Lehman Formation	50*-480	
		Garden City Fm	400*-1050	

MAPS WITH TEXT - Miller & schneyer, 1985; Miller, 1985; Miller et al, 1982; TECTONICS - Allmendinger et al, 1984.

Chart
7 PILOT PEAK AREA, NEVADA

Age	Unit	FEET	Notes
Q	Alluvium & colluvium	0-100	not exposed; accompanied regional extensional faulting
Q	Lake Bonneville deposits	0-200	
M-P	Valley-fill deposits	0-2000 ?	
	ANGULAR UNCONFORMITY		
OLIG	Quartz porphyry dikes	intrusive	30 m.y.?
OLIG	Granodiorite dikes	intrusive	30-37 m.y. K-Ar biot
EO	Bettridge Creek Granodiorite	intrusive	39 m.y. U-Pb zircon
	JURASSIC ROCKS DEFORMED BY LATE JURASSIC-EARLY CRETACEOUS METAMORPHISM		
J	Miners Spring syenogranite	intrusive	155-165 m.y. U-Pb zirc
	PRE-MESOZOIC STRATA HAVE BEEN INTRUDED & TECTONICALLY TRANSPORTED		
P	Strathearn Formation	200	
P	Ely Limestone	800	fusulinids
M	Diamond Peak and Chainman Fms, undiff	750	-major low-angle fault zone, locally silicified
DEV	Guilmette Formation	700	
DEV	Simonson Dolomite	600	
S	Lone Mountain Dolomite & Roberts Mtn Fm undivided	1000	
ORDOVICIAN	Ely Springs Dolomite	500?	severely fractured
ORDOVICIAN	Eureka Quartzite	500	
ORDOVICIAN	Lehman Formation	200	
ORDOVICIAN	Kanosh Shale	300	
ORDOVICIAN	Garden City Limestone	600*	silty limestone
Є	Notch Peak Fm -	100*	*thinned by faulting present in fault slivers
Є	Orr ? Fm - ls & shale	300*	
	DECOLLEMENT SEPARATES BRITTLE DEFORMATION ABOVE FROM PLASTIC DEFORMATION BELOW. INITIAL HEATING IN JURASSIC		
CAMBRIAN	Marble - metamorphosed Toano & Clifside Ls	1100	lower part silty
CAMBRIAN	Phyllite of Killian Springs	1000	graphitic schist
CAMBRIAN	Prospect Mountain Quartzite	3100	white & light-gray cross-laminated
PROTEROZOIC	McCoy Creek Group, Unit G, Siltstone Member	1500	metasiltstone with some marble
PROTEROZOIC	McCoy Creek Group, Unit G, Cg mbr	600	
PROTEROZOIC	McCoy Creek Group, Unit F	1000-1200	light-gray metaquartzite, cross-laminated
PROTEROZOIC	McCoy Creek Group, Unit E	230	phyllite
PROTEROZOIC	McCoy Creek Group, Unit D	700-1300	feldspathic metaquartzite & cg
PROTEROZOIC	McCoy Creek Group, Unit C	100	phyllite
PROTEROZOIC	McCoy Creek Group, Unit B ?	500	coarse white marble
PROTEROZOIC	McCoy Creek Group, Unit A ?	1500	schist; schistose quartzite

MAP WITH TEXT - Miller & Lush, 1981; Miller et al, 1982; GROUNDWATER - Stephens & Wood, 1973; Lines, 1979; STRATIGRAPHY - Miller, 1983, 1984. STRUCTURE - Allmendinger et al, 1984; Miller et al, 1987; SILURIAN - Sheehan, 1979.

Chart
8 WENDOVER - LEPPY HILLS

Age	Unit	FEET	Notes
Q	Alluvium, playa & dune deps.	0-100	
Q	Lake Bonneville deposits	0-200	
MIOCENE - PLIO?	Andesite porphyry flows	0-50	
MIOCENE - PLIO?	Tuffaceous lacustrine siltstone, sandstone, & limestone	500 plus	Salt Lake Fm or Gp
	Contact not exposed		
MIOCENE - PLIO?	Upper rhyolite porphyry	up to 500	
MIOCENE - PLIO?	Lower rhyolite porphyry	400	12 m.y. angular unconformity
	BEGINNING OF REGIONAL EXTENSIONAL FAULTING		
Φ	Granodiorite	intrusive	30 m.y. ? based on correlation with similar rocks in Pilot Range
PERMIAN	Pequop Formation	2000	interbedded bioclastic limestone & sandstone silicified in NW Leppy Hills. *Schwagerina* *Pseudofusulina*
PERMIAN	Ferguson Mountain and Strathearn Fms undivided	230	*Schwagerina*; slight angular unconform.
PENNSYLVANIAN	Hogan Formation	500	brown & white banded with pink shaly partings
PENNSYLVANIAN	Ely Limestone (restricted)	1100	fossiliferous, cherty; low-angle fault zone
MISS	Chainman & Diamond Peak Formations, undivided	430	upper half carbonate
MISS	Needle Siltstone Member	70	
MISS	Joana Limestone	0-20	angular unconformity
DEVONIAN	Pilot Shale	420	only lower part of Pilot present here; major low-angle fault zone locally silicified
DEVONIAN	Guilmette Formation	2000	*Amphipora*; formation tectonically thinned by low-angle bedding faults; *Stringocephalus*
DEVONIAN	Simonson Dolomite	400 plus	light to dark gray, finely laminated

MAP WITH TEXT - Schneyer, 1984; Schaeffer, 1960. DEVONIAN - Sandberg & Poole, 1977; MISSISSIPPIAN - C. A. Sandberg, letter, 1986; Note - Schneyer showed that Tertiary brittle extensional deformation of as much as 70 percent has thinned rocks here. Intraformational low-angle faults are confined to Chainman-Diamond Peak and older units. Section measurers must be aware of this pervasive faulting.

Chart
9 SILVER ISLAND MTS

Period	Formation	Feet	Notes
Q	Alluvium, playa & dune deps	0-100	
Q	Lake Bonneville deposits	0-200	freshwater snails & diatoms
Angular unconformity			
MIOCENE	Salt Lake Group	2800 plus	freshwater snails and bivalves; vitric tuff interbedded with calcareous silt-stone, mudstone & cg, some volcanic rocks
UNCONFORMITY - BEGINNING OF REGIONAL EXTENSIONAL FAULTING			
J	Monzonite & granodiorite stocks on Crater Island	plutons	152 m.y.
PALEOZOIC STRATA WERE PERVASIVELY BROKEN BY MESOZOIC COMPRESSIONAL & CENOZOIC EXTENSIONAL FORCES			
PERM	Undivided Grandeur, Pequop, Ferguson Mtn, & Strathearn Fms	2500	fusulinids
PENN	Ely Limestone	1740	measured at Rishel Pk in northern Leppy Rg
M	Diamond Peak and Chainman Fms	1140	Tetzlaff Peak area
M	Joana Limestone equivalent	0-270	absent at Tetzlaff Canyon
DEVONIAN	Pilot Shale	0-110	absent at Graham Peak
DEVONIAN	Guilmette Limestone	2200	stromatoporoids
DEVONIAN	Simonson Dolomite	1300	
S	Roberts Mountains Fm and Laketown Dolomite	1070	Sheehan, 1979, assigned the upper 850 ft of Schaeffer's Laketown to the Roberts Mtn Fm
ORDOVICIAN	Fish Haven Dolomite	580	*Catenipora, Streptelasma*
ORDOVICIAN	Eureka Quartzite	160-540	
ORDOVICIAN	Crystal Peak Dolomite	120	3 ft Swan Pk Qtzt at top
ORDOVICIAN	Lehman Formation	770	*Bivia, Cybelopsis*
ORDOVICIAN	Kanosh Shale	170	*Orthambonites*
ORDOVICIAN	Garden City Fm — Upper cherty mbr	1800	*Lachnostoma Hesperonomia*
ORDOVICIAN	Garden City Fm — Lower member	1300	intraformational cg
CAMBRIAN	Notch Peak Formation	1900?	faulted section; *Billingsella*
CAMBRIAN	Orr Fm — Sneakover Ls Member	80	
CAMBRIAN	Orr Fm — Corset Spring Sh Mbr	50	*Elvinia* zone
CAMBRIAN	Orr Fm — Johns Wash Ls Mbr	470	
CAMBRIAN	Orr Fm — Candland Shale Mbr	280	becomes sandy to north; *Dicanthopyge*
CAMBRIAN	Orr Fm — Big Horse Limestone M	570 plus	*Tricrepicephalus*
FAULT CONTACT IN JENKINS PEAK AREA			
CAMBRIAN	Lamb ? Dolomite	350 plus	
CAMBRIAN	Trippe & Pierson Cove Formations, Undivided	1800 ?	faulted section
MAJOR FAULT SEPARATES SECTION ABOVE FROM THAT BELOW			
CAMBRIAN	Killian Springs Fm	400	Tetzlaff Peak area
CAMBRIAN	Prospect Mountain Qtzt	1400 plus	base concealed

MAP WITH TEXT - Schaeffer, 1960; Anderson, 1960; GREAT SALT LAKE - Gwynn, 1980; JURASSIC TECTONICS - Miller, 1985; Allmendinger et al, 1984; DEVONIAN - Sandberg & Poole, 1977; SILURIAN - SHEEHAN, 1979; ORDOVICIAN - Oaks et al, 1977; CAMBRIAN - Robison & Palmer, 1968; Hintze & Palmer, 1976; McCollum & McCollum, 1984.

Chart
10 NEWFOUNDLAND MOUNTAINS

Period	Formation	Feet	Notes
Q	Alluvium & Lake Bonneville sediments	0-200	
MIOCENE- PLIO	Valley Fill in Great Salt Lake Desert	0-3000?	valley filling accompanied regional extension
J	Quartz monzonite stock		144-150 m.y. K-Ar
PERM	Meade Peak Mbr Phosphoria Fm	100?	
PERM	Oquirrh Group ?	1300 ±	*Parafusulina*; *Schwagerina*
IP	Conglomerate	0-130	gradational contact; MAJOR UNCONFORMITY
DEVONIAN	Pilot Shale	525-800	*Eoparaphorhynchus* only lower Pilot present
DEVONIAN	Guilmette Formation	1800-2000	*Calvinaria*; Units B and C of Allmendinger & Jordan
DEVONIAN	Simonson Dolomite	600-800	Unit A of Allmendinger & Jordan
SIL	Laketown Dolomite	1100 ±	Laketown - Fish Haven field mapping contact differs from regional stratigraphic assignment; thickness is here adjusted to regional norm of Budge & Sheehan, 1980
ORDOVICIAN	Fish Haven Dolomite	300 ±	
ORDOVICIAN	Eureka Quartzite	490	
ORDOVICIAN	Crystal Peak Dolomite	70-150	
ORDOVICIAN	Swan Peak Quartzite	220	(restricted usage)
ORDOVICIAN	Lehman Formation	300	
ORDOVICIAN	Kanosh Shale	200-380	
ORDOVICIAN	Garden City Limestone	2500 ±	Paddock's Garden City thickness of 3600 feet has been reduced on the basis of cross-sections from Allmendinger & Jordan (in press)
CAMBRIAN	St. Charles Dolomite	1000 -	
CAMBRIAN	Worm Creek Qtzt Mbr	60	
CAMBRIAN	Undivided Upper & Middle Cambrian carbonate rocks	1000 ±	

MAPS WITH TEXT - Allmendinger & Jordan (in press); Paddock, 1956; Doelling, 1980; GREAT SALT LAKE DESERT - Stephens, 1974; Gates, 1984; QUARTZ MONZONITE & STRUCTURE - Allmendinger & Jordan, 1984; PENNSYLVANIAN & DEVONIAN - C. A. Sandberg, letter, 1986; SILURIAN - Sheehan, 1974; ORDOVICIAN - Oaks et al, 1977.

Chart

11 LEMAY ISLAND AREA

Age	Unit	FEET	Notes
Q	Alluvium, eolian & playa dep	0-100	
Q	Lake Bonneville deposits	0-200	
TERT	Conglomerate & sandstone	500?	Valley filling sediments accompany regional extensional faulting
TERT	Rhyodacite flows	100?	33 m.y. K-Ar biotite
CENOZOIC ROCKS UNCONFORMABLE ON ALL OLDER ROCKS			
J	Porphyritic granodiorite of Crater Island stock	3 sq mi exposed	151 m.y. K-Ar biotite
PALEOZOIC STRATA DEFORMED BY LATER TECTONISM & INTRUSION			
PERMIAN	Gerster Limestone	790	brachiopods, bryozoans, & crinoids abundant; some brown chert
PERMIAN	Murdock Mountain Formation	2340	chert; tongue of Gerster Ls; chert & sandstone; black chert & ss & gray dolomite
PERMIAN	Meade Peak Phosphatic Shale	180	Phosphoria tongue
PERMIAN	Grandeur Formation	2500	sandy dolomite; Late Leonardian conodonts; cherty sandstone; conodonts
PERMIAN	Trapper Creek Fm	500	some coquinas of brachiopods, bryozoans
PERMIAN	Badger Gulch Formation*	1080	dark-gray, platy, silty limestone; large fusulinids
PERMIAN	Third Fork Formation*	580-820	*in Little Pigeon Mts
P	Strathearn ? Fm*	200?	
MISS	Tripon Pass Limestone*	1400	Kinderhookian conodonts; thin-bedded, black, silty limestone
DEVONIAN	Guilmette Formation*	1200	highly fault-fractured in Little Pigeon Mts; *Amphipora* common

MAP WITH TEXT - Miller & Glick, 1986; PERMIAN NOMENCLATURE - Miller, Martindale & Fedewa, 1984; STRUCTURE - Allmendinger et al, 1984.

Chart

12 HOGUP & TERRACE MOUNTAINS

Age	Unit	Subunit	FEET	Fossils
Q	Alluvium & colluvium		0-100	
Q	Lake Bonneville deposits		0-200	
T	Valley fill includes the Salt Lake Formation and earlier buried deposits		0-3000	
CENOZOIC ROCKS LIE WITH ANGULAR UNCONFORMITY ON OLDER STRATA				
TRIASSIC	Thaynes Formation		330	*Meekoceras*, *Neospathodus*
TRIASSIC	Dinwoody Formation		1650	*Neogondolella*, *Hindeodella*
LATE PERMIAN	Gerster Limestone		2200	*Timaniella*, *Kuvelousia*
LATE PERMIAN	Plympton Formation	Murdock Mountain Formation	2800	
EARLY PERMIAN	Meade Peak Tongue of Phosphoria Formation		690	
EARLY PERMIAN	Grandeur Formation		1840	
EARLY PERMIAN	Arcturus Formation	Loray ? Formation	3400	
EARLY PERMIAN	Arcturus Formation	"Diamond Creek" Sandstone	1500 -2500	
EARLY PERMIAN	Pequop Formation		2000	
EARLY PERMIAN	Oquirrh Group	Limestone and siltstone unit	4000	*Schwagerina*
PENNSYLVANIAN	Oquirrh Group	Sandstone unit	6000	*Pseudofusulina*, *Triticites*, *Triticites*
PENNSYLVANIAN	Oquirrh Group	Sandy & cherty limestone unit	3000 plus	

MAP WITH TEXT - Stifel, 1964; Doelling, 1980; VALLEY FILL - Cook et al, 1980; TRIASSIC - Clark et al, 1977; Clark & Carr, 1984; Paull & Paull, 1982; PERMIAN - Wardlaw et al, 1979; Maughan, 1979; Bissell, 1973; OQUIRRH - Jordan, 1979, 1981; Jordan & Douglass, 1980.

Chart
13 PORTAGE, UTAH - SAMARIA, IDAHO

Age	Unit		FEET	Remarks
Q	Alluvium & colluvium		0-100	
Q	Lake Bonneville deposits		0-200	
T	Basalt flows & tuffs		0-100	
T	Salt Lake Formation		0-1500	
CENOZOIC ROCKS LIE UNCONFORMABLY OVER PALEOZOIC STRATA				
PERMIAN	Sandstone unit of Allmendinger		1600	may be partly equivalent to Hudspeth Cutoff Ss in Sublette Range
PERMIAN	Oquirrh Group Murphy's units given to the right total 14,450 ft. Allmendinger in the adjacent quadrangle to the west notes total Oquirrh thickness of 11,340 feet. Allmendinger (1985, letter) notes that the lower Oquirrh is locally strained by 60 percent extension west of the North Canyon fault.	Unit 6 of Murphy	1650	cherty ls and ss
PERMIAN		Unit 5	1400	*Pseudoschwagerina*
PERMIAN		Unit 4	1450	cherty siltstone
PERMIAN		Unit 3	4000	*Triticites* *Pseudofusulina*
PENNSYLVANIAN		Unit 2	2450	bioclastic ls and ss *Triticites* *Pseudofusulina*
		PORTAGE CANYON FAULT SEPARATES SEQUENCE		
PENNSYLVANIAN		Unit 1	2200	silty ls & calcareous ss *Pseudostaffella*
	NORTH CANYON FAULT SEPARATES SEQUENCE			
PENNSYLVANIAN	Wells Formation		1300	*Profusulinella* bryozoans, brachiopods conodonts, crinoids
MISSISSIPPIAN	Manning Canyon Shale		260-700	*Caninia* *Rhipidomella*
MISSISSIPPIAN	Great Blue Limestone		1200	*Faberophyllum* *Acrocyathus* *Ekvasophyllum*
MISSISSIPPIAN	Little Flat Formation		900-950	no fossils
MISSISSIPPIAN		Delle Phosphatic Mbr	60	
MISSISSIPPIAN	Lodgepole Limestone*		330-400	*Vesiculophyllum* *includes 15 ft Mission Canyon Ls at top
DEVONIAN	Beirdneau Formation		650-900	
DEVONIAN	Hyrum Dolomite	Upper member	500-550	
DEVONIAN		Samaria Ls Member	550-600	*Allanaria* *Rensellandia*
DEVONIAN	Water Canyon Formation		450-550	fish fragments
SIL	Laketown Dolomite		1200-1300	*Halysites*
O	Fish Haven Dolomite		200-300	*Streptelasma*
O	Swan Peak Quartzite		400	

MAP WITH TEXT - Platt, 1977; Murphy, 1983; Beus, 1963, 1968; Adams, 1962; Jordan 1985; Allmendinger, 1983; OQUIRRH CONODONTS - Akers & Davis, 1985; MISSISSIPPIAN - Isaacson et al, 1985; Wickwire, 1985; C. A. Sandberg, letter, 1986; STRUCTURE - Allmendinger & Platt, 1980; VALLEY FILL - Bolke & Price, 1972; Peterson, 1974.

Chart
14 CLARKSTON MT - MALAD RANGE

Age	Unit		FEET	Remarks
Q	Alluvium & colluvium		0-100	
Q	Lake Bonneville deposits		0-700	
Q	High-level boulder deposits		0-100	high on hillside
TERTIARY	Salt Lake Formation		1000+	includes volcanic rocks and petroliferous ls
TERTIARY	Buried valley fill		several thousand	
TERTIARY	Wasatch Fm deposited before Late Cenozoic Basin & Range faulting			
TERTIARY	Wasatch Formation ?		0-100	bouldery red soil
CENOZOIC ROCKS LIE UNCONFORMABLY OVER PALEOZOIC STRATA				
P	Wells Formation		1300+	PALEOZOIC ROCKS ARE SO PERVASIVELY FAULTED IN THIS RANGE THAT MOST OF THE THICKNESSES SHOWN HERE MAY NOT REPRESENT THE DEPOSITIONAL VALUE
	INTERVAL NOT EXPOSED			
M	Lodgepole Limestone		300	
	INTERVAL NOT EXPOSED			
DEV	Water Canyon Formation		500?	
SIL	Laketown Dolomite and Fish Haven Dolomite Undivided		1600?	light & dark gray, banded *Halysites*
ORDOVICIAN	Swan Peak Quartzite		570	
ORDOVICIAN	Garden City Formation	Upper cherty member	420	*Anomalorthis* *Hesperonomia*
ORDOVICIAN		Lower member	1340	*Trigonocerca* intraformational cg *Hystricurus* *Bellefontia*
CAMBRIAN	St. Charles Fm		700-1100	upper St. Charles thins northward; Worm Creek doubles in thickness north of here.
CAMBRIAN	Worm Creek Qtzt Mbr		100-400	feldspathic ss
CAMBRIAN	Nounan Formation		800-1400	*Aphelaspis* Nounan thins toward the north
CAMBRIAN	Bloomington Formation		500-1000	green shale & nodular ls & intraformational cg Bloomington thickens northward.
CAMBRIAN	Blacksmith Formation		350-450	oolitic
CAMBRIAN	Ute Formation		440	ls & olive shale
CAMBRIAN	Langston Fm	Upper member	30+	
CAMBRIAN		Spence Shale M	60	
CAMBRIAN		Naomi Peak Mbr	40	
CAMBRIAN / p€	Geertsen Canyon Quartzite (formerly Brigham Qtzt)		1100+	

MAP WITH TEXT - Hanson, 1949; Axtell, 1967; Burton, 1973; Gray, 1975; Prammani, 1957; Sprinkel, 1976; Wach, 1967; SOILS - Chadwick et al, 1975; BONNEVILLE - Maw, 1968; SALT LAKE - Adamson et al, 1955; Williams, 1962; BOUGUER GRAVITY - Peterson, 1974; ORDOVICIAN - Ross, 1951.

FIGURE 126 — Raft River Mountains, an east-west domal metamorphic core complex just south of the Utah-Idaho border, has a flat summit held up mostly by Precambrian Elba Quartzite, a strained rock with greatly elongated quartz grains. Quartz mica schists from the Raft River Mountains make excellent flat building stones, prized for their sheen and their toughness in large thin slabs.

FIGURE 127 — Bonneville Salt Flats near Wendover. Paleozoic rocks, largely Cambrian through Devonian, are exposed on the east front of the Silver Island Range.

FIGURE 128 — Wasatch Mountains at north edge of Ogden. Prominent light unit is Cambrian Tintic Quartzite about 1200 feet thick and underlain at the mountain base by early Precambrian granitic gneiss of the Farmington Canyon Complex. The Tintic Quartzite is repeated on the Ogden thrust plate which places broken-up Tintic (the less prominent light band in the middle hillslope) on jumbled slivers of Cambrian limestone and shale (the dark band between the two Tintics). The dark tree-covered area at the rear of the mountain is late Proterozoic strata of the Huntsville sequence on the upper plate of the Willard thrust. These older rocks rest on various Cambrian through Mississippian carbonate units which do not show in this view.

FIGURE 129 — Wasatch Mountains (Wellsville Mtn.) north of Brigham City. Hamlet of Deweyville is at left third of photo beyond the Bear River meanders. Thin-bedded unit on left half of Wellsville Mountain is Pennsylvanian Oquirrh Formation. More massive units at photo center towards base of mountain are Mississippian limestones. Rugged outcrops in mountains at right third of picture are Ordovician through Devonian strata. See Chart 15. Note Lake Bonneville shoreline above Deweyville.

Chart

15 WELLSVILLE MTN - BRIGHAM CITY

System	Unit		FEET	Notes
Q	Alluvium and Lake Bonneville deposits		0-300	
TERT	Upper Tertiary alluvial lacustrine and valley fill deposits		0-2000	
PENN & PERM?	Oquirrh Formation		at least 2600 feet possibly as much as 4500	*Triticites*
		West Canyon Ls Mbr	400	*Idiognathodus*
MISS	Manning Canyon Shale		0-570	
	Great Blue Ls	Upper member	470	cochliodont teeth
		Lower member	550	*Faberophyllum*
	Little Flat Formation		300-800	*Ekvasophyllum*
		Delle Phosphatic Mbr	90	phosphatic
	Lodgepole Ls		840-970	*Siphonodella*
DEV	Beirdneau Fm		0-345	
	Hyrum Dolomite		490	
	Water Canyon Fm		1280	*Protaspis*
S	Laketown Dolomite		1110	
ORD	Fish Haven Dolomite		200	*Halysites*
	Swan Peak Fm		50-300	*Didymograptus*
	Garden City Fm	Upper cherty mbr	425	
		Lower member	965	intraformational cg
CAMBRIAN	St. Charles Fm		1100	conodonts *Matthevia* *Ptychaspis*
		Worm Creek Qtzt M	70	*Elvinia*
	Nounan Fm		1275	*Crepicephalus*
	Bloomington Fm	Calls Ft. Sh Mbr	300	*Bolaspidella*
		Middle ls mbr	650	
		Hodges Sh Mbr	330	
	Blacksmith Dolo		800	
	Ute Fm		650	-*Ehmaniella*
	Langston Fm	Upper mbr	110	-*Glossopleura*
		Spence Sh Mbr	175	*Albertella*
		Naomi Pk M	25	*Scolithus* tubes
	Brigham Group	Geertsen Canyon Quartzite	3500	
?		Browns Hole Fm	350	white qtzt
PROTEROZOIC		Mutual Formation	2200-2600	maroon to purplish-red quartzite
		Inkom Fm	150	
		Caddy Canyon Quartzite	1000	vitreous quartzite tan, green, purple
	Papoose Creek Formation		750- 1500	
	Kelley Canyon Fm		700	black argillite

GENERAL REFERENCE - Oviatt, 1985, 1986; QUATERNARY - Bjorklund & McGreevy, 1974; U.S. Dept. Agriculture, 1975; MISSISSIPPIAN - Dutro, 1979; ORDOVICIAN - Ross, 1951; Oaks et al, 1977; CAMBRIAN - Maxey, 1958; Palmer, 1971; Taylor & Landing, 1981; Podrebarac, 1976; PROTEROZOIC - Sorensen & Crittenden, 1976b; OIL & GAS - Clem & Brown, 1985; GUIDEBOOK - Link et al, 1985.

Chart

16 LOGAN - BEAR RIVER RANGE

System	Unit		FEET	Notes
Q	Alluvium, surficial deposits		0-200	
	Lake Bonneville & glacial dep		0-500	
TERT	Salt Lake Fm	Upper cg & ss unit	1000-2000	thickness figures from west side Cache Valley
		Tuff unit	1200	
		Lower cg unit	2500 ?	
	Wasatch Conglomerate		0-400	
CENOZOIC STRATA ARE UNCONFORMABLE ON ALL OLDER ROCKS				
IP	Wells Formation		600+	top eroded
MISSISSIPPIAN	Monroe Canyon Ls	Cherty ls mbr	70	*Amplexizaphrentis*
		Thickbedded ls m	480	*Faberophyllum*
	Little Flat Fm	Sandy ls mbr	120	*Ekvasophyllum*
		Sandstone mbr	830	*Ekvasophyllum* *Leiorhynchus*
		Delle Phosphatic M	60	platy siltstone
	Lodgepole Limestone		600-700	Cottonwood Canyon Mbr
DEVONIAN	Leatham Formation		0-90	thins southward Contact Ledge
	Beirdneau Sandstone		1100-250	
	Hyrum Dolomite		980-220	thins southward Samaria Ls Mbr
	Water Canyon Formation		210-600	fish plates
SILURIAN	Laketown Dolomite		1200-1500	light and dark gray
ORDOVICIAN	Fish Haven Dolomite		120-490	dark gray
	Swan Peak Formation		400-0	thins southward graptolites
	Garden City Fm	Upper cherty mbr	300-370	*Anomalorthis*
		Lower member	860-1040	silicified trilobites *Hystricurus* *Bellefontia*
CAMBRIAN	St. Charles Formation		700-950	
		Worm Creek Qtzt Mbr	70-120	*Elvinia*
	Nounan Dolomite		800-1200	some limestone
	Bloomington Fm	Calls Fort Member	180	*Elrathia*
		Middle member	720	
		Hodges Shale Mbr	600	
	Blacksmith Formation		330-700	oolitic
	Ute Formation		400-800	*Ehmaniella*
	Langston Fm	Upper member	200-460	
		Spence Shale M	0-210	*Glossopleura*
		Naomi Peak Mbr	0-40	*Albertella* zone
PROT	Brigham Gp	Geertsen Canyon Quartzite	2550	
		Mutual Formation	2600	

CENOZOIC - Bjorklund & McGreevy, 1971; DeGraff, 1976; Schaffer, 1984; Williams, 1962; MISSISSIPPIAN - Isaacson et al, 1985; Newman, 1980; Sando, per. comm., 1985; Mullens & Izett, 1963; DEVONIAN - Sandberg & Gutschick, 1979; Sandberg & Poole, 1977; Williams, 1971; Eliason, 1969; Williams & Taylor, 1964; SILURIAN - Budge & Sheehan, 1980; ORDOVICIAN - Leatham, 1985; Oaks et al, 1977; VanDorston, 1970; Ross, 1951; CAMBRIAN - Taylor & Repetski, 1985; Haynie, 1957; Gardiner, 1974; Hay, 1982; Deputy, 1984; Buterbaugh, 1982; Galloway, 1970; PROTEROZOIC - Blau, 1975. GEOLOGIC MAP - Dover, 1985; Mullens & Izatt, 1963; Berry, 1986. OIL & GAS - Clem & Brown, 1985; GUIDEBOOK - Crook et al, 1985.

Chart
17 PRESTON - POCATELLO, IDAHO

Period	Formation / Member	FEET	Remarks
Q	Alluvium & Lake Bonneville dep	0-200	Bonneville flood 15,000 years ago
Q	Basalt in Portneuf & Gem Valleys	0-200?	0.6 m.y.
T	Salt Lake Fm — Mink Creek Member	1000	cg, tuffaceous matrix
T	Salt Lake Fm — Cache Valley Member	7700	tuff, ls, ss, cg; diabase (Winter, 1985)
CENOZOIC ROCKS REST ON OLDER STRATA IN ANGULAR UNCONFORMITY			Ladd-Bennington 3-12 S-43 E 1980
P	Phosphoria Formation	400	
IP	Wells Formation	800	
MISS	Monroe Canyon Limestone	900	*Lower Mississippian & Devonian thin to absent in Ladd well. Cambrian Ordovician & Silurian strata all thin rapidly to the east
MISS	Little Flat Formation	1000-0*	
MISS	Lodgepole Limestone	800-0*	
DEV	Beirdneau Formation	900-0*	
DEV	Hyrum Dolomite	1600-400*	stromatoporoids; includes Water Cyn Fm
SIL	Laketown Dolomite	1300-300*	*Pentamerus*; light gray
ORDO	Fish Haven Dolomite	1000-500*	dark gray
ORDO	Swan Peak Quartzite	1400-300*	*Thalassinoides* - like "fucoidal" trace fossils
ORDO	Garden City Formation — Upper cherty mbr	370	*Hesperonomia*
ORDO	Garden City Formation — Lower member	1070	intraformational cg; *Symphysurina*
CAMBRIAN	St. Charles Formation — Upper member	1000-500*	
CAMBRIAN	St. Charles Formation — Worm Creek Quartzite Mbr	900-100*	*Elvinia*
CAMBRIAN	Nounan Formation	650	*Tricrepicephalus*
CAMBRIAN	Bloomington Formation	1500-900*	*Ehmania*
CAMBRIAN	Blacksmith Limestone (Elkhead Ls near Pocatello)	550-850	Note - various relations between Brigham Group and Cambrian carbonate units are outlined by Link et al. 1985, p. 100
CAMBRIAN	Ute Limestone (Bancroft Ls) / Lead Bell Shale	500	
CAMBRIAN	Langston Dolo / Lead Bell Shale	300-600	
CAMBRIAN (Brigham Group)	Gibson Jack Formation / Twin Knobs Ls	900-1300	argillaceous siltstone with some qtzt; near Pocatello this fm includes all units to Elkhead
CAMBRIAN / LATE PROTEROZOIC (Brigham Group)	Camelback Mountain Quartzite	1000-2000	*Skolithus*; vitreous white qtzt with feldspathic & pebble cg beds
LATE PROTEROZOIC (Brigham Group)	Mutual Formation	1000-2000	grayish-red qtzt, cg, and thin argillite
LATE PROTEROZOIC (Brigham Group)	Inkom Formation	200-800	green & red argillite
LATE PROTEROZOIC (Brigham Group)	Caddy Canyon Quartzite	500-2000	vitreous, variable pebble cg; siltite in lower part, dolomite locally in middle
LATE PROTEROZOIC (Brigham Group)	Papoose Creek Formation	1000-2000	mottled siltstone & fine-grained qtzt
LATE PROTEROZOIC	Blackrock Canyon Limestone	0-300	present near Pocatello
LATE PROTEROZOIC (Pocatello Formation)	Upper member (phyllitic argillite & qtzt)	1000-2000	Entire Pocatello Fm has been metamorphosed to greenschist facies
LATE PROTEROZOIC (Pocatello Formation)	Scout Mountain	1000-2000	glacial diamictites (Crittenden et al, 1983)
LATE PROTEROZOIC (Pocatello Formation)	Bannock Volc Mbr	100-1000	pillow lava, metadiabase
LATE PROTEROZOIC (Pocatello Formation)	Member	1000 plus	argillite, diamictite, minor sandstone

MAP WITH TEXT - Oriel & Platt, 1980; Oriel, 1968; Armstrong, 1969; Trimble, 1976; Corbett, 1978; Link, 1982; PLEISTOCENE - Scott et al, 1982; Papson, 1977; TERTIARY - Willard, 1972; Winter, 1985; Danzl, 1985; MISSISSIPPIAN - Sando et al, 1981; Isaacson et al, 1985; ORDOVICIAN - Ross, 1951; Oaks et al, 1977; Taylor & Repetski, 1985; CAMBRIAN - Palmer, 1971; Zelazek, 1981; PRECAMBRIAN - Oriel & Armstrong, 1971; Crittenden et al, 1983; Link, 1983; Link & LeFebre, 1983; Link et al, 1985; Lindsey, 1982; Walker, 1983; SUBSURFACE - Graham Campbell, letter, 1985; Crook et al, 1985; STRATIGRAPHIC NAMES -MacLachlan, 1985; STRUCTURE - Sacks & Platt, 1985; Link et al, 1985; Scheu, 1985; GUIDEBOOK - Link et al, 1985.

Chart
18 LAKETOWN CANYON

Period	Formation / Member	FEET	Remarks
Q	Alluvium, colluvium	0- 500	includes lake beds and landslides
M	Salt Lake Formation	0-200	tuff, ls, mudstone
EO	Wasatch Fm	1000+	shale, ss, cg and oncolitic ls; unconformity
JURASSIC	Twin Creek Formation	2300	thickness from well data
JURASSIC	Nugget Ss	1300	cross-bedded
TRIASSIC	Ankareh Fm	940+	
TRIASSIC	Thaynes Fm	1060	*Meekoceras*
TRIASSIC	Dinwoody and Woodside Fms undivided	1180	linguloid brachiopods; pelecypod fragments
P	Phosphoria Fm — Rex Chert M	130	
P	Phosphoria Fm — phosphatic sh	90	phosphate
IP	Wells Fm	580	=Weber Ss; unconformity
MISS	Monroe Canyon Ls	290	
MISS	Little Flat Fm	700	
MISS	Delle Phosphatic Mbr	76	phosphate
MISS	Lodgepole Ls	700	bioclastic cherty
DEV	Beirdneau Fm*	230	salt casts; *equivalent to Logan Gulch Mbr, Three Fks Fm
DEV	Hyrum Dolomite	700-900	
DEV	Water Canyon Fm	250	fish plates
SIL	Laketown Dolomite	1120	light gray
ORD	Fish Haven Dolo	170	unconformity
ORD	Garden City Ls	780	intraformational limestone cg
CAMBRIAN	St. Charles Fm	650	
CAMBRIAN	Nounan Fm	400?	thickness from Valenti (1980) cross sections
CAMBRIAN	Bloomington Fm	700?	
CAMBRIAN	Blacksmith Dolo	500?	
CAMBRIAN	Ute Fm	300?	*Ehmaniella*
CAMBRIAN	Langston Fm	590	
PROT	Brigham Group	several thousand possible	subsurface data shown on Valenti (1980) cross-sections
	Mesozoic strata in lower plates are underlain by thinner Paleozoic strata than shown above		MEADE THRUST & CRAWFORD THRUST AT GREATER DEPTH

PRINCIPAL REFERENCE - Valenti, 1980, 1982; PERMIAN - Peterson, 1977, p. 147; MISSISSIPPIAN -Welsh & Bissell, 1979; Sandberg and Gutschick, 1979, 1984; DEVONIAN - Williams and Taylor, 1964; OIL & GAS - Clem & Brown, 1985; GUIDEBOOK - Chidsey et al, 1985.

Chart
19 CRAWFORD MOUNTAINS AREA

Age	Formation	FEET	Remarks
Q	Alluvium	0-1000	
EOCENE	Fowkes Fm	200-400	48 m.y. K-Ar
EOCENE	Green River Fm	0-200	
EOCENE	Wasatch Fm	1000+	algal limestone; unconformity
EARLY CRETACEOUS	Sage Junction Formation	2000	Aspen Shale Equivalent; coal
EARLY CRETACEOUS	Cokeville Fm	1600	Bear River Fm equivalent (Cokeville Fm through Smiths Fm)
EARLY CRETACEOUS	Thomas Fork Fm	200	
EARLY CRETACEOUS	Smiths Fm	400	
EARLY CRETACEOUS	Gannett Group: Upper part	1400	
EARLY CRETACEOUS	Gannett Group: Ephraim Cg	220	intraformational unconformity
JURASSIC	Stump Ss	300	
JURASSIC	Preuss Formation	•1000	redbeds
JURASSIC	Twin Creek Formation	2000	Giraffe Creek Member; Leeds Creek Member; Watton Canyon Mbr; Boundary Ridge Mbr; Rich Member; Slickrock Member; Gypsum Spring Mbr
JURASSIC	Nugget Ss	1100	
?			
TRIASSIC	Ankareh Fm	1000	
TRIASSIC	Thaynes Ls	1500	
TRIASSIC	Woodside Sh	560	
TRIASSIC	Dinwoody Fm	580	Hogback Ridge Gas Field 12 mi N. of Randolph (Dinwoody Fm through Phosphoria Fm)
P	Phosphoria Fm: Rex Chert M	220	
P	Phosphoria Fm: phosphatic sh	210	
IP	Wells (?) Fm	500	
MISS	Humbug Formation	120	
MISS	Brazer Dolomite	850	conodonts see Norris, 1981; cherty, fossiliferous
MISS	Delle Phosphatic Mbr	14	
MISS	Lodgepole Ls	600	dark gray, fossiliferous
DEV	Leatham Formation	10-50	orange soil
DEV	Beirdneau Formation*	190	*equiv to Logan Gulch Mbr of Three Forks Fm
DEV	Hyrum Dolomite	370	
DEV	Water Canyon Fm	0-65	fish plates
O	Bighorn Dolomite	770	Christmann Fed. 1-20, T10N, R8E TD 18,054 (1981)
Є	Undivided ls & sh	1200	CRAWFORD THRUST
	Mesozoic strata beneath thrust		

QUATERNARY - Hurst 1982; U.S. Dept. Agriculture, 1982; TERTIARY - Nelson, 1971; Oriel & Tracey, 1970; Rubey et al. 1975; CRETACEOUS - Rubey, 1973; JURASSIC - Imlay, 1967; Pipiringos & Imlay, 1979; Jordan, 1985; TRIASSIC - Kummel, 1954; PERMIAN - Smith et al, 1952; Peterson, 1977, p. 147; PENNSYLVANIAN - Ott, 1980; Welsch & Bissell, 1979; MISSISSIPPIAN - Sando et al, 1959, 1976; Sandberg & Gutschick, 1984; DEVONIAN & ORDOVICIAN - Ott, 1980; Chamberlain, 1980; Sandberg & Poole, 1977; Sandberg & Gutschick, 1979; STRUCTURE & GAS PRODUCTION - Ver Ploeg & DeBruin, 1980; Ott et al, 1985; SUMMARY - Baer, 1985; GUIDEBOOK - Chidsey et al, 1985.

Chart
20 ANSCHUTZ RANCH - PINEVIEW

Age	Formation	FEET	Remarks
Q	Alluvium	0-500	
TERT	Wasatch Fm	1000-2000	some algal ls
TERT	Evanston Fm	1000-1500	
CRETACEOUS	Hilliard Shale	0-500	present in footwall
CRETACEOUS	Frontier Fm	400-4000	oil at Pineview Field beneath thrust
CRETACEOUS	Aspen Shale	400-1000	
CRETACEOUS	Bear River Fm	0-3000	
CRETACEOUS	Kelvin Fm	0-3000	gas at South Lodgepole Field
JURASSIC	Stump Fm	200	oil at Pineview
JURASSIC	Preuss Formation	400-1200	salt flows into masses up to 3500 ft thick; SALT
JURASSIC	Twin Creek Ls: Giraffe Creek M	360	
JURASSIC	Twin Creek Ls: Leeds Creek Mbr	310	oil and gas at Pineview, Lodgepole, Elkhorn Ridge, Yellow Creek
JURASSIC	Twin Creek Ls: Watton Canyon M	170	
JURASSIC	Twin Creek Ls: Boundary Ridge Sh	40	
JURASSIC	Twin Creek Ls: Rich Member	230	
JURASSIC	Twin Creek Ls: Sliderock Member	100	
JURASSIC	Twin Creek Ls: Gypsum Spr Mbr	35	
JURASSIC	Nugget Ss	1100	oil and gas at Pineview, Lodgepole, Anschutz Ranch, Anschutz Ranch East, Chicken Creek; Chinle Equivalent
TRIASSIC	Ankareh Fm	1000	
TRIASSIC	Thaynes Ls	1000-1400	sour gas & condensate Yellow Creek, Whitney Canyon
TRIASSIC	Woodside Shale	600-800	
TRIASSIC	Dinwoody Fm	200	sour gas at Hogback Ridge
P	Phosphoria Fm	550	sour gas at Cave Creek
PENN	Weber Ss	600	sour gas at Cave Creek Field
PENN	Amsden Fm	350	
M	Madison Group	600-1000	sour gas at Cave Creek Field
D	Darby Fm	350	Jefferson - Three Forks
S	Laketown Dolomite	0-500	thickens westward
O	Bighorn Dolomite	350	sour gas at Whitney Canyon & Red Canyon
CAMB	mostly limestone, some dolomite and shale	1000-2000	ABSAROKA THRUST at Anschutz Ranch cuts lower Paleozoic section
pЄ	Precambrian basement rocks (Uinta Mtn Group) not yet reached by drilling but may underlie autochthon	several thousand ?	

GENERAL REFERENCES - VerPloeg & DeBruin, 1982; West & Lewis, 1982; Lelek, 1982; STUMP - Pipiringos & Imlay, 1979; MORGAN - Picard, 1977; MADISON - Rose, 1977; PINEVIEW FIELD - Connor & Covlin, 1977; Loucks, 1975.

Chart
21 PROMONTORY RANGE

Age	Formation	FEET	Remarks
Q	Alluvium & colluvium	0-100	
Q	Lake Bonneville deposits	0-200	
TERTIARY	Pre-Bonneville ashes & soils	0-200	Pearlette O & Bishop ash
TERTIARY	Salt Lake Formation and older buried valley fill	0-4000	Rozel Hills basalt flows; maximum in Rozel Graben and Great Salt Lake Graben
CENOZOIC ROCKS LIE WITH ANGULAR UNCONFORMITY ON OLDER STRATA			
PENN	Oquirrh Group	7500	Note: formations shown here are taken mostly from Olson but some of the names have been modified to fit present usage better. The Cambrian units of Olson especially need to be reexamined.
MISSISSIPPIAN	Manning Canyon Shale	1100	
MISSISSIPPIAN	Great Blue Limestone	1500	*Fenestella*
MISSISSIPPIAN	Woodman Formation	700	Delle phosphatic sh
MISSISSIPPIAN	Lodgepole Formation	300	many fossils
DEV	Beirdneau Formation	200	
DEV	Hyrum Dolomite	900	
DEV	Water Canyon Formation	600	
SIL	Laketown Dolomite	1200?	*Halysites*
ORDO	Fish Haven Dolomite	600?	
ORDO	Swan Peak Quartzite	790	*Anomalorthis*
ORDO	Garden City Formation	1170	*Trigonocerca*; *Parahystricurus*
CAMBRIAN	St. Charles Formation	1080	dark, cherty
CAMBRIAN	Worm Creek Qtzt Mbr	80	brown, quartzite
CAMBRIAN	Nounan Formation	2200	light-gray massive dolo & blue-gray ls
CAMBRIAN	Upper shale member	150	shale with ls nodules, intraformational cg & fauna *Bolaspidella*
CAMBRIAN	"Marjum" Limestone	1750	equivalent to upper Bloomington Fm; *Bolaspidella*
CAMBRIAN	"Wheeler" Formation	200	*Ptychagnostus gibbus*
CAMBRIAN	"Swasey" Limestone	200	phyllite & limestone
CAMBRIAN	Whirlwind equivalent	300	*Ehmaniella*
CAMBRIAN	Dome equivalent	100	
CAMBRIAN	Chisholm equivalent	30	*Glossopleura*
CAMBRIAN	Howell equivalent	370	pisolites common; *Ptarmigania*
CAMBRIAN	Pioche Formation	300-380	mostly qtzt & siltstone
CAMBRIAN / PROTEROZOIC (Brigham Group)	Geertsen Canyon Qtzt	3900	
PROTEROZOIC (Brigham Group)	Mutual Formation; Inkom Shale; Caddy Canyon Quartzite; Papoose Creek Formation; Kelley Canyon Formation; Maple Canyon Formation; Fm of Perry Canyon	8000	The late Max Crittenden remapped the Promontory Range and identified the units of his Late Proterozoic Huntsville Sequence as listed. His map and descriptions are yet to be published.

MAP WITH TEXT - Olson, 1960; QUATERNARY - Morrison. 1965; Nash & Smith, 1977; Hood, 1972; TERTIARY - Greenman, 1982; Slentz & Eardley, 1956; Eardley, 1963; Patton & Lent, 1980; Cook et al, 1980. GENERAL - Doelling, 1980; OQUIRRH - Jordan, 1977; ORDOVICIAN - Oak et al, 1977; Hintze, 1951; CAMBRIAN - Crittenden notes. PRE-CAMBRIAN - Christie-Blick, 1985.

Chart
22 HUNTSVILLE AREA ALLOCHTHON

Age	Formation	FEET	Remarks
Q	Alluvial & lake deposits	0-500	
T	Norwood Tuff	0-1100	37.5 m.y. K-Ar
K	Wasatch Fm - Evanston Fm	0-3400	MAJOR UNCONFORMITY
P	Franson Mbr-Park City F	400	cherty
P	Meade Pk Mbr-Phosphoria	230	phosphate
P	Grandeur Mbr-Park City F	270	
P	Wells Formation	600	
P	Round Valley Ls	250-300	
MISS	Monroe Canyon Ls	600	
MISS	Little Flat Fm	900	Delle Phosphate Mbr
MISS	Lodgepole Ls	900	fossiliferous
DEV	Beirdneau Ss	400-500	
DEV	Hyrum Dolomite	400-500	
DEV	Water Canyon Fm	150-250	fish plates
S	Laketown Dolomite	200-1500	
ORD	Fish Haven Dolomite	0-300	
ORD	Garden City Fm	400-1200	flat-pebble ls intraformational cg
CAMBRIAN	St. Charles Fm	900	
CAMBRIAN	Worm Creek Qtzt Mbr	0-50	
CAMBRIAN	Nounan Dolomite	500-700	
CAMBRIAN	Bloomington Fm: Calls Ft Sh Mbr		intraformation cg
CAMBRIAN	Bloomington Fm: Middle ls mbr		
CAMBRIAN	Bloomington Fm: Hodges Sh Mbr	200-450	
CAMBRIAN	Blacksmith Dolomite	500-1200	limestone in places
CAMBRIAN	Ute Formation	400-750	
CAMBRIAN	Langston Dolomite	0-170	
CAMBRIAN / PROTEROZOIC (Brigham Group)	Geertsen Canyon Quartzite	3400-4400	
PROTEROZOIC (Brigham Group)	Browns Hole Fm: Quartzite mbr	100-300	
PROTEROZOIC (Brigham Group)	Browns Hole Fm: Volcanic mbr	0-450	590 m.y. basalt-andesite
PROTEROZOIC (Brigham Group)	Mutual Fm	400-2000	purple to pink cross bedded some feldspar
PROTEROZOIC (Brigham Group)	Inkom Formation	20-450	tuffaceous shale
PROTEROZOIC (Brigham Group)	Caddy Canyon Quartzite	1500-2000	
PROTEROZOIC	Papoose Creek Fm	0-1500	olive siltstone
PROTEROZOIC	Kelley Canyon Fm	600-2000	argillitic, includes chloritic dikes
PROTEROZOIC	Maple Canyon Fm: Conglomerate mbr	100-500	
PROTEROZOIC	Maple Canyon Fm: Green arkose mbr	500-1000	thin limestone locally
PROTEROZOIC	Maple Canyon Fm: Argillite mbr	500	olive-gray
PROTEROZOIC	Fm of Perry Canyon: Graywacke-siltstone member	0-1500	
PROTEROZOIC	Fm of Perry Canyon: Diamictite mbr	0-400	glacial
PROTEROZOIC	Facer Formation	0-2500	varicolored qtzt & schist; 1.6 b.y. Rb-Sr; WILLARD THRUST
A	Farmington Canyon Complex	gneiss	metamorphosed 2.6 b.y.

MAP REFERENCES - Crittenden, 1972; Sorensen and Crittenden, 1976, 1979; Mullens, 1969; SOILS - U.S. Dept. Agriculture, 1980; NORWOOD TUFF - Evernden et al, 1964; MISSISSIPPIAN - Sandberg & Gutschick, 1986; CAMBRIAN -Rigo, 1968; PROTEROZOIC - Crittenden et al, 1971, 1983; Crittenden & Sorensen, 1980; Crittenden & Wallace, 1973; ARCHEAN - Bryant, 1984.

Chart

23 OGDEN - FARMINGTON CANYON

Age	Formation	FEET	Remarks
PLEIST - RECENT	Alluvium, colluvium, fan, landslide, glacial, and Lake Bonneville deposits	0-100?	
MIO - PLIO	Unnamed gravel southeast of Bountiful (Bryant, 1984)	1300	
EOCENE	Norwood Tuff	1100-4000	37.5 m.y. K-Ar late Eocene fauna tuff, tuffaceous ss, cg and clay
PALEOCENE - EOCENE	Wasatch Fm	1300-4000	reddish ss slts, cg, clay; late Paleocene pollen near base
MISS	Humbug Fm	800	
MISS	Deseret Ls	250	
MISS	Lodgepole Ls	400- 500	corals, brachiopods
DEV	Beirdneau Ss	250-300	
DEV	Hyrum Dolomite	0-350	
DEV	Water Canyon Fm	0-90	
O	Fish Haven Dolomite	160-270	
O	Garden City Fm	200-300	
CAMBRIAN	St. Charles Fm	400-650	
CAMBRIAN	Worm Ck Qtzt M	20	linguloid brachiopods
CAMBRIAN	Nounan Dolomite	300-700	
CAMBRIAN	Calls Ft Mbr Bloomington	0- 160	intraformational cg
CAMBRIAN	Maxfield Ls	350-850	boundstone
CAMBRIAN	Ophir Fm	450-650	*Elrathia* *Ehmaniella* *Glossopleura*
CAMBRIAN	Tintic Quartzite	1100-1500	
EARLY PROT	Farmington Canyon Complex: quartz monzonite gneiss - 1790 m.y. Rb-Sr & U-Pb date. Material melted from upper mantle		1790 m.y.
ARCHEAN	Farmington Canyon Complex: migmatite-grades to quartz monzonite and other gneissic rocks		2600-3000 m.y. is age of protoliths which were sandstone, shale, basalt, and, perhaps, siliceous igneous rock
ARCHEAN	Farmington Canyon Complex: schist, gneiss quartzite and some amphibolite		

MAP REFERENCES - Bryant, 1984; Sorensen & Crittenden, 1972; Crittenden & Sorensen, 1985; PRECAMBRIAN - Hedge et al, 1983; CAMBRIAN - Rigo, 1968; NORWOOD TUFF - Nelson, 1972; SOILS -U.S. Dept. Agriculture, 1968, 1980; STRUCTURE - Schirmer, 1985; GUIDEBOOK - Chidsey et al, 1985.

Chart

24 WEBER CANYON NEAR DEVILS SLIDE

Age	Formation	FEET	Remarks
Q	Alluvium, lake beds	0- 500	
P	Unnamed conglomerate	0-1000	
PALEO - EOCENE	Norwood Tuff	5000	gray tuffaceous ss; *Protoreodon*; 37.5 m.y. K-Ar
PALEO - EOCENE	Wasatch Fm	4500	Knight Fm of some authors see Mann, 1974
CRETACEOUS	Evanston Fm	80-600	early Maastrichtian - late Campanian palynomorphs
CRETACEOUS	Echo Canyon ? Conglomerate	1400-3100	Echo Canyon and later beds mostly dip less than 20°; angular unconformity
CRETACEOUS	Henefer Fm	2500	Henefer dips steep to vertical
CRETACEOUS	Frontier Formation	7500	early Coniacian to Cenomainian marine and non-marine palynomorphs
CRETACEOUS	Kelvin Fm	2200	
JURASSIC	Stump Ss	250	
JURASSIC	Preuss Ss	400-800	salt
JURASSIC	Twin Creek Ls: Giraffe Creek M	110	
JURASSIC	Twin Creek Ls: Leeds Creek Mbr	1290	
JURASSIC	Twin Creek Ls: Watton Canyon M	380	
JURASSIC	Twin Creek Ls: Boundary Ridge M	100	
JURASSIC	Twin Creek Ls: Rich Member	540	
JURASSIC	Twin Creek Ls: Sliderock Member	100	
JURASSIC	Twin Creek Ls: Gypsum Spring M	210	
JURASSIC	Nugget Ss	1300	
TRIASSIC	Ankareh Fm	1550	Chinle equiv
TRIASSIC	Thaynes Fm	2600	
TRIASSIC	Woodside Shale	800	
TRIASSIC	Dinwoody Fm	500	
PERM	Franson Mbr Park City	242	
PERM	Meade Pk M Phosphoria	303	
PERM	Grandeur Mbr Park City	476	
PENN	Weber Sandstone	2500-3000	
PENN	Morgan Ss	0-1000	
PENN	Round Valley Ls	400	
MISS	Doughnut Fm	700	
MISS	Humbug Fm	700-900	
MISS	Deseret Ls	500	Delle Phosphatic Mbr
MISS	Lodgepole Ls	800	fossiliferous
D	Beirdneau Ss	230	- Three Forks
D	Hyrum Dolomite	800-900	*Amphipora* unconformity
CAMB	Maxfield Ls	700-1400	
CAMB	Ophir Fm	200-400	
CAMB	Tintic Quartzite	800-1500	*Skolithus*
A	Farmington Canyon Complex		

MAP REFERENCES - Mullens & Laraway, 1964, 1973; Mullens & Cole, 1972; SOILS - U.S. Dept. Agriculture, 1980; WATER - Gates et al, 1984; NORWOOD TUFF - Nelson, 1971, 1982; CRETACEOUS - EOCENE - Mullens, 1971; Nichols & Warner, 1978; Ryer, 1976; Mann, 1974, Crawford, 1979; Hale, 1960; Eardley, 1944; JURASSIC - Stokes, 1959; Imlay, 1967; PERMIAN - Cheney et al, 1953; CARBONIFEROUS - Welsh & Bissell, 1979.

Chart
25 LAKESIDE & GRASSY MOUNTAINS

Period	Unit	FEET	Remarks
Q	Alluvium & Lake Bonneville seds.	0-300	
MIO-P	Valley fill	0-1000	
MIO-P	Salt Lake Formation	0- 160	
E	Andesites of Grayback Hills	0-500	38 m.y.
₸	Thaynes Formation	400	*Prohungarites*
PERMIAN	Gerster Limestone	260	
PERMIAN	Rex Chert Member	1640	
PERMIAN	Meade Peak ? Ls Mbr	230	
PERMIAN	Grandeur Mbr, Park City Fm	1000	cherty ls & dolomite
PERMIAN	Unnamed formation	3350	top not exposed; cherty
PERMIAN	Oquirrh Group — Map unit 6 (Leonardian)	1300	*Schwagerina*
PERMIAN	Oquirrh Group — Map units 3-5 (Wolfcampian)	5700	*Schwagerina* *Pseudofusulina* *Pseudoschwagerina*
PENN	Oquirrh Group — Map units 1-2 (Virgilian, Desmoinesian, and Atokan)	3600	*Triticites* *Pseudofusulinella* *Pseudofusulina*
	Interval covered or faulted out		
MISSISSIPPIAN	Manning Canyon Shale	500 +	poorly exposed
MISSISSIPPIAN	Great Blue Limestone	4080	
MISSISSIPPIAN	Humbug Formation	1550	
MISSISSIPPIAN	Deseret Limestone	370	
MISSISSIPPIAN	Delle Phosphatic Mbr	200	
MISSISSIPPIAN	Lodgepole Limestone	450	
DEV	Pinyon Peak Fm	160	Fitchville equiv
DEV	Beirdneau Fm	1000	
DEV	Hyrum Dolomite	1000	Simonson on maps
DEV	Water Canyon Fm	340	
SIL	Laketown Dolomite	1020	
ORDO	Fish Haven Dolomite	220	
ORDO	Swan Peak Quartzite	0-230	Mid-Ordovician unconformities
ORDO	Kanosh Shale	0-150	
ORDO	Garden City Limestone	1620	intraformational ls cg
CAMBRIAN	St. Charles Formation	1400	hiatus
CAMBRIAN	Worm Creek Qtzt M	40-180	sandy dolomite
CAMBRIAN	Nounan Formation	1900	
CAMBRIAN	Upper shale member	170	ls nodules in green sh
CAMBRIAN	"Marjum" Formation	980	
CAMBRIAN	Wheeler Formation	440	
CAMBRIAN	Swasey Limestone	600	*Glyphaspis* coquina
CAMBRIAN	Whirlwind Formation	70	*Ehmaniella*
CAMBRIAN	Dome Limestone	250	silty limestone
CAMBRIAN	Chisholm Shale	200	*Glossopleura*
CAMBRIAN	Howell Limestone	300	
CAMBRIAN	Pioche Formation	200	green qtzt, trace fossils
CAMBRIAN	Tintic Quartzite	600+	

MAP WITH TEXT - Doelling, 1964, 1980; Young, 1955; TERTIARY & TRIASSIC - Davies, 1980; PERMIAN - Doelling, 1980; Jordan, 1979; Jordan & Douglass, 1980; DELLE - Sandberg & Gutschick, 1984; DEVONIAN - Sandberg & Dreesen, 1984; SILURIAN - Budge & Sheehan, 1980, ORDOVICIAN - Oaks et al, 1977, Taylor & Repetski, 1985.

Chart
26 CEDAR MOUNTAINS

Period	Unit	FEET	Remarks
Q	Alluvium & Lake Bonneville sediments	0-200	
MI-PLI	Valley fill in Skull Valley	0-5000?	
EO-Φ?	Rhyolite plug	-	
EO-Φ?	Basalt	50-100	
EO-Φ?	Basaltic andesite	0-1000	tuffs & flows
K?-T	Unnamed sandstone	300 +	
K?-T	North Horn ? Formation	800 +	conglomerate with red matrix; MAJOR UNCONFORMITY
PERMIAN	Gerster Limestone	320	
PERMIAN	Plympton Formation	2400	
PERMIAN	Meade Pk Mbr, Phosphoria Fm	230	
PERMIAN	Grandeur Formation	1850	
PERMIAN	Unnamed unit	3950	Pequop-Diamond Creek equivalent; cherty
PERMIAN	Oquirrh Group — Wolfcampian — Unit 5	1950-2750	interfingering of Freeman Peak Fm (sandstone facies) and Curry Peak Fm (siltstone facies)
PERMIAN / PENNSYLVANIAN	Oquirrh Group — Wolfcampian / Virgilian — Unit 4	2760-3000	*Triticites*
PENNSYLVANIAN	Oquirrh Group — Mo / Desmoines — Unit 3	2560-3000 +	*Triticites* *Fusulina* *Fusulinella* *Millerella*
PENNSYLVANIAN	Oquirrh Group — Mor-Atok — Unit 2	715-1400	*Chaetetes*
PENNSYLVANIAN	Oquirrh Group — Mor-Atok — Unit 1	435	many fossils
MISSISSIPPIAN	Manning Canyon Shale	1500-2000	gray to black
MISSISSIPPIAN	Great Blue Limestone	2440 +	corals bryozoans; corals
MISSISSIPPIAN	Woodman Formation	1015 +	

MAP WITH TEXT - Maurer, 1970; QUATERNARY - Hood & Waddell, 1968; PERMIAN - Wardlaw et al, 1979; OQUIRRH - Jordan, 1979; Chamberlain & Clark, 1973.

Chart
27 STANSBURY MOUNTAINS

Period	Group	Formation	FEET	Notes
Q		Alluvial, glacial & Lake Bonneville sediments	0 -300	
M-P		Salt Lake Formation	0-1300	
M-P		Basalt flows & dikes	0-130	
E - Φ		Andesite flows, breccias tuff	0-1800	
E - Φ		post-thrusting conglomerate	0-400	red matrix
TRIAS		Thaynes Limestone	1100	*Meekoceras*
TRIAS		Woodside Shale	100	
PERMIAN		Park City / Phosphoria Fm	600	
PERMIAN		Kirkman Limestone & Diamond Creek Ss undivided	500 +	
PERMIAN	Oquirrh Group	Wolfcampian Oquirrh	3000 ±	*Pseudofusulinella* *Schwagerina*
PENNSYLVANIAN	Oquirrh Group	Virgilian & Missourian Oquirrh - Bingham Mine Formation equivalent	8700 ±	*Triticites* *Triticites*
PENNSYLVANIAN	Oquirrh Group	Desmoinesian & Atukan Oquirrh - Butterfield Peaks Formation equivalent	6000 ±	*Wedekindellina* *Fusulinella*
PENNSYLVANIAN	Oquirrh Group	West Canyon Ls	0-840	*Millerella* FAULTED CONTACT
MISSISSIPPIAN		Manning Canyon Shale	200-1600	*Dictyoclostus*
MISSISSIPPIAN		Great Blue Limestone	980-1300	*Ekvasophyllum*
MISSISSIPPIAN		Humbug Formation	710-900	*Siphonodendron*
MISSISSIPPIAN		Deseret Formation	650-750	Delle Phosphatic Mbr
MISSISSIPPIAN		Gardison Limestone	700-1100	
MISSISSIPPIAN		Fitchville Formation	130-650	conodonts
DEV		Pinyon Peak Limestone	0-215	
DEV		Stansbury Formation	0-1700	derived from local uplift
DEV		Simonson ? Dolomite	0-230	
DEV		Sevy Dolomite	0-80	
S		Laketown Dolomite	0-660	*Pentamerus*
ORD		Fish Haven Dolomite	0-260	Ordovician unconformity
ORD		Kanosh Shale	0-270	*Didymograptus*
ORD		Garden City Limestone	1100-1300	*Phyllograptus* *Symphysurina*
CAMBRIAN		Ajax Dolomite	750-910	
CAMBRIAN		Corset Spring Shale	20-150	*Housia varro*
CAMBRIAN		Opex Formation	450-500	
CAMBRIAN		Cole Canyon & Bluebird Dolo	320-450	
CAMBRIAN		Bowman-Herkimer-Dagmar Fms	100-600	
CAMBRIAN		Teutonic Limestone	800-1100	*Elrathia*
CAMBRIAN		Ophir Formation	800-1200	*Ehmaniella* *Glossopleura*
CAMBRIAN		Pioche Formation	300	
CAMBRIAN		Tintic Quartzite	4200	

MAPS WITH TEXT - Rigby, 1958; Tooker & Roberts, 1971; Armin & Moore, 1981; Sorensen, 1982; VOLCANIC ROCKS - Davis, 1959; Hogg, 1972, TRIASSIC - Jordan & Allmendinger, 1979; OQUIRRH -Wright, 1961; Jordan, 1979; Stevens et al, 1983; DESERET - Sandberg & Gutschick, 1984; DEVONIAN - Rigby, 1959; Sandberg & Gutschick, 1979.

Chart
28 NO. OQUIRRH MTS - BINGHAM

Period	Series	Group	Formation	FEET	Notes
Q			Alluvium & Lake Bonneville deposits	0-1000	0.6 m.y. Pearlette O ash 0.7 m.y. Bishop ash
Q			Harkers Alluvium	0-250	
M-P			Valley fill in Tooele Valley	0-12,000	
Φ			Shaggy Peak rhyolite plug		33 m.y.
EOC			Quartz latite porphyry dikes		37 m.y. Traverse volc's
EOC			Bingham & Last Chance qtz monzonite		38-39 m.y. stocks
EOC			post-trusting conglomerate	0-500?	no volcanic clasts
PALEOZOIC STRATA WERE FOLDED AND THRUSTED DURING JURASSIC - CRETACEOUS SEVIER OROGENIC EVENTS					
PERMIAN			Grandeur Limestone, Park City Group	760	
PERMIAN			Diamond Creek Sandstone	2000?	
PERMIAN			Kirkman Limestone	400	
PERMIAN	WOLFCAMPIAN	Oquirrh Group	Freeman Peak Formation (Clinker-rings when struck)	2400	Kesler Canyon Fm of Tooker & Roberts, 1970, is a faulted combination of parts of the Freeman Peak, Bingham Mine, & upper Butterfield Peaks Fms (Welsh, 1985) *Schwagerina* *Pseudoschwagerina*
PERMIAN	WOLFCAMPIAN	Oquirrh Group	Curry Peak Formation	2450	calcareous siltstone & very fine sandstone *Schwagerina* *Triticites*
PENNSYLVANIAN	MISSOURIAN - VIRGILIAN	Oquirrh Group	Bingham Mine Formation	7300	*Triticites* *Linoproductus* *Syringopora*
PENNSYLVANIAN	MISSOURIAN - VIRGILIAN	Oquirrh Group	Commercial LsM	(100-200)	
PENNSYLVANIAN	MISSOURIAN - VIRGILIAN	Oquirrh Group	Jordan Ls Mbr	(100-300)	*Eowaeringella*
PENNSYLVANIAN	ATOKAN - DESMOINESIAN	Oquirrh Group	Butterfield Peaks Formation	9100	Erda Fm of Tooker & Roberts, 1970, is junior synonym for part of the Butterfield Peaks Formation *Chaetetes* *Millerella* *Beedeina*
PENNSYLVANIAN	MOR	Oquirrh Group	West Canyon Limestone	1060-1450	Green Ravine Fm of Tooker & Roberts is junior synonym for West Canyon Ls

MAP WITH TEXT - Swensen, 1975; Tooker & Roberts, 1971; QUATERNARY - Eardley et gl, 1973; Nash & Smith, 1977; VALLEY FILL - Cook et al, 1966; TERTIARY - Moore, 1973; KIRKMAN - Schuerer, 1979; OQUIRRH - Welsh et al, 1961; Welsh, 1983; Welsh letter, 1985; Jordan, 1979; Tooker & Roberts, 1970.

Chart
29 DUGWAY RANGE

Period	Unit		FEET	Remarks
Q	Alluvium, L. Bonneville seds		0-300	Old River Bed 14,000 y.b.p.
T	Valley fill in Dugway V.		0-1000	subsurface
T	Topaz Mtn Rhyolite		0-600	6-8 m.y.
T	Rhyodacite		0-300	36 m.y.?
P	Unnamed Lower Permian ls		1200	Permian on Devonian in Black Hills; see Hintze 1980 c.
MISS	Ochre Mountain Limestone		700	*Inflatia*
MISS	Woodman Formation	Needle Siltstone Member	740	*Zaphrentis*
MISS	Woodman Formation	Delle Phosph Mbr	50	
MISS	Joana Limestone		40	*Cyrtospirifer*
DEVONIAN	Guilmette Formation °Hanauer, Gilson & Goshoot Fms		2200	Upper 280 ft of sandy Guilmette (Hanauer) equivalent to Crystal Pass Fm of SE Nevada; spaghetti stromatoporoids
DEVONIAN	Simonson Dolomite °Englemann Fm		1080	stromatoporoids; *Stringocephalus*
DEVONIAN	Sevy Dolomite		1100?	
SIL	Laketown Dolomite	*Thursday Dol	330?	*Halysites*
SIL	Laketown Dolomite	*Lost Sheep Dol	270	*Pentamerus*
SIL	Laketown Dolomite	*Harrisite Dol	150	
SIL	Laketown Dolomite	*Bell Hill Dol	340	*Virgiana*
ORDOVICIAN	Ely Springs Dolomite	Floride Mbr	100	
ORDOVICIAN	Ely Springs Dolomite	Lower mbr	230	*Palaeophyllum*
ORDOVICIAN	Swan Peak Fm	Quartzite mbr	205	
ORDOVICIAN	Swan Peak Fm	Kanosh Sh M	270	*Orthambonites*
ORDOVICIAN	Pogonip Group	Wah Wah-Juab	380	*Lachnostoma*
ORDOVICIAN	Pogonip Group	Fillmore Fm	1200	*Protopliomerella*
ORDOVICIAN	Pogonip Group	House Ls	150	*Symphysurina*
CAMBRIAN	°Dugway Ridge Fm	Notch Peak Formation	880	*Eurekia*
CAMBRIAN	°Fera Ls	Orr Fm: Sneakover Mbr	200	*Eoorthis*
CAMBRIAN	°Fera Ls	Orr Fm: Johns W-Corset S	80	30 ft green shale
CAMBRIAN	Straight Canyon Fm	Orr Fm: Candland Sh & Big Horse Ls Ms	500	phosphatic brachiopods; *Crepicephalus*; *Tricrepicephalus*; pinkish sandy ls
CAMBRIAN	Lamb Dolomite		890	gray, sugary with pisolitic algae
CAMBRIAN	°Fandangle Ls	Tripp Ls: Fish Sp Mbr	110	*Eldoradia*
CAMBRIAN	°Fandangle Ls	Tripp Ls: Lower member	480	white boundstone
CAMBRIAN	°Fandangle Ls	Pierson Cove Formation	1140	*Modocia*
CAMBRIAN	°Trailer Ls	Wheeler Fm	260	*Elrathia*
CAMBRIAN	°Trailer Ls	Swasey Ls	130	
CAMBRIAN	°Shadscale Fm	Whirlwind Dome Chis	310	*Ehmaniella*; *Glossopleura*
CAMBRIAN	°Shadscale Fm	Howell Limestone	200	
CAMBRIAN	°Shadscale Fm	Pioche Fm: Tatow Mbr	150	
CAMBRIAN	°Bus-Cab	Pioche Fm: Lower mbr	300	
CAMBRIAN	Prospect Mountain Quartzite		5700	white, tan, pink
PROTEROZOIC	Mutual ? Formation		1600	purple
PROTEROZOIC	Inkom Formation		600	
PROTEROZOIC	Caddy Canyon Quartzite		4200+	gray
PROTEROZOIC	Gneissic granite & granitic pegmatite			NOT IN CONTACT; Granite Peak area

*stratigraphic names used by Staatz & Carr

MAPS - Staatz & Carr, 1964; Staatz, 1972; Hintze, 1980c; LAKE BONNEVILLE - Oviatt, 1984; Currey et al, 1983, 1984; VALLEY FILL - Stephens & Sumsion, 1978; VOLCANIC ROCKS - Lindsey, 1979; MISSISSIPPIAN & DEVONIAN - Biller, 1976; Sandberg & Gutschick, 1980; Sandberg, 1985, pers. comm.; CAMBRIAN - Hintze & Robison, 1975; Hintze & Palmer, 1976; PROTEROZOIC - Christie-Blick, 1982b; GNEISS - Fowkes, 1964; Moore & Sorensen, 1979.

Chart
30 SHEEPROCK MOUNTAINS

Period	Unit			FEET	Remarks
Q	Alluvium, L. Bonneville seds.			0-300	Bishop Tuff 700,000 y.b.p.
T	Valley fill in Rush Valley			0-1000	
T	Sheeprock Mts granitic pluton				21 m.y. Rb-Sr
K	Red conglomerate			0-200	MAJOR UNCONFORMITY
PP	Oquirrh Fm			0-10,000	faulted, incomplete
MISSISSIPPIAN	Great Blue Ls	Upper limestone member		1400	*Rhipidomella*; *Caninia*
MISSISSIPPIAN	Great Blue Ls	Chiulos Sh Member		1800	*Calamites*
MISSISSIPPIAN	Great Blue Ls	Lower limestone member		900	*Faberophyllum*
MISSISSIPPIAN	Deseret-Humbug Fms undivided			1150	*Fenestella*; 80' Delle Phosphatic Mbr
MISSISSIPPIAN	Gardison & Fitchville Fms			400- 720*	*in West Tintic Mts
DEV	Pinyon Peak Limestone			250	
DEV	Guilmette Formation			320	includes Victoria Ss
DEV	Simonson Dolomite			500	stromatoporoids
DEV	Sevy Dolomite			350-1000	ant-egg sand grains
S	Laketown Dolomite			900	*Dourillina*
ORD	Ely Springs Dolomite			700	*Catenipora*
ORD	Swan Peak Quartzite			350-450	
ORD	Kanosh Shale			250	*Didymograptus*
ORD	Pogonip Group	Juab Ls, Wah Wah Ls, Fillmore-House		1700	*Hesperonomia*; intraformational cg
CAMBRIAN	Ajax Dolomite			1000	
CAMBRIAN	Opex Formation			940	*Elvinia*
CAMBRIAN	Cole Canyon Dolomite			900	*Tricrepicephalus*
CAMBRIAN	Marjum Formation			1170	
CAMBRIAN	Wheeler Formation			420	*Elrathia*
CAMBRIAN	Swasey-Whirlwind-Dome Fms			330	*Ehmaniella*
CAMBRIAN	Chisholm Shale			270-380	*Glossopleura*
CAMBRIAN	Howell Limestone			150-200	*Albertella* zone
CAMBRIAN	Pioche Shale			430	
CAMBRIAN	Brigham Group	Prospect Mountain (Tintic) Quartzite		2600-4500	light gray cross-bedded some feldspar
PROTEROZOIC	Brigham Group	Mutual ? Fm		1400-1650	maroon, pink, purple, feldspathic
PROTEROZOIC	Brigham Group	Inkom Fm		100-475	green, maroon, gray
PROTEROZOIC	Brigham Group	Caddy Canyon Quartzite		2300-6500 +	white, light gray or tan
PROTEROZOIC	Sheeprock Group	Kelley Canyon Fm		500-1900	olive, tan, gray shale & siltstone
PROTEROZOIC	Sheeprock Group	Dutch Peak Fm	7- slate & argillite	0-800	black
PROTEROZOIC	Sheeprock Group	Dutch Peak Fm	6- quartzite	0-1000	light gray
PROTEROZOIC	Sheeprock Group	Dutch Peak Fm	5- graywacke & diamictite	1000-3000	olive, brown, or black, phyllitic matrix
PROTEROZOIC	Sheeprock Group	Dutch Peak Fm	4- diamictite	0-3000	
PROTEROZOIC	Sheeprock Group	Dutch Peak Fm	3- conglomerate & diamictite	0-3100	dolomite, quartzite & granite boulders in a phyllitic matrix
PROTEROZOIC	Sheeprock Group	Dutch Peak Fm	2- grit & cg	0-1300	maroon
PROTEROZOIC	Sheeprock Group	Dutch Peak Fm	1- qtzt, siltstone	0-160	
PROTEROZOIC	Sheeprock Group	Otts Canyon Fm	Upper member	650-3200	diabase sills in quartzite
PROTEROZOIC	Sheeprock Group	Otts Canyon Fm	Middle member	0-1600	diamictite
PROTEROZOIC	Sheeprock Group	Otts Canyon Fm	Lower member	2300	slate & phyllite

MAPS WITH TEXT - Cohenour, 1959; Christie-Blick, 1982, 1983; Pampeyan, 1986; RUSH VALLEY - Hood et al, 1969; GRANITE DATE - Armstrong, 1970; MISSISSIPPIAN & DEVONIAN - Sandberg & Gutschick, 1984; Sandberg, pers. comm., 1985; CAMBRIAN - Robison, 1960; Hintze & Robison, 1975.

See Christie-Blick, 1983, for correlation of Precambrian rocks in the Simpson Mountains with those in the Sheeprock Mountains.

FIGURE 130 — Wasatch Mountains at the south end of Salt Lake Valley. Big Cottonwood Creek at left center cuts through its own Lake Bonneville deltaic deposits to form the modern tree-covered flood plain. Little Cottonwood Canyon at photo center. Triangular faceted spurs along mountain front document Wasatch fault movements.

FIGURE 131 — Nearly vertical strata of the Pennsylvanian-Permian Oquirrh Group at the north end of the Oquirrh Mountains. Lake Bonneville shorelines are prominent here. See Figure 108 and Chart 28.

FIGURE 132 — Looking down glaciated Little Cottonwood Canyon across Salt Lake Valley. Waste dumps around Bingham pit visible on distant Oquirrh Mountains. Bedded strata on right ridge are quartzites and slates of late Proterozoic Big Cottonwood Formation, here intruded by quartz monzonite of the Little Cottonwood stock (24 m.y.), which forms ragged outcrops on the canyon walls.

FIGURE 133 — Bingham pit in foreground mines disseminated copper in both Oquirrh Group sedimentary rocks and the Bingham quartz monzonite stock. Wasatch Mountains along horizon rise abruptly out of valley because of uplift along the Wasatch fault during the last 10 million years.

Chart 31 SO. OQUIRRH MTS. & SOUTH MTN.

Period	Subdivision	Group	Unit	FEET	Notes
Q			Alluvium & Lake Bonneville sediments	0-1000	
M-P			Valley fill in Cedar Valley	0-2000	
OLIG			Tickville rhyolite flows	0-300	31 m.y.
OLIG			Rhyolite plug near Mercur		32 m.y.
OLIG			Latite flows & lahars	500-1000	
OLIG			Nepheline basalt	0-100	
E			post-thrusting conglomerate	0-200	
PERMIAN	LEON		Park City Group	200	Phosphoria Fm
PERMIAN	LEON		Diamond Creek Ss	2300	
PERMIAN			Kirkman Limestone	100	
PERMIAN	WOLFCAMPIAN	Oquirrh Group	Freeman Peak Formation	1900	
PERMIAN	WOLFCAMPIAN	Oquirrh Group	Curry Peak Formation	3100	
PENNSYLVANIAN	MISSOURI-VIRGIL 'N	Oquirrh Group	Bingham Mine Formation	6500	*Triticites*
PENNSYLVANIAN	MISSOURI-VIRGIL 'N	Oquirrh Group	Commercial Ls	(20-200)	
PENNSYLVANIAN	MISSOURI-VIRGIL 'N	Oquirrh Group	Jordan Ls Mbr	(50-300)	
PENNSYLVANIAN	ATOKAN-DESMOINESIAN	Oquirrh Group	Butterfield Peaks Formation (Bissell's 1959, Pole Canyon, Lewiston Peak, Cedar Fort & Meadow Canyon members are all part of this formation)	8500	*Beedeina*, *Wedekindellina*, cyclical cherty ls, sandy limestone, calcareous sandstone, *Wedekindellina*, *Fusulinella*, *Millerella*
PENNSYLVANIAN	MOR	Oquirrh Group	West Canyon Limestone	900-1200	bioclastic limestone with Morrowan fossils, conodonts & brachiopods
MISSISSIPPIAN			Manning Canyon Shale	1330-1610	abundant land plants and shallow-water marine invertebrate fossils
MISSISSIPPIAN		Great Blue Limestone	Upper limestone member	1950	
MISSISSIPPIAN		Great Blue Limestone	Long Trail Sh. M.	90	
MISSISSIPPIAN		Great Blue Limestone	Lower ls mbr	550	
MISSISSIPPIAN			Humbug Formation	640	
MISSISSIPPIAN			Deseret Limestone	580-730	Delle Phosphatic Mbr
MISSISSIPPIAN			Gardison Limestone	720	cherty corals & brachiopods
MISSISSIPPIAN			Fitchville Dolomite	180-340	
D			Stansbury Fm	10-200	35 ft Pinyon Peak Ls
CAMBRIAN			Opex Dolomite	180	
CAMBRIAN			Cole Canyon Dolomite	210	boundstone
CAMBRIAN			Bluebird Dolomite	310	
CAMBRIAN			Bowman Limestone	250	
CAMBRIAN			Herkimer Limestone	470	
CAMBRIAN			Teutonic Limestone	280	
CAMBRIAN			Ophir Formation	300	*Glossopleura*
CAMBRIAN			Pioche Formation	300	phyllitic sh & qtzt
CAMBRIAN			Tintic Quartzite	subsurface	

MAPS WITH TEXT - Gilluly, 1932; Bissell et al, 1959; Moore, 1973; VALLEY FILL - Hood et al, 1969; Larsen, 1960; OLIGOCENE - Moore, 1973; UPPER PALEOZOIC - Douglass et al, 1974; Bissell, 1959; Welsh, 1983; MANNING CANYON SHALE - Moyle, 1959; Tidwell et al, 1974; Webster, 1984; DESERET PHOSPHATE - Sandberg & Gutschick, 1984; LOWER PALEOZOIC - Rigby, 1959; MERCUR GOLD MINE - Kornze, 1984.

Chart 32 EAST TINTIC MINING DISTRICT

Period	Group	Formation	Unit	FEET	Notes
Q			Alluvium & Lake Bonneville seds	0-200	
PLI			Valley fill-Goshen Valley	0-1000	
PLI			Salt Lake ? Fm	0-2000?	Huckleberry Ridge ash 2 m.y.
MIO			Silver Shield Qtz Latite	0-125	17 m.y.
MIO			Pinyon Creek Cg	0-1000	
OLIGOCENE			Silver City monzonite intrusion		31 m.y.
OLIGOCENE	Laguna Springs Volcanic Group		Tintic Delmar Latite	0-400	32 m.y. flows
OLIGOCENE	Laguna Springs Volcanic Group		Pinyon Queen Latite	0-1100	tuff
OLIGOCENE	Laguna Springs Volcanic Group		North Standard Lat	0-600	flows tuff
OLIGOCENE			Sunrise Peak Monzonite Porphyry intrusion		
OLIGOCENE	Tintic Mtn Volcanic Group		Big Canyon Latite	0-200	flows
OLIGOCENE	Tintic Mtn Volcanic Group		Latite Ridge Latite	0-1500	tuff
OLIGOCENE	Tintic Mtn Volcanic Group	Copper-opolis Latite	Upper flow & tuff brec	0-1500	
OLIGOCENE	Tintic Mtn Volcanic Group	Copper-opolis Latite	Middle flow & agglomerate	0-5000	
OLIGOCENE	Tintic Mtn Volcanic Group	Copper-opolis Latite	Lower flow	0-3000	
OLIGOCENE	Tintic Mtn Volcanic Group	Copper-opolis Latite	Lower tuff	0-2000	
OLIGOCENE	Tintic Mtn Volcanic Group	Copper-opolis Latite	Lower agglomerate	0-3500	
OLIGOCENE			Packard Quartz Latite	0-3000	Swansea Qtz Monzonite "intrusion" is altered Packard Qtz Latite 33 m.y.
OLIGOCENE			Apex Conglomerate	0-500	MAJOR UNCONFORMITY
MISSISSIPPIAN		Great Blue Formation	Poker Knoll Limestone Mbr	600	
MISSISSIPPIAN		Great Blue Formation	Chiulos Member	900	
MISSISSIPPIAN		Great Blue Formation	Paymaster Member	620	*Ekvasophyllum*
MISSISSIPPIAN		Great Blue Formation	Topliff Limestone Mbr	460	*Endothyra*
MISSISSIPPIAN			Humbug Formation	650	
MISSISSIPPIAN		Deseret Limestone	Uncle Joe Member	550	*Rhipidomella*
MISSISSIPPIAN		Deseret Limestone	Tetro Member	475	
MISSISSIPPIAN		Deseret Limestone	Delle Member	5-150	phosphatic shale
MISSISSIPPIAN			Gardison Limestone	450-550	
MISSISSIPPIAN			Fitchville Formation	300	
DEV			Pinyon Peak Limestone	270	
DEV			Guilmette Formation	600	includes Victoria Fm & upper Bluebell as mapped
S			Bluebell Dolomite	300	*Virgiana* *Streptelasma*
ORD			Fish Haven Dolomite	200-350	
ORD			Opohonga Limestone	300-900	*Pseudocybele* intraformational cg *Symphysurina*
CAMBRIAN			Ajax Dolomite	550-660	*Eoorthis*
CAMBRIAN			Opex Formation	140-350	*Elvinia* zone
CAMBRIAN			Cole Canyon Dolomite	830-1300	*Eldoradia*
CAMBRIAN			Bluebird Dolomite	150-220	
CAMBRIAN			Herkimer Limestone	350-430	
CAMBRIAN			Dagmar Dolomite	65-200	white boundstone
CAMBRIAN			Teutonic Limestone	390-420	*Kootenia* *Ehmaniella*
CAMBRIAN			Ophir Formation	300-430	*Glossopleura*
CAMBRIAN			Tintic Quartzite	2300-3200	basalt flow 1500 ft below top
PRO			Big Cottonwood Formation	2600+	mostly phyllitic slate, some quartzite

STRATIGRAPHIC SUMMARY - Morris & Lovering, 1961, 1979; MAPS - Morris, 1964a, b, 1975; Jensen, 1985; AJAX AGE - Taylor & Repetski, 1985; DEVONIAN - Greenhalgh, 1980; Biller, 1976; Sandberg & Poole, 1977; MISSISSIPPIAN - Sandberg & Gutschick 1984.

Chart
33 WASATCH RANGE NEAR SLC

Period	Unit	FEET	Notes
Q	Alluvial, glacial and Lake Bonneville deposits	600-4800	valley-fill
TERTIARY	Hooper Canyon Fm	0-50	
TERTIARY	Conglomerate No. 2	0-1500	"Wasatch" of some maps
TERTIARY	Little Cottonwood qtz monzonite	stock	24-31 m.y.
TERTIARY	Alta granodiorite	stock	32-33 m.y.
TERTIARY	Volc breccia - north SLC	0-1000	35-38 m.y. Clayton Peak stock
TERTIARY	Tuffaceous deposits	0-1000	37-45 m.y.
TERTIARY	Conglomerate No. 1	0-1200	Wasatch Cg; MAJOR UNCONFORMITY
CRETACEOUS	Frontier Formation	7000-8000	not all shown; coal
CRETACEOUS	Kelvin Fm	1600	
CRETACEOUS	Parleys Ls Mbr	100	white, nodular
JURASSIC	Preuss Ss	1000	pale red ss, sh
JURASSIC	Twin Creek Ls: Giraffe Creek Mbr	200	greenish-gray marine beds
JURASSIC	Twin Creek Ls: Leeds Creek Member	1520	*Gryphaea*
JURASSIC	Twin Creek Ls: Watton Canyon Mbr	350	
JURASSIC	Twin Creek Ls: Boundary Ridge Mbr	100	
JURASSIC	Twin Creek Ls: Rich Member	390	cement rock
JURASSIC	Twin Creek Ls: Sliderock Member	150	
JURASSIC	Twin Creek Ls: Gypsum Spring Mbr	140	orangish-pink
JURASSIC–TRIASSIC	Nugget Ss	1300	
TRIASSIC	Ankareh Fm: upper member	700	reddish brown
TRIASSIC	Ankareh Fm: Gartra Grit Mbr	100	
TRIASSIC	Ankareh Fm: Mahogany Mbr	900	reddish brown
TRIASSIC	Thaynes Fm	1800-2300	conodonts; *Meekoceras*
TRIASSIC	Woodside Shale	400	maroon sh, silts
P	Park City Fm	900-1700	phosphorite
PENN	Weber Quartzite	1200-1500	
PENN	Round Valley Ls	600-1000	pink chert
MISS	Doughnut Fm	400-900	dark sh, ls
MISS	Humbug Fm	800	
MISS	Deseret Ls	500-900	Delle Phosphatic M
MISS	Gardison Ls	400-600	
D	Fitchville Fm - south end	0-120	at Beck's Spur
D	Stansbury-Pinyon Pk Fms -north	0-700	
CAMB	Maxfield Ls	0-1200	
CAMB	Ophir Fm	400 -600	*Glossopleura*
CAMB	Tintic Quartzite	1000-2000	
PROTEROZOIC	Mutual Fm	0-4000	reddish-purple feldspathic quartzite
PROTEROZOIC	Mineral Fork Tillite	0-3000	microfossils
PROTEROZOIC	Big Cottonwood Formation	0-16,000	quartzite and argillitic siltstone; not all shown; cross-bedding; ripple marks; mud cracks; rain-drop prints
A ?	Little Willow Fm - Cottonwood area		protolith - 2.6-3.0 b.y.
A ?	Farmington Canyon Complex - north SLC		metamorphism 1.8 b.y. and 2.6 b.y.

GENERAL - Davis, 1983; Crittenden, 1976; QUATERNARY - Van Horn, 1982; Arnow & Mattick, 1968; Hely et al, 1971; U.S. Dept. Agriculture, 1974; TERTIARY - Van Horn, 1981; Naeser et al, 1983; CRETACEOUS - Crittenden, 1963; Ryer, 1976; JURASSIC - Imlay, 1967; TRIASSIC - Solien, 1979a-b; PERMIAN - Cheney et al, 1953; MISSISSIPPIAN - Crittenden, 1959; DEVONIAN - Gosney, 1982; CAMBRIAN - Lochman-Balk, 1976; PRECAMBRIAN - Bryant, 1984; Crittenden et al, 1971, 1983; Knoll et al, 1981; Christie-Blick, 1982.

Chart
34 COALVILLE - PARK CITY - KAMAS

Period	Unit	FEET	Notes
Q	Alluvial & glacial deposits	0-200	
OLIG	Keetley Volcanics	0-2000	andesite and rhyodacite 34-35 m.y.
OLIG	Ontario, Flagstaff Mtn, Pine Creek Valeo, Glencoe, and Mayflower stocks		granodiorite to quartz monzonite porphyries
PA-EO	Wasatch Fm	0- 2000	
PA-EO	Evanston Fm	0-600	
CRETACEOUS	Echo Canyon Cg	1400-3100	angular unconformity
CRETACEOUS	Henefer Fm	0-2500	
CRETACEOUS	Frontier Formation: Upton Ss Mbr	250-450	*Baculites*
CRETACEOUS	Frontier Formation: Judd Shale Mbr	300	marine shale
CRETACEOUS	Frontier Formation: Grass Creek Mbr	950	*Crassostrea*
CRETACEOUS	Frontier Formation: marine ss	200-230	*Cardium*
CRETACEOUS	Frontier Formation: Dry Hollow Member	1100	
CRETACEOUS	Frontier Formation: Oyster Ridge Ss M	150 -200	
CRETACEOUS	Frontier Formation: marine ss	60-150	*Collignoniceras*
CRETACEOUS	Frontier Formation: Allen Hollow Sh M	600-1400	marine shale
CRETACEOUS	Frontier Formation: Coalville Mbr	230	Wasatch coal seam
CRETACEOUS	Frontier Formation: Chalk Creek Member	2500-3100	nonmarine tan ss and reddish sh
CRETACEOUS	Frontier Formation: Spring Canyon Mbr	320-400	
CRETACEOUS	Frontier Formation: Longwall Ss Mbr	40-250	bony coal
CRETACEOUS	Aspen Shale	100-400	fish scales
CRETACEOUS	Kelvin Formation	3000	
CRETACEOUS	Parleys Ls Mbr	250	white, nodular
JURASSIC	Stump Fm	50-220	
JURASSIC	Preuss Ss	1000	salt
JURASSIC	Twin Creek Ls	1380	
JURASSIC	Nugget Ss	1000-1400	fossil sand dunes
TRIASSIC	Ankareh Fm	1230	
TRIASSIC	Thaynes Fm	1150	conodonts; *Meekoceras*
TRIASSIC	Woodside Shale	600-700	
P	Park City Fm	650	phosphorite
ℙ	Weber Ss	1300	
ℙ	Round Valley Ls	300	
MISS	Doughnut Fm	500	
MISS	Humbug Fm	350	
MISS	Brazer Fm	750	Delle Phosphatic M
MISS	Lodgepole Ls	400	
D	Fitchville Fm	150	
CAM	Maxfield Ls	200-800	
CAM	Ophir Fm	400	
CAM	Tintic Quartzite	600	
PROTEROZOIC	Uinta Mountain Group: Red Pine Shale	0-1800	brown to gray micaceous; 950 m.y. Rb-Sr
PROTEROZOIC	Uinta Mountain Group: unnamed quartzite siltstone and shale units	25,000	brick-red to maroon; oldest beds about 1400 m.y.

QUATERNARY - **Baker**, 1970, U.S. Dept. Agriculture, 1976, 1977; OLIGOCENE - Bromfield et al, 1970, 1971; EOCENE - Mann, 1974; CRETACEOUS - Crawford, 1979; Ryer, 1976; Bryant, 1985; Doelling & Graham, 1972b; JURASSIC - Imlay, 1967; Albers, 1975; TRIASSIC - Clark & Rosser, 1976; PERMIAN -Cheney et al, 1953; PALEOZOIC - Bromfield, 1968; Welsh & **Bissell**, 1979; MISSISSIPPIAN - Sandberg & Gutschick, 1980; DEVONIAN - Spreng, 1979; PRECAMBRIAN - Williams, 1953; Crittenden et al, 1967; Crittenden & Peterman, 1975.

Chart

35 PROVO - UTAH VALLEY - WALLSBURG

Age	Group	Unit	FEET	Notes
Q		Alluvium, colluvium, soil	0-50	0-10,000 yrs old
Q		Lake Bonneville deposits	0-200	10,000-30,000 yrs old
MIO-PLIO		Buried valley fill in Utah Valley includes pre-Bonneville alluvial & lake deposits with some interbeds of volcanic tuff	0-13,000	Gulf Oil-Banks #1 23-8S-2E; 13,000 ft deep, bottomed in beds containing Upper Miocene pollen; 10 m.y. +
WASATCH FAULT ZONE SEPARATES UTAH VALLEY FILL FROM PALEOZOIC BEDROCK OF WASATCH MOUNTAINS SHOWN BELOW				
TR		Woodside Shale	300	dark red
PERMIAN	Park City Group	Franson Fm	800	Gerster Ls equiv
PERMIAN	Park City Group	Meade Pk Tongue	280	Plympton Fm equiv; Phosphoria tongue
PERMIAN	Park City Group	Grandeur Fm	310	
PERMIAN		Diamond Creek Ss	670	
PERMIAN		Kirkman Limestone	310	
PERMIAN	Oquirrh Group	Granger Mountain Formation (Wolfcampian)	10,300	1000' ss; 1200' Curry Pk siltstone facies; *Pseudoschwagerina*; mostly fine ss with some siltstone interbeds; *Schwagerina*; fine grained ss with heavy minerals from Uncompahgre source; 1400' Curry Peak siltstone facies
PENNSYLVANIAN	Oquirrh Group	Wallsburg Ridge Formation (Missourian-Virgilian)	5300	*Triticites*; mostly fine grained ss with fewer limestone interbeds than same interval in Oquirrh Mts
PENNSYLVANIAN	Oquirrh Group	Shingle Mill Ls (Lower Missourian)	1100	*Eowaeringella*; equivalent to Jordan & Commercial Ls at Bingham
PENNSYLVANIAN	Oquirrh Group	Bear Canyon Formation (Atokan-Desmoinesian)	7900	*Fusulina*; *Profusulinella*
PENNSYLVANIAN	Oquirrh Group	Bridal Veil Falls Limestone	1050	Morrowan fossils are abundant - brachiopods corals, bryozoans
PENNSYLVANIAN / MISSISSIPPIAN		Manning Canyon Shale	1650	*Dictyoclostus*; *Cravenoceras*; *Lepidodendron*
MISSISSIPPIAN	Great Blue Ls	Upper limestone member	1800	deep water facies
MISSISSIPPIAN	Great Blue Ls	Long Trail Sh Mbr	300	forms strike valley
MISSISSIPPIAN	Great Blue Ls	Topliff Ls Mbr	700	*Apatognathus*
MISSISSIPPIAN		Humbug Formation	540	*Fenestella*
MISSISSIPPIAN		Deseret Limestone	840	320' Uncle Joe Member; 320' Tetro Member; 200' Delle Phosphatic M
MISSISSIPPIAN		Gardison Limestone	600	*Syringopora*
D		Fitchville Dolomite	50-270	regional unconformity
CAMB		Maxfield Limestone	0-850	white dolo in Rock Cyn
CAMB		Ophir Formation	300-500	*Glossopleura*
CAMB		Tintic Quartzite	1200-1300	orange; 25' diabase 160' above base
PROT		Mutual Formation	0-1300	reddish-brown; only in Am Fk Canyon
PROT		Mineral Fork Tillite	0-200	olive-drab
PROT		Big Cottonwood Formation	1350 +	purple & green argillite in Slate Canyon

MAP WITH TEXT - Baker, 1964a, b, 1972, 1973, 1976; Baker & Crittenden, 1961; Hintze, 1978; Davis, 1983b; SOILS - Swenson et al, 1972; LAKE BONNEVILLE - Currey et al, 1983; Bissell, 1963; Hunt et al, 1953; UTAH VALLEY - Brimhall & Merritt, 1981; Cook & Berg, 1961; Davis & Cook, 1984; WASATCH FAULT - Hintze, 1971; Swan et al, 1981; OQUIRRH - Larson & Clark, 1979; Chamberlain & Clark, 1973; Jordan, 1981; Konopka & Dott, 1982; Maxfield, 1957; Murphy, 1954; Alexander, 1978; Welsh & Bissell, 1979; Welsh, letter, 1985; MISSISSIPPIAN - Welsh & Bissell, 1979; Bissell & Barker, 1977; Pinney, 1965; Livingston, 1955; Woodland, 1958; Sandberg & Gutschick, 1979, 1984; CAMBRIAN - Migliaccio, 1958; Hall, 1981; Abbott, 1951.

Chart

36 SPANISH FORK CANYON - THISTLE

Age	Group	Unit	FEET	Notes
Q		Alluvium & Lake Bonneville beds	0-100	
TERTIARY		Volcanic ash	0-100	Miocene ?
TERTIARY		Wanrhodes Fm (chiefly volcaniclastics)	0-500	Oligocene ? equal to Moroni or Tibble Fms
TERTIARY		Uinta Formation	300-500	Eocene
TERTIARY		Green River Formation	400-1800	Gar fish; green
TERTIARY		Colton Formation	0-400	red to green mudstone
TERTIARY		Flagstaff Formation	100-600	oncolites, tar beds; pantodont mammals
TERTIARY	Red Narrows Cg of Young	North Horn Formation	50-2000	forms slumps and landslides near Thistle
CRETACEOUS	Red Narrows Cg of Young	Price River Fm	200-300	tan cg clasts to 3 ft; ANGULAR UNCONFORMITY
CRETACEOUS	Indianola Group	Funk Valley Fm ?	1000 +	Indianola Group thicker and better exposed to the south near Indianola
CRETACEOUS	Indianola Group	Allen Valley Shale	100	
CRETACEOUS	Indianola Group	Sanpete Ss	300	
CRETACEOUS		Cedar Mountain Formation	1000-1800	unstable, slumps; "Morrison?" of old reports; gastroliths common in lower Cedar Mtn
JURASSIC		Twist Gulch Formation	700-1500	called Summerville & Curtis at Monks Hollow by Young, equivalent to Preuss Fm
JURASSIC		Arapien Shale (Imlay's Upper Twin Creek members included in Arapien at Billies Mountain section)	1600	upper Arapien equivalent to Entrada Fm of Young at Monks Hollow; *290' Giraffe Creek Mbr; *275' Leeds Creek Mbr
JURASSIC	Twin Creek Ls	Watton Canyon Mbr	360-345*	*thickness of Twin Creek Limestone members of Imlay at Monks Hollow
JURASSIC	Twin Creek Ls	Boundary Ridge Mbr	40-60*	
JURASSIC	Twin Creek Ls	Rich Member	210-120*	
JURASSIC	Twin Creek Ls	Sliderock Member	70	*Pentacrinus*
JURASSIC	Twin Creek Ls	Gypsum Spring Mbr	10-50*	red beds
JURASSIC		Nugget (or Navajo) Sandstone	1400-1500	white ss irregularly in upper part; red ss in lower part
TRIASSIC	Ankareh Fm	Sandstone unit	310	Kayenta equivalent ?
TRIASSIC	Ankareh Fm	Bentonite unit	370	Chinle equivalent ? slumps
TRIASSIC	Ankareh Fm	Siltstone unit	850	Diamond Fork Jct
TRIASSIC		Thaynes Formation	1200	Moenkopi equivalent unstable, slumps; *Platyvillosus* fauna
TRIASSIC		Woodside Shale	50-150	
PERMIAN	Park City Group	Franson Fm	0-900	cut out in Sp Fk Can by pre-Triassic erosion
PERMIAN	Park City Group	Meade Pk Phosph	0-230	
PERMIAN	Park City Group	Grandeur Ls	200-1170	
PERMIAN		Diamond Creek Sandstone	900	anticline at Castilla
PERMIAN		Kirkman Limestone	75-280	sinkholes in Covered Bridge (Pole) Canyon
PERMIAN	Oquirrh Group	Granger Mountain Formation	5000	Cold Springs; *Pseudofusulina*; *Pseudoschwagerina*; tan fine-grained ss; *Schwagerina*
P	Oquirrh Group	Wallsburg Ridge Fm	500 +	*Millerella*, *Triticites*

MAP WITH TEXT - Young, 1976; Baker, 1976; Witkind and Page, 1983; Pinnell, 1972; Merrill, 1972; Runyon, 1977; Rawson, 1957; CRETACEOUS - Witkind et al, 1986; Lawton, 1982; JURASSIC - Imlay, 1967; Bullock, 1965; Bordine, 1965; THAYNES - James, 1980; PERMIAN - Welsh et al, 1979; LaRell Nielson letter Sept, 85; SUBSURFACE - Irwin, 1971, 1976; Witkind, 1988.

Chart
37 SOLDIER SUMMIT

Period	Formation / Member	FEET	Remarks
Q	Soil, alluvium & colluvium	0-100	
EOCENE	Uinta Fm — Fluviatile seds	200+	
EOCENE	Uinta Fm — Sandstone and limestone facies	700	
EOCENE	Uinta Fm — Saline facies	900	
EOCENE	Green River Fm — Evacuation Creek Member	970	upper tuff zone; E-5 palynomorphs
EOCENE	Green River Fm — Parachute Creek Member	630	tuff zone; E-4 palynomorphs
EOCENE	Green River Fm — "Delta" facies marlstone — Unit D	1000	
EOCENE	Green River Fm — "Delta" facies marlstone — Unit C	600	marlstone & siltstone
EOCENE	Green River Fm — "Delta" facies marlstone — Unit B	1075	E-3 palynomorphs; greenish-gray marlstone, ss, ls
EOCENE	Green River Fm — "Delta" facies marlstone — Unit A	575	fossil bird tracks
EOCENE	Green River Fm — Soldier Summit Member	460	oil shale; fossil tracks; E-2 palynomorphs
EOCENE	Green River Fm — Tabbyune Creek Tongue	260-310	Tongue of Colton Fm
EOCENE	Green River Fm — Middle Fork Tongue	250-280	lacustrine paper sh; ozokerite veins; SOLDIER SUMMIT
EOCENE	Colton Formation	1000-1500	E-1 palynomorphs
PALEOC	Flagstaff Limestone	300-500	P-3 palynomorphs
PALEOC	North Horn Formation	1600	P-2 palynomorphs; red layers locally
CRETACEOUS	Bennion Creek Fm	550	conglomerate & ss
CRETACEOUS	Mesaverde Group — Price River Fm	600	
CRETACEOUS	Mesaverde Group — Castlegate Sandstone	700	
CRETACEOUS	Mesaverde Group — Blackhawk Formation	700	thin coal beds; Pauley #1 well 35-11S-6E 1958 TD 8560
CRETACEOUS	Mesaverde Group — Star Point Sandstone	430	
CRETACEOUS	Mancos Shale — Upper Shale	520	
CRETACEOUS	Mancos Shale — Emery Sandstone Member	1500	
CRETACEOUS	Mancos Shale — Blue Gate Shale Member	1500	
CRETACEOUS	Mancos Shale — Ferron Sandstone Member	600	
CRETACEOUS	Mancos Shale — Tununk Shale M	400	
CRETACEOUS	Dakota - Cedar Mtn Fms	300+	

MAP WITH TEXT - Moussa, 1965; Henderson, 1964; UINTA - Dane, 1954, 1955; GREEN RIVER - Moussa, 1969; Newman, 1974; COLTON - Peterson, 1976; Zawiskie et al, 1982; Smith, 1986; CRETACEOUS - Hale & Van de Graaff, 1964; Heylmun et al, 1965.

Chart
38 DUCHESNE RIVER - ALTAMONT

Period	Formation / Member	FEET	Remarks
Q	Alluvial & glacial deposits	0-300	
Φ	Bishop Cg	0-200	29 m.y.
Φ	Duchesne River Fm	0-3000	
EOCENE	Uinta Fm	0-5000	surface / subsurface
EOCENE	Green River Fm in subsurface only — Parachute Cr and Douglas Cr Members interfinger in subsurface	2000-5700	oil producing mud-supported carbonate interbedded with ss & bright green claystone
PALEO	Currant Creek Fm-about 4000 ft thick as exposed west of Duchesne River — Colton Fm	0-1500	ss, red & green clay
PALEO	Flagstaff Fm or Mbr of Green River Fm	500-2000	tan, gray carbonate, clay, and ss
PALEO	North Horn Formation	0-2500	includes red clays
CRETACEOUS	Mesaverde Group	600-4000	coal
CRETACEOUS	Mancos Shale	1700-3000	
CRETACEOUS	Mancos Shale — Frontier Ss Mbr	600-200	
CRETACEOUS	Mancos Shale — Mowry Sh Mbr	100-300	
CRETACEOUS	Dakota Ss	50-200	← Buckhorn Cg Mbr 0-50 feet
CRETACEOUS	Cedar Mountain Fm	100-200	
JURASSIC	Morrison Fm	1100-1500	
JURASSIC	Curtis Mbr-Stump Fm	100-250	
JURASSIC	Preuss or Entrada Ss	100-700	
JURASSIC	Twin Creek Ls	440-750	
JURASSIC	Navajo Sandstone	700-1300	
TRIASSIC	Chinle Fm	250-450	
TRIASSIC	Chinle Fm — Gartra Grit Mbr	0-100	
TRIASSIC	Ankareh Formation	700-1000	Moenkopi Fm equivalents
TRIASSIC	Thaynes Formation	250-100	Moenkopi Fm equivalents
TRIASSIC	Woodside Shale	600-800	Moenkopi Fm equivalents
TRIASSIC	Dinwoody Formation	0-80	Moenkopi Fm equivalents
P	Park City Fm	300-600	
PENN	Weber Sandstone	1200-1600	
PENN	Morgan Fm	200-300	
PENN	Round Valley Fm	250-350	
MISS	Doughnut Shale	100-250	
MISS	Humbug Fm	350-500	
MISS	Deseret Ls	650	
MISS	Lodgepole Ls	250	"Madison" of old reports
Є	Maxfield Ls	0-300	
Є	Ophir Fm	0-50	
Є	Tintic Quartzite	100-300	
PROTEROZOIC	Uinta Mountain Group — Red Pine Shale	0-3000	
PROTEROZOIC	Uinta Mountain Group — Hades Pass Fm of Wallace	5000-6000	maroon to purplish-red
PROTEROZOIC	Uinta Mountain Group — Mount Watson Formation	1800-2300	light gray to white
PROTEROZOIC	Uinta Mountain Group — Red Castle Fm of Wallace	2000+	arkosic

MAPS - Huddle & McCann, 1947; Huddle et al, 1951; Kinney, 1951; PLEISTOCENE - Grogger, 1974; BISHOP CG - Hansen, 1984; DUCHESNE RIVER - Andersen & Picard, 1972, 1974; Warner, 1966; Kay, 1957; GREEN RIVER - Ryder et al, 1976; Fouch, 1975, 1976; Pitman et al, 1982; Picard et al, 1973; Lucas & Drexler, 1975; CURRANT CREEK - Garvin, 1969; Isby & Picard, 1983, 1985. CRETACEOUS - Doelling & Graham, 1972b; TWIN CREEK - Imlay, 1967; Thaynes - Kummell, 1954; PARK CITY - Smith et al, 1952; CAMBRIAN - Crittenden et al, 1967; PRECAMBRIAN - Sanderson, 1984; SUBSURFACE STRUCTURE - Gries, 1983.

FIGURE 134 — Park City in foreground, Mount Timpanogos on left center horizon. Beds at base of ski runs in Park City gulch are Pennsylvanian Weber Quartzite (wb), overlain by Permian Park City Formation (p), in turn overlain by Woodside Shale (w), and Thaynes Formation (t). Ankareh Formation (a) and Nugget Sandstone (n) to right.

FIGURE 135 — Deer Creek Reservoir in upper Provo Canyon barely visible at bottom center of photo. Layers on Mt. Timpanogos are the lowest 4,000 feet of the 25,000-foot Oquirrh Group (see Chart 35). Aspen Grove graben, a basin-range structure, drops Oquirrh strata down along this side of mountain base, covered by glacial moraines. Hummocky low ground at lower left along Provo River is slump area on Manning Canyon Shale exposed in window in the Deer Creek thrust.

FIGURE 136 — West end of Uinta Mountains along the Summit-Wasatch County border. Lost Lake in left foreground, Trial Lake in right center, Washington Lake beyond to left. View to southwest; Wasatch Mountains faint on distant horizon. Utah Highway 150 winds through view. Bedrock is Proterozoic Uinta Mountain Group. Valleys are choked with glacial debris.

FIGURE 137 — Uinta Basin about 8 miles north of Vernal near Steinaker Reservoir. State Highway 44 (dark) leaves right edge of photo going north. Prominent hogback curving to right through photo center is capped by resistant Frontier Sandstone, Mowry Shale, and Dakota Sandstone. Light-colored softer strata beneath caprocks are Cedar Mountain and Morrison Formations. Dirt road at right edge follows non-resistant Carmel Formation with Navajo (Glen Canyon) Sandstone to the right and on middle distance dip slopes; Stump Formation is left of dirt road. See Chart 40. Left half of foreground exposes Mancos Shale badlands.

Chart

39 FLAMING GORGE - NE UINTA MTS

Age	Unit	FEET	Notes
Q	Alluvial, eolian dep	0-200	
Q	Older alluvial dep	0-150	
M	Browns Park Fm	800	tuffaceous sandstone
Φ	Bishop Conglomerate	0-300	
EOCENE	Bridger Fm	500-2000	
EOCENE	Laney Mbr - Green River Fm	0-1000	45 m.y.
EOCENE	Cathedral Bluffs Mbr	0-1000	evaporites
EOCENE	Wilkins Peak Mbr	300-500	49 m.y.
EOCENE	Tipton Mbr	150	oil shale
EOCENE	Wasatch Fm main body	2000	red and gray sandstone and mudstone
PAL	Ft Union Fm	1200-2300	
CRETACEOUS	Mesaverde Group: Ericson Ss	300-900	
CRETACEOUS	Mesaverde Group: Rock Springs Fm	0-1100	coal
CRETACEOUS	Mesaverde Group: Blair Fm	0-360	
CRETACEOUS	Hilliard Shale	6200	marine shale
CRETACEOUS	Frontier Fm	170-190	
CRETACEOUS	Mowry Shale	200-270	fish scales
CRETACEOUS	Dakota Ss	130-250	Cedar Mtn or Cloverly equivalents in upper part
JURASSIC	Morrison Fm and younger rocks	800-940	
JURASSIC	Curtis Mbr of Stump Fm	145-180	
JURASSIC	Entrada Ss	200-245	
JURASSIC	Carmel Fm	175-330	
JURASSIC	Navajo Ss	815-840	Nugget Sandstone equivalent
TRIASSIC	Chinle Fm	300-380	
TRIASSIC	Moenkopi Fm	725	
TRIASSIC	Dinwoody Fm	360-530	
PERM	Park City Fm: Franson Mbr	175	
PERM	Park City Fm: Meade Pk Phosp	45	phosphorite
PERM	Park City Fm: Lower member	100	
PERM / IP	Weber Sandstone	1550	
IP	Morgan Fm	80	Desmoinesian fusulinids
IP	Round Valley Ls	260	Morrowan & Atokan
MISS	Doughnut Shale	200-500	
MISS	Humbug Fm	360	
MISS	Madison Ls	600	
PROTEROZOIC	Uinta Mountain Group	24,000	red sandstone and quartzite; some red argillite and conglomerate
PROTEROZOIC	Jesse Ewing Canyon Fm	500-1000	alluvial fan deposit
PROTEROZOIC	Red Creek Quartzite	20,000	quartzite with some pelitic schist; 2.3 b.y.
A	Owiyukuts Complex		2.7 b.y. gneiss

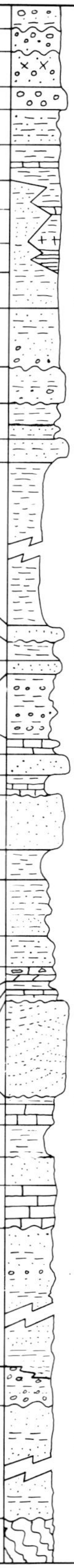

GENERAL REFERENCE - Hansen, 1965; TERTIARY - Roehler, 1972; Hansen, 1984; Mauger, 1977; Winterfeld, 1982; JURASSIC - Otto & Picard, 1975; Lowrey, 1976; Picard, 1975; PERMIAN - Jado, 1978; PROTEROZOIC - Sears et al, 1982; Swayze & Holden, 1984; Sanderson & Wiley, 1986.

Chart

40 DINOSAUR MONUMENT - SE UINTA MTS

Age	Unit	FEET	Notes
Q	Alluvial and eolian deposits	0-100	
M	Browns Park Fm	0-900	tuffaceous ss
Φ	Bishop Conglomerate	0-500	
CRETACEOUS	Mancos Shale	5000	
CRETACEOUS	Frontier Sandstone	200-250	
CRETACEOUS	Mowry Shale	80-120	fish scales
CRETACEOUS	Dakota Sandstone	30-100	
CRETACEOUS	Cedar Mountain Fm	100-200	similar to Morrison
JURASSIC	Morrison Fm	800	varicolored mudstone; DINOSAUR QUARRY
JURASSIC	Stump Fm: Redwater Mbr	130	green glauconitic shale, siltstone
JURASSIC	Stump Fm: Curtis Mbr	30-100	
JURASSIC	Entrada Ss	80-160	
JURASSIC	Carmel Fm	50-130	red sandy sh, ls
JURASSIC	Glen Canyon Sandstone	650-700	Nugget-Navajo equivalent
TRIASSIC	Chinle Fm	200-250	
TRIASSIC	Gartra Mbr	30-100	
TRIASSIC	Moenkopi Fm	700-800	reddish-brown; ripple marks
PERMIAN	Park City Fm: Upper unit	50-100	phosphate
PERMIAN	Park City Fm: Lower unit	50-100	
PERMIAN	Weber Sandstone	1000-1200	
PENN	Morgan Fm: Upper member	600-650	Desmoinesian fusulinids
PENN	Morgan Fm: Lower mbr	150-250	
PENN	Round Valley Ls	200-350	
MISS	Doughnut Shale	100-180	
MISS	Humbug Fm	150-200	
MISS	Madison Ls	600	*Clydagnathus*
CAM	Lodore Fm	400-600	
PROT	Uinta Mountain Group	24,000	cut by alkalic dike dated 483 ± 30 m.y.

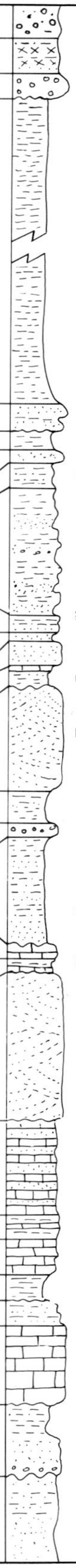

MAP REFERENCES - Hansen et al, 1983; Rowley et al, 1979; Carrara, 1980; TERTIARY - Hansen, 1984; CRETACEOUS - Young, 1975; Doelling & Graham, 1972b; Maione, 1971; MORRISON - Dawson, 1971; ENTRADA - Otto & Picard, 1975; NUGGET - High & Picard, 1975; PERMIAN - Whitaker, 1975; MORGAN - Driese, 1983; MADISON - Dockal, 1980; Welsh & Bissell, 1979; DIKE - Ritzma, 1974; LODORE - Herr et al, 1981, 1982; PHOSPHATE - Garrand, 1985.

Chart
41 UINTA BASIN

		FEET	
Q	Recent alluvium	0-100	
	Outwash, older alluvium	0-200	
OLIG	Duchesne River Formation	0-3000	*Epihippus* subsurface
EOCENE	Uinta Formation	0-5000	
	Green River Formation	2000-6000	42 m.y. K-Ar saline tuff 44 m.y. wavy tuff Mahogany oil-shale marker bed middle marker on well logs carbonate marker
	Colton Formation	0-1500	red ss, silt & shale
PALEOCENE	Flagstaff Formation (or member of the Green River Fm)	0-1500	gray to green clay and ss and limestone
	North Horn Formation	0 -2000	channel ss and red clay and silt
CRETACEOUS	Mesaverde Group	1200-2700	thins northeastward coal
	Mancos Shale	3200-5000	thickens northeastward
	Frontier-Mowry Fms	300-350	Note - Pipiringos & O-Sullivan (1978) show formations listed below under part of Uinta Basin
	Dakota-Cedar Fms	50-200	
JURASSIC	Morrison Fm	500-600	Summerville Fm
	Curtis Fm	0-300	
	Entrada Sandstone	180-220	
	Carmel Fm	0 -100	
	Navajo Sandstone	500-650	Page Sandstone
TRI	Chinle Fm & Gartra Mbr	70-150	
	Moenkopi Fm	100-760	Pre-Triassic rocks
P	Weber and Park City Fms	0-900	were all eroded from the Uncompahgre
M-IP	Limestones, undivided	0-1500	block
D	Devonian undivided	0-160	
C	Ls & ss undivided	0-200	
pC	Granite	basement	

QUATERNARY - Hood, 1976; Hood, & Fields, 1978; UINTA-DUCHESNE RIVER - Kay, 1957; Warner, 1966; EOCENE-PALEOCENE - Fouch, 1976; Ryder et al, 1976; Mauger, 1977; Johnson, 1985; PRE-TERTIARY - Heylmun et al, 1965; STRUCTURE - Smith & Cook, 1985; Bryant, 1985; OIL & GAS -Clem, 1985.

Chart
42 UINTA BASIN MAPS & CROSS SECTION

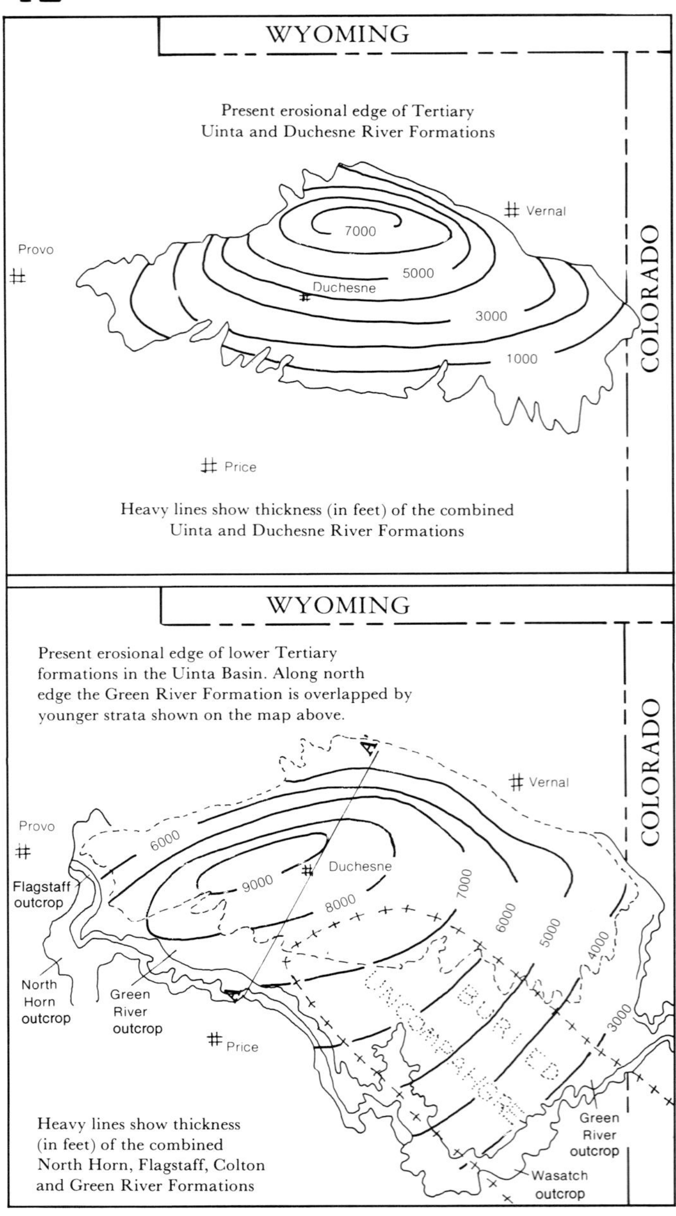

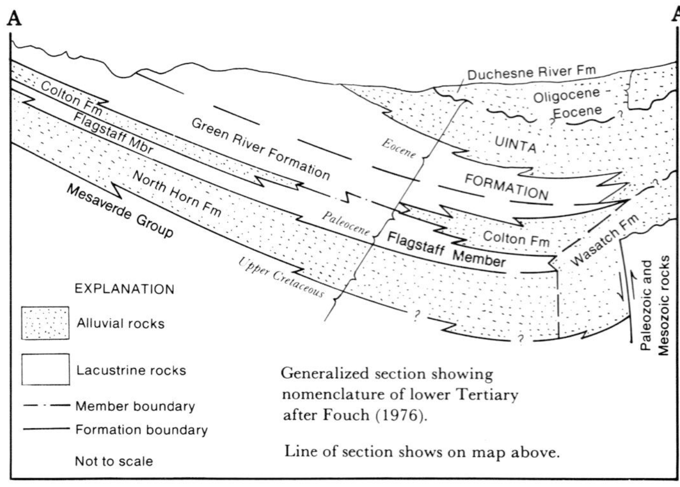

Generalized section showing nomenclature of lower Tertiary after Fouch (1976).

Line of section shows on map above.

REFERENCES - Fouch, 1976; Johnson, 1985.

Chart
43 BONANZA, UTAH - RANGELY, COLO

Age	Formation	Member	FEET	Remarks
Q	Alluvium - dunes		0-200	
EOCENE	Uinta Fm	Unit B	100-600	Gilsonite veins
EOCENE	Uinta Fm	Unit A	400-700	
EOCENE	Green River Fm	Parachute Creek Mbr	500-1200	Mahogany Bed-oil shale
EOCENE	Green River Fm	Garden Gulch M	0-200	
EOCENE	Green River Fm	Cow Ridge Member	400-1800	
PALEO	Wasatch Formation		300-3000	reddish-gray many slumps and landslides
				unconformity
CRETACEOUS	Mesaverde Group	Upper Unit	1100	
CRETACEOUS	Mesaverde Group	Main coal unit	500-700	coal
CRETACEOUS	Mesaverde Group	Minor coal unit	700-750	
CRETACEOUS	Mesaverde Group	Sego Ss	0-250	
CRETACEOUS	Buck Tongue - Mancos Sh		250-300	
CRETACEOUS	Castlegate Ss Tongue		50-200	part of Mesaverde Group
CRETACEOUS	Mancos Shale - main body		4500-5300	City of Rangely; subsurface olive-gray mudstone
CRETACEOUS		Frontier Ss Mbr	50-200	oil in fractured sh
CRETACEOUS		Mowry Sh Mbr	0-60	
CRETACEOUS	Dakota Ss		50-100	fish scales
CRETACEOUS	Cedar Mountain Fm		50-100	
CRETACEOUS		Buckhorn Cg Mbr	0-80	
JURASSIC	Morrison Fm		550-650	dinosaurs
JURASSIC	Curtis Mbr of Stump Fm		100	green
JURASSIC	Entrada Ss		120-150	
JURASSIC	Carmel Fm		70	
JURASSIC	Navajo Ss		650	
TRIAS	Chinle Formation		100-200	
TRIAS		Gartra Grit Mbr	40-80	
TRIAS	Moenkopi Fm		500	
P	Park City Fm		150	
PENN	Weber Sandstone		850-1150	principal Rangely Field reservoir rock
PENN / MISS	Morgan Fm, Round Valley Ls, Doughnut Shale, Humbug Fm undivided		1400-1700	
MISS	Madison Ls		600	
D-O?	Chaffee? Manitou? Dotsero? undivided		0-500	
Є	Lodore - Sawatch Qtzt		100-500	

MAP REFERENCES - Cullins, 1971; Pipiringos, 1977; TERTIARY - Ryder et al, 1976; Picard & High, 1972; Surdam & Stanley, 1980; Cashion & Donnell, 1974; Johnson, 1984, 1985; CRETACEOUS - Young, 1975; PALEOZOIC - Campbell, 1955; Welsh & Bissell 1979; GILSONITE - Jackson, 1985.

Chart
44 GRAND JUNCTION AREA, COLORADO

Age	Formation	Member	FEET	Remarks
Q	Alluvium & colluvium		0-200	
Q	Glacial, pediment, & land slide deposits		0-200	
MIO	Basalt flows of Grand Mesa		0-1000	10 m.y. K-Ar
EOCENE	Uinta Formation		400+	formerly called Evacuation Creek Mbr of Green River Fm
EOCENE	Green River Fm	Parachute Creek Member	300-1100	dolomitic oil shale; includes Mahogany oil-shale bed 5-10 feet thick
EOCENE	Green River Fm	Douglas Creek Member	300-700	Kimball Mtn Tuff Bed
EOCENE	Green River Fm	Garden Gulch M	100-300	Long Point marker bed 1-40 ft thick, fossiliferous
EOCENE / PALEO	Wasatch Formation		200-2000	slump-prone reddish-gray mudstone and sandstone
CRETACEOUS	Mesaverde Group	Ohio Creek Cg	0-100	
CRETACEOUS	Mesaverde Group: Hunter Canyon Formation		1100	Hunter Canyon Fm is coarser, grayer, and has more massive ss beds than Mt. Garfield
				gradational contact
CRETACEOUS	Mesaverde Group: Mount Garfield Fm	Upper part	600	
CRETACEOUS	Mesaverde Group: Mount Garfield Fm	Rollins Ss Mbr	0-150	Cameo coal zone
CRETACEOUS	Mesaverde Group: Mount Garfield Fm	Cozzette Ss M	0-200	*Didymoceras*
CRETACEOUS	Mesaverde Group: Mount Garfield Fm	Corcoran Ss M	0-180	Palisade coal zone
CRETACEOUS	Mesaverde Group: Sego Sandstone	Upper Ss	0-120	Anchor coal zone
CRETACEOUS	Mesaverde Group: Sego Sandstone	Anchor Mine Tongue	50-200	Mancos tongue
CRETACEOUS	Mesaverde Group: Sego Sandstone	Lower Ss	0-100	
CRETACEOUS	Mancos Shale	Early Pierre fauna; Eagle fauna; Telegraph Creek fauna; Niobrara fauna; Carlile fauna; Greenhorn fauna	4000	*Baculites perplexus*; *Baculites obtusus*; *Inoceramus*; *Scaphites*; *Placenticeras*; *Baculites*; *Desmoscaphites*; *Ostrea*; *Placenticeras*; *Ostrea*; *Inoceramus*; *Pycnodonte newberryi*
CRETACEOUS	Dakota Sandstone		150	
CRETACEOUS	Burro Canyon Fm		50-120	
JURASSIC	Morrison Fm	Brushy Basin Mbr	260-340	varicolored mudstone
JURASSIC	Morrison Fm	Salt Wash Mbr	190-310	ss & cg
JURASSIC	Morrison Fm	Tidwell Member	10-100	
JURASSIC	Wanakah Formation		20-50	formerly Summerville
JURASSIC	Entrada Sandstone		60-200	
JURASSIC	Kayenta Formation		0-130	J-2 unconformity
JURASSIC / TR	Wingate Sandstone		210-370	
TR	Chinle Formation		80-120	Uncompahgre unconformity
pЄ	Vernal Mesa-type quartz monzonite; gneiss and schist			qtz mon- 1480 m.y.; gneiss- 1700 m.y.

MAPS WITH TEXT - Cashion, 1973; Lohman, 1965a, b; QUATERNARY - Whitney, 1981; GREEN RIVER - Johnson, 1984; WASATCH - Donnell, 1969; CRETACEOUS - Young, 1955, 1966, 1983b; Fisher et al, 1960; Gill & Hail, 1975; Craig, 1981; Johnson & May, 1980; Johnson et al, 1980; JURASSIC - Cadigan, 1967; O'Sullivan, 1984; O'Sullivan & Pipiringos, 1983; TRIASSIC - Stewart, 1969; PRECAMBRIAN - Bickford & Cudzilo; Tweto, 1977.

Chart
45 GOLD HILL AREA

Period	Formation		FEET	Remarks
Q	Alluvium-Great Salt Lake seds.		0-200	
Q	Lake Bonneville sediments		0-200	
M-P	Pre- Bonneville valley fill		0-1000	subsurface only
M-P	Salt Lake Fm		0-300	
M-P	Rhyolite-basalt volcs		0-300	
EO	Latite & andesite volcs		0-200	
EO	Gold Hill quartz monzonite		intrusive	38 m.y. north part
EO	White Sage Fm		0-600	
EO	Ibapah granite stock		intrusive	39 m.y.
J	Gold Hill stock-main body		intrusive	152 m.y.
ꞦR	Thaynes Formation		0-50	*Meekoceras*
PERMIAN	Gerster Fm		400	*Punctospirifer*
PERMIAN	Park City Group	Plympton Fm	600	
PERMIAN	Park City Group	Meade Pk Mbr	150	
PERMIAN	Park City Group	Kaibab Ls	500	gypsum
PERMIAN	Arcturus Formation		1300	gypsum
PERMIAN	Ferguson Mountain Formation		1900	*Parafusulina* *Syringopora* *Pseudoschwagerina* *Dunbarinella* *Triticites*
ℙ	Ely Limestone		1400	*Wedekindellina* *Millerella*
MISSISSIPPIAN	Chainman Shale		500 ±	← common level of detachment for attenuated Paleozoic rocks above
MISSISSIPPIAN	Ochre Mountain Limestone (equivalent to Great Blue Ls of central Utah)		2000	Herat Shale Member Lower Ochre Mtn 1000 ft (Harmala, 1982)
MISSISSIPPIAN	Woodman Fm		1500 ±	
MISSISSIPPIAN		Delle Phosphatic M		conodonts
MISSISSIPPIAN	Joana Limestone		25-370	
DEV	Guilmette Formation		720-1200	*Cladopora* stromatoporoids
DEV	Simonson Dolomite		1100	stromatoporoids *Stringocephalus*
DEV	Sevy Dolomite		450-670	light-gray
S	Laketown Dolomite		660-970	*Halysites* *Virgiana*
ORD	Ely Springs Dolomite		250	MID-ORDOVICIAN EROSION
ORD	Pogonip Group		0-1500	*Lachnostoma* *Symphysurina*
CAMBRIAN	Notch Peak Fm		1200	
CAMBRIAN	Orr Formation (Hicks Fm of Nolan)		1000	*Eurekia* *Dunderbergia* *Crepicephalus*
CAMBRIAN	Lamb Dolomite		800	
CAMBRIAN	Trippe Limestone		700	*Eldoradia* white boundstone
CAMBRIAN	Young Peak Dolomite		0- 600	
CAMBRIAN	Abercrombie Formation McCollum & McCollum (1984) apply Pierson Cove Fm Wheeler Shale Swasey Ls Whirlwind Fm Dome Ls Chisholm Shale Howell Fm		1760-2700	*Bolaspidella* *Ehmaniella* *Glossopleura*
CAMBRIAN	Pioche Fm	Busby Qtzt M	410-450	*Olenellus*
CAMBRIAN	Pioche Fm	Cabin Sh Mbr	500	micaceous
CAMBRIAN	Prospect Mountain Quartzite		3000	weathers reddish-brown; minor feldspar
PROT	Goshute Canyon Fm	d- sh	270	phyllitic, purple
PROT	Goshute Canyon Fm	c- quartzite	730	
PROT	Goshute Canyon Fm	b- sh, slts, qtzt	1500	phyllitic, micaceous
PROT	Goshute Canyon Fm	a- quartzite	200+	

MAPS WITH TEXT - Nolan, 1935; Bick, 1966; QUATERNARY - Gates & Kruer, 1981; IGNEOUS DATES - Stacey & Zartmann, 1978; Thomson 1973; PERMIAN - Hodgkinson, 1961; Zabriskie, 1970; Wardlaw et al, 1979; Wardlaw & Collinson, 1979; MISSISSIPPIAN - Sandberg & Gutschick, 1984; Harmala, 1982; CAMBRIAN - Hintze & Palmer, 1976; Hintze & Robison, 1975; PROTEROZOIC - Woodward, 1965; Bick, 1966; Christie-Blick, 1982; QUEEN OF SHEBA MINE - James, 1984.

Chart
46 FISH SPRINGS RANGE

Period	Formation		FEET	Remarks
Q	Alluvium, L. Bonneville seds.		0-200	
PLIO / MIO	Valley fill - may be several thousand feet thick at south end of Great Salt Lake Desert		0- 5000?	concealed alluvial deposits, lake deposits and volcanic rocks from Topaz-Keg volcanic area
MIO	Rhyolitic dikes & plugs		intrusion	Fish Springs mines
MIO	Dacite flow		0-50	dark red 30 m.y.
Φ	Skull Rock Pass Cg		0-50	
DEVONIAN	Simonson Dolomite		1000 ±	dark gray, contains stromatoporoids
DEVONIAN	Sevy Dolomite		1000 ±	light gray finely crystalline unfossiliferous
SILUR	Laketown Dolomite	Thursday Mbr	160	
SILUR	Laketown Dolomite	Lost Sheep M	180	
SILUR	Laketown Dolomite	Harrisite Mbr	130	*Favosites*
SILUR	Laketown Dolomite	Bell Hill Mbr	530	*Virgiana*
ORDOVICIAN	Ely Springs Dolomite	Floride Mbr	120-140	
ORDOVICIAN	Ely Springs Dolomite	Lower member	140-260	
ORDOVICIAN	Eureka Quartzite		200-300	
ORDOVICIAN	Watson Ranch Qtzt		110-150	
ORDOVICIAN	Pogonip Group	Deadman Spring Dolo	0-160	*Orthambonites*
ORDOVICIAN	Pogonip Group	Kanosh Shale	0-130	
ORDOVICIAN	Pogonip Group	Juab Limestone	70-150	pliomerid trilobites
ORDOVICIAN	Pogonip Group	Wah Wah Limestone	110-270	
ORDOVICIAN	Pogonip Group	Fillmore Formation	1100 ±	intraformational conglomerates
ORDOVICIAN	Pogonip Group	House Limestone	110-190	*Symphysurina* algal heads
CAMBRIAN	Notch Peak Formation		1000-1250	
CAMBRIAN	Orr Fm	Sneakover Ls Mbr	80-100	
CAMBRIAN	Orr Fm	Corset Spring Sh M	40-60	
CAMBRIAN	Orr Fm	Johns Wash Ls Mbr	170-240	trilobite coquinas
CAMBRIAN	Orr Fm	Candland Shale M	200-280	*Crepicephalus*
CAMBRIAN	Orr Fm	Big Horse Limestone Member	440	
CAMBRIAN	Lamb Dolomite		840	oncolites
CAMBRIAN	Trippe Ls	Fish Springs Mbr	130-140	*Eldoradia*
CAMBRIAN	Trippe Ls	Lower member	630	boundstone
CAMBRIAN	Pierson Cove Formation		1100-1350	very dark gray
CAMBRIAN	Wheeler Shale		590	*Elrathia*
CAMBRIAN	Swasey Limestone		120	
CAMBRIAN	Whirlwind Fm		160	*Ehmaniella*
CAMBRIAN	Dome Limestone		140	*Glossopleura*
CAMBRIAN	Chisholm Fm		230	
CAMBRIAN	Howell Limestone		500 ±	
CAMBRIAN	Pioche Fm	Tatow Member	170	
CAMBRIAN	Pioche Fm	Lower member	420	
CAMBRIAN	Prospect Mountain Quartzite		2000+	pinkish-gray to reddish brown

MAPS - Hintze, 1980a, b; Oliveira, 1975; WATER - Gates et al, 1981; CAMBRIAN - Hintze & Robison, 1975; Hintze & Palmer, 1976; Kopaska-Merkel, 1983; Koepnick, 1976; STRUCTURE - Hintze, 1978; Piekarski, 1980.

Chart
47 SO. DEEP CREEK - KERN MTS

Period	Unit	FEET	Remarks
Q	Alluvium, L. Bonneville seds.	0-200	
M-P	High-level alluvium	0-500	
M-P	Valley fill	0-2000?	alluvial, lake, & volcanic deposits
Φ	Intermediate lavas & tuffs	0-200	34.5 m.y. Heavenly Hills area
E	Skinner Canyon Granite	intrusion	35 m.y.
E	Ibapah Granite	intrusion	39 m.y.
K	Tungstonia Granite	intrusion	75 m.y. two-mica
PERM	Park City Group	500	
PERM	Arcturus Fm	2000*	yellow sandstone and dolomite
P	Ely Limestone	1200	cherty, fossiliferous
MISS	Chainman Shale	1100	top fossiliferous
MISS	Joana Limestone	340	
DEVONIAN	Pilot Shale	260	weathers orange
DEVONIAN	Guilmette Formation	2300	stromatoporoids
DEVONIAN	Simonson Dolomite	700*	stromatoporoids
DEVONIAN	Sevy Dolomite	800*-1100	light-gray
S	Laketown Dolomite	800	
ORDOVICIAN	Ely Spring Dolomite	250*-400	
ORDOVICIAN	Eureka Quartzite	200*-325	
ORDOVICIAN	Crystal Peak Dolomite	30*-120	
ORDOVICIAN	Watson Ranch Qtzt	30*- 150	
ORDOVICIAN	Pogonip Group: Lehman Fm	180*-300	*Orthambonites*
ORDOVICIAN	Pogonip Group: Kanosh Shale	300*-500	
ORDOVICIAN	Pogonip Group: Lower Pogonip undivided	1800-2000	
CAMBRIAN	Notch Peak Formation	310*	
CAMBRIAN	Corset Spring Sh	40	Names of Cambrian carbonate units are perhaps inappropriate here
CAMBRIAN	Orr Formation	100*-600	
CAMBRIAN	Lamb Dolomite	1000	
CAMBRIAN	Trippe Limestone	0*-600	
CAMBRIAN	Young Peak Dolomite	500	
CAMBRIAN	Abercrombie Formation	2500	
CAMBRIAN	Pioche Fm: Busby Qtzt Mbr	500	
CAMBRIAN	Pioche Fm: Cabin Sh Mbr	500	phyllitic
CAMBRIAN	Prospect Mountain Quartzite	3300	
PROTEROZOIC	McCoy Creek sequence: G–phyllite, metasiltstone	300	
PROTEROZOIC	McCoy Creek sequence: F–feldspathic quartzite	1200	
PROTEROZOIC	McCoy Creek sequence: E–phyllite & quartzite	1500	
PROTEROZOIC	McCoy Creek sequence: D–feldspathic qtzt	1100	
PROTEROZOIC	McCoy Creek sequence: C–micaceous quartzite	1500	
PROTEROZOIC	McCoy Creek sequence: B–pelitic schist	2000	FAULTED CONTACT
PROTEROZOIC	Trout Creek sequence: 7–feldspathic qtzt	4500	
PROTEROZOIC	Trout Creek sequence: 6–metasiltstone, micaceous quartzite, & pelitic schist	4000	
PROTEROZOIC	Trout Creek sequence: 5–metadiamictite, cg, qtzt	1200	
PROTEROZOIC	Trout Creek sequence: 4–micaceous quartzite	600	
PROTEROZOIC	Trout Creek sequence: 3–metasiltstone, qtzt, cg	1000	
PROTEROZOIC	Trout Creek sequence: 2–marble	160	diamictites
PROTEROZOIC	Trout Creek sequence: 1–pelitic schist	140	

*attenuated thickness associated with younger-on-older faults that cut out strata.

MAPS WITH TEXT - Nelson, 1966; Ahlborn, 1977; SNAKE VALLEY - Hood & Rush, 1966; GRANITES - Fest et al, 1974; Thomson, 1973; PROTEROZOIC - Christie -Blick, 1982; Murphy, 1975; Rodgers, 1984; STRUCTURE - Rodgers, 1985; Rodgers & Sutter, 1986; Gans et al, 1986; Miller et al, 1986.

Chart
48 NORTHERN SNAKE RANGE, NEVADA

Period	Unit	FEET	Remarks
Q	Alluvium, L. Bonneville seds	0-300	Snake Valley only
MIOCENE-PLIOCENE	Buried valley fill in Spring Valley (This buried sequence may be partly equivalent to Tertiary rocks exposed in Sacramento Pass as listed below.)	0- 9000	upper three-quarters mostly conglomerate lacustrine limestone and sandstone. Basal quarter may be mostly volcanogenic
MIOCENE ? OLIGOCENE	Sacramento Pass sequence: Younger Conglomerate	5000	Large landslide masses of semi-coherent Ordovician & Silurian strata form part of the upper two units of Sacramento Pass sequence
OLIGOCENE	Sacramento Pass sequence: Lacustrine deposits	1600	
OLIGOCENE	Sacramento Pass sequence: Rhyolite tuff	300	
OLIGOCENE	Sacramento Pass sequence: Latite flows	0-500	35 m.y.
OLIGOCENE	Sacramento Pass sequence: Older conglomerate	130	
OLIGOCENE	Silver Creek two-mica granitic pluton		25-30 m.y. reset? K
K	Pole Canyon granitic pluton		79 m.y. (near Lehman Cave)
J	Osceola granitic pluton		145 m.y.
PERM	Arcturus Fm	1000	yellow dolomitic ss
P	Ely Limestone	2000	cherty, fossiliferous
MISS	Chainman Shale	800	fossiliferous
MISS	Joana Limestone	200-400	
DEVONIAN	Pilot Shale	300- 500	
DEVONIAN	Guilmette Formation	2600	
DEVONIAN	Simonson Dolomite	800	
DEVONIAN	Sevy Dolomite	400	light gray
S	Laketown and Ely Springs Dolomites, undivided	1600	
ORD	Eureka Quartzite	10-300	usually brecciated
ORD	Pogonip Group	2100	Strata above the decollement have undergone as much as 500 percent brittle extension along normal faults that sole into the decollement. Thickness figures shown here are maximum values found in these fault blocks and do not necessarily represent the original depositional thickness.
CAMBRIAN	Notch Peak Formation	2300	
CAMBRIAN	Candland-Johns Wash-Corset Sp	300	
CAMBRIAN	Lincoln Peak Formation	650	
CAMBRIAN	Pole Canyon Limestone	800	↑ BRITTLE ↑
CAMBRIAN	marble tectonite	100 ±	SNAKE RANGE DECOLLEMENT
CAMBRIAN	Pioche Shale	0-150	↓ DUCTILE ↓
CAMBRIAN	Prospect Mountain Qtzt	400-700	
PROTEROZOIC	McCoy Creek Group: 1- schist	60	Strata below the decollement have been stretched plastically to one-tenth of their original thickness. Deformation was probably contemporaneous with Oligocene volcanism.
PROTEROZOIC	McCoy Creek Group: 2- quartzite	650	
PROTEROZOIC	McCoy Creek Group: 3- feldspathic qtzt	300	
PROTEROZOIC	McCoy Creek Group: 4- quartzite	150	
PROTEROZOIC	McCoy Creek Group: 5- schist	300	
PROTEROZOIC	McCoy Creek Group: 6- quartzite	300	
PROTEROZOIC	McCoy Creek Group: 7- schist	300	
PROTEROZOIC	McCoy Creek Group: 8- quartzite	350	

MAPS WITH TEXT - Gans et al, 1985, 1986; Miller et al, 1983, 1986; Gans & Miller, 1983; TERTIARY OF SACRAMENTO PASS - Grier, 1983.

Chart 49 THOMAS RANGE - KEG MTS

Age	Unit		FEET	Notes
Q	Alluvial, dune & playa deps		0-100	
Q	Deposits of Lake Bonneville		0- 100	
Q	Basalts of Crater Bench		0-600	0.95 m.y.
PLIO	Pre-Bonneville valley fill		0-1000	
PLIO	Basalt of The Hogback		0-150	5 m.y.
MIOCENE	Rhyolite of The Hogback		0-100	7 m.y.
MIOCENE	Conglomerate of The Hogback		0-750	includes clasts of Joy Tuff and Topaz Mtn Rhyolite
MIOCENE	Topaz Mountain Rhyolite		0-2000	TOPAZ CRYSTALS 8 m.y. flows & extrusive domes of three eruptive phases
MIOCENE	Rhyolite of Keg Mountains		0-1000	10 m.y.
MIOCENE	Spor Mountain Formation	Porphyritic rhyolite member	0-1500	flows, domes & plugs 21 m.y.
MIOCENE	Spor Mountain Formation	Beryllium tuff m	0-200	
OLIG	Tuff of Red Knolls		0-100	30 m.y. "Needles Range" Tuff
OLIG	Quartz latite stocks of Keg Mts			
OLIG	Dell Tuff		0- 500	32 m.y. qtz bipyramids
OLIG	Breccia at Spor Mtn		0-250	dolomite & quartzite blocks
EOC	Joy Tuff		0-700	38 m.y.
EOC	Mt. Laird Tuff		250-1600	39 m.y.
EOC	Keg Spring andesite and latite		0-300	39 m.y.
EOC	Drum Mountains Rhyodacite		0-100	42 m.y. MAJOR UNCONFORMITY
DEV	Sevy Dolomite		1100?	
SILURIAN	Laketown Dolomite	Thursday Member	330	
SILURIAN	Laketown Dolomite	Lost Sheep Mbr	230	*Pentamerus*
SILURIAN	Laketown Dolomite	Harrisite Mbr	140	*Halysites*
SILURIAN	Laketown Dolomite	Bell Hill Member	410	*Virgiana*
ORDOVICIAN	Ely Springs Dolomite	Floride Mbr	130	
ORDOVICIAN	Ely Springs Dolomite	Lower member	300	*Favosites*
ORDOVICIAN	Eureka Quartzite		590	
ORDOVICIAN	Pogonip Group	Kanosh Sh		*Orthambonites*
ORDOVICIAN	Pogonip Group	Juab Ls Wah Wah Ls		*Lachnostoma*
ORDOVICIAN	Pogonip Group	Fillmore Formation	2500 ±	*Trigonocerca* intraformational cg
ORDOVICIAN	Pogonip Group	House Ls		*Symphysurina* algal heads
CAMB	Notch Peak Formation		1000 -	

QUATERNARY - Galyardt & Rush, 1981; Holmes, 1984; TERTIARY - Lindsey, 1979, 1982; Staub 1975; PALEOZOIC - Staatz & Carr, 1964; modified by Dommer, 1980 and Budge & Sheehan, 1980; BERYLLIUM - Davis, 1984; MAP WITH TEXT - Pampeyan, 1986.

Chart 50 DRUM MOUNTAINS

Age	Unit		FEET	Notes
Q	Alluvium, L. Bonneville seds.		0-300	
T	Valley fill		0- 500 +	
OLIG	Tuff of Red Knolls		0-100	30 m.y. FT "Needles Range"
OLIG	Megabreccia of N. Drum Mt		0-200	
OLIG	Joy Tuff		0-600	38 m.y.
EOCENE	Mt. Laird Tuff		0-1500	39 m.y.
EOCENE	Diorite plugs near Joy		intrusive	
EOCENE	Shoshonite & latite of Little Drum Mts		0-2000	flows, flow-breccias laharic breccias & tuffs
EOCENE	Drum Mts Rhyodacite		0-100	42 m.y.
ORDOVICIAN	Eureka Quartzite		480	
ORDOVICIAN	Pogonip Group	Kanosh Shale	150	*Orthambonites*
ORDOVICIAN	Pogonip Group	Juab & Wah Wah Fms undivided	550	*Pseudocybele*
ORDOVICIAN	Pogonip Group	Fillmore Formation	1700	intraformational cg
ORDOVICIAN	Pogonip Group	House Limestone	240	*Symphysurina* algal stromatolites
CAMBRIAN	Notch Peak Formation		1300	
CAMBRIAN	Orr Fm	Sneakover Ls Mbr	220	*Eoorthis* *Kindbladia*
CAMBRIAN	Orr Fm	Corset Spr Sh Mbr	30	
CAMBRIAN	Orr Fm	Light purple member	270	
CAMBRIAN	Orr Fm	Big Horse Limestone Member	1000	silty bioclastic
CAMBRIAN	Lamb Dolomite		620	
CAMBRIAN	Trippe Ls	Fish Springs M	70	*Eldoradia*
CAMBRIAN	Trippe Ls	Lower member	360	white boundstone beds
CAMBRIAN	Pierson Cove Formation		800	
CAMBRIAN	Wheeler Formation		780	*Bolaspidella* *Elrathia* *Peronopsis*
CAMBRIAN	Swasey Limestone		180	
CAMBRIAN	Whirlwind Formation		140	*Ehmaniella*
CAMBRIAN	Dome Limestone		300	
CAMBRIAN	Chisholm Formation		150	*Glossopleura*
CAMBRIAN	Howell Limestone		160	
CAMBRIAN	Pioche Fm	Tatow Member	140	
CAMBRIAN	Pioche Fm	Lower member	230	
CAMBRIAN	Prospect Mountain Quartzite		4000	
PROTEROZOIC	Mutual Formation		3000	red and maroon quartzite
PROTEROZOIC	Inkom Formation		520	
PROTEROZOIC	Caddy Canyon Quartzite		550+	

TERTIARY - Lindsey, 1979, 1982; Leedom, 1974; Pierce, 1974; PALEOZOIC - Dommer, 1980; CAMBRIAN - Rowell et al, 1982; PROTEROZOIC - Christie-Blick, 1982.

Chart

51 HOUSE RANGE AREA

Age	Unit		FEET	Remarks
Q	Alluvium, L. Bonneville seds.		0-500	
PLIO	Tule Valley fill		0-2000	1500' drill hole showed 800' of lake marls above alluvial sand and gravel
OLIG	Needles Range Tuff		0-50	29 m.y.
OLIG	Skull Rock Pass Cg		0-200	
OLIG	Freshwater ls & breccia		0-300	fossil leaves
OLIG	Tunnel Spring Tuff		0-50	33 m.y.
OLIG	Dacitic ash-flow tuff		0-200	36 m.y. Whirlwind Valley
K	Breccia & mylonite along SE trending tear faults of Sevier Orogeny			
J	Notch Peak quartz monzonite intrusion			170 m.y.
ORD	House Limestone		500	*Hystricurus*, *Symphysurina*
LATE CAMBRIAN	Notch Peak Fm	Lava Dam Member	380	stromatolites, *Missisquoia*
LATE CAMBRIAN	Notch Peak Fm	Red Tops Mbr	110	*Saukiella*
LATE CAMBRIAN	Notch Peak Fm	Hellnmaria Member	1200	stromatolites, *Matthevia*, *Idahoia*
LATE CAMBRIAN	Orr Fm	Sneakover Ls M	110-155	
LATE CAMBRIAN	Orr Fm	Corset Spring Sh M	100-110	*Taenicephalus*, *Elvinia*
LATE CAMBRIAN	Orr Fm	Johns Wash Ls M	140-200	
LATE CAMBRIAN	Orr Fm	Candland Sh M	210-270	*Dunderbergia*, *Aphelaspis*
LATE CAMBRIAN	Orr Fm	Big Horse Limestone Member	720	*Crepicephalus*, bioclastic ls, *Cedaria*
MIDDLE CAMBRIAN	Weeks Fm (found only in central House R) 1200'	Lamb Dolomite	450	
MIDDLE CAMBRIAN	Trippe Ls	Upper Mbr	320	white boundstone common in U. Trippe
MIDDLE CAMBRIAN	Trippe Ls	Lower Mbr	520	*Lejopyge*
MIDDLE CAMBRIAN	Marjum Formation of central House Range 1360'-1410' shaly with many fossils	Pierson Cove Formation of eastern flank of House Range 1200' few fossils	1200-1410	very dark gray, *Marjumia*, *Bolaspidella*
MIDDLE CAMBRIAN	Wheeler Shale		420-490	*Asaphiscus*, *Elrathia*, *Peronopsis*
MIDDLE CAMBRIAN	Swasey Limestone		250	
MIDDLE CAMBRIAN	Whirlwind Formation		140	*Ehmaniella*
MIDDLE CAMBRIAN	Dome Limestone		320	
MIDDLE CAMBRIAN	Chisholm Formation		220	*Glossopleura*
MIDDLE CAMBRIAN	Howell Ls	Upper member	380	light gray
MIDDLE CAMBRIAN	Howell Ls	Millard Member	310	dark gray
EARLY CAMBRIAN	Pioche Fm	Tatow Member	180	
EARLY CAMBRIAN	Pioche Fm	Lower member	420	*Olenellus*, phyllitic siltstone and quartzite
EARLY CAMBRIAN	Prospect Mountain Quartzite		2000+ exposed; regional thickness 4000+	pink, vitreous, minor cross bedding

MAPS - Hintze, 1981a, b,c; TULE VALLEY - Stephens, 1975; JURASSIC - Granath et al, 1983; UPPER CAMBRIAN - Hintze et al, 1980; Hintze & Palmer, 1976; Rees et al, 1976; MIDDLE CAMBRIAN - Hintze & Robison, 1975; Robison, 1976; 1982; Gunther & Gunther, 1981; Brady & Koepnick, 1979; COCORP SUBSURFACE-- Von Tish et al, 1985

Chart

52 CRICKET MOUNTAINS AREA

Age	Unit		FEET	Remarks
Q	Alluvium, L. Bonneville seds.		0-200	
Q	Basalt of Black Rock		0-100	1.3 m.y.
PLIOCENE	Lacustrine sediments		0-200+	2.0 m.y. ash included
PLIOCENE	Rhyolite at Mid-Dome		0-200	2.5 m.y.
PLIOCENE	Freshwater limestone		0-100	
PLIOCENE	Rhyolite of Cudahy Mine		0-500	2.4 ? m.y. snowflake obsidian
PLIOCENE	Rhyodacite of Coyote Hills		0-600	2.7 m.y.
PLIOCENE	Plio-Pleistocene shale, marl, limestone and sandstone in upper 1500 feet of well		1500	1982 Cominco wall - Sec 28 T20S, R8W penetrated Precambrian and Cambrian rocks below 3200 feet
MIO	Miocene ? marl, dolomite, shale and sandstone in well		1000	
PALEOC	Conglomerate of Red Pass		0-300	–Cg of High Rock Pass of Lemmon & Morris
PALEOC	Limestone of Fillmore Canyon		0-150	
PALEOC	Breccia of Cat Canyon		0-150	€ Carbonate blocks
CAMBRIAN	Notch Peak Formation		1600	regional thickness; incomplete section in Cricket Mts.
CAMBRIAN	Orr Fm	Sneakover Ls Mbr	100	
CAMBRIAN	Orr Fm	Corset Spring Sh M	40	*Elvinia*
CAMBRIAN	Orr Fm	Johns Wash Ls M	100	
CAMBRIAN	Orr Fm	Candland Shale M	160	*Dunderbergia*
CAMBRIAN	Orr Fm	Big Horse Ls Mbr	650	*Crepicephalus*
CAMBRIAN	Wah Wah Summit Fm	White Marker Mbr	160-230	
CAMBRIAN	Wah Wah Summit Fm	Ledgy member	400	
CAMBRIAN	Trippe Ls	Fish Springs Mbr	100	*Eldoradia*
CAMBRIAN	Trippe Ls	Lower member	650	white algal boundstone beds
CAMBRIAN	Limestone of Cricket Mountains		2000	unfossiliferous mottled limestone, dolomite and algal boundstone
CAMBRIAN	Whirlwind Formation		200-260	*Ehmaniella*
CAMBRIAN	Dome Limestone		230-320	
CAMBRIAN	Chisholm Formation		160-260	*Glossopleura*
CAMBRIAN	Howell Ls	Upper member	100-160	
CAMBRIAN	Howell Ls	Millard Member	200-250	
CAMBRIAN	Pioche Fm	Tatow Member	100	hematite bed
CAMBRIAN	Pioche Fm	Lower member	600-700	*Olenellus*
CAMBRIAN	Prospect Mountain Quartzite		7300	pink vitreous quartzite; basalt flow 5600 ft below Pioche Fm; white to pink arkosic quartzite
PROTEROZOIC	Mutual Quartzite		2100	maroon to purplish-red coarse-grained quartzite
PROTEROZOIC	Inkom Fm		500	
PROTEROZOIC	"Caddy Canyon Qtzt"		300	names in quotes may all be facies of Caddy Canyon Qtzt of SE Idaho
PROTEROZOIC	"Blackrock Canyon Limestone"		800	
PROTEROZOIC	"Pocatello Formation"		150	

FRISCO THRUST AT 8340 FT DEPTH COMINCO WELL PUTS p€/ €

MAPS & STRATIGRAPHY - Hintze, 1984; Lemmon & Morris, 1984; VALLEY FILL - Case & Cook, 1979; unpublished Cominco well log, 1982; CENOZOIC - Crecraft et al, 1981; Mehnert et al, 1978; Whelan & Bowdler, 1979; PROTEROZOIC - Woodward, 1968; Christie-Blick, 1982b.

Chart

53 GOSHEN - NORTHERN LONG RIDGE

Period	Formation		FEET	Notes
Q	Alluvium, Utah Lake, and Lake Bonneville deposits		0-1000	
M-P	Buried valley fill in Goshen Valley		several thousand	
OLIGOCENE	Goldens Ranch Fm		0-70	
	Laguna Springs Latite		0-300	32 m.y. welded tuff
	Latite Ridge Latite		0-200	
	Copper-opolis Latite	Upper flow	0-400	
		Upper tuff	0-20	these volcanic rocks were all erupted from the Tintic volcanic center between 32 and 33 m.y. ago
		Upper tuff-breccia	0-1000	
		Middle flow	0-950	
		Mid. tuff-breccia	0-200	
K-T	North Horn Formation		0-100+	ANGULAR UNCONFORMITY
MISSISSIPPIAN	Great Blue Limestone		200+	
	Humbug Formation		650	
	Deseret Limestone		600-700	Delle Phosphatic Mbr 100'
	Gardison Limestone		400	many fossils
D	Fitchville Dolomite		230	
	Pinyon Peak Limestone		50-100	Victoria included with Pinyon Pk by some
	Victoria Formation		40-120	
S	Bluebell Dolomite		230	
O	Opohonga Limestone		250	intraformational cg
CAMBRIAN	Ajax Dolomite		300	
	Opex Formation		200	
	Cole Canyon Dolomite		440	includes white beds of laminated boundstone
	Bluebird Dolomite		180	white twiggy bits
	Herkimer Limestone		240-320	*Lingulella*
	Dagmar Dolomite		50	white boundstone
	Teutonic Limestone		370	pisolitic at base
	Ophir Formation		300	*Ehmaniella* *Glossopleura*
	Tintic Quartzite		1400	*Skolithos*
		Diabase flow	50	
			160	
PROTEROZOIC	Mutual Formation		2700	purple and red slate and argillite inter-bedded with pink and orange quartzite
	Big Cottonwood Formation		130+	pink quartzite

MAP WITH TEXT - Jensen, 1985; Clark, 1953; Peacock, 1953; Petersen, 1953; Peterson, 1953; Sirrine, 1953, MISSISSIPPIAN-DEVONIAN - Rigby & Clark, 1962; Sandberg & Gutschick, 1984; LOWER PALEOZOIC - Hintze, 1962; Brady, 1965; PRECAMBRIAN - Woodward, 1972.

Chart

54 MT. NEBO - SANTAQUIN AREA

Period	Formation		FEET	Notes
Q	Alluvium & colluvium		0-5000	0.7 m.y. Bishop Ash
T	Moroni Formation		0-800	volcaniclastics
	Flagstaff Limestone		0-50	
CRET	North Horn Fm		0-600	red, post-orogenic ANGULAR UNCONFORMITY
	Indianola Group		0-50+	
	Cedar Mountain Fm		600	Rees Flat area
JURASSIC	Twist Gulch Fm		500	
	Arapien Shale		2000+	gypsiferous
	Twin Creek Ls	Watton Canyon M	100	intensely contorted
		Boundary Ridge M	100	
		Rich Member	100-200	*Pentacrinus*
		Sliderock Member	50-80	
		Gypsum Spring M	10	Triassic-Jurassic rocks are generally attenuated near the Nebo Thrust where they are not as thick as listed here
	Nugget Sandstone (Navajo)		1400	
TRIASSIC	Ankareh Formation		900	
	Thaynes Formation		800-1300	
	Woodside Shale		200-1000?	
PERMIAN	Park City Group	Franson Fm	430	
		Meade Peak Phosp	160	cherty, phosphatic Mbr of Phosphoria Fm
		Grandeur Fm	600	
	Diamond Creek Ss		330-800	
	Kirkman Limestone		130-170	
PENNSYLVANIAN	Oquirrh Group Undivided		12,000?	Mt. Nebo crest
MISSISSIPPIAN	Manning Canyon Shale		1000+	slumps & landslides
	Great Blue Limestone		900+	thin-bedded
	Humbug Formation		700	
	Deseret Limestone		800	
	Gardison Limestone		300	
D	Fitchville Formation		230	
CAMBRIAN	Ajax Dolomite		0-340	cut out to north
	Opex Formation		200	
	Cole Canyon Dolomite		500	
	Bluebird Dolomite		80-150	
	Herkimer Limestone		230-440	
	Dagmar Dolomite		40	white boundstone
	Teutonic Limestone		350-410	
	Ophir Formation		240	
	Tintic Quartzite		800-850	
		Diabase flow	50	amygdaloidal
			70-90	
pC	Big Cottonwood Fm		860-1230	
	Farmington Canyon Complex-schist			

MAP WITH TEXT - Black, 1965; Brady, 1965; Biek, 1986; Levot, 1985; Sorensen et al, 1983; Metter, 1955; Demars, 1956; Foutz, 1960; Smith, 1965; DEVONIAN - Rigby, 1959; KIRKMAN - LaRell Nielson letter, 1985; BISHOP ASH - Nash & Smith, 1977. QUATERNARY - Zoback, 1987.

Chart

55 MORONI - CEDAR HILLS

Period	Unit		FEET	Remarks
Q	Younger alluvial deposits		0-200	
Q	Older alluvial deposits		0-500?	
TERTIARY	Salt Creek Fanglomerate		0-500	
TERTIARY	Moroni Formation		200-2000	erroneously called Goldens Ranch Fm in old reports; 34-38 m.y.
TERTIARY	Crazy Hollow Fm		200	varicolored muds
TERTIARY	Green River Formation		0-1200	44 m.y.
TERTIARY	Colton Formation		0-700	
TERTIARY	Flagstaff Limestone		0-700	white ls with interbedded clastics
TERTIARY / CRETACEOUS	North Horn Formation		750-1500	poorly exposed, forms red soil with pebble & cobble float, some local limestone lenses
				ANGULAR UNCONFORMITY
CRETACEOUS	Indianola Group	Sixmile Canyon equivalent	7100	fluvial deposits; massive alluvial fan conglomerate
CRETACEOUS	Indianola Group	Funk Valley equivalent	1600-2000	*Inoceramus* marine ss & shale with interbedded cg
CRETACEOUS	Indianola Group	Allen Valley eq	0-360	marine shale & ss *Collignoniceras*
CRETACEOUS	Indianola Group	Sanpete Fm equivalent	1400	braided fluvial pebbly sandstone
CRETACEOUS	Cedar Mountain Formation (characteristically contains calcareous nodules, gastroliths in lower third, interbedded cg, ss, grit, and locally pink oncolitic limestone beds)		1300	NOTE: Indianola, Cedar Mountain, and Twist Gulch units thin drastically in the vicinity of folds with Arapien cores. See Witkind, 1983.
JURASSIC	Twist Gulch Formation		600-1000	poorly exposed
JURASSIC	Arapien Shale		several thousand	gypsiferous, with salt locally; incompletely exposed

MAP WITH TEXT - Jefferson, 1982; Hawks, 1980; Banks, 1986; Fograsher, 1956; Cooper, 1956; STRATIGRAPHY - Hardy, 1962; Witkind, 1983; GROUNDWATER - Robinson, 1971.

Chart

56 FAIRVIEW - SCOFIELD AREA

Period	Unit		FEET	Remarks
Q	Alluvium, valley fill		0-500	thickest in Sanpete Valley
TERTIARY	Alkalic diabase dikes & sills		1-5' wide	5 m.y.
TERTIARY	Mica-peridotite dikes & sills		1-5' wide	24 m.y.
TERTIARY	Moroni Formation		0-700	agglomerate, tuff & welded tuff, qtzt conglomerate at base
TERTIARY	Green River Fm	Upper member	1400-2000	formerly called Crazy Hollow; calcareous shale & ls, minor ss & cg
TERTIARY	Green River Fm	Middle member	320	algal, ostracodal, oolitic
TERTIARY	Green River Fm	Lower member	580	green and brown shale & ss, lesser ls, cg
TERTIARY	Colton Formation		200+	red sh, ss, mostly covered
TERTIARY	Flagstaff Limestone		300+	
TERTIARY / CRETACEOUS	North Horn Formation		1400-2000	prone to slumping thin lignite beds
CRETACEOUS	Mesaverde Group	Price River Formation	240-320	
CRETACEOUS	Mesaverde Group	Castlegate Sandstone	220-320	
CRETACEOUS	Mesaverde Group	Blackhawk Formation	1100-1300	coal
CRETACEOUS	Mesaverde Group	Star Point Sandstone	300-600	Three States Utah Fuel #8 18-13S-7E and Pacific-Western Gordon Creek #1 24-14S-7E and Skelly #5 Unit 25-14S-7E
CRETACEOUS	Mancos Shale	Upper Blue Gate Shale Member (formerly incorrectly called Masuk Mbr)	1050	
CRETACEOUS	Mancos Shale	Emery Sandstone Member	1400-1700	
CRETACEOUS	Mancos Shale	Lower Blue Gate Shale Member	1850	dark gray marine shale
CRETACEOUS	Mancos Shale	Ferron Ss Mbr	490	
CRETACEOUS	Mancos Shale	Tununk Sh Mbr	330	dark gray marine shale
CRETACEOUS	Dakota Sandstone		160	
CRETACEOUS	Cedar Mountain Formation		1440	formerly identified as "Morrison?"
JURASSIC	Summerville Formation		450-520	equivalent to the Twist Gulch Fm in Sanpete Valley
JURASSIC	Curtis Formation		170-240	
JURASSIC	Entrada Formation		1080-1150	equivalent to the Arapien Shale in Sanpete Valley
JURASSIC	Carmel Formation		1220	Twin Creek Ls equiv (lower part)
JURASSIC	Navajo Sandstone		340	
JURASSIC	Kayenta Formation		130	
JURASSIC	Wingate Sandstone		350	
TRIASSIC	Chinle Shale		280-330	
TRIASSIC	Moenkopi Fm	Upper member	750	
TRIASSIC	Moenkopi Fm	Sindbad Ls Mbr	100	
TRIASSIC	Moenkopi Fm	Lower member	360	
PERM	Park City Formation		230	
PERM	Diamond Creek Sandstone		750	
	Older strata not penetrated by drill			

MAP WITH TEXT - Oberhansley, 1980; Knowles, 1985; Jensen, 1986; DIKES - Thomas, 1976; Tingey, 1986; SOILS - Swenson et al, 1981; CRETACEOUS - Doelling, 1972; Peterson et al, 1980; SUBSURFACE - Walton, 1954; Edson et al, 1954; Wells, 1954; Irwin, 1971, 1976; GROUNDWATER - Robinson, 1971.

Chart 57 GILSON - S. EAST TINTIC MTS

Age	Unit		Feet	Remarks
Q	Alluvium, L. Bonneville seds		0-300	12,000 y.b.p. dates on shells
OLIGOCENE	Copperopolis Latite	Middle agglomerate member	0-1000	
		Sage Valley Ls Mbr	0-300	algal limestone
	Fernow Quartz Latite		0-1500	welded ash-fall tuff
	Volcanic gravel		0-300	
T	Flagstaff Limestone		200+	Paleocene; cg with ls clasts
PERMIAN	Franson Member of the Park City Formation		540	
	Meade Peak Shale Tongue of the Phosphoria Fm		250-330	phosphate
	Grandeur Member of the Park City Formation		750-960	
	Diamond Creek Sandstone		680-880	
	Oquirrh Group	Furner Valley Limestone	5000-6000	*Pseudofusulina*, *Schwagerina*, cherty, silty ls, *Triticites*
PENNSYLVANIAN		Bingham Mine Formation	3200-3400	*Triticites*, *Eowaeringella*
		Butterfield Peaks Formation	5800	*Beedeina*, *Wedekindellina*, *Fusulinella*
		West Canyon Limestone	960+	*Composita*, *Rhipidomella*
MISSISSIPPIAN	Manning Canyon Shale		(1000 ±)	not exposed, faulted out?
	Great Blue Formation	Chiulos Member	625	black shale & brown quartzite
		Lower member	600	*Turbophyllum*
	Humbug Formation		570-750	*Lithostrotionella*, *Syringopora*
	Deseret Limestone		560-1100	bryozoans
		Delle Phosphatic Mbr	70	phosphatic shale
	Gardison Limestone		360-570	*Syringopora*
	Fitchville Formation		260-400	*Composita*
DEV	Pinyon Peak Limestone		130-180	
	Guilmette Formation		200-250	10 ft quartzite bed
	Simonson Dolomite		140	
	Sevy Dolomite		360	
S	Laketown Dolomite		870	*Halysites*
ORD	Fish Haven Dolomite		240	*Favosites*
	Opohonga Limestone		1000	*Kirkella*, intraformational conglomerate
Є	Ajax Dolomite		590	

MAPS WITH TEXT - Costain, 1960; Morris, 1977; Higgins, 1982; Wang, 1970; Pampeyan, 1986; OQUIRRH GROUP - Morris et al, 1977.

Chart 58 MILLS - SOUTHERN LONG RIDGE

Age	Unit			Feet	Remarks
Q	Younger alluvium-colluvium			0-500	pediments, alluvial fans, landslides, slumps
	Older alluvium & valley fill			0-1000	
TERTIARY	Cazier Canyon Agglomerate			0-750	
	Copperopolis Latite			0-50	flow
	Fernow Quartz Latite			0-50	33 m.y.
	Goldens Ranch Fm	Sage Valley Ls M		0-260	abundant fossil leaves
		Hall Canyon Cg Mbr	V unit	400	many volcanic clasts
			Q unit	400	quartzite & ls clasts
		Chicken Cr Tuff M		0-100	33-34 m.y.
	Green River Fm			820	Flagstaff-Colton-Green River become conglomeratic to the north
	Colton Fm	Orme Spring Cg		260	
	Flagstaff Fm			320	UNCONFORMITY
CRETACEOUS	North Horn Formation			1200	Red Narrows Cg of Meibos
	Cg of Spring Canyon (Upper Indianola?)			1100	MAJOR UNCONFORMITY Rocks beneath are folded
	Undivided Cretaceous strata - Indianola and Cedar Mountain Fms?			2000	Placid Oil WXC #1A Howard 5-14S-1W
JURASSIC	Twist Gulch Formation			2000	reddish brown siltstone, ss, grit
	Arapien Shale			2800	Arapien contains many salt & gypsum beds, some as much as 100 feet thick
	Twin Creek Limestone			800	
	Nugget Sandstone			1400	bottom of well
TRIASSIC	Ankareh Formation			950?	Triassic strata inferred by Meibos from regional exposures east of Nephi, and from interpretation of subsurface
	Thaynes Formation			1300?	
	Woodside Shale			1000?	
PERMIAN	Park City Group	Franson Fm		540	Paleozoic strata exposed in hills in Dog Valley and on Furner Ridge
		Meade Peak Phos		300	
		Grandeur Limestone		850	
	Diamond Creek Ss			500-800	
PENNSYLVANIAN	Oquirrh Group	Furner Valley Limestone		5400	
		Bingham Mine Formation		3200	
		Butterfield Peaks Formation		5800	

MAP WITH TEXT - Meibos, 1983; Morris, 1977; Clark, 1986; TERTIARY - Lambert, 1976; Vorce, 1979; TWIN CREEK - Sprinkel, 1982; GROUNDWATER - Bjorklund & Robinson, 1968.

FIGURE 138 — West face of House Range as seen from mile 28 on U.S. Highway 6-50. Top of Notch Peak is type section for Late Cambrian Notch Peak Formation. Tree-covered bench below the peak is made by shaly members of the upper Orr Formation. Jurassic granitic intrusion forms knobby slopes left of joint-controlled canyon. See Chart 51.

FIGURE 139 — North end of Canyon Range looking east to distant Mt. Nebo. Utah Highway 132 and railroad follow Sevier River through Leamington Canyon at left side. Light-colored beds on lower left slope are Lake Bonneville deposits. Near-vertical beds on spur just to right of railroad-highway merger is Precambrian Mutual Formation (m) and Cambrian Tintic Quartzite (t). Pioche Formation (p) forms first saddle. Middle Cambrian limestones and shales (ls) form strike ledges. Hill top is capped by Upper Cretaceous Canyon Range Conglomerate (cr). See Chart 59.

FIGURE 140 — Fossil Mountain at Ibex in the southern Confusion Range. Ordovician Juab Limestone ledges in foreground. White peak top is Eureka Quartzite; gray ledges beneath are Crystal Peak Dolomite; light ledges beneath are Watson Ranch Quartzite; slopes to base of hill are Lehman Formation and Kanosh Shale. See Chart 70.

FIGURE 141 — Blue Mountain thrust near Lund. Arrow points down thrust surface which separates dark gray Middle Cambrian carbonates and shales from underlying light-colored sandstones and siltstones of the Jurassic Temple Cap Formation. See Chart 74.

Chart
59 LEAMINGTON - CANYON RANGE

Age	Formation	FEET	Notes
Q	Alluvium & Lake Bonneville sediments	0-600	L. Bonneville 16-14,000 y.b.p.
OLIG ? - MIO - PLIO	terrestrial shale with some sandstone and siltstone	1940	Argonaut well near IPP site
	halite and minor anhydrite and bentonitic claystone	5150	
	pink to red conglomerate	30	
OLIGOCENE	Andesitic rocks of Sage Valley	0-200	
	Fool Creek Conglomerate	3800	unconsolidated clasts of quartzite and limestone
U.CRETACEOUS-PALEOC	Red beds of Wide Canyon	1200-2500	=Flagstaff Fm?
	Canyon Range Formation (Conglomerate of Leamington Pass of Higgins)	3600	Slide blocks of limestone breccia
CAMBRIAN	Undifferentiated limestone & dolomite	1200	MAJOR UNCONFORMITY boundstone
	Wheeler Shale	100	*Elrathina*
	Swasey Limestone	600	
	Whirlwind Formation	140	*Ehmaniella*
	Dome Limestone	180	
	Chisholm Formation	250	*Glossopleura*
	Howell Limestone	300	
	Pioche Formation	750	*Skolithus*
	Tintic Quartzite	2700	pale orange
PROTEROZOIC	Mutual Formation	1900-2400	
	Inkom Formation	300	
	"Caddy Canyon Quartzite"	950	Names in quotes have been taken from SE Idaho, perhaps inappropriately, because of lack of better formal Utah nomenclature.
	"Papoose Creek Formation"	750	
	"Blackrock Canyon Limestone"	150-550	
	"Pocatello Formation" Upper member	720-900	
	"Pocatello Formation" Middle member	1300	
	"Pocatello Formation" Lower member	600	

CANYON RANGE ALLOCHTHON RESTS ON PAVANT ALLOCHTHON

MAPS WITH TEXT - Higgins, 1982; Holladay, 1984; Michaels, 1986; QUATERNARY - Varnes & VanHorn, 1984; Oviatt, 1984; Currey et al, 1984; ARGONAUT WELL - Lindsey et al, 1984; TERTIARY - Campbell, 1979; CRETACEOUS - Stolle, 1978; Higgins, 1982; CAMBRIAN - Higgins, 1982; PROTEROZOIC - Christie-Blick, 1982b; COCORP SUBSURFACE - Van Tish et al, 1985.

Chart
60 SCIPIO PASS - PAVANT ALLOCHTHON

Age	Formation	FEET	Notes
Q	Alluvium & colluvium	0-100	
	Older fanglomerate	0-250	
MIO	Oak City Formation	0-1600	unconsolidated pinkish soil
	Goldens Ranch Fm ?	150+	
OLIGO	Fool Creek Conglomerate	0-3800	mostly unconsolidated clasts of quartzite and limestone
U. CRETACEOUS - PALEOCENE	Red beds of Wide Canyon	1200+	=Flagstaff Fm ?
	Canyon Range Formation Upper member	4000	=North Horn Fm ? oncolites gastropod impressions
	Canyon Range Formation Middle member	840	
	Canyon Range Formation Lower member	590	=Price River Fm ?

OROGENIC & LATER DEPOSITS LIE ABOVE MAJOR UNCONFORMITY

Age	Formation	FEET	Notes
DEVONIAN	Cove Fort Quartzite	250	PAVANT ALLOCHTHON INCLUDES DEVONIAN TO LATE PROTEROZOIC STRATA
	Simonson Dolomite	770	
	Sevy Dolomite	2700	
S	Laketown Dolomite	1100	orthid brachiopod rugose coral cherty stromatolites
ORD	Fish Haven Dolomite	140	
	Eureka Quartzite	220	
	Kanosh Shale	370	*Orthambonites*
	Pogonip Group	1300	intraformational conglomerate
CAMBRIAN	Ajax Dolomite	960	interbedded light & dark gray, algal heads
	Opex Formation	670	*Tricrepicephalus*
	Cole Canyon Dolomite	540	
	Herkimer Limestone	350	
	Dagmar Dolomite	85	boundstone
	Teutonic Limestone	560	*Glossopleura*
	Pioche Formation	260-400	*Skolithus*
	Tintic Quartzite	3100	forms pale orange cliffs
PROT	Mutual Formation	1500	red and pinkish-purple quartzite

MESOZOIC STRATA BENEATH PAVANT ALLOCHTHON

MAPS WITH TEXT - Millard, 1983; Holladay, 1984; Michaels, 1986; TERTIARY - Campbell, 1979; CRETACEOUS - Stolle, 1978; CAMBRIAN - Michaels, 1986.

Chart
61 GUNNISON PLATEAU

Period	Unit		FEET	Remarks
Q	Alluvium, colluvium, mass movement, & terrace deposits		0-500	
TERTIARY	Salt Creek Fanglomerate		0-50	reddish, crudely bedded
	Moroni Formation		0-150	34-38 m.y. volcanics
	Leucomonzonite intrusive bodies			Levan stock
	Monzonite porphyry intrusions			dikes & small stocks
	Crazy Hollow Formation		0-100	black chert pebbles
	Green River Formation		300-1400	Early Tertiary formations thin over loci of Arapien upwelling
	Colton Formation		0-800	
	Flagstaff Limestone		0-900	
CRETACEOUS	North Horn Formation		0-840	
	Price River Formation		0-500	may be partly Indianola as presently mapped
	South Flat Formation Lawton(1987, letter) says South Flat as mapped includes parts of Indianola and North Horn		0-950	Blackhawk Fm equivalent fossil leaves abundant ANGULAR UNCONFORMITY between Price River and Indianola at Wales
	Indianola Formation or Group	Member IV	1430	gray quartizte cg
		Member III	0-800	tan conglomerate
		Member II	0-200	greenish sandstone
		Member I	1800-5560	some fresh-water ls Paleozoic limestone and dolomite clasts are abundant in cg of member I
	Cedar Mountain Formation		250+	60' thick at Wales Gap thicker on west side
JURASSIC	Twist Gulch Formation		460-530	Summerville-Curtis equivalent
	Arapien Shale		5150-5470	greatly contorted in surface outcrops; diapiric Chevron #1 Chriss Canyon 33-16S-1E Dixel #1 Gunnison 15-16S-1E 800-1000 feet of salt 1700 feet below top 50 feet of anhydrite 50 feet of salt 50 feet of salt Arapien is equivalent to Imlay's Leeds Creek and Giraffe Creek members
	Twin Creek Limestone		310	includes Imlay's lowest five members
	Navajo Sandstone		1000	

MAP WITH TEXT - Auby, 1986; Banks, 1986; Mattox, 1986; McDermott, 1986; HYDROGEOLOGY - Gates, 1982; TERTIARY - Weiss, 1969, 1982; Millen, 1982; Fograsher, 1956; Stanley & Collinson, 1979; Marcantel & Weiss, 1968; CRETACEOUS - Thomas, 1960; Hays, 1960; JURASSIC - Standlee, 1982; Sprinkel, 1982.

Chart
62 REDMOND - VALLEY MOUNTAINS

Period	Unit	FEET	Remarks
Q	Alluvium, terrace deposits	0-300	Axtell Fm is locally deformed by diapiric movement of Arapien Shale.
	Axtell Formation	0-50	
TERTIARY	Valley fill - Sevier Valley	0-800	
	Sevier River Fm	0-1500	5 m.y FT
	Formation of Aurora	0-600	39 m.y.
	Crazy Hollow Formation	0-1000	
	Green River Formation	0-1200	yellow fresh-water limestone green shale beds
	Colton Formation	0-450	varicolored sh, ss
	Flagstaff Limestone	200-1400	
CRET	North Horn Formation	100-1600	
	Price River Formation	0-50	ANGULAR UNCONFORMITY
	Indianola Conglomerate	500+	
JURASSIC	Twist Gulch Formation	200+	red siltstone
	Arapien Shale	5000±	gray & red mudstone, siltstone, ss & ls including gypsum and salt. Arapien occurs chiefly as large diapiric masses that affect all Tertiary units, not only by causing depositional thinning but also by producing post-depositional tilting

Well below is about 8 miles west of the nearest exposure of Arapien Shale and did not encounter any Arapien.

Placid Oil Co WXC-USA 1-2 24-19S-2W

Period	Unit	FEET	Remarks
T	North Horn Formation	5600	
K	Indianola Group	1200	
ORDOVICIAN	Fish Haven Dolomite	200	
	Eureka Quartzite	130	Eureka repeated by faults in well
	Lehman & Kanosh Fms	210	
	Juab, Wah Wah, Fillmore & House fms undivided	1700	intraformational cg
CAMBRIAN	Ajax Dolomite	620	
	Opex Formation	390	
	Cole Canyon & Bluebird Dol	600	
	Herkimer & Dagmar Fms	370	
	Teutonic Limestone	640?	
	Ophir Formation	300	
	Tintic Quartzite	750	
pC	Crystalline basement	-	1.1 b.y. K-Ar

MAP WITH TEXT - Gilliland, 1951; Witkind, 1981; STRUCTURE - Witkind & Page, 1984; SUBSURFACE - J. L. Baer unpublished; HYDROGEOLOGY - Gates, 1982; Young & Carpenter, 1965.

Chart

63 WASATCH PLATEAU EAST OF MANTI

Period	Group / Formation	Member	FEET	Notes
Q	Joes Valley graben fill		0-200	
PALEOC	Flag staff Ls	Musina Peak Mbr	120	mudstone, dolomite, gypsum
		Cove Mtn Mbr	560	limestone & mudstone
		Ferron Mtn Mbr	300-600	
	North Horn Formation		1300-1600	red mudstone beds; oncolitic limestone
CRETACEOUS	Mesaverde Group	Price River Fm	400-700	Upper Cretaceous Units become coarser westward; channel sands and muds inter-finger complexly so that recognition of surface map units in well logs is uncertain
		Castlegate Sandstone	100-300	
		Blackhawk Formation	500-800	
		Star Point Ss	200-300	
	Mancos Shale	Upper Blue Gate Shale Member	500-600	formerly Masuk Mbr
		Emery Sandstone Member	1400-1600	coal
		Lower Blue Gate Shale Member	2100	Phillips Petroleum 27-195-3E 1975; subsurface; marine
		Ferron Sandstone Mbr	660	coal
		Tununk Shale Member	960	marine
	Dakota Sandstone		110	
	Cedar Mountain Fm		250	
JURASSIC	Morrison Fm ?		440	lower Cedar Mtn Fm ?; J-5 unconformity
	Twist Gulch Formation		1120	Summerville equiv 930'; Curtis equiv 190'; Entrada equiv 540'
	Arapien Shale		1540	salt & gypsum beds equivalent to upper two members of Twin Creek Ls in north-central Utah
	Twin Creek Limestone		500	Twin Creek here used for lowest 5 members of Imlay
	Navajo Sandstone		740	
	Kayenta Fm		120	
	Wingate Sandstone		360	
TRIASSIC	Chinle Fm	Upper Chinle Sh	30	
		Moss Back Ss M	35	TR-3 unconformity
	Moen-kopi Fm	"Upper red" mbr	310	
		"Shnabkaib" Mbr	480	anhydrite
		"Middle red" mbr	450	
		Sinbad Ls Mbr	540	TR-1 unconformity; Welch et al, 1979
PERM	Black Box Dolomite		140	
	Toroweap Fm		460	
	Elephant Canyon Fm		100	EMERY UPLIFT
MISS	Redwall Ls	Upper member	180	
		Thunder Springs M	470	chert
		Whitmore Wash M?	180	
D	Unnamed Devonian rocks		390	
CAMBRIAN	"Lynch" Dolomite		710	
	Maxfield Limestone		580	
	Ophir Shale		240	
	Tintic Quartzite		130+	

MAP WITH TEXT--Kitzmiller, 1982; PALEOCENE--Fadagau, 1949; Tomida & Butler, 1980; Stanley & Collinson, 1979; CRETACEOUS--Sanchez & Brown, 1983; Flores & Marley, 1979; Doelling, 1972; Peterson et al, 1980; SUBSURFACE--Phillips Petroleum No. 1 United States E well log as interpreted by J. L. Baer.

Chart

64 HUNTINGTON - FERRON - EMERY

Period	Group / Formation	Member	FEET	Notes
Q	Alluvium & terrace deposits		0-200	
TERTIARY	Alkalic diabase dikes & sills		1-5 ft wide	8 m.y. 18 m.y. 24 m.y.
	Green River Fm		100 –	
	Colton Formation		300-1500	
	Flagstaff Limestone		200-1500	freshwater ls
	North Horn Formation		500-2500	prone to slump
CRETACEOUS	Mesaverde Group	Price River Formation	600-1000	
		Castlegate Sandstone	150-500	
		Blackhawk Formation	700-1000	coal; Hiawatha coal beds
		Star Point Sandstone	100-1000	
	Mancos Shale	Upper Blue Gate Shale Member	300-1300	*Scaphites hippocrepis*
		Emery Ss Mbr	800-50	*Clioscaphites*
		Lower Blue Gate Shale Member	1600-2400	*Baculites, Placenticeras*
		Ferron Ss Member	750-150	coal; Phillips Petroleum, Huntington 1, 15-17S-8E
		Tununk Sh Mbr	400-650	
	Dakota Sandstone		0-60	
	Cedar Mtn Fm	Upper member	100-200	
		Buckhorn Cg Mbr	0-50	
JURASSIC	Morrison Fm	Brushy Basin Member	570	varicolored
		Salt Wash Member	510	
	Summerville Fm		260-420	red
	Curtis Formation		200-280	green or white
	Entrada Formation		520-800	
	Carmel Formation		1300-400	gypsum
	Navajo Sandstone		320	
	Kayenta Formation		250	
	Wingate Sandstone		400	
TRIASSIC	Chinle Fm	Church Rock Mbr	200	
		Moss Back Ss M	80	
		Temple Mtn Mbr	60	
	Moenkopi Fm	Moody Canyon Mbr	200	
		Torrey Member	220	
		Sinbad Ls Mbr	60	
		Black Dragon Mbr	300	
PER	Black Box Dolomite		160	
	Cedar Mesa Ss		450	Diamond Creek Ss
P	"Hermosa" Fm		450	
MISS	"Madison" or "Redwall" limestone & dolomite		1200	

MAP WITH TEXT - Stokes & Cohenour, 1956; Doelling, 1972; Spieker, 1931; COLTON - Zawiskie et al, 1982; SOILS - Swenson et al, 1970; COAL FIELD TRIP - Cross et al, 1975; Mercier, 1982; Ryer, 1982; COAL - Doelling, 1972; Johnson, 1978; Parker, 1975; Flores et al, 1982; MANCOS - Cobban, 1975; FERRON - Cotter, 1975; Ryer, 1981; SUBSURFACE - Heylmun et al, 1965; Welsh & Bissell, 1979; Irwin, 1971, 1976; TRIASSIC - Stewart, 1972a; Blakey, 1974; NORTH HORN MAMMALS - Robison, 1986.

Chart

65 HELPER - PRICE - WELLINGTON

Period	Unit	Member	FEET	Remarks
Q	Alluvium & pediment deps		0-100	
TERT	Alkalic diabase dikes & sills		0-5 wide	8 m.y. 18 m.y. 24 m.y.
TERT	Flagstaff Limestone		220-280	
TERT / CRETACEOUS	North Horn Formation		1040-2170	North Horn & Price River Fms thin eastward as they pass over the northern end of the San Rafael Swell thus dating its rise
CRETACEOUS	Mesaverde Group	Price River Fm	270-1150	
CRETACEOUS	Mesaverde Group	Castlegate Ss	180-290	
CRETACEOUS	Mesaverde Group	Blackhawk Formation	1000-1500	Sunnyside Ss & coal Kenilworth Ss & coal Aberdeen Ss & coal Spring Canyon Ss
CRETACEOUS	Mancos Shale	Shale tongue	120-350	
CRETACEOUS	Mancos Shale	Star Point Ss Mbr	300-20	Storrs Ss Panther Ss
CRETACEOUS	Mancos Shale	Shale tongue	800-1300	*Baculites*
CRETACEOUS	Mancos Shale	Emery Ss Member	120-180	*Desmoscaphites*
CRETACEOUS	Mancos Shale	Shale tongue	350-850	*Placenticeras*
CRETACEOUS	Mancos Shale	Upper Garley Cyn M	30-60	*Baculites codyensis*
CRETACEOUS	Mancos Shale	Shale tongue	80-110	*Placenticeras*
CRETACEOUS	Mancos Shale	Lower Garley Cyn M	40-70	
CRETACEOUS	Mancos Shale	Blue Gate Shale Member	2500±	Shell Oil-North Springs 27-15S-9E
CRETACEOUS	Mancos Shale	Ferron Ss Member	100-10	Shell Oil - Miller Creek 26-15S-10E
CRETACEOUS	Mancos Shale	Tununk Sh Mbr	200-300	
CRETACEOUS	Dakota Sandstone		0-30	Pan Am-Farnham Dome 7-15S-12E
CRETACEOUS	Cedar Mtn Fm	Upper member	150-750	Pure Oil-Washboard 12-16S-9E
CRETACEOUS	Cedar Mtn Fm	Buckhorn Cg Mbr	0-50	Pure Oil-Desert Lake 1-17S-10E
JURASSIC	Morrison Formation		800±	
JURASSIC	Summerville Formation		120-180	J-5 unconformity
JURASSIC	Curtis Formation		140-180	J-3 unconformity
JURASSIC	Entrada Formation		150-950	
JURASSIC	Carmel Formation		300-700	
JURASSIC	Page Sandstone		70	J-2 unconformity
JURASSIC	Navajo Sandstone		150-300	
JURASSIC	Kayenta Formation		120-200	
JURASSIC	Wingate Sandstone		300-400	
TRIASSIC	Chinle Fm	Upper member	200-300	bentonitic
TRIASSIC	Chinle Fm	Moss Back Mbr	20-60	Tr-3 unconformity
TRIASSIC	Moenkopi Fm	Upper member	550-700	
TRIASSIC	Moenkopi Fm	Sinbad Ls Mbr	50	
TRIASSIC	Moenkopi Fm	Black Dragon Mbr	250-350	Tr-1 unconformity
PERM	Black Box Dolomite		170	
PERM	White Rim Sandstone		500-700	
PERM	Pakoon Dolomite		650-800	
P	Callville Limestone		250-300	
MISS	Doughnut Formation		600-700	
MISS	Humbug Formation		400-500	
MISS	Redwall Dolomite		750-970	
D	Pinyon Peak Ls		20	
D	Ouray Formation		110-160	
€	Cambrian dolomite		350	
€	Ophir Shale		200	
€	Tintic Quartzite		210	
p€	Crystalline basement		-	

MAP WITH TEXT - Nethercott, 1985; Witkind, 1979; Carroll, 1986; Russon, 1986; CRETACEOUS - Cobban, 1975; Cross et al, 1975; Doelling, 1972; Frey & Howard, 1985; Lawton, 1983; Mercier, 1982; Young, 1975; SUBSURFACE - Heylmun et al, 1965; JURASSIC-TRIASSIC - Pipiringos & O'Sullivan, 1978; CARBONIFEROUS - Welsh & Bissell, 1979; Irwin, 1971, 1976; Welsh, pers. comm., 1985.

Chart

66 SUNNYSIDE - WOODSIDE AREA

Period	Unit	Member	FEET	Remarks
Q	Alluvial, colluvial, pediment and terrace deposits		0-200	
TERTIARY	Green River Formation		3200+	Edge of Uinta Basin Tar sands
TERTIARY	Colton Formation (formerly Wasatch)		900-3000	Roan Cliffs varicolored ss, sh thicken to southeast
TERTIARY	Flagstaff Limestone		0-30	
TERTIARY / CRETACEOUS	North Horn Formation		100-500	
CRETACEOUS	Price River Fm	Bluecastle Ss Mbr	100-300	
CRETACEOUS	Price River Fm	Mudstone member	100-400	Book Cliffs
CRETACEOUS	Castlegate Sandstone		80-300	
CRETACEOUS	Blackhawk Fm	Upper mudstone	100-200	
CRETACEOUS	Blackhawk Fm	Sunnyside Member	100-190	Principal coal beds
CRETACEOUS	Blackhawk Fm	Lower mudstone	150-200	
CRETACEOUS	Blackhawk Fm	Kenilworth Mbr	110-220	coal
CRETACEOUS	Blackhawk Fm	Mancos tongue	200	
CRETACEOUS	Blackhawk Fm	Aberdeen Mbr	0-10	
CRETACEOUS	Mancos Shale (main body)		3400	*Scaphites hippocrepis* *Baculites aquilaensis*
CRETACEOUS	Mancos Shale	Ferron Ss Mbr	10-50	*Scaphites warreni*
CRETACEOUS	Mancos Shale	Tununk Sh Mbr	400	*Collingnoniceras*
CRETACEOUS	Dakota Sandstone		0-60	*Pycnodonte newberryi*
CRETACEOUS	Cedar Mtn Fm	Upper mudstone	110-350	
CRETACEOUS	Cedar Mtn Fm	Buckhorn Cg Mbr	10-80	
JURASSIC	Morrison Fm	Brushy Basin M	150-250	
JURASSIC	Morrison Fm	Salt Wash Ss M	50-250	
JURASSIC	Summerville Formation		150-200	chocolate torte beds
JURASSIC	Curtis Formation		130-180	
JURASSIC	Entrada Sandstone		150-450	
JURASSIC	Carmel Formation		200-500	gypsum
JURASSIC	Page Sandstone		50±	J-2 unconformity
JURASSIC	Navajo Sandstone		320-350	
JURASSIC	Kayenta Formation		50-100	
JURASSIC	Wingate Sandstone		310-440	Pan-Am Dragerton 11-15S-13E
TRIASSIC	Chinle Fm	Upper member	200-230	
TRIASSIC	Chinle Fm	Moss Back Ss Mbr	40-50	Reserve Oil-Cedar Siding 21-16S-13E
TRIASSIC	Moenkopi Fm	Upper member	450-600	Forest-Arnold-Little Park 25-16S-14E
TRIASSIC	Moenkopi Fm	Sinbad Ls Mbr	50±	
TRIASSIC	Moenkopi Fm	Black Dragon Mbr	300-450	
PERM	Black Box Dolomite		140	"Kaibab"
PERM	White Rim Sandstone		600-650	"Coconino"
PERM	Pakoon Dolomite		800	"Oquirrh" of some well logs
P	Callville Limestone		550	
M	Doughnut Formation		850-250	
M	Humbug Formation		400-180	
M	Redwall Dolomite		800-0	
D	Ouray-Elbert Fms		120-0	formations truncated on southeast flank of Uncompahgre Uplift
€	Cambrian carbonate		350-0	
€	Ophir Shale		150-0	
€	Tintic Quartzite		170-0	
	Precambrian crystalline basement			

MAP WITH TEXT - Osterwald et al, 1974, 1981; Witkind, 1980; COLTON - Zawiskie et al, 19081; TAR SAND - Holmes et al, 1948; CRETACEOUS - Cobban, 1975; Doelling, 1972; Lawton, 1983; Mercier, 1982; Young, 1975; JURASSIC - Pipiringos & O'Sullivan, 1978; SUBSURFACE - Heylmun et al, 1965; Welsh & Bissell, 1979, Irwin, 1971, 1976; John E. Welsh, pers. comm., 1985.

Chart
67 GREEN RIVER AREA

Period	Group / Formation	Unit	FEET	Remarks
Q		Alluvium, pediment dep	0-100	
T		Tuscher Fm	400	
CRETACEOUS	Mesaverde Group	Farrer Fm	700	barren
		Bluecastle Ss Tongue	100-130	tongue of Castlegate Ss
		Neslen Fm	200-250	thin coal at base
		Sego Sandstone	50-150	
		Buck Tongue-Mancos Sh	0-50	
		Castlegate Ss	80	
		Blackhawk Fm	140	thin coal beds
		Grassy Ss	0-100	
		Sunnyside Ss	100-0	
		Mancos Shale tongue	100-200	
		Kenilworth Ss	50-150	
	Mancos Shale	Pierre fauna; Telegraph Creek fauna; Niobrara fauna	3000	*Scaphites*; *Inoceramus*; *Baculites*; *Desmoscaphites*; *Ostrea*
		Ferron Ss M	10-30	*Collignoceras*
		Tununk Mbr	350-400	*Pycnodonte newberryi*
		Dakota Sandstone	0-30	
	Cedar Mountain Fm	Cedar Mountain Fm	150-180	
		Buckhorn Cg M	0-30	
JURASSIC	Morrison Fm	Brushy Basin M	240-420	
		Salt Wash Mbr	160-290	J-5 unconformity
		Tidwell Member	20-50	red siltstone
		Summerville Fm	100-400	
		Curtis Formation	130-230	green
		Entrada Ss	410-470	stone goblins
		Carmel Formation	220-300	gypsum; J-2 unconformity
		Page Sandstone	0-80	
		Navajo Ss	430-510	
		Kayenta Fm	190-240	
		Wingate Ss	300-400	
TRIASSIC	Chinle Fm	Church Rock Mbr	200-400	
		Moss Back Cg Mbr	60-100	
		Temple Mountain M	0-40	Tr-3 unconformity
	Moenkopi Fm	Moody Canyon & Torrey Members	470-650	
		Sinbad Limestone M	30-50	
		Black Dragon Mbr	170-210	
PERMIAN		Black Box Dolomite	60-160	"Kaibab" in old reports
	Cutler Group	White Rim Ss	300-500	"Coconino" in old reports
		Organ Rock Shale	0-300	subsurface
		Elephant Canyon Formation	1000-1200	*Dunbarinella*; *Schwagerina*
PENNSYLVANIAN	Hermosa Group	Honaker Trail Fm	500-1000	
		Paradox Fm	1000-2500	gypsum & anhydrite & halite
		Pinkerton Trail Fm	0-500	
		Molas Fm	50	
M		Leadville Limestone	600-800	corals; endothyrid forams
D		Ouray Ls	0-100	
		Elbert Fm	150-250	
CAMB		"Lynch" Dolomite & "Maxfield" Ls	800-1000	
		Ophir Fm	150-200	
		Tintic Quartzite	160-200	
pC		Granitic & metamorphic rocks		1800 m.y. Rb-Sr

MAPS WITH TEXT--Trimble & Doelling, 1978; Williams & Hackman, 1971; CRETACEOUS--Young, 1955, 1966; McGookey, 1972; Fisher et al, 1960; Lawton, 1983; Lessard, 1973; Harris, 1980; JURASSIC--O'Sullivan, 1980b; Wright & Dickey, 1963; Derr, 1974; TRIASSIC--Stewart et al, 1972a, 1972b; Blakey, 1974; Orgill, 1971; Lupe, 1979; PERMIAN--Baars, 1962; Rascoe & Baars, 1972; PENNSYLVANIAN--Mallory, 1972; MISSISSIPPIAN-Craig, 1972; DEVONIAN--Baars, 1972; Parker & Roberts, 1966; CAMBRIAN--Lochman-Balk, 1972.

Chart
68 CISCO - HARLEY DOME

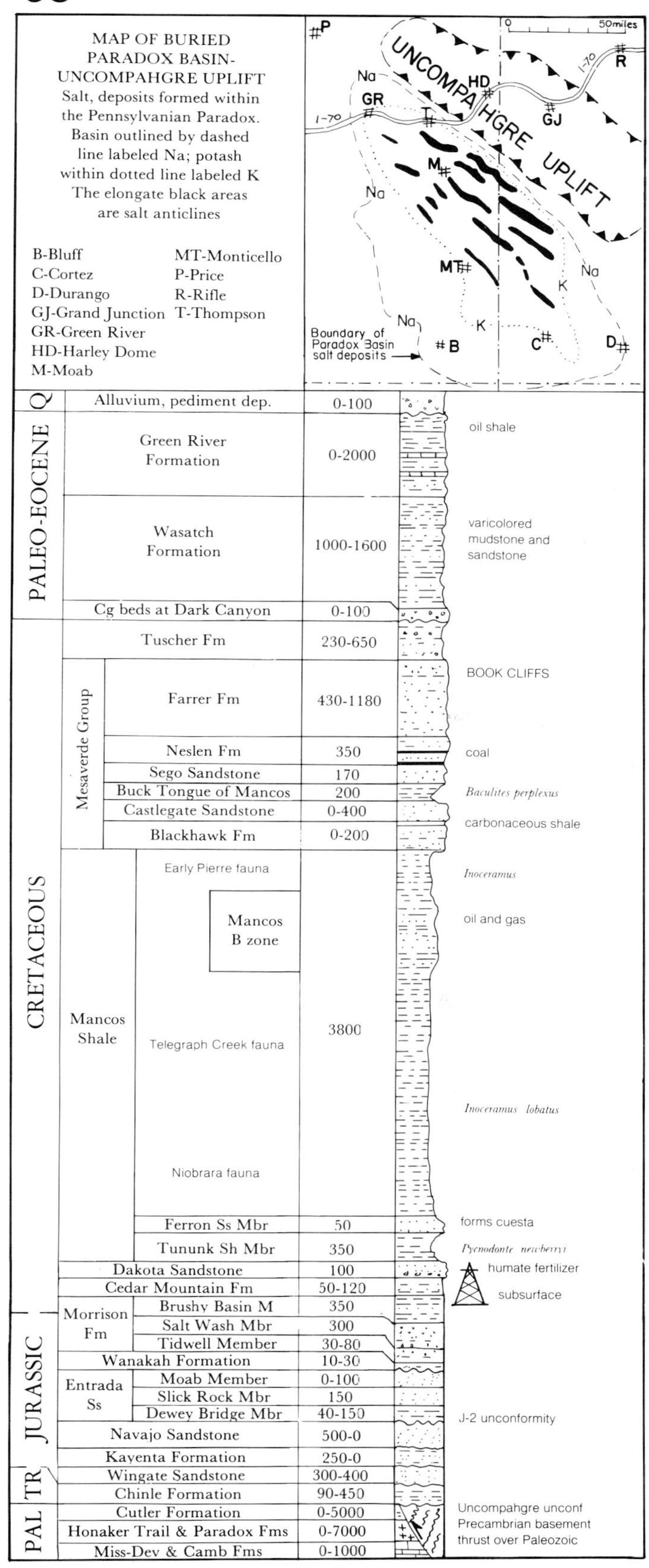

MAP OF BURIED PARADOX BASIN-UNCOMPAHGRE UPLIFT
Salt, deposits formed within the Pennsylvanian Paradox. Basin outlined by dashed line labeled Na; potash within dotted line labeled K
The elongate black areas are salt anticlines

B-Bluff
C-Cortez
D-Durango
GJ-Grand Junction
GR-Green River
HD-Harley Dome
M-Moab
MT-Monticello
P-Price
R-Rifle
T-Thompson

Period	Group / Formation	Unit	Thickness	Remarks
Q		Alluvium, pediment dep.	0-100	
PALEO-EOCENE		Green River Formation	0-2000	oil shale
		Wasatch Formation	1000-1600	varicolored mudstone and sandstone
		Cg beds at Dark Canyon	0-100	
CRETACEOUS		Tuscher Fm	230-650	
	Mesaverde Group	Farrer Fm	430-1180	BOOK CLIFFS
		Neslen Fm	350	coal
		Sego Sandstone	170	
		Buck Tongue of Mancos	200	*Baculites perplexus*
		Castlegate Sandstone	0-400	carbonaceous shale
		Blackhawk Fm	0-200	
	Mancos Shale	Early Pierre fauna; Mancos B zone; Telegraph Creek fauna; Niobrara fauna	3800	*Inoceramus*; oil and gas; *Inoceramus lobatus*
		Ferron Ss Mbr	50	forms cuesta
		Tununk Sh Mbr	350	*Pycnodonte newberryi*
		Dakota Sandstone	100	humate fertilizer
		Cedar Mountain Fm	50-120	subsurface
JURASSIC	Morrison Fm	Brushy Basin M	350	
		Salt Wash Mbr	300	
		Tidwell Member	30-80	
		Wanakah Formation	10-30	
	Entrada Ss	Moab Member	0-100	
		Slick Rock Mbr	150	
		Dewey Bridge Mbr	40-150	J-2 unconformity
		Navajo Sandstone	500-0	
		Kayenta Formation	250-0	
		Wingate Sandstone	300-400	
TR		Chinle Formation	90-450	
PAL		Cutler Formation	0-5000	Uncompahgre uncont Precambrian basement thrust over Paleozoic
		Honaker Trail & Paradox Fms	0-7000	
		Miss-Dev & Camb Fms	0-1000	

MAP WITH TEXT - Cashion, 1973; Williams, 1964; Willis, 1985; Stokes, 1952; GREEN RIVER - Cashion, 1967b; WASATCH - Fouch et al, 1982; CRETACEOUS - Young, 1955, 1966; Willis, 1985; Fisher et al, 1960; Craig, 1981; Doelling & Graham, 1972b; Gill & Hail, 1975; Kellogg, 1977; Jackson, 1983; JURASSIC - O'Sullivan, 1981b, 1984; TRIASSIC - Stewart et al, 1972a; PRECAMBRIAN - Tweto, 1977; UNCOMPAHGRE FAULT - White & Jacobson, 1983; Heyman, 1983; Frahme & Vaughn, 1983.

Chart
69 BURBANK HILLS

Age	Group	Formation / Member	FEET	Remarks
Q		Alluvium & colluvium	0-100	
T		Valley fill	0-600	
OLIGOCENE		Tuffaceous sandstone	0-100	
OLIGOCENE		Post-Needles Range Cg	0-600	
OLIGOCENE	Needles Range Gp	Wah Wah Spr Fm	0-300	29.5 m.y.
OLIGOCENE	Needles Range Gp	Cottonwood Wash Tf	0-200	30.6 m.y. big biotite
OLIGOCENE		Cg of Skull Rock Pass	0-300	Eureka boulders
OLIGOCENE		Tunnel Spring Tuff	up to 1900 at Cowboy Pass	small doubly-terminated quartz crystals abundant; 33.7 m.y.
STRATA BELOW OVERLAIN UNCONFORMABLY BY VARIOUS ROCKS ABOVE				
PERMIAN		Arcturus Formation	2000 plus	
PERMIAN	Ely Ls	Riepe Spring Mbr	500	*Pseudoschwagerina*, *Triticites*
PENN	Ely Ls	Cherty member	2000	*Chaetetes*; bryozoans, brachiopods, gastropods, corals abundant
PENN	Ely Ls	Chertless member	200	*Zellerina*
MISSISSIPPIAN	Chainman Shale	Jensen Member	470	many brachiopods
MISSISSIPPIAN	Chainman Shale	Willow Gap Ls Mbr	0-300	
MISSISSIPPIAN	Chainman Shale	Camp Canyon Member: Upper unit	670	
MISSISSIPPIAN	Chainman Shale	Camp Canyon Member: Black shale	320	Chainman units from Sadlick, 1965, except Delle Phosphatic Member
MISSISSIPPIAN	Chainman Shale	Skunk Spring Ls M	0-5	
MISSISSIPPIAN	Chainman Shale	Needle Siltstone Mbr	320	
MISSISSIPPIAN	Chainman Shale	Delle Phosphatic Mbr	30 ?	
MISSISSIPPIAN		Joana Limestone	500	
MISSISSIPPIAN		Pilot Shale	470	upper member, Leatham Member, lower member
DEVONIAN		West Range Limestone	630	
DEVONIAN	Guilmette Fm	Main member	3000	brown-weathering ss; *Manticoceras*
DEVONIAN	Guilmette Fm	Breccia member	300-600	paleokarst cave collapse
DEVONIAN		Simonson Dolomite	600	dark gray, light gray
DEVONIAN		Sevy Dolomite	600 plus	American Quasar Antelope Valley 36-235-18W 1981
Є-O-S		Laketown & Ely Springs Dol	1350	
Є-O-S		Eureka Quartzite	300 plus	FAULTS
Є-O-S		Pogonip Group	3200	
Є-O-S		Notch Peak Formation	1900	FAULTS
Є-O-S		Orr Formation	900	

OLIGOCENE - Bushman, 1973; Hintze, 1974c; ELY - St. Aubin-Hietpas, 1983; CHAINMAN - Sadlick, 1965; Gordon, 1984; Tynan, 1980; Flower & Gordon, 1959; Karklins, 1986; DEVONIAN - Biller, 1976; Hose, 1966.

Chart
70 CONFUSION RANGE

Age	Group	Formation / Member	FEET	Remarks
Q		Alluvium, L. Bonneville seds.	0-200	
M-P		Valley fill	0-1000	concealed, may include volcanics
M-P		Rhyolite of Honeycomb Hills	0-100	3-5 m.y.
M-P		Basalt of Honeycomb Hills	0-200	6 m.y.
OLIG		Post-Needles Range Cg	0-600	
OLIG	Needles Range Gp	Wah Wah Springs T	0-300	29 m.y.
OLIG	Needles Range Gp	Cottonwood Wash T	0-300	
OLIG		Skull Rock Pass Cg	0-300	Eureka boulders
OLIG		Tunnel Spring Tuff	0-1000	33 m.y. Crystal Peak
TRIAS		Thaynes Formation	1000-1900	platy siltstone; thin limestones contain conodonts
PERMIAN		Gerster Limestone	800-1100	siliceous nodules; *Punctospirifer*
PERMIAN		Plympton Formation	550-700	some gypsum
PERMIAN		Kaibab Limestone	500-600	chert nodules
PERMIAN		Arcturus Formation	3000 ±	yellow sandstone; gypsum; limy dolomite
P		Ely Limestone	1800-2000	cyclic detrital limestone; chert and fossils abundant
MISS	Chainman Shale	Jensen Member	190	very fossiliferous
MISS	Chainman Shale	Willow Gap Ls M	340	see Sandberg et al, 1980, for paleoenvironments
MISS	Chainman Shale	Camp Canyon Member	1050	
MISS	Chainman Shale	Skunk Sp Ls M	5-15	*Goniatites*
MISS	Chainman Shale	Needle Siltstone M	170	Delle Phos Mbr 15' at base
MISS		Joana Limestone	10-300	thins to north
DEVONIAN		Pilot Shale	700-1090	see Sandberg et al, 1980, for conodont zones
DEVONIAN		Guilmette Formation	2000-2650	"spaghetti" and "cauliflower" stromatoporoids; 600' massive solution breccia
DEVONIAN		Simonson Dolomite	540-700	dark dolomite with relicts of fossils
DEVONIAN		Sevy Dolomite	1100-1300	barren light gray dolomite with floating quartz sand grains
SIL		Laketown Dolomite	1000-1200	silicified corals, brachiopods at top; dasycladacean algae; *Pentamerus*
ORDOVICIAN		Ely Springs Dolomite	450-600	*Streptelasma*
ORDOVICIAN		Eureka Quartzite	450-540	
ORDOVICIAN		Crystal Peak Dolomite	90	
ORDOVICIAN		Watson Ranch Quartzite	175-250	*Eofletcheria*
ORDOVICIAN	Pogonip Group	Lehman Formation	170-210	ostracodes
ORDOVICIAN	Pogonip Group	Kanosh Shale	500-560	*Orthambonites*
ORDOVICIAN	Pogonip Group	Juab Limestone	160	
ORDOVICIAN	Pogonip Group	Wah Wah Limestone	250	*Pseudocybele*, *Presbynileus*
ORDOVICIAN	Pogonip Group	Fillmore Formation	1800	*Trigonocerca*, *Leiostegium*
ORDOVICIAN	Pogonip Group	House Limestone	420-500	*Symphysurina*
CAMB	Notch Peak Fm	Lava Dam Mbr	320	*Missisquoia*
CAMB	Notch Peak Fm	Red Tops Mbr	120-130	*Saukiella*; stromatolites
CAMB	Notch Peak Fm	Hellnmaria Mbr	1000-1300	*Matthevia*

MAPS - Hintze, 1974a, b, c; Hose, 1977; VALLEY FILL - Stephens, 1977; VOLCANIC ROCKS - Hogg, 1972; Bushman, 1973, Best & Grant, 1986; TRIASSIC - Carr & Paull, 1983; Collinson et al, 1976; PENN-PERMIAN - Hose & Repenning, 1959; Wardlaw & Collinson, 1979; MISSISSIPPIAN - Webster et al, 1984; Sadlick, 1965; Spinosa et al, 1977; Karklins, 1986; DEVONIAN - Hose, 1966; Gutschick & Rodriguez, 1979; Sandberg, Poole & Gutschick, 1980; SILURIAN - Budge & Sheehan, 1980; ORDOVICIAN - Hintze, 1973, 1979; Rigby & Hintze, 1977; Ethington & Clark, 1981; Chamberlin, 1975; CAMBRIAN - Hintze et al, 1986.

Chart
71 MOUNTAIN HOME RANGE

Age	Group	Formation	Member	Feet	Remarks
Q		Alluvium & colluvium		0-200	
T		Older alluvium & valley fill		0-1000 ?	
OLIGOCENE	Needles Range Group	Lund Fm-Tuff mbr		up to 300	27.9 m.y. crystal-rich
OLIGOCENE	Needles Range Group	Wah Wah Springs Fm	Outflow tuff member	up to 1000	29.5 m.y. crystal-rich
OLIGOCENE	Needles Range Group	Cottonwood Wash Tuff		900-1300	30.6 m.y. big biotite phenocrysts
OLIGOCENE		Escalante Desert Fm	Beers Spring M	0-30	
OLIGOCENE		Escalante Desert Fm	Lamerdorf Tuff M	0-50	crystal-poor
OLIGOCENE		Escalante Desert Fm	Andesite mbr	0-40	32.3 m.y.
OLIGOCENE		Sawtooth Peak Fm		0-160	
OLIGOCENE		Pre-volcanic sandstone and conglomerate		0-2000	thickest on west flank of range south of pass lacks volcanic clasts
STRATA BELOW OVERLAIN UNCONFORMABLY BY VARIOUS ROCKS ABOVE					
PERMIAN		Arcturus Formation		2200	orange dolomitic ss
PERMIAN		Ely Ls	Riepe Spring Ls Mbr	500	*Pseudofusulina* *Pseudoschwagerina*
PENN		Ely Ls	Cherty member	2030	*Chaetetes* cyclic cherty ls, dolo, ss fossils abundant *Profusulinella* *Endothyra*
MISSISSIPPIAN		Ely Ls	Chertless member	200	*Zellerina*
MISSISSIPPIAN		Chainman Shale	Jensen Member	260-400	many invertebrates
MISSISSIPPIAN		Chainman Shale	Willow Gap Ls Mbr	300-350	
MISSISSIPPIAN		Chainman Shale	Camp Canyon Member: Rusty limestone	650	
MISSISSIPPIAN		Chainman Shale	Camp Canyon Member: Clayey limestone	300	brachiopods and mollusks common
MISSISSIPPIAN		Chainman Shale	Camp Canyon Member: Black shale	400	*Cravenoceras*
MISSISSIPPIAN		Chainman Shale	Skunk Spring Ls Mbr	5-10	
MISSISSIPPIAN		Chainman Shale	Needle Siltstone Mbr	600	
MISSISSIPPIAN		Chainman Shale	Delle Phosphatic Mbr: Limestone U	80-100	
MISSISSIPPIAN		Chainman Shale	Delle Phosphatic Mbr: Siltstone Unit	40-65	
MISSISSIPPIAN		Joana Limestone		450	horn corals
MISSISSIPPIAN / DEVONIAN		Pilot Shale		360	upper member Leatham Member lower member
DEVONIAN		Guilmette Fm	Upper ls mbr	260-390	equivalent to West Range Ls of Pioche, Nv
DEVONIAN		Guilmette Fm	Sandstone mbr	100-130	
DEVONIAN		Guilmette Fm	Dolomite and sandstone mbr	1500 plus	spaghetti-type stromatoporoids very common

MAP WITH TEXT - Hintze, 1986; Hintze & Best, 1986; VALLEY - Stephens, 1976; ELY - St. Aubin-Hietpas, 1983; DEVONIAN-MISSISSIPPIAN - Sandberg & Gutschick, 1984; Sandberg et al, 1980.

Chart
72 INDIAN PEAK RANGE

Age	Group	Formation	Member	Feet	Remarks
Q		Alluvium & colluvium		0-100	
T		Volcanic ss & cg		350-1400	
MIOCENE		Steamboat Mountain Fm	Rhyolite member	0-1500	12-13 m.y. flows & domes
MIOCENE		Blawn Fm	Rhyolite member	0-500	flows, domes, dikes
MIOCENE		Blawn Fm	Mafic flow member	0-1500	18 m.y.
MIOCENE		Bauers Tuff Mbr Condor Canyon		0-400	22.3 m.y.
MIOCENE		Dacite of Spanish George Spring		0-800	lava dome
OLIGOCENE		Isom Formation		0-2500	26 m.y. flat vesicles
OLIGOCENE		Hornblende andesite flows		0-800	
OLIGOCENE	Needles Range Group	Lund Fm	Andesite member	0-1300	
OLIGOCENE	Needles Range Group	Lund Fm	Tuff member	0-1800	27.9 m.y. COLLAPSE OF CALDERA
OLIGOCENE		Ryan Spring Fm	Rhyolite flow mbr	0-600	28.4 m.y.
OLIGOCENE		Ryan Spring Fm	Mackleprang Tuff M / Andesite flow mbr	0-1500	
OLIGOCENE		Ryan Spring Fm	Greens Canyon Tuff / Andesite flow mbr	0-1500	
OLIGOCENE	Needles Range Group	Wah Wah Springs Fm	Intracaldera member	100-7000	crystal-rich dacite tuffs, breccias & caldera wall collapse deposits
OLIGOCENE	Needles Range Group	Wah Wah Springs Fm	Outflow tuff member	0-1600	COLLAPSE OF THE INDIAN PEAK CALDERA 29.5 m.y.
OLIGOCENE	Needles Range Group	Cottonwood Wash Tuff		0-1000	30.6 m.y. big biotite
OLIGOCENE		Escalante Desert Fm	Beers Spring Mbr	0-500	green to brown ss
OLIGOCENE		Escalante Desert Fm	Lamerdorf Tuff M	0-200	32.3 m.y.
OLIGOCENE		Escalante Desert Fm	Andesite flow mbr	0-200	COLLAPSE OF THE PINE VALLEY CALDERA
OLIGOCENE		Escalante Desert Fm	Marsden Tuff Mbr	0-1000?	
OLIGOCENE		Sawtooth Peak Formation		0-1000	34.6 m.y. ash-flow tuff
PALEOZOIC STRATA BENEATH VOLCANIC ROCKS HAVE BEEN FOLDED AND THRUSTED PRIOR TO BEING FURTHER DEFORMED BY TERTIARY CALDERA COLLAPSE AND VOLCANOGENIC ALTERATION					
P		Callville Limestone		850	
M		Joana Limestone		400	
M		Pilot Shale		0-350	
DEV		Guilmette Formation		50-1300	
DEV		Simonson Dolomite		150 plus	dark gray
DEV		Sevy Dolomite		350 plus	light gray
S		Laketown Dolomite		1100	THICKNESSES SHOWN FOR PALEOZOIC FORMATIONS ARE MAXIMUMS FOUND IN THIS AREA. THEY DO NOT NECESSARILY REPRESENT THE ORIGINAL FULL THICKNESS.
ORDOVICIAN		Ely Springs Dolomite		500	
ORDOVICIAN		Eureka Quartzite		500	
ORDOVICIAN		Crystal Peak Dolomite		240	
ORDOVICIAN		Lehman & Kanosh Fms undiv		500	
ORDOVICIAN		Juab & Wah Wah Ls undivided		500	
ORDOVICIAN		Fillmore Formation		1500	
ORDOVICIAN		House Limestone		400	
CAMBRIAN		Notch Peak Formation		2000	algal stromatolites
CAMBRIAN		Orr Formation		1300	
CAMBRIAN		Wah Wah Summit Fm	White member	160	boundstone
CAMBRIAN		Wah Wah Summit Fm	Lower member	600	
CAMBRIAN		Trippe Limestone		100 plus	

MAP WITH TEXT - Best, Hintze & Holmes, 1986; Best et al, 1986. TERTIARY - Best & Grant, 1986; Best, Mehnert, Keith & Naeser, 1986.

Chart
73 WAH WAH MOUNTAINS

System	Group / Formation	Unit	Feet	Remarks
Q		Alluvium & colluvium	0-100	
		Valley fill	0-3000	
MIOCENE		Basalt Mbr, Steamboat Mtn Fm	0-600	12-13 m.y. exposed at Brimstone Reservoir
	Blawn Fm	Rhyolite member	0-1300	20 m.y.
		Tuff member	1200	crystal-poor welded ash-flow tuff
		Mafic flow member	600	
		Garnet tuff member and source pluton	0-500 intrusion	few microgarnets 24 m.y.
OLIGOCENE		Isom Formation	0-90	26 m.y.
		Three Creeks Tuff Mbr	0-600	27 m.y. Bullion Canyon Volc
	Needles Range Group	Lund Formation	up to 1500	27.9 m.y.
	Needles Range Group	Wah Wah Springs Fm	up to 1500	29.5 m.y.
	Needles Range Group	Cottonwood Wash Tuff	0-50	30.6 m.y.
	Needles Range Group, Escalante Desert Fm	Beers Spring M	0-130	Wah Wah Pass equivalents:
		Lamerdorf Mbr tuff & andesite	0-2000	Andesite of Long Valley Andesite of Kelley's Place Dacite of Wah Wah Cove
		Marsden Tuff M	0-200	33.6 m.y.
		Small intrusive bodies at Wah Wah Pass		diorite; rhyolite porphyry
		Older andesite	0-50	
		Conglomerate	0-250	

STRATA BELOW OVERLAIN UNCONFORMABLY BY VARIOUS ROCKS ABOVE. ALL PRE-TERTIARY ROCKS ARE PART OF THE UPPER PLATE OF THE WAH WAH THRUST

System	Group / Formation	Unit	Feet	Remarks
ORD	Pogonip Group	Fillmore Fm	1800	
		House Limestone	450	
CAMBRIAN	Notch Peak Fm	Lava Dam Mbr	320-430	
		Red Tops Mbr	50-130	
		Hellnmaria Member	1340	algal stromatolites *Matthevia*
	Orr Fm	Sneakover Ls M	170	
		Steamboat Pass M	270	abundant trilobites
		Big Horse Ls Mbr	700	*Tricrepicephalus*
	Wah Wah Summit Fm	White marker mbr	130-170	bioclastic ls white boundstone
		Ledgy member	700-780	
	Trippe Ls	Fish Springs Mbr	120	*Eldoradia*
		Lower member	500-540	much white boundstone
		Pierson Cover Formation	1220-1280	dark gray
		Eye of Needle Limestone	210-240	
		Swasey Limestone	440-520	
		Whirlwind Formation	40-80	*Ehmaniella*
		Dome Limestone	280-310	
		Peasley Limestone	210-230	
		Chisholm Shale	40-80	*Glossopleura*
	Howell Ls	Upper member	190	light gray
		Millard Member	180	dark gray
	Pioche Fm	Tatow Member	200	
		Lower member	640	
	Prospect Mountain Quartzite	Upper member	2400	pink to tan
		Basalt member	100-190	
		Lower member	1100	
PROT		Mutual Formation	1200	purplish-red
		Inkom Formation	230	purple phyllitic shale
		Caddy Canyon Qtzt	1600	tan to pink
		Papoose Creek ? Fm	600	brown-weathering
		Blackrock Canyon ? Ls	600-800	algal beds
		Pocatello ? Fm	2000	tan argillaceous quartzite

MAP WITH TEXT - Hintze, 1974d; Wheeler, 1980; Abbott et al, 1983; NEEDLES RANGE - Best & Grant, 1986; CAMBRIAN - Hintze et al, 1980; Hintze & Palmer, 1976; Hintze & Robison, 1975.

Chart
74 BLUE MTN & SO. WAH WAH MTS

System	Group / Formation	Unit	Feet	Remarks
Q		Alluvium & colluvium	0-100	
P		Valley fill	0-500?	
MIOCENE	Steamboat Mtn Fm	Rhyolite mbr	0-1600	12-13 m.y.
		Basalt mbr	0-500	
	Blawn Fm	Rhyolite member	0-400	18 m.y.
		Tuff member	0-1200	
		Mafic flow member	0-700	20 m.y.
		Garnet-bearing tuff m	0-700	23 m.y.
		Bauers Mbr Condor Cyn F	0-400	23 m.y.
		Andesite flows	0-1000	24 m.y.
OLIGOCENE		Isom Formation	0-500	26 m.y.
		Three Creeks Tuff Member	0-200	27 m.y. Bullion Canyon Volc
	Needles Range Group	Lund Formation	0-1600	27.9 m.y.
	Needles Range Group, Wah Wah Springs F	Intracaldera m	0-300	
		Outflow mbr	0-1000	29.5 m.y.
	Needles Range Group	Cottonwood Wash Tuff	0-1000	30.6 m.y. big biotite
	Needles Range Group, Escalante Desert Fm	Beers Spring Mbr	0-1300	
		Upper rhyolite flows	0-1200	
		Lamerdorf Tuff M	0-300	32.3 m.y.
		Lower rhyolite flows	0-500	
		Andesite flow mbr	0-1200	
		Marsden Tuff Mbr	0-100	

ROCKS ABOVE OVERLIE ROCKS BELOW WITH ANGULAR UNCONFORMITY. MESOZOIC ROCKS ARE AUTOCHTHONOUS AT BLUE MOUNTAIN

System	Group / Formation	Unit	Feet	Remarks
JUR		Temple Cap Formation	1200	sandstone & red siltstone
		Navajo Sandstone	2000	metamorphosed to quartzite at Blue Mtn
TR	Chinle Fm	Upper member	280	pillow lavas in upper 150 feet
		Shinarump Cg Mbr	120	fossil wood
		Moenkopi Formation	300 plus	hornfels

STRATA BELOW ARE BENEATH THE WAH WAH THRUST. ROCKS ABOVE THIS THRUST ARE SHOWN ON COLUMN 73, WAH WAH MTS

System	Group / Formation	Unit	Feet	Remarks
P		Callville Limestone ? Great Blue Ls ?	1900	Identity of upper unit uncertain. Abbott et al called it Callville, but John Welsh (letter, 1985) prefers Great Blue. Woodman through Guilmette thicknesses from C. A. Sandberg, letter, 1986.
MISS		Woodman Formation	850	
		Delle Phosphatic Mbr	50	
		Gardison Limestone	150	
		Fitchville Formation	60	
DEV		Pinyon Peak Limestone	570	
		Guilmette Formation	120	
		Simonson Dolomite	100-500	dark gray
		Sevy Dolomite	700	light gray
S		Laketown Dolomite	1100-1300	
ORD		Ely Springs Dolomite	400	
		Eureka Quartzite	150-200	STRATA ATTENUATED BY THRUSTING
		Kanosh Shale	25-100	
		Juab Limestone	200 plus	

STRATA BELOW FORM UPPER PLATE OF THE BLUE MOUNTAIN THRUST

System	Group / Formation	Unit	Feet	Remarks
CAMBRIAN		Notch Peak Formation	200 plus	FAULTED FRAGMENTS
		Orr Formation	200 plus	
		Wah Wah Summit Fm	1000	
		Trippe Limestone	650	white boundstone beds
		Pierson Cove Formation	1600	dark gray
		Eye of Needle Limestone	400-550	
		Swasey Limestone	550-650	
		Whirlwind Formation	120	
		Dome & Peasley Ls undiv	800	
		Chisholm Formation	160	SOLE OF THRUST
		Howell Limestone	100 plus	

MAP WITH TEXT - Abbott et al, 1983; Morris et al, 1982; Best et al, 1986; Weaver, 1980; TERTIARY - Best & Grant, 1986; Best, Mehnert et al, 1986. TRIASSIC - Olmore, 1972; MISSISSIPPIAN - J. E. Welsh per. comm. 1985; ORDOVICIAN - Chamberlain, 1975.

Chart
75 FRISCO - BEAVER LAKE MTS

Period	Unit	Feet	Remarks
Q	Alluvium & colluvium	0-100	
TERTIARY	Fanglomerate & basin-fill in Wah Wah & Milford valleys	0-2000	
	CACTUS STOCK ALTERED SURROUNDING ROCKS		
	Granodiorite of the Cactus Stock	intrusive	28 m.y. INTRUSIVE CONTACT
	Horn Silver Andesite	500-2000	34 m.y.
	Conglomerate of High Rock Pass	0-300	
	CALLVILLE-PAKOON OVERTURNED BENEATH GRAMPIAN THRUST		
P-P	Pakoon Dolomite	460	
	Callville Limestone	340	cherty
	STRATA BELOW ARE CUT BY INTRUSIONS IN BEAVER LAKE MTS		
M	Deseret Limestone	300 plus	"Fm of Rose Spr Canyon"
	Gardison Limestone	600 plus	
DEV	Pinyon Peak Formation	150	
	Cove Fort Qtzt Mbr	100	
	Simonson Dolomite	300 plus	
	Sevy Dolomite	600 plus	
S	Laketown Dolomite	800 plus	
ORDO	Ely Springs Dolomite	70 plus	
	STRATA BELOW ARE ATTENUATED BELOW THE FRISCO THRUST		
	Watson Ranch Quartzite	300	
	Kanosh Shale	130	abundant fossils
	Wah Wah & Juab Fms	350	
	House & Fillmore Fms	600 plus	
CAMBRIAN	Notch Peak Formation	1000	altered near the Cactus Stock
	Orr Fm: Steamboat Pass Sh M	100	
	Orr Fm: Big Horse Ls Mbr	800	
	STRATA BELOW ARE ATTENUATED ALONG BEAVER LAKE THRUST		
	Dome Limestone	50 plus	
	Chisholm Shale	50 plus	
	Howell Limestone	30 plus	
	Pioche Formation	10 plus	phyllitic
	STATA BELOW ARE IN UPPER PLATE OF THE FRISCO THRUST		
	Prospect Mountain Quartzite: Upper member	5600	pink vitreous quartzite
	Prospect Mountain Quartzite: Basalt flow mbr	15-50	
	Prospect Mountain Quartzite: Lower member	1600	
LATE PROTEROZOIC	Mutual Quartzite	2100	maroon quartzite
	Inkom Formation	500	red & green argillite
	"Caddy Canyon Quartzite"	300	
	"Blackrock Canyon Ls": Upper member	0-150	
	"Blackrock Canyon Ls": Limestone member	30-150	names in quotes may all be facies of the Caddy Canyon Qtzt of SE Idaho
	"Blackrock Canyon Ls": Lower member	500-600	
	"Pocatello Formation"	150+	

MAP WITH TEXT - Hintze, Lemmon & Morris, 1984; Lemmon & Morris, 1984, 1979; Welsh, 1973; GUIDEBOOK - Utah Geological Association Publication 3, 1973, Geology of the Milford Area; GROUNDWATER - Stephens, 1974b; DEVONIAN & MISSISSIPPIAN - C. A. Sandberg, letter, 1986.

Chart
76 STAR RANGE - SHAUNTIE HILLS

Period	Unit	Feet	Remarks
Q	Alluvium & colluvium	0-100	
P	Dissected fan deposits in Big Wash	0-500	cemented, tilted
MIOCENE	Steamboat Mountain Fm: Rhyolite mbr; Basalt mbr	0-300	12-13 m.y.
	Felsite & aplite dikes	intrusive	
	Granite near Shauntie	pluton	
	Monzonite	small pluton	
	Quartz monzonite	plutons	21 m.y. Rocky Range
	Blawn Fm: Rhyolite member	0-200	18 m.y.
	Blawn Fm: Mafic lava flow mbr; Dacite of Squaw Pk	0-600	Note: Dacite of Squaw Pk is not a member of Blawn Fm
	Blawn Fm: Tuff member	0-2000	23 m.y.
OLIGOCENE	Isom Formation	20-150	26 m.y.
	Three Creeks Tuff Mbr, BCV	0-50	27 m.y. Bullion Canyon Volc
	Needles Range Group: Lund Tuff	250	27.9 m.y.
	Needles Range Group: Wah Wah Springs Fm	100-700	29.5 m.y.
	Needles Range Group: Cottonwood Wash T	0-30	30.6 m.y.
	Needles Range Group: Escalante Desert F	0-200	32.3 m.y.
	Horn Silver Andesite	0-100	33 m.y.
	Andesite of Shauntie Hills	0-1500	reddish-brown flows
	Conglomerate of High Rock Pass	0-50	ANGULAR UNCONFORMITY
JUR	Navajo Sandstone	600-1000	
TRIASSIC	Chinle Fm: Upper member	400-530	red siltstone, sh, ss
	Chinle Fm: Shinarump Cg M	50-200	petrified wood
	Moenkopi Fm: Upper red member	700-1300	fine-grained ss and siltstone
	Moenkopi Fm: Virgin Ls Member	100-250	
	Moenkopi Fm: Lower red member	300-340	
	Moenkopi Fm: Timpoweap Member	600-720	gypsum *Meekoceras*
PERMIAN	Plympton Formation	100-390	Harrisburg equivalent
	Kaibab Limestone	350-600	cherty, fossiliferous
	Toroweap Formation	380-440	
	Talisman Quartzite	680-730	
	Pakoon Dolomite	270-400	
P	Callville Limestone	200-500	
MISS	Manning Canyon Shale	60+	incomplete exposure
	Great Blue Limestone	100-1000	
	Humbug Formation	350-430	
	Deseret Ls: Uncle Joe Mbr	120	
	Deseret Ls: Tetro Member	500	cherty
	Gardison Limestone	100	Delle Phosphate Mbr
DEVONIAN	Fitchville Formation	30	
	Pinyon Peak Ls: Upper member	180	conodonts
	Pinyon Peak Ls: Lower member	250	Cove Ft Qtzt Bed
	Guilmette Formation	450	stromatoporoids
	Simonson Dolomite	500	
	Sevy Dolomite	340	light gray

MAP WITH TEXT - Baer, 1962; Lemmon & Morris, 1979; Felt, 1981; Haymond, 1981; DEVONIAN - Sandberg & Dreesen, 1984; Sandberg & Poole, 1977; MISSISSIPPIAN - Sandberg, pers comm. 1985; UPPER PALEOZOIC - John Welsh, pers. comm. 1985.

Chart
77 FILLMORE - COVE FORT AREA

Period	Unit		Feet	Remarks
Q	Basalt & L. Bonneville seds.		0-500	0.07-0.007 m.y. flows
Q	Rhyolite of White Mtn		0-50	0.4 m.y.
Q	Basaltic flows of Cove Ft		0-150	0.3-0.5 m.y. flows
Q	Basalts near Kanosh		0-300	0.7 m.y. cones, flows
M-P	Basin-fill deposits		0-300	Sevier River Fm equiv
M-P	Joe Lott Tuff Mbr of Mt. B.V.		0-200	19 m.y.
M-P	Osiris Tuff		0-200	23 m.y.
OLIGOCENE	Tuff of Albinus Canyon		0-100	25 m.y.
OLIGOCENE	Three Creeks Tuff Member of the Bullion Canyon Volcanics		0-2500	27 m.y. caldera east of Cove Fort
OLIGOCENE	Volcanics of Wales Canyon		0-200	andesitic flows
OLIGOCENE	Volcanics of Dog Valley		0-200	andesitic flow breccia
OLIGOCENE	Wah Wah Springs Fm		0-400	29.5 m.y.
K-T	North Horn Fm		250-1500	CRETACEOUS ROCKS ARE SYNOROGENIC
JURA	Navajo Sandstone		1200-1700	JURASSIC - TRIASSIC STRATA ARE AUTOCHTHONOUS PRE-OROGENIC
TRIASSIC	Chinle Fm	Upper member	70-270	
TRIASSIC	Chinle Fm	Shinarump Cg M	150-300	petrified wood
TRIASSIC	Moenkopi Fm	Upper red member	760	
TRIASSIC	Moenkopi Fm	Virgin Ls Mbr	350	"Shnabkaib" formerly
TRIASSIC	Moenkopi Fm	Lower red member	550	"middle red" formlerly
TRIASSIC	Moenkopi Fm	Sinbad Ls Mbr	180	*Meekoceras*
TRIASSIC	Moenkopi Fm	Black Dragon Mbr	280	Woodside Sh equiv
PERMIAN	Kaibab Limestone		880	PALEOZOIC STRATA ARE PART OF THE PAVANT ALLOCHTHON
PERMIAN	Toroweap Fm	Evaporite member	230	gypsum
PERMIAN	Toroweap Fm	Limestone member	240	
PERMIAN	Queantoweap Sandstone		800	Cedar Mesa Ss equiv
PERMIAN	Pakoon Dolomite		450	Elephant Canyon Fm equiv
P	Callville Limestone		500	
MISS	Humbug Formation		0-350	
MISS	Deseret Fm	Uncle Joe Mbr	10-160	
MISS	Deseret Fm	Tetro Member	500	Redwall Ls of Davis, 1983
MISS	Deseret Fm	Delle Phosph Mbr	165-190	
MISS	Gardison Limestone		250	
DEVONIAN	Pinyon Pk Ls	Upper member	135	coralline limestone sandy pelmicrite
DEVONIAN	Pinyon Pk Ls	Cove Ft Qtzt Mbr	80-160	stromatoporoids
DEVONIAN	Guilmette Formation		460-570	dark gray
DEVONIAN	Simonson Dolomite		180	
DEVONIAN	Sevy Dolomite		710	weathers light gray abundant fish plates
S	Laketown Dolomite		250	algal structures
ORD	Ely Springs Dolomite		320	
ORD	Eureka Quartzite		170	
ORD	Pogonip Group		1100	*Orthambonites* intraformational cg
CAMBRIAN	Upper Cambrian dolomite		1000?	
CAMBRIAN	Pierson Cove Formation		1000	algal boundstone dark gray
CAMBRIAN	Middle Cambrian Ls		400	*Glossopleura*
CAMBRIAN	Pioche Fm	Tatow Member	90	*Chancia ebdome*
CAMBRIAN	Pioche Fm	Lower member	200	
CAMBRIAN	Tintic Quartzite		3000 -	
	MESOZOIC ROCKS BENEATH PAVANT THRUST			

QUATERNARY - Dennis et al, 1946; Hoover, 1974; Luedke & Smith, 1978; Mehnert et al, 1978; TERTIARY - Steven, 1979, 1981; Steven & Morris, 1983; MAPS AND PALEOZOIC STRATA - Davis, 1983; George, 1985; PERMIAN - Welsh, 1976; Irwin, 1971, 1976; MISSISSIPPIAN - Sandberg & Gutschick, 1984; J. E. Welsh per. comm., 1985; ORDOVICIAN - Chamberlin, 1975.

Chart
78 RICHFIELD - MONROE AREA

Period	Unit	Feet	Remarks
Q	Alluvium, colluvium, landslide & glacial deposits	0-500	glacial above 9800 ft
PLIO–MIOCENE	Concealed valley fill in Sevier Valley - includes buried Sevier River Fm and possible diapiric masses of Jurassic Arapien Shale and evaporites	6000 ?	thickness estimated from geophysical data
MIOCENE	Sevier River Formation- exposed on Sevier Valley flanks	0-1000	7-14 m.y.
MIOCENE	Basalt flow	20	
MIOCENE	Joe Lott Tuff M, Mt. Belknap V	20	19 m.y.
MIOCENE	Osiris Tuff	100	23 m.y. ash-flow tuff
OLIGOCENE	Bullion Canyon Volcanics: Lava flows; Tuff of Glenwood Mt; of Signal Peak; Antimony Tuff M of Mt. Dutton Fm	400-1500	formerly known as Dry Hollow Fm, now believed to have been derived from several volcanic centers
OLIGOCENE	Tuff of Albinus Canyon	200	25 m.y.
OLIGOCENE	Three Creeks Tuff Mbr / Volcanic rocks of Cliff Canyon	200	27 m.y.
OLIGOCENE	Wah Wah Springs Fm, N R Gp	100	29.5 m.y. Needles Range Gp
OLIGOCENE	Volcanic rocks of Dog Valley / Older volcanic rocks	600	dacitic rocks in Clear Creek Canyon
EOCENE	Formation of Aurora	480	39 m.y.
EOCENE	Crazy Hollow Formation	260	
EOCENE	Green River Formation	0-50	44 m.y.
EOCENE–PALEOCENE	Flagstaff Formation	0-1900	red strata on east flank of Pavant Range NW of Richfield are generally equivalent to scenic early Cenozoic strata at Bryce Canyon and Cedar Breaks. Anderson and Rowley (1975) suggested that a paleohighland near Marysvale separated the two early Cenozoic basins of deposition
PALEOCENE–CRETACEOUS	North Horn Formation	3300	overlies Paleozoic rocks on west side of Pavant Range; basal conglomerates sometimes called Price River--Indianola
J	Arapien Shale	2000 +	occurs as diapiric masses near Glenwood, forms badlands
Mz	Pre-Arapien Mesozoic strata	4000	occurs on west side of Pavant Range and in deeper wells drilled in Sevier Valley
PALEO	Paleozoic limestone, dolomite, and quartzite	5000-10,000	

MAP WITH TEXT - Schneider, 1964; Lautenschlager, 1952; Rowley et al, 1981; Steven, 1979; VALLEY FILL - Young & Carpenter, 1965; Brown & Cook, 1982; VOLCANICS - Cunningham & Steven, 1979; Steven et al, 1979; Steven, 1984; FLAGSTAFF - Schneider, 1967.

FIGURE 142 — Wasatch Plateau east of Manti. Flagstaff Formation dips towards Sanpete Valley. Elevations exceed 10,000 feet and slump topography develops in basins cut into the North Horn Formation.

FIGURE 143 — Badland hills from foreground to middle distance in photo center are Jurassic Arapien Shale on the east side of Sevier Valley between Richfield and Salina. Early Cenozoic strata dip valleyward at right. Structure in Sevier Valley has been complicated by mobility of evaporite-bearing Arapien Shale.

FIGURE 144 — Cretaceous beds on the west side of the San Rafael Swell. Interstate I-70 in foreground. Tununk Shale forms easily eroded lower hillslopes. Caprock ledges are Ferron Sandstone. See Chart 81.

FIGURE 145 — Jurassic beds on the west side of the San Rafael Swell. Interstate I-70 in foreground is on silty maroon beds of the Entrada Formation. Curtis Formation forms thin light-green caprock ledge in middle distance. Prominent cuesta caprock just below horizon is Cretaceous Ferron Sandstone. See Chart 81.

Chart
79 SALINA CANYON MOUTH

Period	Group	Unit	FEET	Notes
Q		Alluvium, colluvium, lake beds	0-300	
TERTIARY		Osiris Tuff	0-350	23 m.y.
TERTIARY		Fm of Black Cap Mtn	0-600	volcaniclastic ss
TERTIARY		Tuff of Albinus Canyon ?	0-250	25 m.y.
TERTIARY		Three Creeks Tuff Mbr	0-30	27 m.y.
TERTIARY		Unnamed ss, cg, mudstone	0-100	30 m.y. tuffaceous ss
TERTIARY		Formation of Aurora	0-1400	bentonitic 39 m.y. formerly Bald Knoll Fm
TERTIARY		Crazy Hollow Formation	500-1000	red & pink mudstone salt & pepper ss
TERTIARY	Green River Fm	Upper member	730	pale yellow algal ls 44 m.y. K-Ar on tuffs
TERTIARY	Green River Fm	Lower member	430	green shale
TERTIARY		Colton Formation	100-530	red mudstone
TERTIARY		Flagstaff Limestone	0-100	ss, cg, ls, mudstone
TERTIARY / CRETACEOUS		North Horn Formation	0-1500	STRATA BELOW THE NORTH HORN FM ARE STEEPLY TILTED
CRETACEOUS	Indianola Group	Sixmile Canyon Formation (not exposed in Salina Canyon) (fluvial sandstone &siltstone in upward-fining cycles)	4500	°Names of equivalent units on east side of Wasatch Plateau; Castlegate Ss°; Blackhawk Fm°; Upper Blue Gate Sh° (formerly Masuk Sh); Emery Ss°
CRETACEOUS	Indianola Group	Funk Valley Formation (only lower third exposed in Salina Canyon) (marine shale and near-shore sandstone)	2400	Lower Blue Gate Sh° *Inoceramus*; Ferron Ss°
CRETACEOUS	Indianola Group	Allen Valley Shale	900	*Collignoniceras*; Tununk Shale°
CRETACEOUS	Indianola Group	Sanpete Formation	1200	Dakota Ss°
CRETACEOUS		Cedar Mountain Fm	850	90-96 m.y. FT zircon from bentonite bed
JURASSIC	Twist Gulch Fm	Summerville equiv?	10	
JURASSIC	Twist Gulch Fm	Curtis Fm equiv	200	yellow ss
JURASSIC	Twist Gulch Fm	Entrada Ss equivalent	1700	reddish-brown siltstone mudstone & sandstone
JURASSIC	Arapien Shale	Unit E	400	salt-bearing shale
JURASSIC	Arapien Shale	Unit D	3000	gypsiferous shale mudstone & sandstone
JURASSIC	Arapien Shale	Unit C	1000	gypsum
JURASSIC	Arapien Shale	Units A & B	400	equivalent to Leeds Creek Mbr, Twin Creek Ls
Mz		Subsurface Mesozoic strata	3300+	pre-Arapien strata
Pz		Subsurface Paleozoic strata	3400+	likely nearly horizontal

PRINCIPAL REFERENCE--Willis, 1986; TERTIARY--Weiss, 1982; Sperry, 1980; CRETACEOUS--Lawton, 1982; Fouch et al, 1982; Spieker, 1949; JURASSIC--Imlay, 1980; Standlee, 1982; STRUCTURE--Witkind & Page, 1982; Stokes, 1982; Lawton, 1985; Hardy, 1952.

Chart
80 UPPER SALINA CANYON I-70

Period	Group	Unit	FEET	Notes
Q		Alluvium and colluvium	0-100	
EO		Green River Formation	1000+	white; green
PALEOCE		Colton Formation	500	red
PALEOCE		Flagstaff Limestone	300-1000	fresh-water ls
PALEOCE		North Horn Formation	900-1200	red and gray mudstone, ss, prone to slumping
CRETACEOUS	Mesaverde Group	Price River Fm	700	
CRETACEOUS	Mesaverde Group	Castlegate Sandstone	210-240	
CRETACEOUS	Mesaverde Group	Blackhawk Formation	800-900	coal; Phillips Petroleum 20-225-3E 1973
CRETACEOUS	Mesaverde Group	Star Point Sandstone	400	
CRETACEOUS	Mancos Shale	Upper Blue Gate Shale Member	500	formerly Masuk Sh M
CRETACEOUS	Mancos Shale	Emery Ss Member	1200	coal
CRETACEOUS	Mancos Shale	Lower Blue Gate Shale Member	1500	*Inoceramus*; *Inoceramus*
CRETACEOUS	Mancos Shale	Ferron Ss Member	720	coal
CRETACEOUS	Mancos Shale	Tununk Shale Member	940	*Inoceramus*
CRETACEOUS		Dakota Sandstone	100	
CRETACEOUS		Cedar Mountain Fm	300	
JURASSIC		Morrison Fm	400	
JURASSIC		Summerville Fm	600	
JURASSIC		Curtis Formation	190	
JURASSIC		Entrada Sandstone	1000	silty redbeds Twist Gulch equivalent
JURASSIC		Arapien Shale	1100	salt & gypsum; Arapien is equivalent to upper 2 members of Imlay's Twin Creek Ls.
JURASSIC		Twin Creek Limestone	550	As used here the Twin Creek includes Imlay's lower 5 members
JURASSIC		Navajo Sandstone	900	
JURASSIC		Kayenta Formation	120	
JURASSIC		Wingate Sandstone	330	
TRIASSIC	Chinle Fm	Upper member	200	bentonitic
TRIASSIC	Chinle Fm	Moss Back Ss M	60	
TRIASSIC	Moenkopi Formation	Upper member	1200	
TRIASSIC	Moenkopi Formation	Sinbad Ls Mbr	270	
TRIASSIC	Moenkopi Formation	Black Dragon Mbr	250	
PERM	Kaibab Fm	Harrisburg Mbr	150	Plympton Fm equivalent
PERM	Kaibab Fm	Fossil Mtn Mbr	110	Black Box Dolo equivalent
PERM		Pakoon Dolomite	800+	
M		Redwall Limestone	800+	
D		Undivided Devonian ls, dolomite, & ss	600+	
CAMB		Undivided Cambrian dolomite, limestone shale & quartzite	1500+	

EXPOSED TERTIARY & CRETACEOUS--Bachman, 1959; Baughman, 1959; Doelling, 1972; Stanley & Collinson, 1979; Spieker & Baker, 1928; SUBSURFACE--Phillips Petroleum No. 1 United States D well log as interpreted by J. L. Baer and J. E. Welsh.

Chart
81 SAN RAFAEL SWELL, WEST FLANK

System	Group / Formation	Unit	FEET	Remarks
Q		Alluvium & surficial dep	0-100	
T		Alkalic diabase dikes & sills	–	3 to 5 m.y.
CRETACEOUS	Mancos Shale	Blue Gate Shale Mbr	1500 +	
CRETACEOUS	Mancos Shale	Ferron Ss Member	150-400	coal beds
CRETACEOUS	Mancos Shale	Tununk Shale Mbr	400-700	*Pycnodonte newberryi*
CRETACEOUS		Dakota Sandstone	0-50	
CRETACEOUS		Cedar Mountain Formation	100-420	
CRETACEOUS		Buckhorn Cg Mbr	0-30	106 m.y. fission track
JURASSIC	Morrison Fm	Brushy Basin Member	250-400	134 m.y.
JURASSIC	Morrison Fm	Salt Wash Mbr	200-300	
JURASSIC	Morrison Fm	Tidwell Member	60	J-5 unconformity
JURASSIC	San Rafael Group	Summerville Fm	200	red
JURASSIC	San Rafael Group	Curtis Fm	100-250	green J-3 unconformity
JURASSIC	San Rafael Group	Entrada Fm (earthy facies)	300-400	resembles Summerville Fm here
JURASSIC	San Rafael Group	Carmel Fm	250-300	gypsum *Camptonectes*
JURASSIC	San Rafael Group	Page Sandstone	100	J-2 unconformity
JURASSIC	Glen Canyon Gp	Navajo Ss	450-550	
JURASSIC	Glen Canyon Gp	Kayenta Fm	50-250	
JURASSIC	Glen Canyon Gp	Wingate Ss	350-400	
TRIASSIC	Chinle Fm	Church Rock M	150-250	
TRIASSIC	Chinle Fm	Moss Back Ss M	80-120	
TRIASSIC	Chinle Fm	Temple Mtn M	0-100	Tr-3 unconformity
TRIASSIC	Moenkopi Fm	Moody Canyon Member	400	
TRIASSIC	Moenkopi Fm	Torrey Mbr	250-300	
TRIASSIC	Moenkopi Fm	Sinbad Ls Mbr	50-150	*Meekoceras*
TRIASSIC	Moenkopi Fm	Black Dragon M	140-200	subsurface
PERMIAN		Black Box Dolomite	100-200	"Kaibab" of old reports
PERMIAN		White Rim Sandstone	600-800	"Coconino" of well logs Standard Oil of Calif -Continental Unit #2 1937 to Precambrian
PERMIAN		Pakoon Dolomite (Elephant Canyon)	500-700	NOTE - This area is on the east flank of the late Paleozoic Piute Platform. The White Rim Ss rests directly on the Redwall Ls in 20S-7E Pan-Am Ferron well
PENN		Honaker Trail Fm ?	200-500	
PENN		Paradox Fm ?	300-500	lower Oquirrh Fm equivalent
PENN		Pinkerton Trail Fm ?	100-200	Manning Cyn equiv
MISS		Redwall Ls	600-1000	Gardison Ls equiv Fitchville Fm equiv
DEV		Ouray Limestone	100-200	
DEV		Elbert Fm	200-400	
CAMBRIAN		Lynch Dolomite & Maxfield Limestone undivided	1000-1300	
CAMBRIAN		Ophir Fm	200	
CAMBRIAN		Tintic Quartzite	150-300	
p€		schist	-	

MAP--Williams & Hackman, 1971; QUATERNARY--Williams, 1972; TERTIARY--Gilluly, 1929; CRETACEOUS--Ryer, 1981; Maxfield, 1976; Tschudy et al, 1984; Doelling, 1972; JURASSIC--O'Sullivan & Pierce, 1983; O'Sullivan, 1984; Stanton, 1976; Smith, 1976; Bagshaw, 1974; Kowallis & Heaton, 1984; TRIASSIC--Sanderson, 1974; Hood & Patterson, 1984; Stewart, 1972a, 1972b; Lupe, 1979; Ash, 1975; Blakey, 1974, 1977; Mitchell 1985; PERMIAN - Rascoe & Baars, 1972; Kiser, 1976; Orgill, 1971; Welsh et al, 1979; SUBSURFACE - Mallory, 1972; Craig, 1972; Baars, 1972; Lockman-Balk, 1972; Welsh & Bissell, 1979.

Chart
82 HANKSVILLE AREA

System	Group / Formation	Unit	FEET	Remarks
Q		Alluvium & dune sand	0-100	sand from Entrada Fm
K	Morrison Fm	Brushy Basin Member	300-320	dinosaur bone varicolored rocks
JURASSIC	Morrison Fm	Salt Wash Ss Member	250-200	
JURASSIC	Morrison Fm	Tidwell Member	100	J-5 unconformity
JURASSIC	San Rafael Group	Summerville Fm	100	
JURASSIC	San Rafael Group	Curtis Formation	60	J-3 unconformity
JURASSIC	San Rafael Group	Entrada Formation	560	stone babies of Goblin Valley St Pk
JURASSIC	San Rafael Group	Carmel Formation	240	gypsum
JURASSIC	San Rafael Group	Page Sandstone	30-70	J-2 unconformity
JURASSIC	Glen Canyon Group	Navajo Sandstone	500-600	rounded cliffs
JURASSIC	Glen Canyon Group	Kayenta Fm	300	ragged tree-covered ledges
JURASSIC	Glen Canyon Group	Wingate Ss	300-350	vertical cliffs J-0 unconformity
TRIASSIC	Chinle Fm	Church Rock Mbr – Upper	150	varicolored
TRIASSIC	Chinle Fm	Church Rock Mbr – Black ledge	40	
TRIASSIC	Chinle Fm	Church Rock Mbr – Lower	70	
TRIASSIC	Chinle Fm	Moss Back Mbr	90	uranium
TRIASSIC	Chinle Fm	Temple Mtn Mbr	40	Tr-3 unconformity
TRIASSIC	Moenkopi Fm	Moody Canyon Mbr	120-170	
TRIASSIC	Moenkopi Fm	Torry Member	260-250	
TRIASSIC	Moenkopi Fm	Sinbad Ls Mbr	30-50	snails & pelecypods
TRIASSIC	Moenkopi Fm	Black Dragon M	80-120	subsurface
PERMIAN		Black Box Dolomite	0-200	Delhi Oil- #1 Russell, Tidewater-Big Flattop Standard of Calif San Rafael Unit # 2 drilled in '53, '49, '37
PERMIAN	Cutler Group	White Rim Sandstone	200-400	
PERMIAN	Cutler Group	Organ Rock Shale	100-200	
PERMIAN	Cutler Group	Cedar Mesa Sandstone (arkosic interbeds)	0-900	
PERMIAN	Cutler Group	Elephant Canyon Formation	700	arkosic interbeds
PENNSYLVANIAN	Hermosa Group	Honaker Trail Formation	900	REGIONAL UNCONFORMITY HERMOSA GROUP thickens eastward towards Paradox Basin
PENNSYLVANIAN	Hermosa Group	Paradox Formation	700-1000	salt & gypsum thin to absent over Nequoia Arch west of Canyonlands
PENNSYLVANIAN	Hermosa Group	Pinkerton Trail Formation	500	
PENNSYLVANIAN		Molas Formation	0-100	red residual deposit
MISS		Redwall Limestone	900	cherty
D		Elbert & Ouray Fms	200	
CAMBRIAN		"Lynch" Dolomite and "Maxfield" Limestone undivided	900	
CAMBRIAN		Ophir Formation	200	
CAMBRIAN		Tintic Quartzite	180	
p€		Metamorphic rocks	-	

MAP WITH TEXT - Hunt et al, 1953; JURASSIC - Craig & Dickey, 1956; Peterson, 1984; Wright & Dickey, 1978; TRIASSIC - Stewart, 1972a, b; Blakey, 1974; Mitchell, 1985; BLACK BOX - Welsh et al, 1979; SUBSURFACE - Wells, 1954; Welsh & Bissell, 1979; Irwin, 1971, 1976.

FIGURE 146 — "Reef" along southeast margin of the San Rafael Swell. Light-colored massive flatirons in upper half of left photo edge are Jurassic Navajo Sandstone; darker and sharper overlying flatirons are Carmel Formation; Entrada Formation forms strike valley with dry wash; Curtis Formation forms low hogback at center foreground; dark slope to right of Curtis is Summerville Formation and Tidwell Member of Morrison Formation. Massive hogback in right third is Salt Wash Member of the Morrison. See Chart 81.

FIGURE 147 — View from inside the south rim of the San Rafael Swell looking southward to Factory Butte. Talus-covered shaly beds in foreground are Church Rock Member, Chinle Formation (c), overlain by Kayenta (k), and Navajo (n) Sandstones. See Figure 146 for overlying strata. Factory Butte exposes Cretaceous beds. See Chart 81.

FIGURE 148 — View southward from Arches National Park toward Abajo laccolithic mountains in distance. Colorado River cuts across Moab (Spanish) Valley at left center. Moab Valley is a collapsed salt anticline. Moab fault extends across photo at base of cliffline capped by Wingate Sandstone (w) near photo center. Arch-forming Entrada Sandstone (e) below photo center rests on jointed Navajo Sandstone in foreground. See Chart 84.

FIGURE 149 — Upheaval Dome in Canyonlands National Park shows Jurassic Navajo Sandstone forming a ring syncline around a circular anticline cored with Permian Organ Rock Shale and White Rim Sandstone. Upheaval Dome was likely formed by meteor impact.

Chart

83 CANYONLANDS PARK AREA

Period	Group / Formation	Member / Unit	FEET	Remarks
Q	Alluvium - dune sand		0-100	
K / JURASSIC	Morrison Fm	Brushy Basin Member	300-400	dinosaur bone
JURASSIC	Morrison Fm	Salt Wash Mbr	200-400	
JURASSIC	Morrison Fm	Tidwell Member	20-60	J-5 unconformity
JURASSIC	San Rafael Group	Wanakah Formation	0-60	
JURASSIC	San Rafael Group	Entrada Sandstone	220-350	ARCHES NAT'L PARK
JURASSIC	San Rafael Group	Dewey Bridge Mbr	20-130	
JURASSIC	San Rafael Group	Page Sandstone	0-100	J-2 unconformity
JURASSIC	Glen Cayon Group	Navajo Sandstone	250-450	rounded cliffs
JURASSIC	Glen Cayon Group	Kayenta Fm	200-260	DEAD HORSE POINT
JURASSIC	Glen Cayon Group	Wingate Sandstone	30-430	vertical cliff face J-0 unconformity
TRIASSIC	Chinle Fm	Church Rock Member	240-310	
TRIASSIC	Chinle Fm	Owl Rock M	70-120	
TRIASSIC	Chinle Fm	Petrified Forest M	50-100	bentonite
TRIASSIC	Chinle Fm	Moss Back Ss M	0-100	Tr-3 unconformity
TRIASSIC	Moenkopi Fm	Moody Canyon M	40-120	
TRIASSIC	Moenkopi Fm	Torrey Member	100-200	
TRIASSIC	Moenkopi Fm	Sinbad Ls Mbr	0-20	
TRIASSIC	Moenkopi Fm	Black Dragon Mbr	0-230	Tr-1 unconformity
TRIASSIC	Hoskinnini Sandstone		0-50	
PERMIAN	Cutler Group	White Rim Ss	0-250	
PERMIAN	Cutler Group	Organ Rock Fm	0-400	
PERMIAN	Cutler Group	Cedar Mesa Sandstone	300-1000	Surface rock in The Needles area. Interfingers eastwarc with arkosic beds within the Park
PERMIAN	Cutler Group	Elephant Canyon and Halgaito Fms (time equivalent)	400-1200	Thickens and becomes less limy towards Uncompahgre Uplift to the east unconformity
PENNSYLVANIAN	Hermosa Group	Honaker Trail Formation	1000-1500	
PENNSYLVANIAN	Hermosa Group	Paradox Formation	500-4000	Oldest rocks exposed in Canyonlands, at river level in the Gypsum Canyon area salt & potash interbedded with black shale and anhydrite
PENNSYLVANIAN	Hermosa Group	Pinkerton Trail Formation	200-300	
PENNSYLVANIAN	Molas Formation		0-100	
MISS	Leadville Limestone		400-600	Pre-Paradox rocks are known from wells and from geophysical measurements
D	Ouray Limestone		100	
D	Elbert Formation		300	
CAMBRIAN	Undivided Upper Cambrian dolomite, limestone, & shale		800-900	
CAMBRIAN	"Bright Angel" Shale		0-200	
CAMBRIAN	Ignacio Quartzite		100-200	
pЄ	Metamorphic rocks		-	

GEOLOGIC MAP - Huntoon et al, 1982; JURASSIC - Craig & Dickey, 1956; Wright & Dickey, 1978; TRIASSIC - Stewart, 1971a; Blakey, 1974; Pipiringos & O'Sullivan, 1978; Lupe, 1984; O'Sullivan & Maclachlan, 1975; PERMIAN - Baars, 1975; PENNSYLVANIAN - Welsh & Bissell, 19779; Szabo & Wengerd, 1975; SUBSURFACE - Rocky Mountain Associatoin of Geologists Atlas, 1972. RIVER GUIDEBOOK - Baars & Molenaar, 1971.

Chart

84 MOAB - ARCHES - LA SAL AREA

Period	Group / Formation	Member / Unit	FEET	Remarks
Q	Alluvium, colluvium		0-100	
Q	Glacial & eolian deposits		0-200	0.6 m.y. ash beds
Q	Geyser Creek Fanglomerate		0-600	
T	Diorite & monzonite porphyry dikes, sills, laccoliths of LaSal Mts			29 m.y. intrusive contact
CRET	Mancos Shale	Upper shale Mbr	0-500 +	
CRET	Mancos Shale	Ferron Ss Mbr	0-100	
CRET	Mancos Shale	Lower Shale Mbr	300-500	*Pycnodonte newberryi*
CRET	Dakota Sandstone		0-200	
CRET	Burro Canyon Fm		80-250	
JURASSIC	Morrison Fm	Brushy Basin Mbr	250-450	dinosaur bones
JURASSIC	Morrison Fm	Salt Wash Member	130-350	
JURASSIC	Morrison Fm	Tidwell Member	20-100	formerly Summerville J-5 unconformity
JURASSIC	Entrada Ss	Moab Tongue	50-180	white caprock
JURASSIC	Entrada Ss	Slick Rock Mbr	200-500	ARCHES NAT'L PARK
JURASSIC	Entrada Ss	Dewey Bridge M	40-240	Page Ss locally J-2 unconformity
JURASSIC	Navajo Sandstone		0-550	
JURASSIC	Kayenta Formation		140-300	DEAD HORSE POINT
JURASSIC	Wingate Sandstone		150-450	vertical cliffs
TRIASSIC	Chinle Fm	Siltstone member	150-300	
TRIASSIC	Chinle Fm	Black ledge	0-150	manganese coating
TRIASSIC	Chinle Fm	Limy member	0-50	
TRIASSIC	Chinle Fm	Claystone mbr	0-100	
TRIASSIC	Chinle Fm	Moss Back Mbr	0-50	Tr-3 unconformity
TRIASSIC	Moenkopi Fm	Upper slope mbr	230	MEMBERS IN CASTLE VALLEY 140' - Pariott Mbr
TRIASSIC	Moenkopi Fm	Ledge-forming m	260	380' - Sewemup Mbr
TRIASSIC	Moenkopi Fm	Sinbad Ls Mbr	0-60	220' - Ali Baba Mbr
TRIASSIC	Moenkopi Fm	Lower slope mbr	100-200	220' - Tenderfoot Mbr
TRIASSIC	Hoskinnini Sandstone		0-100	Permian ?
PERMIAN	Cutler Group	White Rim Ss	0-250	tar sands
PERMIAN	Cutler Group	Organ Rock Sh	0-300	
PERMIAN	Cutler Group	Cedar Mesa Sandstone	0-1200	CANYONLANDS NAT'L PARK
PERMIAN	Cutler Group	Arkosic facies Arkosic facies of Cutler Group is derived from the Uncompahgre Uplift to the east and interfingers with both the Cedar Mesa and Elephant Canyon Fms	400-3000	Arkosic facies is not a separately named map unit but forms the undivided lower part of the Cutler Group at Moab
PERMIAN	Cutler Group	Elephant Canyon Fm (formerly called Rico facies of Cutler)	0-1500	Elephant Canyon Fm is exposed along Colorado R south of Potash plant. Fossils include corals, bryozoans, brachiopods and fusulinids
PENNSYLVANIAN	Hermosa Group	Honaker Trail Formation	1500-2000	Honaker Trail Fm exposed on west side of Moab Fault near Arches Hdqtr. Arkose, ss, & maroon mudstone is interbedded with fossiliferous limestone that contains corals & brachiopods
PENNSYLVANIAN	Hermosa Group	Paradox Fm (cyclic sequence of salt,anhydrite, potash, black shale, dolomite and locally arkose)	2000-5000	potash & salt Salt from the Paradox Fm has been mobilized into elongate masses up to 15,000 ft thick which have formed the salt anticlines in this region
PENNSYLVANIAN	Hermosa Group	Pinkerton Trail Fm	0-150	*Anthracospirifer*
PENNSYLVANIAN	Molas Formation		0-100	
M	Leadville Dolomite		300-600	Oil zones in Lisbon and Big Flat oil fields.
DEV	Ouray Limestone		0-150	
DEV	Elbert Formation		100-200	
DEV	Elbert Formation	McCracken Ss M	25-100	unconformity
CAMB	"Lynch" Dolomite		800-1000	
CAMB	"Bright Angel" Shale		0-100	
CAMB	Ignacio Quartzite		100-300	
pЄ	Schist, gneiss, granitic and pegmatite dikes and plugs			1480 m.y. granite 1700 m.y. gneiss

MAPS WITH TEXT--Williams, 1964; Doelling, 1985; Huntoon et al, 1975; Carter & Gualtieri, 1965; QUATERNARY--Biggar et al, 1981; Nash & Smith, 1977; Coleman, 1983; Richmond, 1962; LACCOLITHS--Hunt, 1958, 1983; CRETACEOUS-- Craig, 1981; JURASSIC--O'Sullivan, 1984; O'Sullivan & Pierce, 1983; Kocurek, 1981; TRIASSIC--O'Sullivan & MacLachlan, 1975; Stokes & Madsen, 1979; Gilland, 1979; O'Sullivan, 1970; Lupe, 1979; Shoemaker & Newman, 1959; PERMIAN--Baars, 1975; Bohn, 1977; Terrell, 1972; Mack & Rasmussen, 1984; PENNSYLVANIAN-- Mallory, 1972; Melton, 1972; Welsh & Bissell, 1979; MISSISSIPPIAN--Craig, 1972; DEVONIAN--Baars, 1972; CAMBRIAN--Lochman-Balk, 1972; PRECAMBRIAN--Tweto, 1977; Hedge et al, 1968; SALT ANTICLINES--Huntoon, 1982; McGill & Stranquist, 1979; Reynolds et al, 1985; LISBON FIELD--Parker, 1981.

Chart
85 MINERAL MOUNTAINS

Period	Unit		FEET	Remarks
QUAT	Alluvium, colluvium, & hot spring deposits		0-200	
QUAT	Basalt flows & cones		0-100	0.01-0.07 m.y.
QUAT	Rhyolite domes, flows, tuffs		0-200	0.5-0.8 m.y.
TERTIARY	Basin-fill in Milford Valley		1000-5000	
TERTIARY	Porphyritic quartz latite flow		0-50	7.9 m.y.
TERTIARY	Rhyolite dikes		-	
TERTIARY	Diabase & diorite dikes		-	
TERTIARY	Plutonic complex in the Mineral Mountains	Fine-grained granite dikes		A thermal event about 10 m.y. ago reset all K-Ar biotite dates. Original age of these plutons is probably 20-30 m.y.
		Leucocratic granite		SE 1/4 of pluton
		Syenite stock		
		Coarse-grained biotite granite		SW 1/4 of pluton
		Porphyritic quartz monzonite		
		Coarse-grained biotite quartz monzonite		N 1/2 of pluton
		Diorite breccia		
		Diorite		INTRUSIVE CONTACT
TERTIARY	Claron ? Conglomerate		110	red matrix
JURASSIC	Carmel Formation		570	*Ostrea* *Pentacrinus*
JURASSIC	Navajo Sandstone		1540	
TRIASSIC	Moenkopi Fm	Upper red beds	500-1000	
		Sinbad Ls Mbr	570	*Monotis* *Flemingites*
		Black Dragon Mbr	140	red bed
PERMIAN	Kaibab Fm	Harrisburg Mbr	250	gypsum
		Fossil Mtn Mbr	410	*Dictyoclostus*
PERMIAN	Toroweap Formation		280	
PERMIAN	Talisman Quartzite		1180	erroneously called Coconino in some reports. Equivalent to Queantoweap Ss.
PERMIAN	Pakoon Dolomite		700	aplite dike intrusion
P	Callville Limestone		500	
MISS	Humbug Formation and Deseret Limestone undivided		700	Devonian-Mississippian strata in SW part of range are bleached, silicified, and mostly unfossilfierous
		Tetro Ls Mbr	150	-cherty
MISS	Gardison Limestone		350	unconformity
D	Pinyon Peak Limestone		150	unconformity
D	Simonson Dolomite		400+	
Cambrian strata at north end of range not in contact with rocks above				
CAMBRIAN	Notch Peak and Orr Formations		1300	attenuated and converted to marble
Cambrian rocks below are thrust over Cambrian rocks above				
CAMBRIAN	Howell Limestone ?		190+	
CAMBRIAN	Pioche Formation		25-100	*Glossopleura*
CAMBRIAN	Prospect Mountain Qtzt		800+	
Precambrian rocks not in contact with Paleozoic strata				
pC	Banded gneiss, metaquartzite, schist			1750 m.y. basement

MAP WITH TEXT - Sibbett & Nielson, 1980; Earll, 1957; Liese, 1957; Petersen, 1975; Whelan & Bowler, 1979; VALLEY FILL - Mower & Cordova, 1974; QUATERNARY - Lipman et al, 1978; Mehnert et al, 1978; MESOZOIC-PALEOZOIC - Petersen, 1974; Lee, 1908; J. E. Welsh, per. comm., 1985; PRECAMBRIAN - Stacey & Hedlund, 1983.

Chart
86 BEAVER - MANDERFIELD

Period	Unit		FEET	Remarks
H	Recent floodplain alluvium		0-20	less than 10,000 yrs
PLEISTOCENE	Alluvial terraces of Beaver		0-15	0.012-0.015 m.y.
PLEISTOCENE	Alluvial terraces of Greenville		0-15	0.12-0.14 m.y.
PLEISTOCENE	Old terrace alluvium		0-30	0.25 m.y. (North Creek)
PLEISTOCENE	Gravel of ancestral Indian Cr		10	0.4 m.y.
PLEISTOCENE	Basalt of Red Knoll		0-400	cinder cone
PLEISTOCENE	Basalt of Crater Knoll		0-400	cinder cone
PLEISTOCENE	Gravel of Last Chance Bench		6-15	0.5 m.y.
PLEISTOCENE	Tephra of Ranch Canyon		0-150	0.55 m.y. pumice
PLEISTOCENE	Rhyolite domes & flows		0-200	source of tephra of Ranch Canyon
PLEISTOCENE	Basalt of Cunningham Hill		1-20	1.1 m.y.
PLIOCENE	Upper basin-fill deposits including Lake Beaver beds	Last Chance Bench ash	600-800	1.6 m.y. ?
		Huckleberry Ridge ash		2.0 m.y. ostracodes
		Middle ash		2.1 m.y.
		Indian Creek ash		2.3-2.4 m.y.
		Hogsback ash		ostracodes
		Blancan bones		fossil zebra
PLIOCENE	Lower basin-fill deposits	Piedmont fluvial & deltaic deposits	1000?	ostracodes
		Conglomerate of Maple Flats	800	
MIOCENE		Bolson deposits	1000-3000	gypsiferous 7.6 m.y. basalt near Minersville Reservoir
MIOCENE	Rhyolite & rhyolite tuff of Gillies Hill		0-1000	9 m.y.
MIOCENE	Mount Belknap Volcanics	Mount Baldy Rhyolite Member	300	exposed in Tushar Mountains
		Joe Lott Tuff Mbr	30	19 m.y.
MIOCENE	Potassium-rich mafic lava flows		500	22 m.y.
MIOCENE	Formation of Lousy Jim		0-1000	22.5 m.y. rhyodacite lavas & flow-breccia
MIOCENE	Tuff of Lion Flat		0-300	rhyolite ash-flow tuff
MIOCENE	Osiris Tuff		0-100	23 m.y.
MIOCENE / OLIGOCENE	Bullion Canyon Volcanics 22-29 m.y.	Delano Pk Tuff Mbr		23 m.y. exposed in the Tushar Mountains
OLIGOCENE		Mount Dutton Formation 22-27 m.y.	1000+	locations of calderas and stocks shown by Cunningham et al, 1983

MAP WITH TEXT - Machette, 1983; Machette & Steven, 1983; Cunningham et al, 1983; LATE CENOZOIC - Machette, 1982, 1985; Forester & Bradbury, 1981; Evans & Steven, 1982.

Chart 87 SOUTHERN TUSHAR MOUNTAINS

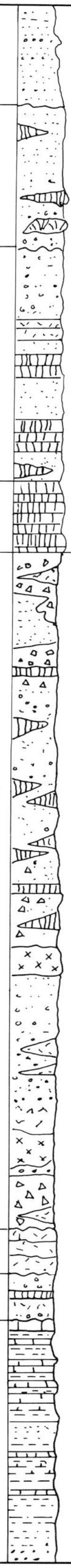

Period	Unit	Feet	Notes
QUAT	Alluvium, colluvium, river terrace deposits, piedmont & landslide dep	0-200	0.015 m.y. terrace; 0.14 m.y. terrace; 0.25 m.y. piedmont deposits
P	Sedimentary & volcanic basin-fill in west includes basalt and rhyolite / Sevier River Fm in east part of area	0-500	See Beaver-Manderfield column for more deta l in upper basin-fill; 6.4 m.y. basalt; 8 m.y. rhyolite dome & flow
MIOCENE	Rocks on west side near Beaver include tuff of Lion Flat 22 m.y. / Rocks on east side include: Mafic gravels of Gunsight Flat 22-23 m.y. Mafic lava flows of Circleville Mtn. Mafic lafa flows of Birch Creek Mtn 22-23 m.y.	1000	22-23 m.y.
MIOCENE	Lava flows of Kents Lake	0-800	andesite porphyry flows
MIOCENE	Osiris Tuff*	100	23 m.y.
MIOCENE	Sandstone Mbr of Mt Dutton Fm	0-500	*Certain widespread distinctive tuff beds from more distant source volcanoes are interlayered with the Mount Dutton rocks of local origin
MIOCENE–OLIGOCENE	Mount Dutton Formation, alluvial & vent facies	3000-8000	Mount Dutton Formation, mostly andesite and dacite, came from a series of east-west stratovolcanoes located where the present southern Tushar Mts are
OLIGOCENE	Spry intrusion		25-26 m.y.
OLIGOCENE	Bear Valley Fm	1000	
OLIGOCENE	Antimony Tuff M	0-30	25 m.y.
OLIGOCENE	Kingston Can T M	0-10	26 m.y.
OLIGOCENE	Volcanic and intrusive rocks of Bull Rush Creek		26 m.y.?
OLIGOCENE	Buckskin Breccia	0-600	
OLIGOCENE	Isom Fm*	30-60	26 m.y.
OLIGOCENE	Wah Wah Springs Fm, Needles Range Group	50-80	29.5 m.y.
OLIGOCENE	Local volcanic and tuffaceous sedimentary rocks	0-100	32 m.y.
PALEOCENE	Claron Formation	900	cut by Spry intrusion; top third- white Claron fresh-water limy beds; bottom two-thirds red Claron

MAP WITH TEXT - Anderson et al, 1986 a, b, in preparation.

Chart 88 ANTIMONY - JOHNS VALLEY

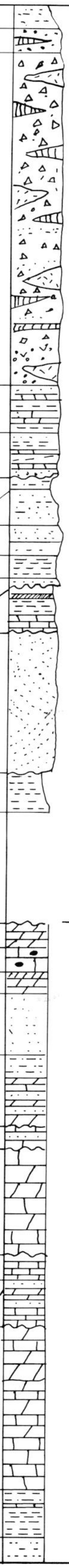

Period	Unit	Feet	Notes
Q	Alluvium & surficial deposits	0-200	
P	Sevier River Formation	0-200	
MIO	*Osiris Tuff	30	23 m.y. *Distinctive ash-flow tuff bed from a distant source
OLIGOCENE	Mount Dutton Formation mostly dacitic volcanic mudflow breccia, minor ss, cg, flows, and flow breccia	1500-2500	
OLIGOCENE	Kingston Can, Tuff M	30	26 m.y.
OLIGOCENE	*Three Creeks Tuff M	100	27 m.y. - Member of Bullion Canyon Volcanics
OLIGOCENE	*Wah Wah Sp Fm, NRGp	130	29.5 m.y.
PALEO	Flagstaff Fm / Claron Fm White m	up to 1100	Antimony ore host
PALEO	Flagstaff Fm / Claron Fm Red mbr	up to 1100	
CRET	Kaiparowits Formation	0-50	ANGULAR UNCONFORMITY Flagstaff-Claron lies across all tilted Cretaceous units here
CRET	Wahweap Formation	0-500	
CRET	Straight Cliffs Formation	0-1200	
CRET	Tropic Formation	0-600	
CRET	Dakota Sandstone	0-200	Dakota coal in Johns Valley
JURASSIC	Carmel Formation	560 plus	Overlain by Flagstaff at Antimony Canyon
JURASSIC	Navajo Sandstone (undivided Glen Canyon Group)	1650 plus	
TRIASSIC	Chinle Formation	120 plus	fault bounded
TRIASSIC	Moenkopi Formation	1000 ? not exposed	Tenneco Oil Antimony Canyon # 1 30-30S0-2W 1966
PERMIAN	Plympton Formation	120	
PERMIAN	Kaibab Limestone	130	
PERMIAN	Toroweap Formation	120	
PERMIAN	Queantoweap Sandstone	600	Cedar Mesa equivalent
PERMIAN	Pakoon Formation	670	70 ft red siltstone
P	Callville Limestone	130	
MISS	Deseret Limestone and Gardison Limestone undivided	1100	Redwall of well logs
DEV	Pinyon Peak Limestone	220	
DEV	Cove Ft Qtzt Mbr	20	
DEV	Guilmette Formation	200	
CAMBRIAN	Dolomite & limestone undivided	1800	
CAMBRIAN	Bright Angel Shale	550	
CAMBRIAN	Tapeats Quartzite	150 plus	

MAP WITH TEXT - Rowley et al, 1986 in press; Smith, 1957; Traver, 1949; Doelling, 1975; SUBSURFACE - John E. Welsh, letter, 1985; Doelling & Davis, 1978.

Chart
89 MARYSVALE AREA

Age	Formation	Member	FEET	Remarks
Q	Alluvium, colluvium, fan, landslide, and glacial deposits		0-300	
MIOCENE	Sevier River Fm	Rhyolite, rhyodacite flows	0-1500	5-8 m.y. Piute Res. area
		Basalt flows	0-400	13 m.y.
		sediments	0-1000	7-14 m.y.
		Basalt flows	0-100	
	Mount Belknap Volcanics	Gray Hills Rhyo M	0-400	Erupted from 2 main source areas of flows, domes and tuffs
		Crystal-rich member	0-800	19 m.y.
		Red Hills Tuff Mbr	0-400	19 m.y.
		Joe Lott Tuff Mbr	0-800	19 m.y.
		Blue Lake Rhyo M	0-800	21-16 m.y.
		Intrusive rhyo plugs, domes, dikes		
	Volcanic & intrusive rocks, Monroe Pk caldera		0-2000	21-23 m.y. Monroe Peak caldera source 1 mile east of Marysvale
	Osiris Tuff		0-1500	23 m.y.
OLIGOCENE	Bullion Canyon Volcanics 22-29 m.y.	Delano Pk Tuff Mbr		23 m.y.
		Mt Dutton Fm 22-27 m.y. crystal-poor dacitic rocks predominant to the south	(0-500)	in the Sevier Plateu east of Marysvale equivalent rocks include volcanic rocks of Signal Peak and volcanic rocks of Little Table
			1000-5000	
		Antimony Tuff Mbr	(0-50)	25 m.y.
		Three Cks Tf M	(0-600)	27 m.y.
	Wah Wah Springs Fm, Needles Range Group		0-100	29.5 m.y.
	Conglomerate		0-60	
	Older volcanic rocks		0-300	
JURASSIC	Twist Gulch Fm		700	red siltstone Summerville, Curtis, Entrada equivalent
	Arapien Shale		1145	550' gray shale with mud cracks, salt casts 170' red ss 285' gray shale 140' red ss
	Twin Creek Limestone (included with Arapien on maps)		1000	medium to thin-bedded gray shaly limestone *Ostrea, Camptonectes Pentacrinus*
	Navajo Sandstone (undivided Glen Canyon Group)		2000	
TRIASSIC	Chinle Fm	Upper member	200-400	
		Shinarump Cg M	130-220	
	Moenkopi Fm	Upper member	1400	
		Sinbad Ls M	350	*Meekoceras*
		Black Dragon M	120	
PERMIAN	Kaibab Fm	Harrisburg Mbr	170	
		Fossil Mtn Mbr	165	*Dictyoclostus*
	Toroweap Formation		480	
	Queantoweap Sandstone		580	
	Pakoon Dolomite		400	exposed in mine
P	Callville Limestone		300+	

QUATERNARY - Young & Carpenter, 1965; TERTIARY - Anderson & Rowley, 1975; Rowley et al, 1979, 1981b, 1986a, b; Cunningham & Steven, 1979; Steven et al, 1979; Rowley et al, 1981; Steven, 1984; Steven & Rowley, 1984; JURASSIC - Callaghan & Parker, 1962; TRIASSIC - Kerr et al, 1957; PALEOZOIC - Kennedy, 1960; Kerr et al, 1957; Welsh et al, 1979; GEOLOGIC MAP - Cunningham et al, 1983.

Chart
90 FISH LAKE PLATEAU

Age	Formation	Member	FEET	Remarks
Q	Alluvium, moraines		0-200	
TERT	Latitic lavas of the Fish Lake Plateau		400-1000	flows, tuff, agglom 25 m.y.
	Flagstaff Limestone		580	Marathon Oil 1-26S-3E; some North Horn?
CRETACEOUS	Mancos Shale	Ferron Ss Mbr	140+	coal
		Tununk Shale Member	740	
	Morrison Fm	Brushy Basin Member?	510	Cedar Mountain Fm? (Cretaceous)
JURASSIC		Salt Wash Ss M?	50	Dakota ? Ss
	Summerville Fm		230	red
	Curtis Fm		310	
	Entrada Sandstone		1010	Desert Wash well 14-25S-5E
	Carmel Formation		880-1150	anhydrite; anhydrite
	Navajo Sandstone		950-1140	
	Kayenta Formation		150	
	Wingate Sandstone		300-350	
TRIASSIC	Chinle Fm	Upper member	330-400	bentonite
		Moss Back Ss M	20-140	
	Moenkopi Fm	Upper member	650-870	
		Sinbad Ls Mbr	160	
		Lower member	150	
PERMIAN	Kaibab Formation		210	gypsum
	Toroweap Formation		500	
	Cedar Mesa Sandstone		410	
	Elephant Canyon Fm		140	Pakoon
P	Callville (Hermosa) Fm		160	
MISS	Redwall Limestone	Horsehoe Mesa Member	540	
		Mooney Falls M	30	
		Thunder Springs Member	400	cherty
		Whitmore Wash M	290	
D	Ouray Limestone		130	
	Elbert Formation		270	
Є	Cambrian dolomite and limestone		top only reached	

SURFACE - Hardy & Muessig, 1952; McGookey, 1958; SUBSURFACE - Marathon Oil South Mt. Terrill No. 1A-1;p GROUNDWATER - Carpenter et al, 1967.

FIGURE 150 — Mount Hillers, elevation 10,650 feet, is part of the Henry Mountains cluster of laccoliths. Upturned Mesozoic strata encircle Mt. Hillers which lies in the middle of the Henry Mountains structural basin. Beds in foreground are Cretaceous Blue Gate Shale. See Chart 92.

FIGURE 151 — View northward along Waterpocket Fold towards Capitol Reef Park in distant left. Burr Trail dirt road shows as winding path near left edge of photo. Rugged reef rocks are Navajo, Kayenta, and Wingate Sandstones. Burr Trail is on Moenkopi Formation on east flank of Circle Cliffs anticline. Formations east of reef are Carmel Formation, Jca; Entrada Sandstone, Je; Summerville Formation, too thin to label; Morrison Formation, Jm; Dakota Sandstone, Kd; Tununk Shale, Kt; Ferron Sandstone, Kf; Blue Gate Shale, Kb; Muley Canyon Sandstone, Kmc; Masuk Shale, Km; Tarantula Mesa Sandstone, Ktm.

FIGURE 152 — Over St. George looking north to Pine Valley Mountains laccolith. Thin beds at bottom of photo are middle member of Jurassic Kayenta Formation. Massive cliffs just above are upper Kayenta. Base of Navajo Sandstone is in middle of picture where eolian cross-bedded sandstones of the Navajo overlie the horizontally-bedded Kayenta. White cliffs behind are upper Navajo. Black late Cenozoic basalts cap benches of "inverted topography." The lavas originally followed paleostream valleys.

FIGURE 153 — Over Bloomington looking at St. George and Pine Valley Mountains in distance. Two levels of "inverted topography" show at left with airport mesa representing a lower and younger inverted valley than the higher West Black Ridge paleovalley on the left. Soft light-colored sediments in foreground are Petrified Forest Member of Chinle Formation. See Figure 152 for younger strata beyond St. George.

Chart
91 CAPITOL REEF AREA

System	Group / Formation	Member	Feet	Remarks
Q	Alluvium, valley, fill		0-200	
Q	Glacial deposits		0-100	
T	Basalts & tuffs on Thousand Lake Mtn		500+	5-6m.y.
T	Flagstaff ? Ls tuffaceous seds		500	
				unconformity
CRETACEOUS	Mancos Shale	Blue Gate Shale Member	1400	
CRETACEOUS	Mancos Shale	Ferron Ss Mbr	250	thin coal beds
CRETACEOUS	Mancos Shale	Tununk Shale Member	575	*Pycnodonte newberryi*
CRETACEOUS	Cedar Mt Fm & Dakota Ss		0-50	100 m.y. fission track
JURASSIC	Morrison Fm	Brushy Basin M	100-300	120 m.y.
JURASSIC	Morrison Fm	Salt Wash Ss Mbr	30-300	140 m.y.
JURASSIC	Morrison Fm	Tidwell Member	50-100	160 m.y.
JURASSIC	San Rafael Group	Summerville Fm	150-300	J-5 unconformity
JURASSIC	San Rafael Group	Curtis Formation	0-80	green
JURASSIC	San Rafael Group	Entrada Sandstone	480-780	
JURASSIC	San Rafael Group	Carmel Formation	300-1000	*Pentacrinus*
JURASSIC	San Rafael Group	Page Sandstone	0-100	J-2 unconformity
JURASSIC	Glen Canyon Gp	Navajo Sandstone	800-1100	
JURASSIC	Glen Canyon Gp	Kayenta Fm	350	
JURASSIC	Glen Canyon Gp	Wingate Sandstone	320-370	J-0 unconformity
TRIASSIC	Chinle Fm	Upper member	100-200	
TRIASSIC	Chinle Fm	Middle member	250-350	fossil wood
TRIASSIC	Chinle Fm	Shinarump Cg M	0-90	Tr-3 unconformity
TRIASSIC	Moenkopi Fm	Moody Canyon M	320-430	
TRIASSIC	Moenkopi Fm	Torrey Member	250-320	
TRIASSIC	Moenkopi Fm	Sinbad Ls Mbr	70-140	*Hemiprionites*
TRIASSIC	Moenkopi Fm	Black Dragon M	50-110	Tr-1 unconformity
PERMIAN	Black Box Dolomite		70	subsurface
PERMIAN	White Rim Ss		420	Pacific Western Unit No. 1 Teasdale 1949, to Redwall
PERMIAN	Organ Rock Shale		330	
PERMIAN	Cedar Mesa Ss		800+	
PERMIAN	Elephant Canyon Fm		200-500	
PENN	Hermosa Fm		700-1000	gypsum
PENN	Molas Fm		0-150	
MISS	Redwall Ls		940	
DEV	Ouray Limestone		140	
DEV	Elbert Fm		300	
CAMBRIAN	"Lynch" Dolomite & "Muav" Limestone undivided		1200-1600	
CAMBRIAN	Ophir Formation		200-300	
CAMBRIAN	Tintic Quartzite		200	
p€	"granite"		-	

MAP WITH TEXT--Smith et al, 1963; QUATERNARY--Flint & Denny, 1958; Bjorkland, 1969; Marine, 19€2; TERTIARY--Schneider, 1967; CRETACEOUS--Hill, 1982; Uresk, 1979; Lessard, 1973; Bagshaw, 1977; Doelling & Graham, 1972b; JURASSIC--Pipiringos & Sullivan, 1978; Petersen & Roylance, 1982; Kowallis & Heaton, 1984; Petersen & Pack, 1982; TRIASSIC--Ash, 1975; Lupe, 1984; Blakey, 1974; Dean, 1981; Friz, 1980; PERMIAN--Welsh et al, 19"9; SUBSURFACE--Rascoe & Baars, 1972; Mallory, 1972; Craig, 1972; Baars, 1972; Lochman-Balk, 1972; Irwin, 1971, 197€.

Chart
92 CIRCLE CLIFFS - HENRY MTS

System	Group / Formation	Member	Feet	Remarks
Q	Alluvial, surficial seds		0-100	
T	Diorite, porphyry laccoliths, stocks, bysmaliths of Henry Mts			23-29 m.y.
CRETACEOUS	Beds on Tarantula Mesa		60+	
CRETACEOUS	Tarantula Mesa Ss		230-400	Mesaverde Ss
CRETACEOUS	Mancos Shale	Masuk Sh Mbr	450-560	
CRETACEOUS	Mancos Shale	Muley Canyon Ss M	180-360	coal "Emery" of old reports
CRETACEOUS	Mancos Shale	Blue Gate Shale Member	1300-1500	
CRETACEOUS	Mancos Shale	Ferron Ss Mbr	190-250	minor coal
CRETACEOUS	Mancos Shale	Tununk Sh M	440-600	*Pycnodonte newberryi*
CRETACEOUS	Dakota Sandstone		0-50	
CRETACEOUS	Cedar Mt Fm	Upper member	0-20	
CRETACEOUS	Cedar Mt Fm	Buckhorn Cg M	20-50	
JURASSIC	Morrison Fm	Brushy Basin M	120	
JURASSIC	Morrison Fm	Salt Wash Ss M	280-450	
JURASSIC	Morrison Fm	Tidwell Mbr	0-90	J-5 unconformity
JURASSIC	San Rafael Gp	Summerville Formation	200	red
JURASSIC	San Rafael Gp	Curtis Fm	0-30	J-3 unconformity
JURASSIC	San Rafael Gp	Entrada Sandstone	450	
JURASSIC	San Rafael Gp	Carmel Fm	300	gypsum
JURASSIC	San Rafael Gp	Page Ss	100	J-2 unconformity
JURASSIC	Glen Canyon Gp	Navajo Sandstone	550-1100	
JURASSIC	Glen Canyon Gp	Kayenta Formation	250-350	
JURASSIC	Glen Canyon Gp	Wingate Ss	200-400	J-0 unconformity
TRIASSIC	Chinle Fm	Owl Rock M	150-250	
TRIASSIC	Chinle Fm	Petrified For't M	150-350	petrified wood
TRIASSIC	Chinle Fm	Monitor Butte M	0-200	
TRIASSIC	Chinle Fm	Shinarump Cg M	0-60	Tr-3 unconformity
TRIASSIC	Moenkopi Fm	Moody Canyon M	0-300	
TRIASSIC	Moenkopi Fm	Torrey Mbr	250-350	
TRIASSIC	Moenkopi Fm	Sinbad Ls M	40-60	
TRIASSIC	Moenkopi Fm	Lower member	0-50	Tr-1 unconformity
PERMIAN	Black Box Dolomite		0-100	
PERMIAN	Cutler Group	White Rim Ss	100-400	
PERMIAN	Cutler Group	Organ Rock Shale	200-300	
PERMIAN	Cutler Group	Cedar Mesa Sandstone	1000-1500	subsurface
PENN	Hermosa Fm		700-1000	gypsum
PENN	Molas Fm		0-50	red silty shale
MISS	Redwall Limestone		760	
D	Ouray & Elbert Fms undivided		500	Pre-Mississippian thicknesses estimated from regional data
CAMBRIAN	"Lynch" Dolomite		750	
CAMBRIAN	Unnamed shale		200	
CAMBRIAN	Muav Ls		300	
CAMBRIAN	Bright Angel Shale		300	micaceous
CAMBRIAN	Tapeats Sandstone		300	
p€	"granite"		-	

MAPS WITH TEXT--Hunt et al, 1952; Doelling, 1975; Huntoon et al, 1986; Morton, 1984; Whitlock, 1984; Davidson, 1967; Steed, 1954; CENOZOIC--Hunt, 1958; Affleck & Hunt, 1980; CRETACEOUS--Smith, 1983; Law, 1979, 1980; Doelling & Graham, 1972b; Peterson et al, 1980; Lessard, 1973; Maxfield, 1976; Lawyer, 1972; Molenaar, 1983; JURASSIC--Stokes, 1980; Peterson, 1980, 1984; Pipiringos & O'Sullivan, 1978; TRIASSIC--Lupe, 1984; Blakey, 1974; Friz, 1980; PERMIAN-- Gridley, 1974; PALEOZOIC--Doelling, 1975; Munger et al, 1965; Wells, 1954; Irwin, 1971, 1976.

Chart

93 BULL VALLEY MTS - BERYL JCT

Age	Unit	FEET	Notes
Q	Alluvium, dunes	0-100	
Q	Basalt of Black Hills	0-600	0.5 m.y.
PLI	Valley fill under Escalante Desert	0-3000	may include other basalts
MIOCENE	Muddy Creek? Fm 10-6 m.y.: Basalt of Enterprise	0-300	7.7 m.y.
MIOCENE	Muddy Creek? Fm 10-6 m.y.: Basalt of Modena	0-200	10.8 m.y.
MIOCENE	Muddy Creek? Fm 10-6 m.y.: Mine Series Volcaniclastics of Siders (1985)	0-1200	Mineralization in Escalante Silver Mine 11.6 m.y.
MIOCENE	Rhyolitic rocks of Flattop Mtn Suite - Hogsback, Shinbone, Cow Creek, Little Pine, Creek, & Pilot Peak	0-1000	Rhyolite of Beryl Jct of Siders, 1985 10.8 m.y. Rhyolite of Pinon Pt 12.8 m.y.
MIOCENE	Ox Valley Tuff	0-400	12-15 m.y.
MIOCENE	Cove Mtn Fm: Mbr of Cedar Spr	0-300	tuffs
MIOCENE	Cove Mtn Fm: Basalt of Pilot Pk	0-400	basalt cobbles
MIOCENE	Cove Mtn Fm: Racer Canyon Tuff	0-1500	19 m.y. source caldera may be near Caliente, Nevada
MIOCENE	Cove Mtn Fm: Mbr of Willow Sp	0-50	
MIOCENE	Porphyry of Maple Ridge	0-300	
MIOCENE	Breccia of Shoal Creek	0-200	
MIOCENE	Rencher Fm	0-1000	Quartz latite tuff and associated lavas and intrusive rocks of local origin
MIOCENE	Quichapa Group: Harmony Hills Tuff	0-500	21 m.y. Dacite of Pinon Park Wash of Siders (1985) 21.7 m.y.
MIOCENE	Quichapa Group: Breccia of Little Creek	0-1200	
MIOCENE	Quichapa Group: Condor Canyon Fm	0-200	23-24 m.y. Andesite of Enterprise of Siders,
MIOCENE	Quichapa Group: Leach Canyon Fm	0-350	25 m.y.
Φ	Isom Fm	0-150	26 m.y.
Φ	WahWah Springs Tuff	0-50	29.5 m.y. Needles Range Group
PAL	Claron Formation	0-600	
CRETACEOUS	Iron Springs Formation	2000	thin coaly beds 80 m.y.
JURASSIC	Carmel Fm: Crystal Creek Mbr	0-100	160 m.y. "Entrada" Ss
JURASSIC	Carmel Fm: Judd Hollow Mbr	400	"Homestake" Ls
JURASSIC	Temple Cap Fm	200	red siltstone
JURASSIC	Navajo Sandstone	2000	subsurface only
	SQUARE TOP MTN THRUST		
M-IP-P	Pakoon Dolomite, Callville Ls, & Chainman Shale in allochthon at Goldstrike	2000	marble at Mineral Mtn

MAPS - Blank, 1959; Siders, 1985; TERTIARY - Best et al, 1980; Rowley et al, 1979; MINES - Siders, 1985; Morris, 1980; Perry, 1976; Thomson & Perry, 1975; ESCALANTE DESERT - U.S. Dept. Agriculture, 1960; Mower, 1982.

Chart

94 BEAVER DAM MOUNTAINS

Age	Unit	FEET	Notes
Q	Alluvium, dune sand	0-100	
Q	Gunlock basalt flows	0-60	1.0-1.6 m.y.
P	Alluvium, landslides, breccia	0-150	6-10 m.y.
MIOCENE	Muddy Creek Fm	0-1000+	
MIOCENE	Ox Valley Tuff	0-500	12-15 m.y
MIOCENE	Cove Mountain Fm	0-700	19 m.y.
MIOCENE	Rencher Formation	0-300	
MIOCENE	Quichapa Volcanic Group	0-1500	21-25 m.y.
Φ	Isom Tuff & sediments	0-400	26 m.y.
Φ	Wah Wah Springs Tuff & seds	0-200	29.5 m.y.
PAL?	Claron Formation	0-400	red & white
?	Grapevine Wash Formation	0-2000	
CRET	Iron Springs Formation	2400-3000	yellowish- gray unfossiliferous calcareous ss
CRET	Dakota Conglomerate	0-50	80 m.y. FT zircon major unconformity
JURASSIC	Carmel Fm	450- 550	160 m.y. K-Ar biotite
JURASSIC	Temple Cap Fm	450	red siltstone, sh J-1 unconformity
JURASSIC	Navajo Sandstone	2000-2500	white & red cross-bedded calcareous ss
JURASSIC	Kayenta Fm: Upper member	800	pink ss and maroon mudstone
JURASSIC	Kayenta Fm: Middle member	1200	maroon siltstone and sandstone
JURASSIC	Kayenta Fm: Lower member	150	
JURASSIC	Moenave Fm	380	
TRIASSIC	Chinle Fm: Petrified Forest Member	700	varicolored mudstone
TRIASSIC	Chinle Fm: Shinarump Cg M	100-260	petrified wood
TRIASSIC	Moenkopi Fm: Upper red mbr	500-600	maroon ss & slts
TRIASSIC	Moenkopi Fm: Shnabkaib Member	800-870	
TRIASSIC	Moenkopi Fm: Middle red mbr	300-350	gypsiferous red sh
TRIASSIC	Moenkopi Fm: Virgin Ls Mbr	180-300	marine sh & ls
TRIASSIC	Moenkopi Fm: Lower red mbr	0-150	Rock Canyon Cg locally
PERMIAN	Kaibab Fm: Harrisburg Mbr	80-350	gypsum, ls
PERMIAN	Kaibab Fm: Fossil Mountain M	250-300	cherty ls
PERMIAN	Toroweap Fm: Woods Ranch Mbr	150-350	gypsum
PERMIAN	Toroweap Fm: Brady Canyon M	200-300	cherty ls
PERMIAN	Toroweap Fm: Seligman Mbr	100-200	gypsum & shale
PERMIAN	Queantoweap Sandstone	1500-2000	yellowish- gray sandstone
PERMIAN	Pakoon Dolomite	700-900	gypsum
P	Callville Limestone	1500-2000	cyclic beds *Lithostrotionella* cherty
M	Redwall Limestone	700-900	Thunder Sp Chert M
D	Muddy Peak Dolomite	700	stromatoporoids
CAMBRIAN	Nopah Dolomite	1300	stromatolites
CAMBRIAN	Bonanza King Formation	2500	*Crepicephalus* laminated white boundstone
CAMBRIAN	Bright Angel Shale	250	mottled ls
CAMBRIAN	Tapeats Quartzite	1200	often attenuated
p€	gneiss, schist, pegmatite	-	

GENERAL--Hintze, 1985a,b, 1986; Hintze et al, 1986; Steed, 1980; CENOZOIC--Rowley et al, 1979; Blank, 1959; Nations et al, 1985; CRETACEOUS-- Johnson, 1984; CARMEL--Blakey et al, 1983; MOENKOPI--Jenson, 1986; Lambert, 1984; KAIBAB-TOROWEAP--Nielson, 1981; Clark, 1981; REDWALL--McKee & Gutschick, 1969; Welsh & Bissell, 1979.

FIGURE 154 — West of Cedar City looking at Iron Springs Gap between The Three Peaks hills on the right and Granite Mountain on the left. Both are quartz monzonite laccoliths, about 20 million years old, that have domed adjacent Jurassic, Cretaceous, and early Cenozoic strata. Magnetite (iron ore) contact replacement ore bodies occur chiefly in the Jurassic Carmel limestones. See Chart 95.

FIGURE 155 — Looking southwest at the Harrisburg Dome, a late Mesozoic anticline between St. George and Hurricane. Utah Highway 9 to Zion Park is at bottom photo center. Low hills in core of dome are Harrisburg Member of the Permian Kaibab Formation; encircling banded beds are Early Triassic Moenkopi Formation; hogback cut by highway at right is Shinarump Conglomerate Member of the Upper Triassic Chinle Formation. See Chart 94.

FIGURE 156 — Northwestern portion of Zion National Park is seen along Interstate I-15 between Cedar City and St. George. Monoliths near top are Jurassic Navajo Sandstone. Thinner beds beneath are Kayenta Formation. Beds near base of hill are Early Triassic Moenkopi Formation. Hilltop at left is capped by late Cenozoic black basalt, underlain by light beds of the Jurassic Carmel Formation. See Chart 96.

FIGURE 157 — Triassic-Jurassic strata west of Zion. Pk, Permian Kaibab Formation; Trmlt, Timpoweap and lower red members of the Moenkopi Formation; Trmv, Virgin Limestone Member; other Moenkopi members as shown on Chart 96. Shinarump Conglomerate is rimrock for Test Sled Mesa. Springdale Sandstone Member of Moenave Formation, Jm, holds up the next higher bench. Kayenta and Navajo Sandstones form cliffs in background.

Chart
95 CEDAR CITY - IRON MINES

Age	Formation	Member	FEET	Remarks
Q	Holocene alluvium		0-100	
Q	Basalt cones & flows		0-200	0.3 to 1.7 m.y.
PLIO	Valley fill - alluvial fan and channel deposits, lake beds and older basalt flows		0-6000	known from well records in Cedar Valley
MIOCENE	Quartz monzonite porphyry intrusions			20 m.y. - iron ores
MIOCENE	Harmony Hills Tuff		0-350	21 m.y.
MIOCENE	Volcanic mudflows		0-150	
MIOCENE	Mt Dutton Fm		0-1000	Black Mountains
MIOCENE	Condor Canyon Fm	Bauers Tuff M	0-200	22 m.y.
MIOCENE	Condor Canyon Fm	Volcanic breccia	0-250	
MIOCENE	Condor Canyon Fm	Swett Tuff Mbr	0-50	23 m.y.
MIOCENE	Leach Canyon Fm	Table Butte Tuff M	100-700	
MIOCENE	Leach Canyon Fm	Narrows Tuff Mbr	100-300	24 m.y.
MIOCENE	Flows of Mud Spring		0-150	
Φ	Isom Fm	Hole-in-the-Wall Tf	0-120	member of Isom Fm
Φ	Isom Fm	Bald Hills Tuff M	100-400	26 m.y.
Φ	Wah Wah Springs Fm, N R Gp		0-300	29.5 m.y.
PALEOC	Claron Fm	Member E	300-450	white ss & shale
PALEOC	Claron Fm	Member D	20-120	white ls
PALEOC	Claron Fm	Member C	100-500	white ss, ls, sh, cg
PALEOC	Claron Fm	Member B	100-500	red sh, ss, cg, ls
PALEOC	Claron Fm	Member A	200-750	purple-red ls red cg
CRET	Iron Springs Fm west of Cedar City only	"Straight Cliffs & Wahweap Ss" = Sandstone facies of Tropic Shale	1600-2500	Kaiparowits area names in quotes improperly applied near Cedar City
CRET	Iron Springs Fm west of Cedar City only	"Tropic Fm" of Averitt = Dakota Fm of Lawrence	1100	Culver coal beds
JURASSIC	Carmel Fm	Winsor Member	300-320	pre-K unconformity
JURASSIC	Carmel Fm	Paria River M	140	"Curtis" gypsum
JURASSIC	Carmel Fm	Crystal Creek M	150-200	"Entrada"
JURASSIC	Carmel Fm	Co-op Creek Mbr	540-650	*Pentacrinus*
JURASSIC	Temple Cap Fm		0-50	
JURASSIC	Navajo Sandstone		1700	fossil sand dunes large cross-beds ZION PARK
JURASSIC	Kayenta Fm	Kayenta - Cedar City Tongue	400-800	
JURASSIC	Kayenta Fm	Shurtz Ss Tongue of Navajo	60-350	
JURASSIC	Kayenta Fm	- Lower member	270-420	
JURASSIC	Moenave Fm	Springdale Ss M	110-140	Silver Reef
JURASSIC	Moenave Fm	Dinosaur CanyonM	240-400	reddish-brown
TRIASSIC	Chinle Fm	Petrified Forest M	200-400	fossil wood
TRIASSIC	Chinle Fm	Shinarump Cg M	40-90	
TRIASSIC	Moen-kopi Fm	Upper red mbr	515	
TRIASSIC	Moen-kopi Fm	Shnabkaib Mbr	320	
TRIASSIC	Moen-kopi Fm	Middle red mbr	370	
TRIASSIC	Moen-kopi Fm	Virgin Ls Mbr	130	marine fossils
TRIASSIC	Moen-kopi Fm	Lower red mbr	450	
TRIASSIC	Moen-kopi Fm	Timpoweap Mbr	100	
PERMIAN	Kaibab Fm	Harrisburg Mbr	120	gypsum
PERMIAN	Kaibab Fm	Fossil Mtn Mbr	280	cherty, fossiliferous
PERMIAN	Toroweap Fm	Woods Ranch M	100	gypsum
PERMIAN	Toroweap Fm	Brady Canyon M	230	
PERMIAN	Toroweap Fm	Seligman Mbr	60	
PERMIAN	Queantoweap Ss		1000-1200	subsurface only
PERMIAN	Pakoon Dolomite		700	
P	Callville Limestone		600	cyclic ls, ss cherty
M	Redwall Limestone		1100	cherty, fossilferous
D	Devonian dolomite		500	
CAMB	Nopah Dolomite and Muav Ls undivided		1300	Supra-Muav
CAMB	Bright Angel Shale		250	
CAMB	Tapeats Sandstone		500	
pC	gneiss & schist		-	

MAP WITH TEXT - Averitt, 1962, 1967; Averitt & Threet, 1973; Rowley, 1975, 1976; Rowley & Threet, 1976; Mackin & Rowley, 1975, 1976; Mackin et al, 1976; IRON ORE - Bullock, 1973; CENOZOIC - Rowley et al, 1979; CRETACEOUS - Lawrence, 1965; Moir, 1974; Cashion, 1967; JURASSIC - Thompson & Stokes, 1970; Blakey et al, 1983; TRIASSIC - Stewart et al, 1972; PERMIAN - Nielson, 1981.

Chart
96 ZION PARK - CEDAR BREAKS AREA

Age	Formation	Member	FEET	Remarks
QUAT	Alluvium, terrace deposits		0-100	
QUAT	Cinder cones & basalt flows	Stage IV	0-700	few thousand years
QUAT	Cinder cones & basalt flows	Stage III		less than 0.5 m.y.
QUAT	Cinder cones & basalt flows	Stage II		1-2 m.y.
QUAT	Cinder cones & basalt flows	Stage I		3-6 m.y.
TERT	Mount Dutton Fm		0-800	
TERT	Claron Formation (also called Wasatch Formation)		1430	White ls at top Pink Cliffs made of fresh-water lake beds basal red cg & ss
CRETACEOUS	Kaiparowits Fm		0-400	duck-bill dinosaurs
CRETACEOUS	Wahweap and Straight Cliffs Sandstones, undivided		600-1200	drab ss, some sh
CRETACEOUS	Tropic Shale		700-800	
CRETACEOUS	Dakota Formation		600-400	coal
JURASSIC	Carmel Fm	Winsor Member	0-150	K unconformity yellow ss
JURASSIC	Carmel Fm	Paria River Mbr	60	gypsum
JURASSIC	Carmel Fm	Crystal Creek M	140	red ss
JURASSIC	Carmel Fm	Co-op Creek M	350-250	*Pentacrinus* J-2 unconformity
JURASSIC	Temple Cap Fm	White Throne M	160-100	
JURASSIC	Temple Cap Fm	Sinawava Mbr	30	J-1 unconformity
JURASSIC	Navajo Sandstone		*1600-2000	*thickness near Rockville
JURASSIC	Kayenta Fm	Tenney Canyon T	*800-200	
JURASSIC	Kayenta Fm	Lamb Point T	0-20	Tongue of Navajo Ss
JURASSIC	Kayenta Fm	Kayenta main body	*600 -150	forms slope
JURASSIC	Moenave Fm	Springdale Ss M	140-190	
JURASSIC	Moenave Fm	Whitmore Point M	60-70	
JURASSIC	Moenave Fm	Dinosaur Canyon M	150-240	J-0 unconformity
TRIASSIC	Chinle Fm	Petrified Forest M	450	varicolored
TRIASSIC	Chinle Fm	Shinarump Cg M	50	Tr-3 unconformity
TRIASSIC	Moen-kopi Fm	Upper red member	250	
TRIASSIC	Moen-kopi Fm	Shnabkaib Mbr	400-430	gypsum
TRIASSIC	Moen-kopi Fm	Middle red member	340	
TRIASSIC	Moen-kopi Fm	Virgin Ls Mbr	120	
TRIASSIC	Moen-kopi Fm	Lower red member	300	
TRIASSIC	Moen-kopi Fm	Timpoweap Mbr	50-120	
TRIASSIC	Moen-kopi Fm	Rock Canyon Cg M	0-20	Tr-1 unconformity
PERMIAN	Kaibab Fm	Harrisburg Mbr	120-130	gypsum
PERMIAN	Kaibab Fm	Fossil Mountain M	300-330	cherty, fossiliferous
PERMIAN	Toroweap Fm	Woods Ranch M	140-170	gypsum
PERMIAN	Toroweap Fm	Brady Canyon M	180	
PERMIAN	Toroweap Fm	Seligman Mbr	130	gypsum
PERMIAN	Queantoweap Sandstone		1330	tan & red ss
PERMIAN	Pakoon Dolomite		700	
P	Callville Limestone		650	
M	Redwall Limestone		1000	cherty
D	Ouray-Elbert Fms, undivided		230	
CAMBRIAN	Nopah Dolomite?		1000	
CAMBRIAN	Muav Limestone		1000	
CAMBRIAN	Bright Angel Shale		500	
CAMBRIAN	Tapeats Sandstone		500	
pC	Vishnu Schist		-	

MAPS WITH TEXT--Gregory, 1950a, b; Hamilton, 1978; QUATERNARY--Hamblin, 1970, 1986; Holmes, 1979; TERTIARY--Schneider, 1967; CRETACEOUS--Lawrence, 1965; Doelling & Graham, 1972a; JURASSIC--Blakey et al, 1983; Taylor, 1981; Thompson & Stokes, 1970; TRIASSIC - JURASSIC - Peterson & Pipiringos, 1979; Wilson, 1965, 1967; TRIASSIC - Stewart et al 1972a, b; Lambert, 1984; Nielson & Johnson, 1979; PERMIAN - Nielson, 1981; SUBSURFACE - J. E. Welsh, per. comm., 1985.

Chart
97 KANAB - ALTON AREA

Period	Formation	Member	Feet	Remarks
Q	Alluvium, surficial deps		0-100	
Q	Basalt flows & cones		0-300	1.1m.y.
Q	Older terrace deposits		0-100	
TERT	Claron Formation		1000	fresh-water limestone; PINK CLIFFS
CRETACEOUS	Kaiparowits & Wahweap formations undivided		1600-2000	
CRETACEOUS	Straight Cliffs Ss		200-500	coal
CRETACEOUS	Tropic Shale		600-1200	marine; *Inoceramus*
CRETACEOUS	Dakota Formation		200-300	minor coal
JURASSIC	Romana Ss		0-70	K-1 unconformity cuts down through Upper Jurassic units to the west
JURASSIC	Entrada Ss	Escalante Mbr	0-90	
JURASSIC	Entrada Ss	Cannonville M	0-150	
JURASSIC	Entrada Ss	Gunsight Butte M	0-280	
JURASSIC	Carmel Fm	Winsor Mbr	150-300	
JURASSIC	Carmel Fm	Paria River M	60-150	"Curtis" gypsum
JURASSIC	Carmel Fm	Thousand Pockets	0-80	Tongue of Page Ss
JURASSIC	Carmel Fm	Crystal Creek M	150-180	"Entrada" of old reports
JURASSIC	Carmel Fm	Co-op Creek M	200-80	
JURASSIC	Temple Cap Ss	White Throne M	40-0	J-2 unconformity
JURASSIC	Temple Cap Ss	Sinawava Mbr	30-0	J-1 unconformity
JURASSIC	Navajo Sandstone		1700-2000	WHITE CLIFFS
JURASSIC	Tenney Canyon Tongue-Kayenta		120-0	
JURASSIC	Lamb Point Tongue-Navajo		0-400	
JURASSIC	Kayenta Formation		180-280	
JURASSIC	Moenave Fm	Springdale Ss M	120-220	VERMILION CLIFFS
JURASSIC	Moenave Fm	Whitmore Point M	70-120	fossil fish
JURASSIC	Moenave Fm	Dinosaur Canyon M	50-200	J-0 unconformity
TRIASSIC	Chinle Fm	Petrified Forest M	470	fossil wood
TRIASSIC	Chinle Fm	Shinarump Cg M	10-50	
TRIASSIC	Moenkopi Fm	Upper red member	210-240	
TRIASSIC	Moenkopi Fm	Shnabkaib Mbr	390	gypsum
TRIASSIC	Moenkopi Fm	Middle red Mbr	240-320	
TRIASSIC	Moenkopi Fm	Virgin Ls Mbr	20-50	*Tirolites*
TRIASSIC	Moenkopi Fm	Lower red member	230	
TRIASSIC	Moenkopi Fm	Timpoweap Mbr	80-170	
PERMIAN	Kaibab Limestone		320-370	cherty
PERMIAN	Toroweap Formation		350-450	anhydrite
PERMIAN	Hermit Formation		650	Organ Rock equivalent; subsurface
PERMIAN	Cedar Mesa (Queantoweap) Sandstone		580-620	
P	Hermosa Fm (Callville Limestone)		540-590	
MISS	Redwall Ls	Horseshoe Mesa M	170-290	cherty
MISS	Redwall Ls	Mooney Falls Mbr	240	
MISS	Redwall Ls	Thunder Springs M	210	cherty
MISS	Redwall Ls	Whitmore Wash M	100	
D	Ouray Limestone		0-150	
D	Temple Butte Ls		110	Elbert equivalent
CAMBRIAN	"Supramuav"		630	
CAMBRIAN	Muav Limestone		590	
CAMBRIAN	Bright Angel Fm		340	
CAMBRIAN	Tapeats Sandstone		260-330	
pC	Grand Canyon Series		820+	

MAPS & TEXT--Gregory, 1948, 1951; Cashion, 1967a; Goode, 1973a, b; CRETACEOUS--Doelling & Graham, 1972a; Lawrence, 1965; Moir, 1974; JURASSIC--Blakey et al, 1983; Wright & Dickey, 1978; Thompson & Stokes, 1970; NAVAJO--Peterson & Pipiringos, 1979; TRIASSIC--Wilson, 1967; Blakey & Gubitosa, 1979; Marzolf & Dunne, 1983; PERMIAN--Cheevers & Rawson, 1979; Rawson & Turner-Peterson, 1979; SUBSURFACE--Tapp, 1963; Munger et al, 1965; Irwin, 1971, 1976.

Chart
98 BRYCE CANYON - ESCALANTE AREA

Period	Formation	Member	Feet	Remarks
Q	Alluvium, surficial rocks		0-200	
TERTIARY	Younger basalt		0-150	12 m.y. Red Canyon
TERTIARY	Osiris Tuff		0-300	23 m.y.
TERTIARY	Mount Dutton Formation		0-1000	23-26 m.y.
TERTIARY	Claron Formation	Sandstone member	200-400	
TERTIARY	Claron Formation	White limestone M	200-550	
TERTIARY	Claron Formation	Pink limestone member	700-900	BRYCE CANYON NATIONAL PARK; fresh-water lake beds
TERTIARY	Pine Hollow Fm	present on Aquarius Plateau near Escalante	0-450	gray mudstone
TERTIARY / CRETACEOUS	Canaan Peak Formation	present on Aquarius Plateau near Escalante	0-900	includes Cretaceous igneous pebbles
CRETACEOUS	Kaiparowits Formation		1000-3000	latest Cretaceous; duck-bill dinosaurs; turtles; greenish-gray
CRETACEOUS	Wahweap Fm		900-1500	light-gray to white
CRETACEOUS	Straight Cliffs Fm	Drip Tank Mbr	200-350	
CRETACEOUS	Straight Cliffs Fm	John Henry Member	800-1000	coal
CRETACEOUS	Straight Cliffs Fm	Smoky Hollow M	100-300	
CRETACEOUS	Straight Cliffs Fm	Tibbet Canyon M	80-180	
CRETACEOUS	Tropic Shale		600-800	marine fossils
CRETACEOUS	Dakota Formation		100-200	unconformity
CRETACEOUS	Cedar Mountain Fm		0-50	
JURASSIC	Morrison Fm		0-300	light, greenish gray
JURASSIC	Entrada Ss	Escalante Mbr	40-280	banded earthy facies
JURASSIC	Entrada Ss	Cannonville Mbr	150-200	
JURASSIC	Entrada Ss	Gunsight Butte M	250-350	
JURASSIC	Carmel Fm	Upper member	350-500	Wiggler Wash Member; Winsor Member; Paria River Member
JURASSIC	Carmel Fm	Thousand Pockets	0-200	Tongue of Page Ss
JURASSIC	Carmel Fm	Judd Hollow T	40-110	
JURASSIC	Carmel Fm	Lower member	60-80	J-2 unconformity
JURASSIC	Glen Canyon Gp	Navajo Sandstone	900-1700	large-scale festoon crossbedding; subsurface
JURASSIC	Glen Canyon Gp	Kayenta Fm	110-450	
JURASSIC	Glen Canyon Gp	Wingate Ss	75-1000	J-0 unconformity
TRIASSIC	Chinle Fm	Upper member	400-600	
TRIASSIC	Chinle Fm	Shinarump Cg M	60-130	Tr-3 unconformity
TRIASSIC	Moenkopi Fm	Upper member	800-1380	
TRIASSIC	Moenkopi Fm	Timpoweap Mbr	60-220	Tr-1 unconformity
PERMIAN	Kaibab Limestone		120-230	light gray dolomite
PERMIAN	White Rim Sandstone		140-200	
PERMIAN	Toroweap Fm		350-450	anhydrite beds
PERMIAN	Cutler Group	Organ Rock Sh	50-200	
PERMIAN	Cutler Group	Cedar Mesa Sandstone	700-1400	
PERMIAN	Cutler Group	Halgaito Tongue	350	P-P unconformity
P	Hermosa Formation		350-500	
P	Molas Fm		40-70	M-P unconformity
M	Redwall Limestone		900	D-M unconformity
D	Ouray Ls		100-150	conodonts abundant
D	Elbert Fm		250-300	C-D unconformity
CAMB	Muav Limestone		900	
CAMB	Bright Angel Shale		600	
CAMB	Tapeats Sandstone		300	feldspathic
pC	Basement unidentified		-	not drilled yet

COAL MAPS & TEXT--Bowers, 1973a, b, c, 1975; Stephens, 1973; Zeller, 1973a, b, c, d, 1978; Zeller & Stephens, 1973; CRETACEOUS--Doelling & Graham, 1972a; Peterson, 1969; Johnson & Vaninetti, 1982; Ryer, 1983; DeCourten, 1985; TERTIARY--Schneider, 1967; Bowers, 1972; GARFIELD COUNTY--Doelling, 1975; KODACHROME BASIN--Hannum, 1980; Hornbacher, 1985. SUBSURFACE--Munger et al, 1965; Welsh & Bissell, 1979; Irwin, 1971, 1976. JURASSIC-Thompson & Stokes, 1970; Blakey et al, 1983.

FIGURE 158 — Cedar Breaks is an erosional escarpment cut into the early Cenozoic Claron Formation which is overlain by middle and late Cenozoic volcanic rocks.

FIGURE 159 — Zion National Park, south entrance, village of Springdale in foreground along Utah Highway 15. Slopes on both flanks of the Virgin River floodplain are talus-covered Petrified Forest Member of the Chinle Formation. Prominent ledge in middle of slope is Springdale Sandstone Member of the Moenave Formation. Kayenta Formation has a slope-forming member below the main cliffs, and an upper member which makes up the lower third of the cliffs including all the horizontally-bedded stream-laid sandstones. The Navajo is an eolian cross-bedded sandstone that makes up the upper two-thirds of the monoliths whose cap rock is the Temple Cap Formation. See Chart 96.

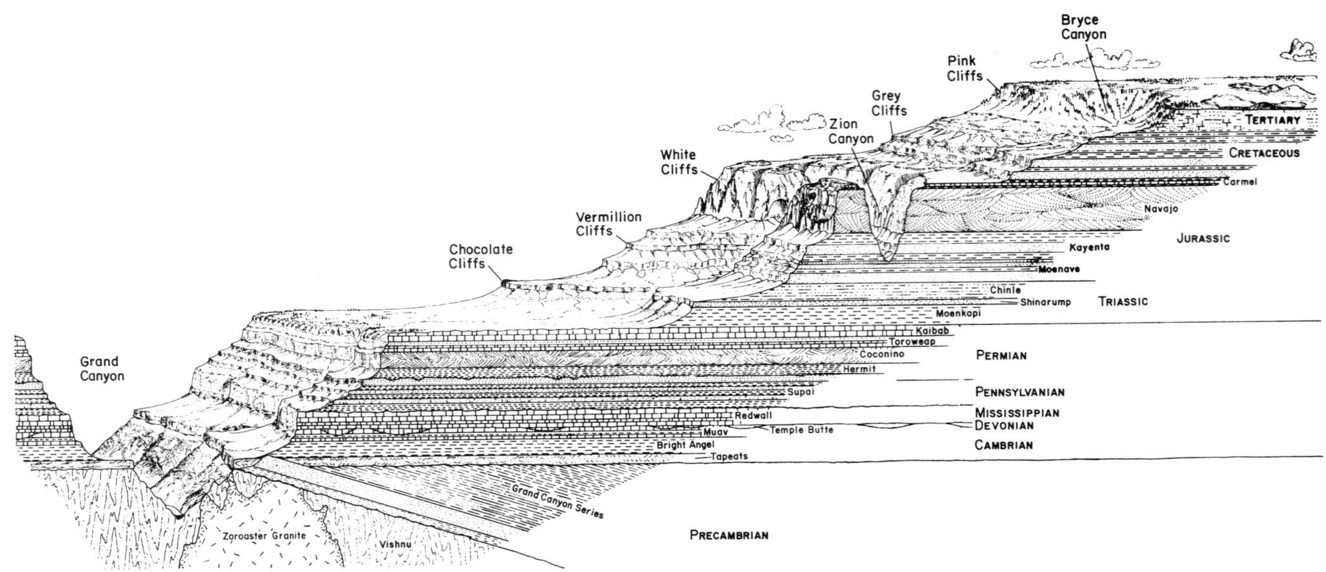

FIGURE 160 — Perspective sketch showing the "Grand Staircase" between Grand Canyon and Bryce Canyon. Vertical scale greatly exaggerated. Compare with Figure 161. Sketch courtesy of W.K. Hamblin.

FIGURE 161 — Grand Staircase east of Kanab. Oldest strata in right foreground are Shnabkaib Member of the Moenkopi Formation, overlain by the upper red member. Shinarump Member of the Chinle Formation caps the nearest bench whose backslope, behind, is on the non-resistant lower part of the Kayenta Formation, Jk. White cliffs are mostly Navajo Sandstone, Jn, overlain by Carmel, Jc. In distance Cretaceous marine beds, Ku, are overlain by early Cenozoic Claron Formation. See Chart 97.

FIGURE 162 — Old Paria townsite about 35 miles east of Kanab and 10 miles north of U.S. Highway 89. Beds dip east on flank of "The Cockscomb" or East Kaibab monocline. Early Triassic Moenkopi Formation in left foreground is overlain by colorful beds of the Late Triassic Chinle Formation. The Moenave Formation is the resistant dark-shadowed caprock. Beyond it can be seen white cliffs of Navajo Sandstone.

FIGURE 163 — U.S. Highway 89 crosses "The Cockscomb," or in structural terms, the East Kaibab monocline, about halfway between Kanab, Utah and Page, Arizona. The Timpoweap Member of the Moenkopi Formation forms the dipslope at left photo edge. Strike valley is on Moenkopi and Chinle Formations, here concealed but much attenuated by faulting. Cockscomb is more resistant sandstones of Moenave, Kayenta, and Navajo Formations badly broken and attenuated in places. Dip slopes right of Cockscomb are Carmel Formation with flat-lying Cretaceous marine strata forming mesas in right middle distance.

FIGURE 164 — Navajo Mountain, elevation 10,400 feet, dominates vistas near Glen Canyon. It is a laccolith with a small part of the igneous core exposed (Condie, 1964). Cretaceous Dakota Formation caps the mountaintop and uptilted Navajo Sandstone rises on its flanks. Rocks in foreground at Lake Powell shoreline are Navajo Sandstone. Lighter, thinner-bedded rocks above are mostly Entrada Sandstone.

FIGURE 165 — Monument Valley's magnificent towers are formed mostly of massive DeChelly Sandstone of Permian age. The monuments are erosional remnants of a once-continuous layer which is underlain by more easily eroded Organ Rock Shale. Some buttes and mesas are capped by thin layers of Triassic Moenkopi Formation and Shinarump Conglomerate.

Chart
99 GRAND CANYON AREA, ARIZONA

Age	Unit	Member	FEET	Remarks
Q	Alluvium, terrace gravel		0-100	1.2 m.y. intracanyon flows
T	Stage I, II, III, IV basalts		0-500	3.6-7.1 m.y.
T	Eocene fluviolacustrine dep		0-200	
₮	Moenkopi Fm		0-400	
PERMIAN	Kaibab Fm	Harrisburg Mbr	0-180	gypsum
PERMIAN	Kaibab Fm	Fossil Mt Mbr	250-350	*Dictyoclostus*
PERMIAN	Toroweap Fm	Woods Ranch M	100- 200	gypsum
PERMIAN	Toroweap Fm	Brady Canyon M	0-200	
PERMIAN	Toroweap Fm	Seligman Mbr	20-50	
PERMIAN	Coconino Sandstone		250-400	fossil sand dunes small reptile tracs
PERMIAN	Hermit Shale		300-350	fossil ferns
P	Supai Group	Esplanade Ss	230-420	
P	Supai Group	Wescogame Fm	100-250	fusulinids, corals brachiopods, conodonts
P	Supai Group	Manakacha Fm	150-320	
P	Supai Group	Watahomigi Fm	80- 200	
MISS	Surprise Canyon Fm		0-400	U. Miss. channel fill
MISS	Redwall Limestone	Horseshoe Mesa M	50-100	*Vesiculophyllum*
MISS	Redwall Limestone	Mooney Falls M	200-400	*Lithostrotionella*
MISS	Redwall Limestone	Thunder Springs M	80-150	*Syringopora*
MISS	Redwall Limestone	Whitmore Wash M	50-100	
D	Temple Butte Ls		0-400	thickens westward
CAMB	Nopah Dolomite		100-300	"Supra-Muav"
CAMB	Muav Limestone		350-800	535 m.y. FT *Glossopleura*
CAMB	Bright Angel Shale		200-500	
CAMB	Tapeats Sandstone		0-300	565 m.y. FT *Olenellus*
LATE PROTEROZOIC	Grand Canyon Supergroup, Chuar Group: Sixtymile Fm		200	GRAND CANYON OROGENY 800 m.y.
LATE PROTEROZOIC	Grand Canyon Supergroup, Chuar Group: Kwagunt Fm	Walcott Member	840	*Melanocyrillium*
LATE PROTEROZOIC	Grand Canyon Supergroup, Chuar Group: Kwagunt Fm	Awatubi Member	1230	biohermal stromatolites
LATE PROTEROZOIC	Grand Canyon Supergroup, Chuar Group: Kwagunt Fm	Carbon Butte	250	
LATE PROTEROZOIC	Grand Canyon Supergroup, Chuar Group: Galeros Fm	Duppa Mbr	570	
LATE PROTEROZOIC	Grand Canyon Supergroup, Chuar Group: Galeros Fm	Carbon Canyon Member	1550	alternating limestone and shale
LATE PROTEROZOIC	Grand Canyon Supergroup, Chuar Group: Galeros Fm	Jupiter Member	1520	
LATE PROTEROZOIC	Grand Canyon Supergroup, Chuar Group: Galeros Fm	Tanner Member	640	massive dolomite 950 m.y.
MIDDLE PROTEROZOIC	Grand Canyon Supergroup: Nankoweap Fm		300+	sandstone 1050 m.y.
MIDDLE PROTEROZOIC	Grand Canyon Supergroup, Unkar Group: Cardenas Basalt and diabase dikes & sills		700-1500	basalt flows 1070 m.y.
MIDDLE PROTEROZOIC	Grand Canyon Supergroup, Unkar Group: Dox Ss	Ochoa Pt M	300	
MIDDLE PROTEROZOIC	Grand Canyon Supergroup, Unkar Group: Dox Ss	Comanche Point Mbr	600	stromatolite
MIDDLE PROTEROZOIC	Grand Canyon Supergroup, Unkar Group: Dox Ss	Solomon Temple Member	900	diabase dikes 1120 m.y.
MIDDLE PROTEROZOIC	Grand Canyon Supergroup, Unkar Group: Dox Ss	Escalante Creek Member	1300	
MIDDLE PROTEROZOIC	Grand Canyon Supergroup, Unkar Group: Shinumo Quartzite		1000-1500	
MIDDLE PROTEROZOIC	Grand Canyon Supergroup, Unkar Group: Hakatai Shale		400-800	diabase sills 1120 m.y.
MIDDLE PROTEROZOIC	Grand Canyon Supergroup, Unkar Group: Bass Limestone		120-300	Hotauta Cg
MIDDLE PROTEROZOIC	Zoroaster Granite		intrusion	1,650 m.y.
MIDDLE PROTEROZOIC	Trinity & Elves Chasm Gneisses		island arc	sedimentary and
MIDDLE PROTEROZOIC	Vishnu Schist & Amphibolite		protolith?	volcanic protolith

MAP WITH TEXT - Huntoon et al, 1976; Maxon, 1961; TERTIARY - Hamblin, 1970; Best et al, 1980; Young & Hartman, 1984; TRIASSIC - McKee, 1954; PERMIAN - Cheevers & Rawson, 1979; Rawson et al, 1979; McKee, 1934; White, 1929; PENNSYLVANIAN - McKee, 1982; MISSISSIPPIAN - Billingsley & Beus, 1985; McKee & Gutschick, 1969; DEVONIAN - Beus, 1969; Ritter, 1983; CAMBRIAN - McKee, 1945; Lochman-Balk, 1971; PROTEROZOIC - Elston, 1979; Elston & McKee, 1982; Bloeser, 1985; Ford & Breed, 1973; Elston & Scott, 1976; Lucchitta & Hendricks, 1983; Stevenson & Beus, 1982; Wassenburg & Lanphyre, 1965; Dalton, 1972.

Chart
100 GLEN CANYON

Age	Unit	Member	FEET	Remarks
Q	Alluvium, dunes, terraces		0-100	
CRETACEOUS	Wahweap Formation		1300	
CRETACEOUS	Straight Cliffs Fm	Drip Tank Mbr	0-20	
CRETACEOUS	Straight Cliffs Fm	John Henry Member	1150	Christensen coal zone
CRETACEOUS	Straight Cliffs Fm	Smoky Hollow M	30-90	
CRETACEOUS	Straight Cliffs Fm	Tibbet Canyon M	70-150	
CRETACEOUS	Tropic Shale		610-730	bentonitic
CRETACEOUS	Dakota Formation		20-170	*Pycnodonte newberryi*
CRETACEOUS	Cedar Mountain Fm		0-50	
JURASSIC	Morrison Fm	Upper member	120-350	
JURASSIC	Morrison Fm	Salt Wash Mbr	180-360	J-5 unconformity
JURASSIC	Romana Sandstone		20-120	J-3 unconformity
JURASSIC	Entrada Ss	Middle member	190-540	red & white banded
JURASSIC	Entrada Ss	Lower member	300-560	reddish-orange cross-bedded
JURASSIC	Carmel Fm		140-260	Winsor Member Paria River Member
JURASSIC	Page Sandstone		30-130	J-2 unconformity
JURASSIC	Glen Canyon Group	Navajo Sandstone	1170-1230	
JURASSIC	Glen Canyon Group	Kayenta Fm	250-330	
JURASSIC	Glen Canyon Group	Wingate Ss	250-310	J-0 unconformity
TRIASSIC	Chinle Fm	Owl Rock Mbr	130-150	
TRIASSIC	Chinle Fm	Petrified Forest Member	450-520	varicolored bentonitic
TRIASSIC	Chinle Fm	Shinarump Cg M	140-200	Tr-3 unconformity
TRIASSIC	Moenkopi Formation		230-270	Tr-1 unconformity
PERMIAN	Kaibab Limestone		0-150	
PERMIAN	Cutler Group	White Rim Ss	0-230	
PERMIAN	Cutler Group	Organ Rock Shale	400-500	
PERMIAN	Cutler Group	Cedar Mesa Sandstone	800-1500	subsurface
PERMIAN	Cutler Group	Elephant Canyon Formation	500-600	
PENN	Hermosa Formation		500-570	
PENN	Molas Fm		150-200	
MISS	Redwall Ls	Horseshoe Mesa M	170	
MISS	Redwall Ls	Mooney Falls M	200	
MISS	Redwall Ls	Thunder Spr M	150	cherty
MISS	Redwall Ls	Whitmore Wash M	110	
D	Ouray Limestone		140	
D	Elbert Formation		150-230	
CAMBRIAN	Muav Limestone		430	
CAMBRIAN	Bright Angel Shale		680	
CAMBRIAN	Tapeats Sandstone		300	
p€	Basement unidentified		-	Lithology not known

COAL MAPS & TEXT--Peterson, 1973, 1975, 1980; Peterson & Barnum, 1973a, b, c, d; Doelling & Graham, 1972a; Waldrop et al, 1967; REGIONAL MAP--Hackman & Wyant, 1973; JURASSIC--Blakey et al, 1983; PALEOZOIC--Munger, et al, 1965; Wells, 1954; Welsh & Bissell, 1979; Irwin, 1971, 1976.

Chart
101 MONUMENT VALLEY, UTAH

Period	Group / Formation	Unit	Feet	Notes
Q		Alluvium & dune sand	0-100	
T		Lamprophyric dike near Oljeto	intrusive	28 m.y.
JURASSIC	Glen Canyon Group	Navajo Sandstone	600 +	
JURASSIC	Glen Canyon Group	Kayenta Formation	200-230	
JURASSIC	Glen Canyon Group	Wingate Sandstone	310-360	
TRIASSIC	Chinle Fm	Church Rock, Owl Rock, Petrified Forest, and Monitor Butte members undivided	770-1020	multicolored bentonitic
TRIASSIC	Chinle Fm	Shinarump Cg M	0-200	Caprock for most of the monuments in this area
TRIASSIC		Moenkopi Formation	100-350	
TRIASSIC		Hoskinnini Sandstone	40-120	Permian? Rocks at Oljeto & buttes of Monument V
PERMIAN	Cutler Group	DeChelly Sandstone	0-400	
PERMIAN	Cutler Group	Organ Rock Formation	300-700	Rocks at Goulding T.P. & lower slopes in Monument Valley
PERMIAN	Cutler Group	Cedar Mesa Sandstone	500-700	Resistant sandstone that is exposed at the surface over much of the northern Monument Upwarp. Grades eastward to gypsiferous redbeds within Monument Valley.
PERMIAN	Cutler Group	Halgaito Formation	400-700	Rocks at Mexican Hat
PENNSYLVANIAN	Hermosa Group	Honaker Trail Formation	400-600	Oldest exposed strata are in San Juan River Canyon. Raplee Anticline near Mexican Hat shows Paradox evaporites. Equivalent rocks in Halgaito Anticline are carbonates.
PENNSYLVANIAN	Hermosa Group	Paradox Formation	500-1000	This is the type area for Honaker Trail Fm Skelly Oil, 1952 27-40S-12E Nokai # 1-A
PENNSYLVANIAN	Hermosa Group	Pinkerton Trail Formation	200-400	
PENNSYLVANIAN		Molas Formation	150	
MISS	Redwall Ls	Horseshoe Mesa M	110	
MISS	Redwall Ls	Mooney Falls Member	220	
MISS	Redwall Ls	Thunder Springs M	140	cherty
MISS	Redwall Ls	Whitmore Wash M	90	
DEV		Ouray Limestone	110	
DEV		Elbert Formation	310	McCracken Ss Mbr
CAMBRIAN		Muav Limestone	630	"Lynch" Dolomite
CAMBRIAN		Bright Angel Shale	290	
CAMBRIAN		Tapeats Sandstone	180 +	

MAP WITH TEXT - Hackman & Wyant, 1973; Lewis & Trimble, 1959; Mullens, 1960; CHINLE - Lupe, 1984; MOENKOPI - Blakey, 1974; PERMIAN - Baars, 1973a; PENNSYLVANIAN - Wengerd, 1958; Wengerd & Matheny, 1958; Baars, 1973b; SUBSURFACE - Munger, 1965; Wells, 1954; Baars, 1958; Welsh & Bissell, 1979; TERTIARY BASALT - Stuart-Alexander et al, 1972; Ellingson, 1973; Delaney et al, 1986.

Chart
102 MONTICELLO - BLUFF - ANETH

Period	Group / Formation	Unit	Feet	Notes
Q		Alluvium, colluvium	0-100	
Q		Loess, pediment gravel	0-100	loess forms red soil
T		Dikes, plugs, diatremes	intrusive	Alhambra Rock
T		Abajo Mts laccoliths	intrusive	diorite 20-29 m.y.
CRET		Mancos Shale	0-700	
CRET	Mancos Shale	Bridge Creek Ls Mbr	0-50	*Pycnodonte* Greenhorn 50 ft of Mancos below
CRET		Dakota Sandstone	30-150	
CRET		Burro Canyon Fm	50-180	
JURASSIC	Morrison Fm	Brushy Basin Mbr	200-440	varicolored mudstone
JURASSIC	Morrison Fm	Westwater Canyon M	0-180	
JURASSIC	Morrison Fm	Recapture Mbr	0-285	
JURASSIC	Morrison Fm	Salt Wash Member	0-550	uranium mines
JURASSIC	Morrison Fm	Bluff Ss Mbr	0-340	J-5 unconformity
JURASSIC		Wanakah Fm	60-200	formerly Summerville
JURASSIC		Entrada Sandstone	50-400	
JURASSIC		Carmel Fm	0-120	J-2 unconformity
JURASSIC		Navajo Sandstone	300-800	thickens westward
JURASSIC		Kayenta Fm	50-250	stream-channel ss
JURASSIC		Wingate Sandstone	250-450	
TRIASSIC	Chinle Fm	Church Rock M	100-500	light brown
TRIASSIC	Chinle Fm	Owl Rock Mbr	100-400	pale red
TRIASSIC	Chinle Fm	Petrified Forest M	0-300	varicolored, bentonite
TRIASSIC	Chinle Fm	Moss Back Cg M	0-200	
TRIASSIC	Chinle Fm	Monitor Butte M	0-200	
TRIASSIC	Chinle Fm	Shinarump Cg M	0-100	Tr-3 unconformity
TRIASSIC	Moenkopi Fm	Moody Canyon M	0-100	slope-forming mbr
TRIASSIC	Moenkopi Fm	Torrey Mbr	200	ledge-forming mbr
TRIASSIC	Moenkopi Fm	Hoskinnini Tongue	0-100	
PERMIAN	Cutler Group	De Chelly Ss	0-550	
PERMIAN	Cutler Group	Organ Rock Fm	100-900	
PERMIAN	Cutler Group	Cedar Mesa Sandstone	500-1200	Rocks at Natural Bridges National Monument In subsurface east of Comb Ridge this unit is gypsiferous redbeds
PERMIAN	Cutler Group	Halgaito Fm	400-500	reddish brown
PENNSYLVANIAN	Hermosa Group	Honaker Trail Formation	500-1500	"Rico Fm"
PENNSYLVANIAN	Hermosa Group, Paradox Fm	Oil zones: Ismay, Desert Creek, Akah, Barker Creek, Alkali Gulch	500-3500	halite anhydrite & potash interbedded with black shale, dolomite and algal reef mounds which yield oil
PENNSYLVANIAN	Hermosa Group	Pinkerton Trail Fm	100-400	Pre-Pennsylvanian in subsurface only
PENNSYLVANIAN		Molas Formation	0-150	
MISS		Leadville Limestone	300-600	= Redwall Ls
DEV		Ouray Formation	50-140	
DEV		Elbert Formation	160-350	
DEV	Elbert Formation	McCracken Ss Mbr	0-120	
DEV		Aneth Formation	0-200	fish plates
CAMB		"Supra-Muav" Dolomite	0-300	"Lynch" Dolo
CAMB		"Bright Angel" Shale	0-300	
CAMB		Ignacio Quartzite	100-300	
pЄ		Schist, gneiss, granitic and pegmatite dikes and plugs		

MAPS WITH TEXT-Haynes et al, 1972; Witkind, 1964; O'Sullivan, 1965;
QUATERNARY-Nielson & Erickson, 1980; Huff & Lesure, 1965; Biggar et al, 1981; TERTIARY-Stuart-Alexander, 1972; Armstrong, 1969; CRETACEOUS-Molenaar, 1975; Craig, 1981; Tschudy et al, 1984; JURASSIC-O'Sullivan & Craig, 1973; O'Sullivan, 1980a; Peterson, 1986; O'Sullivan & Maberry, 1975; Pipiringos & O'Sullivan, 1975, 1978; Tyler & Etheridge, 1983; Lupe, 1983; TRIASSIC-O'Sullivan, 1970; O'Sullivan & Green, 1973; Blakey, 1974; Stewart et al, 1972a, b; O'Sullivan & MacLachlan, 1975; Lupe, 1984; PERMIAN-Baars & Stevenson, 1979; Rascoe & Baars, 1972; Loope, 1984; PENNSYLVANIAN-Mallory, 1972; Hite & Cater, 1972; Choquette, 1983; Welsh & Bissell, 1979; Clem & Brown, 1984; MISSISSIPPIAN-Craig, 1972; DEVONIAN-Baars, 1972; CAMBRIAN-Lochman-Balk, 1972; OIL & GAS-Clem & Brown, 1984.

FIGURE 166 — New sands made from old. Monument Valley.

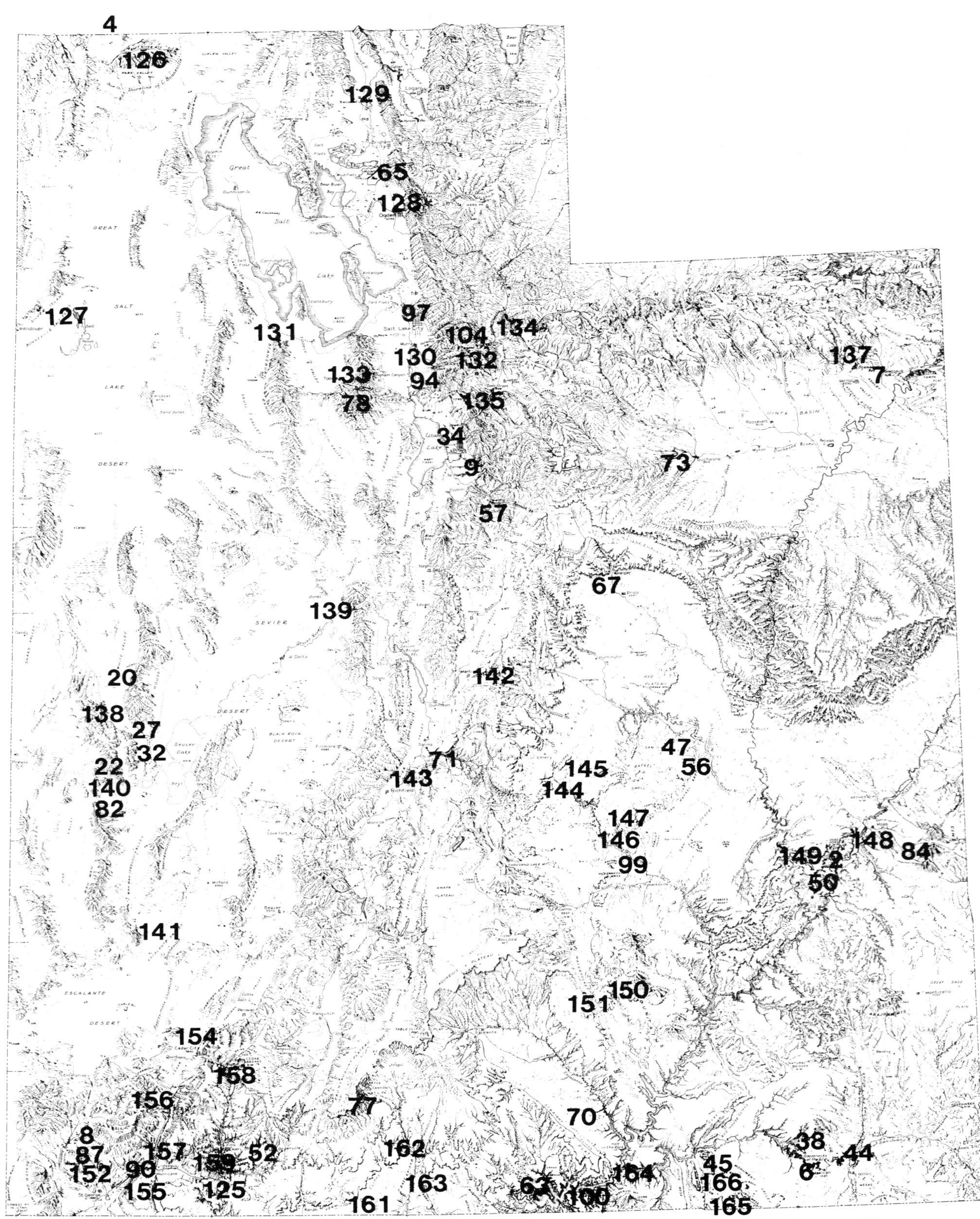

FIGURE 167 —Index map showing locations of photographs in this book. Landforms base map courtesy of Merrill K. Ridd.